DIVINING THE ETRUSCAN WORLD

The Etruscan *Brontoscopic Calendar* is a rare document of omens foretold by thunder. It long lay hidden, embedded in a Greek translation within a Byzantine treatise from the age of Justinian. The first complete English translation of the *Brontoscopic Calendar*, this book provides an understanding of Etruscan Iron Age society as revealed through the ancient text, especially the Etruscans' concerns regarding the environment, food, health, and disease. Jean MacIntosh Turfa also analyzes the ancient Near Eastern sources of the calendar and the subjects of its predictions, thereby creating a picture of the complexity of Etruscan society reaching back to before the advent of writing and the recording of the calendar.

Jean MacIntosh Turfa is Rodney Young Fellow in the Mediterranean Section of the University of Pennsylvania Museum of Archaeology and Anthropology and Adjunct Professor of Classics and Ancient Studies at St. Joseph's University, Philadelphia. She has published catalogs of collections of Etruscan antiquities and articles on Etruscan art, seafaring, votive offerings, and divination and medicine.

apasi atialc alcsentresc,
cen zic zichu turce

DIVINING THE ETRUSCAN WORLD

The *Brontoscopic Calendar* and Religious Practice

JEAN MACINTOSH TURFA
The University of Pennsylvania Museum
of Archaeology and Anthropology

CAMBRIDGE
UNIVERSITY PRESS

CAMBRIDGE
UNIVERSITY PRESS

University Printing House, Cambridge CB2 8BS, United Kingdom

One Liberty Plaza, 20th Floor, New York, NY 10006, USA

477 Williamstown Road, Port Melbourne, VIC 3207, Australia

314-321, 3rd Floor, Plot 3, Splendor Forum, Jasola District Centre, New Delhi - 110025, India

79 Anson Road, #06-04/06, Singapore 079906

Cambridge University Press is part of the University of Cambridge.

It furthers the University's mission by disseminating knowledge in the pursuit of education, learning and research at the highest international levels of excellence.

www.cambridge.org
Information on this title: www.cambridge.org/9781107009073

First published 2012

A catalogue record for this publication is available from the British Library

Library of Congress Cataloging in Publication data
Turfa, Jean MacIntosh, 1947–
Divining the Etruscan world : the *brontoscopic calendar* and religious practice / Jean MacIntosh Turfa.
p. cm.
Includes bibliographical references and index.
ISBN 978-1-107-00907-3 (hardback)
1. Etruscans – Religion. 2. Omens. 3. Calendar, Greek. 4. Astronomy, Greek.
5. Lydus, Johannes Laurentius, 490 ca. 565. De ostentis. I. Title.
BL813.E8T87 2012
299΄.9294–dc23 2011040848

ISBN 978-1-107-00907-3 Hardback

CONTENTS

PREFACE AND ACKNOWLEDGMENTS

In covering the topics of this book, my research paralleled the situation of our knowledge of the Etruscans in general – we do not see them through their own eyes but always through a filter of other cultures' slanted perception, rather like an archaeological excavation in which we must sift through later layers to reach the Iron Age. Our only *literary* sources for Etruria are the Greek and Latin authors, and an unbiased, annalistic history of the Etruscans was never their intention. For the Greeks, Etruscans were hostile aliens who sought to bar their merchants from profits and their colonists from claiming land – and the Etruscans' success at both cemented the negative stereotypes. For the Romans, Etruria was even darker – alien yet not utterly, for even in the eighth to seventh centuries BC, plenty of Latin, Faliscan, and other families had Etruscan in-laws, patrons, or allies.

Denigrate them as treacherous and your own betrayals may be justified, encapsulate them as doughty warriors and your victories over them mean more, or relegate them to a position of alleged respect and usefulness – when their official function may be subjugated to the Roman state. The Etruscans became identified as the people with the best religious techniques, who bequeathed to the Romans their rituals for founding cities, communicating with the gods, and expiating sins. Seen through the eyes of Cicero or Seneca, they have fallen to the level of the imaginary Native Americans who were claimed as "spirit guides" for nineteenth-century parlor séances. And the *Brontoscopic Calendar*, by its oracular phrasing and topic of divination, could support this viewpoint on Etruscan religion.

By the last days of Etruscan political autonomy – or their independent cultural identity – that is, the first half of the first century BC, the *Brontoscopic Calendar* was so little known that Publius Nigidius Figulus, a respected friend of Cicero, felt the need to make it accessible to a broader segment of Roman intellectuals by translating it into Latin. He must have known that a number of worthy men of letters would (a) be interested in it and (b) be incapable of reading it in the original Etruscan. The Latin document he created must have circulated, even after Figulus' untimely death in 45 BC while in exile, after having supported Pompey in the Civil War. It must have been copied, because five centuries later a text of it reached the desk of one of the few Latin secretaries in Justinian's Constantinople, John the Lydian. He translated the Latin document into (Byzantine) Greek, presuming (like Figulus before him) that a wider audience awaited if the language were more accessible. Lydus' treatise, *De ostentis* (*On Portents*), in which the calendar text and other arcane works were embedded, was disseminated and periodically cited by later scholars. In the meantime, pagan apologists were holding up the *etrusca disciplina*, the teachings of Tages, the strange child-prophet of earliest Etruria, as a foil to the Christian scriptures. They were the words of the gods – divinely inspired revelations, brought to men by a supernatural being – and they were scriptures written down at Tages' dictation by a noble scribe, who was also founder of the new city, Tarquinii.

Through the modern period, intellectuals with political goals have often retrieved the Etruscans from perceived oblivion: when fascists held up the prehistoric Terramare culture and the Roman Empire as models of a noble past and justification for modern authoritarian regimes, D. H. Lawrence countered with the color and freedom of the painted Etruscan tombs (of Tarquinia), where smiling couples danced and feasted, in *Etruscan Places*. The alternative of a goose-stepping militaristic state could never compare to that.

Back to antiquity: Figulus, Caecina, Tarquitius Priscus, and other scholars from the first century BC through the second century AD were salvaging and publicizing selected texts or portions of the *etrusca disciplina*, but there were already other texts and systems of divination in circulation. In the affluent and educated circles of the Hellenistic Greek kingdoms, especially the Seleucid and Ptolemaic spheres, and in the Roman Republic and Principate, people were delving into the occult and all sorts of popular divination. The personal horoscope today is probably the best-known legacy of this intellectual world.

But its roots tapped some of the same sources (no doubt via later copies) that the Iron Age Etruscans had experienced in the formulation of the *Brontoscopic Calendar*: the venerable and vast divination literature of Mesopotamia. Today, the best exemplars of these Mesopotamian texts are known from the final dynasty of the Assyrian empire, the clay tablet libraries of the Sargonid kings, especially the so-called library of Ashurbanipal at Nineveh, preserved by the conflagration and collapse of the palace at the end of the seventh century BC. The original cuneiform texts may be traced back through centuries of ongoing editing and augmentation to (at least) the early second millennium BC, and their ideology and topics stem from the third millennium and the foundation of the great Sumerian cities: Uruk, Ur, Sippar, Eridu, Babylon, and others. The first period of intensive codification – crystallization – of these texts was around 1000 BC, and they testify to the development of both scientific astronomy and astrology, of medical diagnosis, and of the systematic study of natural phenomena and wonders.

Several major works of Mesopotamian divination literature are known to us, and portions of the great texts (some extending over dozens of clay tablets) consulted by Neo-Assyrian and Neo-Babylonian scholars can be seen to have surfaced in the Etruscan *Brontoscopic Calendar*: the *Enūma Anu Enlil* on celestial events, the *Šumma ālu* on human activities, the *Šumma izbu* on the omens of deformed births, and the *Iqqur īpuš*, a calendar-based compilation of all the others.

The Etruscan document preserved by Figulus is not a slavish copy or excerpt of the Mesopotamian texts – whoever produced it had thoroughly "digested" the Mesopotamian literature and deleted all the references to phenomena impossible in Italy, such as sesame crops, date palm plantations, scheming crown princes, and war with the Amorites or Elamites. The composer(s) of the Etruscan text grafted on many references to situations and topics that would only have been applicable in Iron Age Etruria at that time, such as the power of the common people or other urban social factions, and the mention of fruits, nuts, pests, and diseases found in Italy rather than the Near East. Other references to "the women and slaves" evoke a society different from those of the Italic peoples or Greeks, but women and slaves certainly do appear in Mesopotamian legal documents. There, however, divination texts do not refer to them with the same tone or situations as found in the Etruscan *Brontoscopic Calendar.*

But the Mesopotamian source texts used by the unknown Etruscan author(s) had already attained, and would continue to enjoy, a life of their own, spawning many divergent successors. Mesopotamian divination literature was still being copied and annotated in the Near East and Levant through the Seleucid period (in cuneiform tablets and Aramaic papyri), and readers of the Hellenistic and Roman worlds delved into those later editions again and again, emerging with the stuff of personal horoscopes, fortune-telling texts, farmers' almanacs, and also *brontologia* and *seismologia*, the terms by which texts of divination by thunder and earthquake, respectively, were cited in the Hellenistic and Roman periods. Sometimes, over the next millennium, the thunder-omens and earthquake lore were attributed to past prophets, from Zoroaster to David and Enoch, and a few works even recalled Tages and scholars with Etruscan names. These brontoscopic documents were also copied during the Byzantine Middle Ages, and a large corpus, showing much cross-fertilization ever since the Seleucid and Ptolemaic period, was compiled over the nineteenth to twentieth centuries and published as the *Catalogus codicum astrologorum graecorum* (*CCAG*) by Cumont, Boll, and subsequent generations of scholars from 1898 to 1953. Yet another such scripture surfaced when scholars analyzed fragments of papyrus scrolls discovered in the Qumran Caves.

It seems that Seleucid and Ptolemaic Greco-Syrians and Greco-Egyptians, Roman scholars with political ambitions, and Palestinian scholars writing in Aramaic all accessed the stream of Mesopotamian divination literature and created from it their own, shorter, and usually more narrowly focused, documents. The lore, especially the astrological/astronomical texts, had a significant impact on thought as far away as India and China – just as, in its first heyday, in the second to first millennia BC, the literature had been copied and translated in the kingdoms of the Hittites, Syrians, Urartians, Canaanites, and Persians. The Mesopotamian tradition was like the fountain of Apsu, living water always surging underground beneath the feet, there to be dipped into when the next generation of sages was ready to appreciate it.

I hope I have offered evidence to prompt a(nother) reassessment of the deep Near Eastern roots, not just of this text, but of political, social, and economic developments in Etruria, for which we have long known the tangible symbols – such as scepters, thrones, and parasols, and a wealth of narrative and floral art.

There is a greater issue here too in the epistemology or metaphysics of the study of ancient religion that cannot be resolved in this book. We moderns today do not follow Etruscan or Roman religion as a personal belief system. We may therefore excuse our cynicism on grounds of our greater enlightenment. But there in the ruins of Ashurbanipal's palace, in the clay tablets painstakingly collected and archived, is the proof that a deep-seated religious tenet had been taken over piecemeal by Etruscans from an alien culture – and so how could it be a direct revelation from the gods? Probably no one in Figulus' day was really aware of that deep Mesopotamian heritage or able to judge it: a Ptolemaic scholar might have noticed similarities between horoscope predictions and the *Brontoscopic Calendar*, but he or she could have ascribed it to the truth of divine revelations. We might see the adoption and adaptation of Mesopotamian texts as a sort of duplicity at the heart of the religion of the Etruscans, the "most religious of all men" (Livy 5.1.6) – but would they have felt the same way?

But the *etrusca disciplina* was also, especially, the province of the elite, kept, as far as we can tell, by aristocratic priestly families and shared at their discretion. Presumably it was cited as the reason why things in the Etruscan city were done the way they were done. Some of the special features of the calendar furnish evidence for determining that it was created in the eighth to seventh centuries BC, in Etruria, but with input from someone skilled in Mesopotamian divination and in the use of Akkadian tablet texts or perhaps Aramaic or other translated versions of the great works of Assyrian and Babylonian religion. It would have taken only a single family, a single man (or woman) to work with a Near Eastern scholar trained in the divination texts, to produce the original of the text used by Figulus.

By making the Etruscan *Brontoscopic Calendar* accessible (in English) to a wider group of readers who are probably not familiar with Byzantine Greek or Golden Age Latin, I acknowledge a responsibility to suggest some sensible modes of interpretation for it.

Chapter 1 is an introduction to the study of this unusual document, and Chapter 2 provides background on Etruscan religion and its scriptures and their impact on Republican Rome. The reader may then want to refer to Appendix A, which offers excerpts from Classical sources (with English translations) for our knowledge of Etruscan and Roman beliefs and rituals. Chapter 3 covers the environment and climate of Iron Age Italy and ancient perceptions on weather phenomena: a text on divination by thunder(storm) could not have appeared at a more opportune time to exert strong influence on the people of early historic Italy. The Greek text and my English translation follow, the first English version of this rare text (the preliminary version of my translation was presented in 1999 at the Sixth Langford Conference of Florida State University and appeared in the resultant book, *The Religion of the Etruscans*, edited by de Grummond and Simon, 2006). Chapters 4 through 7 analyze the content of the *Brontoscopic Calendar* and discuss the environment of Iron Age and Archaic Etruria as related to the concerns of life on the land and the hopes and fears of the people who used this unique divination tool (especially food, safety, health, and society). Chapter 8 discusses the Near Eastern sources of such divination, and Appendix B furnishes samples of the Akkadian divination texts that were the inspiration for many of the omen predictions found in the Etruscan document. Chapter 9 surveys other brontoscopic treatises, all dated later than that recorded by Figulus, and all demonstrably derived from the Hellenistic and Roman-period literature of pop-culture astrology and personal divination. Appendix C gives examples of their Greek texts with English translations. (There are many points of comparison between the society of Late Republican Rome and our own, including a large literate population with leisure time for studies and the curiosity to clamor for convenient fortune-telling as a substitute for cultic wisdom and communal ritual.)

The Etruscans gave the Western world the tie-beam truss without which the basilica and modern housing could not function, the foresail on merchant clippers that enabled the Romans' bread-and-circuses lifestyle, the "Latin" alphabet and "Roman" numerals, and a wealth of art and technology, much of it indeed derived from other cultures but all funneled to *us* through Etruria. We will never know the full extent of the Etruscan legacy – it probably includes many elements of Western statecraft, the status of women, and popular medicine and beliefs. My study here will point out some of their earlier sources and some of their unique situation and perceptions.

ACKNOWLEDGMENTS

I am particularly indebted to Ingrid Edlund-Berry, Stephanie Budin, and my husband, Alexander F. Turfa, for kindness and help in the preparation of this book.

I am indeed grateful to a great number of scholars, colleagues, and friends for help far beyond the call of duty or friendship, editorial assistance, discussions, references, shared results of research, acquisition of books, articles, and more. Each favor – or rescue – is unique. So many have there been that I must name them alphabetically; each contribution has meant so much: Laura Ambrosini, Maria Paola Baglione, Gilda Bartoloni, Hilary Wills Becker, Marshall Becker, Cecilia Beer, Jon Berkin, Claudio Bizzarri, Larissa Bonfante, Dominique Briquel, Ann Blair Brownlee, Jim Burr, Jennifer Chiappardi, Nancy de Grummond, Dick Ellis, Adriana Emiliozzi, Valentina Follo, Erlend Gehlken, Sarah Gettys, Margarita Gleba, Jean Gran-Aymerich, Anne Guinan, Marie-Laurence Haack, Rick Hamilton, Adrian Harrison, Susan Jones, Peter Kuniholm, Lesley Lundeen, Jodi Magness, Lynn Makowski, Eileen Markson, Janet Monge, Albert Nijboer, Carl Nylander, Annette Rathje, Cynthia Reed, David Romano, Brian Rose, Francine Sarin, Phil Schmitz, Darby Scott, Erika Simon, Stephan Steingräber, Alwin G. Steinmayer, Simonetta Stopponi, Judith Swaddling, L. Bouke van der Meer, and Don White.

ABBREVIATIONS

CAD	*The Assyrian Dictionary* of the Oriental Institute of the University of Chicago
CCAG	Cumont, Boll et al. 1898–1953. *Catalogus codicum astrologorum graecorum*. Brussels. (1889–1911)
CIE	*Corpus Inscriptionum Etruscarum*. Leipzig and Florence. (1893–)
ES	E. Gerhard, *Etruskische Speigel*, vols. 1–4. Berlin. (1843–1867)
ET	H. Rix et al., *Etruskische Texte. Editio minor. I: Einleitung, Konkordanz, Indices; II: Texte*. Tübingen. (1991) Provenance: Ad – Adria; Af – Africa; AH – Ager Hortanus; Ar – Arretium; AS – Ager Saenensis; AT – Ager Tarquiniensis; AV – Ager Vulcentanus; Cl – Clusium; CM – Campania; Co – Cortona; Cr – Caere; Cs – Corsica; Cy – Cyprus; Fa – Falerii et Ager Faliscus; Fe – Felsina; Fs – Faesulae; La – Latium; Li – Liguria; LL – *Liber linteus* Zagrabiensis; Na – Gallia Narbonensis; NU – Nummi; OA – Inscriptiones Originis Australis; OB – Inscriptiones Originis Borealis; OI – Inscriptiones Originis Ignotae; Pa – Padana; Pe – Perusia; Po – Populonia; Ru – Rusellae; Sa – Sabina; Si – Sicilia; Sp – Spina; Ta – Tarquinia; TC – Tabula Capuana; Um – Umbria et Ager Gallicus; Vc – Vulci; Ve – Veii; Vn – Vetulonia; Vs – Volsinii; Vt – Volaterrae
REE	*Rivista di Epigrafia Etrusca*, appearing continuously in *Studi Etruschi*

FIGURE 1. Engraved bronze hand mirror from a tomb at Tuscania, c. 300 BC: scene of a haruspicy ritual. A youth labeled *pava tarchies* examines a sheep's liver in company of gods and heroes (Veltune, Ucernei, Avl Tarchunus, Rathlth) as Thesan (Dawn) looks on. Variously identified as either an Etruscan youth's cultic initiation or the prophet Tages teaching the *etrusca disciplina* to Tarchon, founder of Tarquinia. The Etruscan scriptures (*etrusca disciplina*) were attributed to the supernatural Tages, who dictated them during the protohistoric period; by the first century BC, the *Brontoscopic Calendar* would have been understood as part of that body of ritual texts. (Florence, Museo Archeologico Nazionale, inv. 77759. After L. A. Milani, *Il R. Museo Archeologico di Firenze*, vol. 2, Florence, 1912, pl. 39.)

FIGURE 2. Bronze votive figurine, 4th c. BC, depicting an Etruscan haruspex. The costume of the haruspex, consisting of a tall, pointed hat and fringed (or fleece) cloak fastened with an ancient fibula, symbolized the early traditions of Etruria, such as pastoralism and rituals linked with the land. Inscribed (*ET* Vs 3.7) in letters of the region of Volsinii (Orvieto): *tn turce vel sveitus*, "Vel Sveitus gave this." (See Bonfante and Bonfante 2002:165, no. 46.) (Vatican, Museo Etrusco Gregoriano. Sketch after Pfiffig 1975:48, fig. 6.)

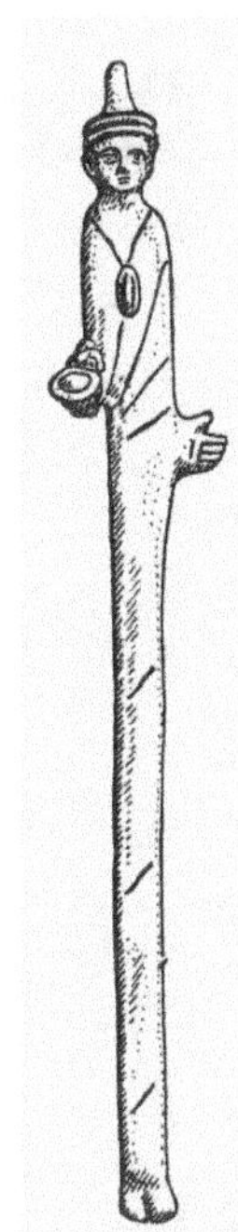

FIGURE 3. Bronze votive figurine depicting a haruspex in elongated style, 3rd c. BC. (H. 0.34 m.) The attenuated format for anthropomorphic figures was used in votive images of gods, goddesses, priests, and votaries, including many deposited at the Sanctuary of Diana at Nemi. Does this design, which may be traced to the Archaic period, reflect some preference for slimness as a healthy condition? See the apodosis in the Etruscan *Brontoscopic Calendar* for December 29: "If it thunders, it signifies the most healthful leanness for the bodies." (See Cristofani 1985: 275, no. 73. On the style, see Haynes 1985:322, no. 199; for Archaic elongated-style figurines, see Terrosi Zanco 1961; Turfa 2006A:99, fig. VI.10.) (Rome, Museo Nazionale Etrusco di Villa Giulia, inv. 24478 [formerly Collezione Kircheriana]. Sketch after Pfiffig 1975:119, fig. 50.)

FIGURE 4. Terracotta inscribed *Tabula Capuana*, the so-called Capua tile, ca. 480 BC, found in the necropolis at Santa Maria di Capua Vetere, loc. Quattordici Ponti. Not actually a roof tile, but a leaf in what was once a stacked set of texts comprising a calendar of dates on which certain religious ceremonies were to be performed. The text may derive from an even older document (see Cristofani 1995). (Berlin, Staatliche Museen, Antikensammlung, inv. 30892. Facsimile drawing after Bonfante 2006B:12, fig. II.3.)

FIGURE 5. Sample section of the *Liber linteus Zagrabensis*, the "Linen Book of Zagreb," section IX, obtained in Egypt in 1848–49, inscribed (in black and red ink) probably in the 2nd c. BC on a much older textile. It was later used to wrap the mummy of a woman buried during the 1st c. BC or AD. For text and explication, see van der Meer 2007:199–210 (transcription). (Zagreb National Museum. Drawing after Roncalli 1985:42.)

FIGURE 6. Stone slab with bilingual epitaph of priest, Lars Cafates, mid-1st c. BC (a contemporary of Nigidius Figulus), found built into the Late Antique walls of Pesaro, the ancient Pisaurum, with other finds from a nearby Roman necropolis. (This is the only bilingual Etruscan inscription found outside Etruria.) Inscription (*ET* Um 1.7): L. Cafatius l(uci) f(ilius) Ste(llatina tribu) haruspex fulguriator/cafates l(a)r(th) l(a)r(is) netśvis. Trutnvt. Frontac ("Lucius Cafatius, son of Lucius, of the Stellatina tribe, haruspex, interpreter of thunderbolts/ Cafates Larth, son of Laris, haruspex, priest and interpreter of thunderbolts"; see Bonfante and Bonfante 2002:69). (Pesaro, Museum Oliverianum, inv. 378. Drawn after photos, Benelli 2000 and Stopponi 2000:186, no. 329.)

FIGURE 7. Relief on bronze sheet from the side panel of a parade chariot found at Castel San Mariano (Umbria), ca. 525–500 BC, depicting Tinia (Jupiter) with a trident-like thunderbolt, in conversation with Hercle (Hercules). One of the Archaic versions of the thunderbolt. (Perugia, Museo Archeologico Nazionale, inv. 453 [Bellucci inv. 1427].) After A. E. Feruglio, 131–34, in *Gens antiquissima Italiae: Antichità dall'Umbria a New York* (1991).

FIGURE 8. Engraved bronze hand mirror, ca. 470 BC. Tinia in a scene with Menrva as Thesan (Eos/ Dawn) and Thetis beseech his aid for their mortal sons (Memnon and Achilles) during the Trojan War. Tinia holds triple thunderbolts. (Vatican: Museo Gregoriano Etrusco, no. 12257. After *ES*, vol. 4:pl. 396.)

FIGURE 9. Engraved bronze hand mirror depicting the gods in Olympos, from Vulci, ca. 325 BC. Tinia, enthroned in the center of the upper register, holds a lightning bolt with floral appearance. (Paris, Bibliothèque Nationale, Cabinet des Medailles, inv. 1287. After *ES*, vol. 2:pl. 181.)

FIGURE 10. Engraved bronze hand mirror, 3rd c. BC. Menrva brandishes a lightning bolt resembling a warrior's spear. (Berlin, Staatliche Museen, Antikensammlung. After *ES*, vol. 3:pl. 246 [see Pfiffig 1975:256, fig. 111; see now Corpus Speculorum Etruscorum Bundesrepublik Deutschland 4:24a].)

FIGURE 11. Engraved bronze hand mirror, ca. 300–275 BC, Lasa Vecu (Vegoia?) holding a small, floral lightning bolt(?), or perhaps a plant, converses with Menrva. The lightning bolt, or perhaps a symbol of the earth, crops, or fertility, is depicted in the hands of a supernatural prophetess (see Appendix A, Text 2). (Rome, Museo Nazionale Etrusco di Villa Giulia. After *ES*, vol. 1:pl. 37.)

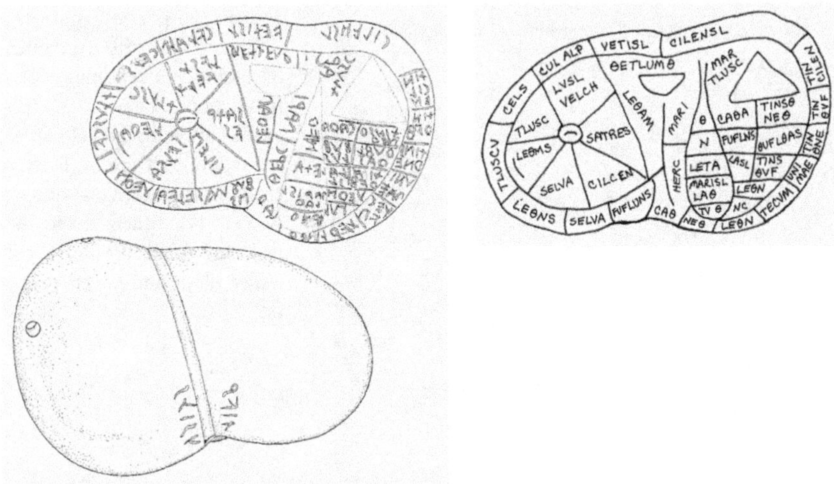

FIGURE 12. Views of the bronze Piacenza liver model found at Settima (territory of Piacenza), end of the 2nd to beginning of the 1st c. BC. The names of gods are inscribed on both surfaces, but the set of deities is not the Greek Olympians, nor is it those usually depicted in art or cultic rituals. A, B = facsimile drawings, dorsal and ventral sides; C = names of gods on ventral side transliterated. (Dorsal side: Usil the sun god, and Tiv/Tiur, the Moon.) (Piacenza, Museo Civico, inv. 1101. After Bonfante and Bonfante 2002:173, no. 51; Turfa and Gettys 2009:46, fig. 5; also see van der Meer 1987:10–12.)

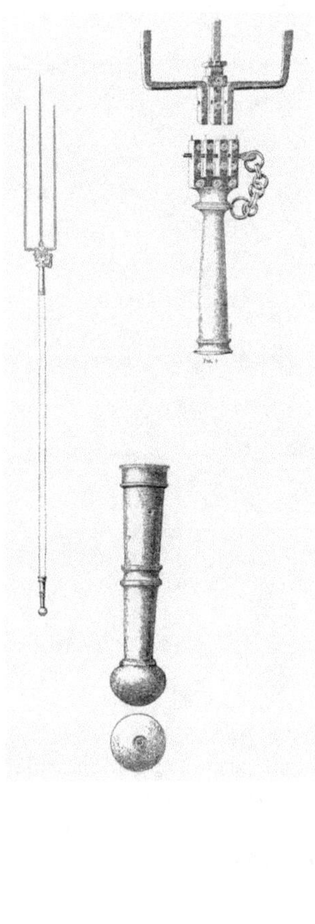

FIGURE 13. Bronze trident from Vetulonia, Tomba del Tridente, first quarter of the 7th c. BC. The beautifully fashioned, two-part prongs are too decorative and fragile to function as spears. (Total length 89.5 cm.) The implement does resemble Near Eastern thunderbolt insignia. (See Cygielman and Pagnini 2006:72–75, nos. 191–192, pl. VII,f,I; Sciacca 2004:271, figs. 2–3.) (Florence, Museo Archeologico Nazionale, inv. 7123, 7124. After Falchi 1908:426–27, figs. 7–9.)

FIGURE 14. A fragmentary Assyrian *kudurru* (ornamental monument reflecting a land transaction), in a rock relief erected by Sennacherib (704–681 BC) at Bavain (Khinnis). To left of center in the lower row of gods' symbols is a trident shape, representing Adad, the storm god. (After Black and Green 1992:17, fig. 10.)

FIGURE 15. Relief from the Temple of Ninurta, Kalḫu (Nimrud), era of Ashurnasirpal II (883–859 BC): the god Ninurta (or perhaps Adad) wields triple-pronged double-sided thunderbolts in confronting a monster. (After Black and Green 1992:142–43, fig. 117 [after Layard].)

FIGURE 16. Relief (upper register) from Room II of the Palace of Sargon II (721–705 BC) at Dur-Sharrukin (Khorsabad). Servants (eunuchs?) dipping lion situlae into a large cauldron for a banquet, probably a victory celebration for Sargon's campaign in the East. Related lion situlae have been found in tombs at Phrygian Gordion and Etruscan Veii. (After Botta and Flandin 1849:pl. 76 [see Ebbinghaus 2008:183, fig. 3].)

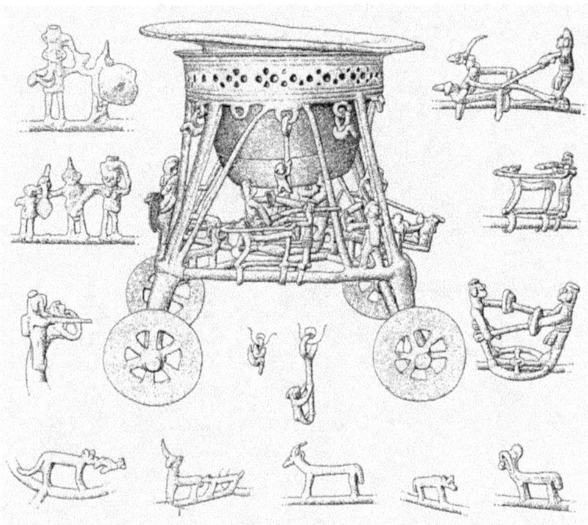

FIGURE 17. Wheeled bronze censer from Tomb 2, Capodimonte (Bisenzio), Olmo Bello necropolis, end of 8th to beginning of 7th c. BC. Figures and scenes relevant to the society of the rulers of Bisenzio decorate it: warrior and woman with water jar, warrior and child with woman, plowman with team of oxen, hunter, archer, swordsmen dueling, dogs, sheep/goats, wolf with prey, pendant birds, monkeys. (See Torelli 1997:37–46.) (Rome, Museo Nazionale Etrusco di Villa Giulia. Drawing after Woytowitsch 1978:pl. 24, and Kossack 1999:41, fig. 25.)

FIGURE 19. The Plikaśna situla, found in 1700 at Chiusi. Gilded silver vase incised with the name "Plikaśnaś," and scenes with men, women, soldiers, and animals, in the hunt, games, procession, and perhaps preparation for sacrifice. Perhaps made in the region of Caere by an immigrant Near Eastern artist, ca. 650 BC. (See Martelli 1973:pl. 31.) (Florence, Museo Archeologico Nazionale, inv. 2594. After Inghirami 1825:pls. XIX–XX.)

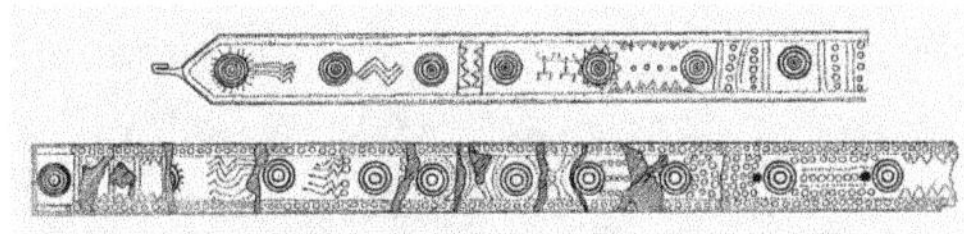

FIGURE 18. Drawing of bronze belt from Vulci, Tomb 42F, early 7th c. BC. One of three belts, deposited in a woman's tomb, this has an unusual set of images, including human figures and perhaps symbols of celestial bodies. (University of Pennsylvania Museum, inv. MS 691. After Dohan 1942:fig. 63; see Turfa 2005:102, no. 31.)

FIGURE 20. Incised scenes from the Etrusco-Corinthian Tragliatella oinochoe, deposited in a chamber tomb in the Tragliatella necropolis (between Cerveteri and Veii), during the last quarter of the 7th c. BC. On the neck: man and woman, flanked by ship and goat, birds and man with goat on lead; feline protomes pendant below the scene (omitted here). On the body, centered on two images of a draped woman: parade of infantry with matched shields converges on family scene of couple with child; matron turns left to enter a stone-cut tomb, two couples copulating, maze game labelled "Truia" and two cavalrymen with matched shields, one with a pet monkey. On a painted band above base, incised hound and hare. Early images expressing Iron Age beliefs and behaviors: symbols of the ruling class include hunting (hound and hare), herds (goat), fowl or wild birds, a ship, a city militia, aristocratic cavalry display, two scenes of *hieros gamos* (ritual marriage), and a family with inscriptions proclaiming "I am –" (*mi mamarce, mi velelia, mi thesathei*) "father Mamarce, mother Thesathei, and daughter Velelia." The matron turns away and seems to enter her rock-cut tomb. (See Small 1986; Torelli 1997:29–31, 181–82, suggests Theseus and Ariadne, with additional comments.) (Rome, Musei Capitolini, inv. 358. After Giglioli 1929; Amann 2000:pl. 15b.)

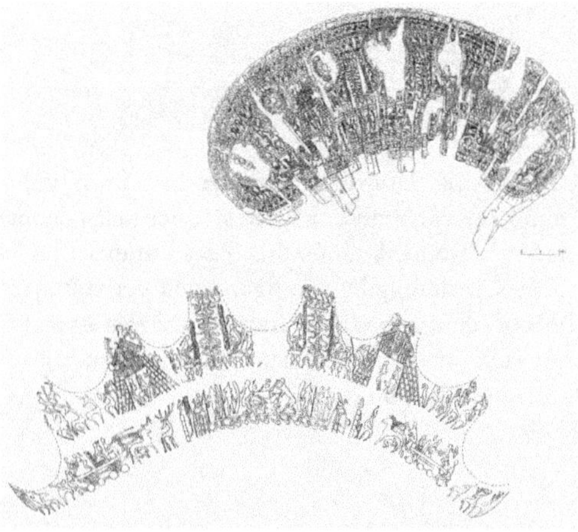

FIGURE 21. The Verucchio throne from Tomb 89, Podere Lippi necropolis, Verucchio, the burial of a distinguished man, ca. 700–680 BC (von Eles 2002). Incised and openwork back of a cylindrical throne of wood. Amid geometric patterns overall, the interior back of the throne bears intaglio figural scenes of life among the elite of Verucchio: women and warriors, animals, a procession of carts (possibly one side male, the other female) converges on houses with women weaving elaborate textiles. Various interpretations have been suggested: daily life, religious ceremonies, wedding or family life (cf. Torelli 1997:52–68). (After Kossack 1999:65–66, figs. 43–44.)

FIGURE 23. The Certosa situla, a bronze, repoussé-decorated vase with scenes from the life and rituals of the ruling class of Felsina-Bologna, ca. 500 BC, believed to preserve traditions that began much earlier in Tyrrhenian Etruria, the ceremonial activities of the Archaic city. The four registers depict fantastic beasts, herders and huntsmen, the elite at leisure, processions of men and women (offering sacrifice?) and a parade of infantry and cavalry. (See Bonfante 1981:36ff.) (Bologna, Museo Civico Archeologico. After Kossack 1999:69, fig. 46.)

FIGURE 22. Sheet-bronze, repoussé-decorated ceremonial rattle (*tintinnabulo*) from Bologna, Arsenale necropolis, Tomba degli Ori, ca. 600 BC. The life and importance of women in Etruscan society: scenes of matrons preparing and weaving elaborate textiles; the rattle, in the shape of a sacrificial ax head, was presumably used for religious rituals. (After Morigi Govi 1971:pls. LII and LIV.)

FIGURE 24. Relief panel from one side of the Sperandio sarcophagus: procession (triumph and sacrifice?), ca. 480 BC. Found in the Sperandio necropolis of Perugia in 1843. L. of sarcophagus 1.90 m.; relief panel 1.575 x 0.58 m. The short sides have small reliefs depicting banqueters served by nude youths; the front side has a procession of a young man leading bearded men who are roped together and carrying goods, also women, men with laden mules or donkeys, a dog and equipment, young men carrying spears and others herding goats and cattle. Possible interpretations range from the initiation of a *ver sacrum* emigration (to found a new town) to the return from a raid, departure of leader and colonists, or the procession and sacrifice of a triumph. In any case, servants of lower degree, captives, beasts, and rituals fit the lifestyle assumed in the *Brontoscopic Calendar.* (Perugia, Museo Archeologico, inv. 195. After Jannot 1984:42–44, no. 340 (Group C,I,1), figs. 158–159.)

IMAGES RELATING TO THE SOCIETY PORTRAYED IN THE ETRUSCAN *BRONTOSCOPIC CALENDAR*

PART I

BACKGROUND

CHAPTER 1

THE *BRONTOSCOPIC CALENDAR* AND ITS TRANSMISSION

A REMARKABLE DOCUMENT, ALBEIT THIRD-HAND

More than once, an entire culture has been eliminated, or overwhelmed by a conqueror, and its literature and memories lost to posterity. To recover some lost remnant of a people's beliefs and expressions is a rare privilege, and it appears that one such window of opportunity has opened onto the religious teachings for which Etruscans were famed. Preserved in the manuscript of a short book on omens, *De ostentis* (*On Portents*), published by Johannes Lydus ("John the Lydian") in Justinian's Constantinople, is a precious example of Etruscan sacred scripture, even though it is not recorded in the Etruscan language. The (Byzantine) Greek edition, Εφήμερος Βροντοσκοπία, which appears at *De ostentis* 27–38, is the only surviving text of an Etruscan divinatory calendar. It had previously been published in Latin by Publius Nigidius Figulus (fr. 83 in Swoboda 1889), a respected contemporary of Cicero. If its original-language text had survived, it would be the longest coherent Etruscan document extant, because so few Etruscan texts even have complete sentences or multiple lines of comment on the same topic. No trace of the Etruscan original has yet been found, but Lydus, taking his cue from Figulus, implied that it was of great antiquity, a part of the Etruscan religious *disciplina* dictated by the legendary prophet Tages, he himself understood to have lived during the protohistoric Iron Age, probably the ninth to eighth centuries BC. Figulus' Latin translation was available in sixth-century AD Constantinople when Lydus converted it into contemporary Greek language, making it much more accessible to his fellow citizens in the Eastern Empire. Today, only his Byzantine Greek text remains.

Organized into twelve months, beginning in June, the calendar had functioned as a reference table for priests interpreting the phenomenon of "thunder," a loud noise heard in the heavens (with or without storms). Embedded in it is a wealth of social, agricultural, religious, and medical information. The stratified society of the calendar is composed of urban factions that include "powerful men," nobles, a "band of youth," common people, women, and a servile class that alternately seems to foment rebellion and be stricken by plagues. The *Brontoscopic Calendar* mentions a king, war, onslaughts of various noxious and "harmless" diseases, and a wide array of meteorological conditions. It predicts an abundance or dearth of certain crops, including barley and wheat, fruit and nuts. Herds, flocks, wild birds, and fish of both river and sea are also noted, along with "reptiles" and pests.

Comparative analysis shows that the *Brontoscopic Calendar*'s sources lie in equal parts in the native, emerging Etruscan culture of Iron Age Italy and in the wisdom of Mesopotamia, as known to us from the great cuneiform tablet libraries of Ashur, Nineveh, and Babylon. I suggest that, at the dawn of urbanized history in Tyrrhenian Etruria, the elite groups who were establishing the cities, stratified societies, and centralized governments availed themselves of any tools that came to hand, and one of the best – already known to the Near Eastern kingdoms for millennia – was control of the calendar and the bonding of religion with the state.

THE MATTER OF TRANSMISSION

Those who study Etruscan civilization need to be experts in the culture of the Etruscans' enemies, for the only histories and commentaries that survive are those written by their Greek rivals and Roman conquerors. Fortunately, archaeology is rectifying the picture of Etruscan society and religion, known to have been distorted, albeit not always deliberately, by foreign authors. I hope that my English version will stimulate among twenty-first-century scholars a renewed interest in this text and in its early Etruscan development.

The original Etruscan text was probably composed early in the seventh century BC, if not slightly before, and may have been the product of a single Etruscan devotee or possibly one group of priestly colleagues conversant with native, Italic, and Etruscan religious traditions. The thunder calendar was created after someone had somehow been introduced to the divination literature of the Near East, either on the home ground of Etruria, or possibly in the Levant or some middle ground, such as Cyprus. The transfer could have been accomplished by a single Assyrian or Levantine diplomat-translator or priest, perhaps traveling with an embassy that participated in an economic and political mission to one of the Etruscan chiefs. The Near Eastern visitor then communicated the lore of Akkadian divinatory reference texts to some counterpart priest (or priestess) in Etruria. The "donor" of the divination texts might have been Levantine – Phoenician or Syrian – and the texts recorded not on bulky cuneiform tablets (which have never yet been found in Italy) but in Aramaic or Phoenician on convenient papyri or scrolls. (The courts of Assyria and its neighbors all had multilingual scribes and archives of organic media – and clay.) The creation and transmission of the Etruscan text would have required only a small coterie of experts: in later centuries, individual (noble) families were recognized as the custodians of cultic lore, and priesthoods characteristically passed from father to son.[1] Their scions, too, were proud of their learning – they continued tradition by writing their own treatises on topics of religion, as proclaimed on Hellenistic monuments, such as the sarcophagus of Lars Pulenas (he names his family and ancestors and claims *ziχ netśrac*, to "have written a work on haruspicy."[2] The precarious time for a rare and privileged religious document would have been the centuries immediately after its creation, but that would have been during the *floruit* of the major Etruscan cities and the heyday of the great families whose names continued to the end of the first millennium BC. It is true that we lack Etruscan literature (as opposed to personal inscriptions) for the Archaic period, and the long texts that have been found are nearly all on nonperishable materials (and nearly all of religious impact – for instance, the gold Pyrgi plaques, the Capua tile, the stone Perugia *cippus*). But the linen book known as the Zagreb Mummy Binding (*Liber linteus*) was preserved in the dry climate of Egypt and is a tantalizing remnant of the religious texts that Etruria once created.

As a written Etruscan text, the *Brontoscopic Calendar* did survive into the early first century BC, when it was made accessible to a Roman statesman and scholar of probable Etruscan ancestry, Figulus. In the days around or just before 50 BC, there were still people in central Italy capable of reading Etruscan (as, later, was Emperor Claudius), but Figulus foresaw the future demise of the Etruscan language and Rome's need for religious inspiration. By then, however, many other texts, of more recent, and questionable, origins were circulating in Ptolemaic Alexandria, in Rome among the literati, and in the Hellenistic East. All of those that survive, however, betray undeniable hallmarks of their post-Assyrian ancestry. Some may even have used the Figulus document as one of their source references (see Chapter 9).

PUBLIUS NIGIDIUS FIGULUS

Whatever form the original Etruscan *Brontoscopic Calendar* took, by the sixth century AD it only survived in the Latin translation of the Roman statesman and scholar, Figulus. He had earned a reputation as a senator, statesman, grammarian, occultist, and natural historian (encyclopedist), yet only the last twenty years of his life are known in any detail. That any information exists is in large part because Figulus was a friend whom Cicero greatly admired (he called him *omnium doctissimo et sanctissimo* ("most learned and pious of all men") in the letter he wrote him in exile in 46 BC (*Ad familiares* 4.13.3)). Integrity seemed to be a hallmark of this scholar, along with a keen intellect;

[1] Cf. Haack 2005 and 2006; Briquel 2002.

[2] *ET* Ta 1.17; Bonfante and Bonfante 2002:149–51, no. 31, 3rd c. BC.

Cicero exclaimed that Figulus was "graced with all the other arts that befit a free man."[3]

Figulus must have been born around 100 BC, to a well-to-do family of plebeian background, probably in Perusia, where Nigidii are attested in epitaphs since the second century BC.[4] Hall suggests that the Nigidii gained favor in Rome during the Second Punic War, and thus they would have maintained a presence in Roman politics from that time.[5] Unwavering integrity and sincerity were Figulus' downfall – and his fame. The friendship with Cicero began around 63, when Figulus participated in the anti-Catilinarian crusade (Cicero *Pro Sulla* 42). He held the magistracy of praetor in 58 BC (the one certain date in his biography), when he was among the *amicissimos et acerrimos civis* ("most friendly and keenest citizens") who supported Cicero (*Epistulae ad Quintum fratrem* 1.2.16). During the Civil War, Nigidius fought with the *optimates* for Pompey at Pharsalos (cf. Cicero *Ad Atticum* 7.24). Figulus, *cui cura deos secretaque caeli nosse fuit* ..., "to whom it has been given to know the cultivation of the gods and the secrets of heaven," delivers a speech in Lucan's *De bello civili* (1.639–72), in which he prophesies "that peace of yours comes with a master" (1.670, *cum domino pax ista venit*), but this means that freedom will end. Clearly, for his contemporaries, divination was his most distinguishing pursuit. The inevitable exile followed in 46, and he died a year later, just as Cicero was preparing a speech to deliver on his behalf. It is not certain that Figulus was the subject of a Suetonian biography,[6] but his fame was further transmitted in Suetonius' comment that it was "common knowledge" (*nota ac vulgata res est*) that Figulus had predicted the Civil War and the birth of Augustus (*Augustus* 94.5). Figulus' place in Cicero's *Timaeus* is understood today as a memorial tribute, perhaps crafted from the text the orator had composed for his undelivered speech to petition Caesar for clemency to Figulus.

Figulus was a prolific writer, and his lengthy *Commentarii grammatici* led to frequent comparison with his contemporary, Varro.[7] A tradition that Figulus revived Pythagoreanism is not supported by his surviving scholarship, fragments of which have been noted by Pliny, Gellius, the scholiasts, and Byzantine authors. The reason for his *cognomen* Figulus, "the Potter," was explained by the scholiast on Lucan (*De bello civili* 1.639) and by Augustine (*De civitate dei* 5.3). When asked how horoscopes for twins could differ, Figulus supposedly demonstrated that in a revolving universe, no two beings can have the same identity or fate, just as two ink splashes on a *potter's* turntable are always distinct from each other.[8]

[3] *In Vatinium* 16, quoted by Barton 1994:37.

[4] Cf. Harris 1971:321–22.

[5] Hall 1996:165n113.

[6] See Della Casa 1962:17–36.

[7] Gellius *Noctes Atticae* 19.14; Servius *Ad Aen.* 10.175.

[8] The possibility of a completely different and Etruscan source of his *cognomen* remains open, however; see Della Casa 1962:9–17.

Titles are known for a number of Figulus' otherwise lost religious and natural historical works, all in Latin:

De diis (*On the Gods*): probably consulted by Cicero for his own *De natura deorum*; cf. Macrobius Saturnalia 3.4.6

De hominum natura (*On the Nature of Men*)

De animalibus (*On Animals*)

De ventis (*On Winds*): contained meteorological, astronomical, and astrological data

Sphaera Graecanica and *Sphaera Barbarica* (*The Dome of the Sky, Greek*, and *The Dome of the Sky, Near Eastern*): discussed names of constellations, astrology, horoscopes

De extis (*On Divination by Organs* [of sacrificed animals])

De augurio privato (*On Private Divination*)

De somniis (*On Dreams*): one quote or paraphrase is preserved in Lydus, *De ostentis* 45

Diarium tonitruale (*Brontoscopic Calendar*): said to be a translation or adaptation of part of the books of Tages

Figulus was distinguished for writing *communes litterae*, in contrast to Varro's reputation for *theologia*, according to Servius (*Ad Aen.* 10.175). Harris suggests that the work cited by Servius, in which Figulus denigrated the reputation of the Ligurians as thieves, was called *De terras* (*de terris*?) and should be considered a separate treatise from the *Sphaera*(*e*) to which it is sometimes attributed.[9] Little is known about the exact timing of his writings, but they cannot have begun "much before the 80s or 70s at the earliest."[10] According to Macfarlane,[11] the output of Figulus the author was "numerous, diverse and certainly at hand" to influence authors and speakers of the Augustan period, but we do not know how long individual treatises circulated after that time: the *Brontoscopic Calendar* is the only work to have survived in near-complete (albeit translated!) form – could that possibly have been influenced by its different, "scriptural" character in contrast to all the personal and philosophical treatises?

Astrology, apart from some popular incidents with the wandering *Chaldei*, "Chaldeans" as Romans called hireling soothsayers, seems only to have been legitimized in Rome during the first century BC, under the aegis of two scholars: Figulus and Tarutius of Firmum (also known to Cicero).[12]

[9] Harris 1971:23.
[10] Harris 1971:182.
[11] Macfarlane 1996:243.
[12] Barton 1994:37.

It seems that Figulus also expressed the Stoic and supposedly Babylonian doctrine of the world's ending in conflagration and of catasterism, the ascension of individual souls to the stars after death (presumably in the *Sphaera Graecanica* and *Sphaera Barbarica*). As so often happens, the public perception of deep philosophical works can reduce them to mere fortune-telling and magic, and so Figulus was credited with having predicted the Civil War between Pompey and Caesar, and also the birth of Augustus (Lucan *De bello civili* 1.649–65). Suetonius said that "everyone believes this story" (*Augustus* 94.5).

Suetonius recorded a portent that obliquely cited lightning and was identified by Figulus as forecasting the greatness of Octavian at the time of his birth in 63 BC. This expressed a long-standing Italic tradition, apparently, for Suetonius adds that "in ancient days, when a part of the wall of Velitrae had been struck by lightning, the prediction was made that a citizen of the town would one day rule the world."[13] On the day of his son's birth, in the midst of the Catilinarian affair, Octavius arrived late at the senate, "then Publius Nigidius, as everyone knows, learning the reason for his tardiness and being informed also of the hour of the birth, declared that the ruler of the world had been born."[14] Figulus' noting the hour of birth sounds more like astrological divination, but the association between divination and ruling power is one that occurs in many cultures.

Figulus' erudition earned him the respect of his contemporaries and presumably furthered the promising public career that his personal political convictions ultimately ruined. He was one of the most successful (or best documented) of the men who turned their sights from Etruria to seek their futures in Late Republican Rome. Ancient scholars seem not to have questioned the authenticity or validity of his research, and the praise of Cicero, that scathing critic of diviners, registers as particularly sincere about a colleague whose personal ethics were as admirable as his arcane studies.[15]

LYDUS (JOHN THE LYDIAN)

Johannes Laurentius Lydus, "John the Lydian," was born in Philadelphia in Lydia in 490 AD, and he received an excellent Classical education. He came

[13] Suetonius *Augustus* 94.2/Rolfe 1998:287 – *Velitris antiquitus tacta de caelo parte muri responsum est eius oppidi civem quandoque rerum potiturum.*

[14] Suetonius *Augustus* 94.5/Rolfe 1998:289 – *nota ac vulgata res est P. Nigidium comperta morae causa, ut horam quoque partus acceperit, affirmasse dominum terrarum orbi natum.*

[15] For references on Nigidius Figulus, see Schmidt, 2000; Hall 1984:316–17, no. 61.2; Liuzzi 1983; Della Casa 1962 (also see review by Thesleff 1965); Kroll 1936; Legrand 1931, all of which have been superseded by references above; Turfa 2006D:174–75. Texts of surviving fragments are Swoboda 1889 and Hakkert 1964.

to Constantinople in 511, where, after studying philosophy (especially the Neoplatonism popular in previous centuries), he received an appointment to the praetorian prefecture. His next forty years were spent in civil service in Constantinople, where one of his distinctions was his ability to translate Latin documents – apparently by the era of Justinian, few officials were so qualified. Lydus' palace career finished with the prestigious appointment to teach in the Imperial School (ca. 543); he retired in 551, but he continued to write; he died in Constantinople in 560 AD.[16]

In his position in the Byzantine government and the palace school, Lydus must have had access to many manuscripts that are now lost, for although Byzantium treasured many elements of past civilizations, under Justinian there were three waves of persecution of pagans. In 528–29, 545–46, and 562 AD even courtiers, civil servants, scholars, and privileged aristocrats were seized in the purges, and one can only assume that questionable texts were destroyed. Lydus' treatise *De ostentis*, like his others written in the contemporary Greek, made otherwise obscure works by Latin writers accessible to the scholars of his day. It seems, however, to have been published without any of the difficulties – namely, charges of heresy – that accrued to some other experts on pagan topics, for Lydus was honored with the chair of Latin Philology at the palace school. Some of his comments on natural phenomena seem to clarify a Christian viewpoint pleasing to his superiors. For instance, in *De ostentis* 9, noting alarming past solar eclipses, he observes that "there is good reason to honor Divine Providence since for the entire duration of this most blessed reign, there has been no eclipse whatever verified under this sign [the royal triangle Aries-Leo-Sagittarius]." At *De ostentis* 54, he notes tendentiously, "the whole universe is organized according to physical laws but above them is the rule of Divine Providence and of the laws of fate."[17] Though he was not likely a "crypto-pagan," Lydus was certainly a curious scholar inquiring into antique Roman curiosities, and, as Michael Maas puts it, "simply someone whose professional life, culture, and intellectual interests were free of Christian decoration."[18]

The survival of his work shows that his audience believed that he was not a pagan adherent of the old works he translated. It also speaks to the perceived authenticity of the works he transmitted, at least those in *De ostentis*, because there could be no reason to alter them for his own personal beliefs or for political propaganda. His high ultimate rank in the palace bureaucracy must also have guaranteed him access to libraries and archives, which is additional

[16] Sources for his life are his own works and a brief entry in Suidas, s.v. *Joannes Philadelpheus Lydos*; see Klotz 1927:2210.

[17] ... προνοίᾳ δὲ θεοῦ καὶ λόγῳ σιγῆς. Other references to Providence are *De ostentis* 16 and 22.

[18] Maas 1992:5.

circumstantial evidence for other Byzantine scholars' confidence in the authenticity of the ancient works preserved in the Latin expert's treatises.

As Ilaria Domenici, Erika Maderna, Dominique Briquel, and others have pointed out,[19] Lydus evinced a firm conviction that what one learns from the past can be valid for the present. Like many of his scholarly and political contemporaries, he saw himself as Italian (although born in Asia Minor), and he saw the power of Constantinople as the logical development of Roman civilization – a culture that had itself grown from Etruscan roots. His selection of natural phenomena for divination or analysis may well have seemed to his readers to be confirmed when the comet of 540 AD presaged the destruction of Antioch by "Rome's" enemy, Khosru of Persia, and he and his contemporaries knew of other events to link to past political disasters.

Briquel, noting the proactively Christian tone of Justinian's Constantinople, explains Lydus' lack of heretical rumor by his reputation for esotericism: "En fait, avec en auteur comme Jean le Lydien, c'est simplement d'un goût pour l'ésotérisme, pour les saveurs mystérieux et cachées, qu'il s'agit."[20] Introducing *De ostentis*, Lydus cites a great variety of authors on the topic of divination: among the Greeks, Egyptians, and Near Easterners named were Petosiris, "the divine Ptolemy" (the 2nd-c. AD Alexandrian astronomer and author of the *Tetrabiblion*), and even Zoroaster. It appears that Tages and the Etruscan scriptures offered an Italian parallel for the Eastern material already so familiar to scholars, and they became a focus for Lydus of the wisdom of Rome, to which he and his fellow Constantinopolitans were the logical heirs.[21] Although eloquent in denouncing public figures (like John of Cappadocia) in his work on contemporary bureaucracy, Lydus offers no criticism of ancient paganism, and, apart from the mild references to Divine Providence, he gives little indication of Christian belief having influenced his writing.

For Lydus, the Etruscan *Brontoscopic Calendar* was just one of many pagan documents on the subject of divination that crossed his desk; it appears to be the oldest of the remaining brontoscopies (calendars for divination by thunder) of Classical antiquity that Lydus preserved, embedded in his antiquarian and learned treatises. These held other texts for divination also, using as omens weather, earthquakes, natural or unnatural phenomena, and of course the zodiac, moon, and stars. The rest of the texts had been created in the heady atmosphere of affluence, encyclopedic scholarship, and popular culture that flourished in the Hellenistic and Roman Near East and Egypt, and they continued to be disseminated, at least among scholars, through the entire Byzantine and medieval period (the last of the late brontoscopic manuscripts was hand-copied in the nineteenth century!).

[19] Domenici and Maderna 2007:11–12; Briquel 1997:197–98.

[20] Briquel 1997:197.

[21] Briquel 1997:197–98; Mastandrea 1979:76; Domenici and Maderna 2007:11–12, 18–19.

In the field of ancient religion, Lydus published one other antiquarian work that remains of great importance – *De mensibus* (*On the Months*), on the Roman calendar and holidays – and it includes some pagan festivals still celebrated under Justinian. In the history of Byzantium, Lydus is actually better known for *De magistratibus* (*On the Magistracies*), a complete historical discussion of Roman bureaucracy from its origins to his own day. There, in addition to making strongly political remarks about his contemporaries, he stressed the continuity of ancient developments with the culture of his own day. (References exist to lost works by Lydus, ranging from panegyrics on his contemporaries to straightforward history, and even poetry.)[22]

SURVIVAL AND TRANSMISSION OF THE TEXT

So precarious was the survival of the written works of antiquity that the Etruscan *Brontoscopic Calendar* would not be known to us today if a single Byzantine manuscript (*Codex Caseolinus*) had not been (re-)discovered in 1784 in the Bibliothèque Nationale in Paris. This held the only complete text of *De ostentis* and has been dated to the end of the ninth to the beginning of the tenth century. (Only three fragments were previously known.) The manuscript was originally published in 1823 by Karl Benedikt Hase, with updated editions appearing in 1837 (edited by August Immmanuel Bekker), 1863, and finally in 1897 (the last two edited by Curt Wachsmuth). This antiquarian compendium of curiosities on the topic of divination usually is known by the Latin *De ostentis*, but it was titled by Lydus Περὶ διοσημείων, which has been rendered recently by Italian translators Domenici and Maderna (2007) as *Sui segni celesti*, or *On Signs in the Heavens*.

In addition to the document attributed to Nigidius Figulus, *De ostentis* purports to (and surely does) furnish Greek word-for-word translations of several other texts on divination. The Figulus document is the oldest of them, followed by a treatise attributed to the first-century BC author Fonteius Capito. Introduced with short discussions by Lydus, other brief texts are associated with third-century AD and later Roman authors, such as Labeo and Claudius Tuscus, another name that hints at the Etruscan background of such lore. (See Chapter 9.)

PAST SCHOLARSHIP AND THEORIES ON AUTHENTICITY

Scholarship during the past century has been divided on the topic of the value and/or authenticity of the brontoscopic text preserved by Lydus at *De ostentis*

[22] References on Lydus, including previous bibliography, are Baldwin 1991; Tinnefeld 1999; Jones et al. 1980; Maas 1992:1–11; Turfa 2006D:174; Domenici and Maderna 2007:7–19, 28–33.

27–38. Carmine Ampolo (1990–91) and Domenici and Maderna (2007) have summarized the main scholarly approaches to the unusual document. The nineteenth-century editor of this material, Curt Wachsmuth,[23] was skeptical of its authenticity and criticized prior scholarship as too credulous. Wachsmuth treated the text as a Byzantine product, suggesting that it lacked the depth of astronomical/astrological expertise that had made Figulus famous among his contemporaries. (I would instead interpret this as evidence that Figulus had faithfully translated an original document rather than forging it, because a product of his own thought would never have been devoid of his brand of astrological data – and all the later *brontoscopia* are full of the zodiac.)

Anton Swoboda (1889), the previous editor of the Lydus manuscript, had greater respect for the integrity of the text, but he felt uncertain as to whether Figulus could be credited with the complete document. He thought that certain elements indicated a concern with the regime of Justinian, namely, a supposed reference to Christianity in the December 3 allusion to fish ("men will excessively consume their flocks because of a dearth of fish"), and, supposedly, a veiled reference to Theodora in the September 6 entry ("there shall be power among the women greater than [what is] appropriate to their nature"). But the flocks/fish entry, and several similar predictions, can be found in Mesopotamian predecessors and also in the Hellenistic-Roman successors. The entry of September 6 refers to women as a group and not to a queen: there is no reason to connect it to the Empress Theodora. In contrast, examples abound of Etruscan women acting independently, or cited in texts of a "public" nature, such as epitaphs, dedications, and even mortgages.[24] A handful of Byzantinisms noted by early commentators are surely the by-products of Lydus' translation skills and the uncertainty of the Latin translation from which he worked – perhaps more than 500 years after its composition in Etruscan. Oscar Thulin appraised the manuscript tradition of the *Brontoscopic Calendar* in his monumental compendium on the *etrusca disciplina*, *Die etruskische Disciplin*, which remains indispensable for the study of Etruscan religion, even though archaeology has augmented the picture for us.[25] Thulin accepted an Etruscan/Late Republican origin for the original text. He developed the opinion that the document preserved by Lydus had a genuine Etruscan substrate that had been contaminated with Chaldean and Egyptian elements, namely, the astrological literature generated by scholars in Egypt under Persian and Ptolemaic rule, and the Near East.

Wilhelm Kroll (1936), in his *Realenzyklopedia* (*RE*) entry on P. Nigidius Figulus, was able to point to discrepancies between the original Babylonian texts and the Figulus document in the apodoses (prediction clauses) that

[23] Wachsmuth 1897:185.

[24] As in the Cortona Tablet; cf. Amann 2000:80–118.

[25] Thulin 1906:vol. 1, 68–78, 94.

cite enemies of the state and also its internal social or political conflicts. The Mesopotamian apodoses are invariably much simpler and seem more "generic" in their predictions. In fact, earlier studies of Babylonian and related texts by Carl Bezold and Franz Boll (1911) already pointed out close links to Mesopotamian literature that would negate some of Wachsmuth's critical, pro-Byzantine points. Kroll suggested that some elements of the text could be attributed to the events of the last years of Figulus' life, but he also saw some sort of intermediate editor or redactor between the work of Figulus and the manuscript generated by Lydus. With no additional literature by Figulus, or by that unnamed editor, surviving, it is not possible to pursue his theory: it is clear that the bulk of the text is unique, and a wealth of Near Eastern scholarship since 1936 has shown us ample Mesopotamian predecessors.

New respect for the authenticity and value of the brontoscopic document came from André Piganiol (1951), who definitively separated the work from its supposed Byzantine context. He considered several elements of the text to be derived from an Etruscan original, while accepting that some details were references inserted after the fact of events in the lifetime of Figulus. He astutely associated the text with the Babylonian traditional hemerologies, but to explain this he had recourse to the Herodotean version of the Etruscans as Lydian immigrants.[26] He maintained that the political and social conditions cited in the text reflected those of Etruscan civilization at its height (thus probably the Late Archaic to Hellenistic periods). Piganiol, like many others, has noted the similarity of some of the language in the *De haruspicum responso* narrative to the language of this, and other, Etruscan religious writings (cf. Cicero *De har. resp.* 19.40). The friction portrayed there between *optimates* and lesser citizens was believed to reflect not only the atmosphere of Late Republican Rome but also social conditions in the Late Etruscan cities. Cicero insisted that he was recording the words of the haruspices unembellished.[27] The syntax and perception of the formulaic response intoned by the haruspices sounds like the archaic language of a cultic document, and the omen of *strepitus cum fremitu* certainly seems (to us) like an allusion to thunder-omens (*De har. resp.* 10.20; see Chapter 2). Predictions such as those for mid-October appear to have language (always at two removes from the original, of course) similar to that of the haruspices in 56 BC. Could it be that the haruspices in Cicero's day had been quoting or mimicking older, Etruscan ritual texts and formulae?

Mid-October *Brontoscopic Calendar* omens:

19. If it thunders, it warns the downfall of a ruler or the overthrow of a king, but it warns both discord among the common people and abundance.

[26] A position now corrected by Briquel 1991.

[27] *Haruspicum verba sunt haec omnia: nihil addo de meo* (19.40); cf. Ampolo 1990–91:186.

20. If it thunders, it warns there will be a festering wound, and for the many, extreme suffering out of the discord.
21. If it thunders, there will be coughing sicknesses and oppression of the heart.
22. If it thunders, it threatens for the people bad conditions and spotted diseases.
24. If it thunders, out of the discord of those in power, the common people will oppress.

At about the same time as Piganiol, Stefan Weinstock also was reaching a level of newfound respect for the brontoscopic text. In "*Libri Fulgurales*" (1951) and an earlier article (1950), he was able to discern between the products of later authors such as Labeo and Capito, whose works on divination could be traced to the Hellenistic (Ptolemaic- or Seleucid-era) texts of the traveling Chaldeans. This was the tradition that arose in the melting pot of Egyptian and Mesopotamian divination – especially astrology – that followed in the wake of the Persian and Macedonian conquests. The works of authors such as Labeo owe everything to the circulation of the Chaldeans and perhaps to Classical philosophy, but nothing in them is really Etruscan. For Weinstock, the *Brontoscopic Calendar* text was Etruscan, and translated into Latin by Figulus, but it was still (in his view) affected by Late Republican intrusions, specifically (supposed) references to Catiline, Caesar, and Pompey. He suggested that Figulus had collected accounts of *prodigia* that had occurred during the time of the Civil War, and inserted these into his new edition of the Etruscan *Brontoscopic Calendar* as he translated it into Latin. (If this is so, then the insertions, massaged through Lydus' Greek, are surprisingly seamless and undetectable.) And the references that have been linked to actors in the drama of Late Republican Rome (e.g., July 14, "…, one man will come to power over many. But this man is most unjust in state affairs," or April 9, "If it thunders, it signifies victory for the kingdom and good cheer for the powerful ones") are already present in the much older Mesopotamian prototypes. Might it instead be possible that the Roman demagogues simply were aware of such documents as the Etruscan *Brontoscopic Calendar* and were consciously courting comparisons with their predictions?

Wilhelm and Hans Georg Gundel in their *Astrologoumena* discussed the Latin authors and Roman texts in the tradition of *Laienastrologie* but saw only the similarities between the *Brontoscopic Calendar* and the later, astrological works.[28] Their flowchart emphasizes the continued cross-fertilization between Mesopotamian literature and the Hellenistic treatises that were constantly being generated afresh.[29] For the Gundels, the Figulus document

[28] Gundel and Gundel 1966:135–39.
[29] Gundel and Gundel 1966:383, fig. 1.

had "mit etruskischen Wahrsagungen nichts zu tun" ("nothing to do with Etruscan soothsaying"), but was rather another example of the regurgitated Graeco-Egyptian *Vulgatalehren*.[30] But of course in 1966, they did not have access to the full panorama of pre-Hellenistic Mesopotamian cuneiform texts that have since been published.

French scholarship has provided a steady stream of analyses of the genre of the *Brontoscopic Calendar.* Legrand (1930), in his monograph on Nigidius Figulus, published a French translation of the text. Jacques Heurgon (1959), like Piganiol, gave credit for the Etruscan character of the document, and (like Piganiol) for its cherchez la femme references that sound distinctly un-Roman. By noting the *Brontoscopic Calendar* in his *Vie quotidienne chez les Étrusques* (1961),[31] Heurgon brought it to the attention of a wider readership (and it was there that I first became aware of it). The French school of thought has in general been especially open-minded about the authenticity of the text and its Etruscan background – and the general topic of Etruscan divination.

Dominique Briquel (1997 and 1999) has examined the epigraphic and literary evidence of the *etrusca disciplina* and noted its repercussions extending for centuries after the end of Etruscan autonomy. He has a series of *tables-rondes* on Etruscan divination and has published almost a millennium of literary sources, treated in chronological order, as *La divination dans le monde étrusco-italique*, supplements to *Caesarodunum* from 1985 through 2005.[32] An article therein by Françoise-Hélène Pairault-Massa (1985) explores Etruscan divination and its sources, especially noting the turning point in the fourth century BC; most of the documents we know were created afterward.

Charles Guittard (1997 and 2003) views the *Brontoscopic Calendar* as having a genuine Etruscan substrate upon which were grafted a number of Roman insertions. Against Wachsmuth's insinuation of a complete Roman or Byzantine forgery, he notes that a Roman work would have used a year beginning in January and emphasizes the significance of meteorological citations of thunder.[33] Adriana Della Casa (1962; reviewed by Thesleff 1965), in her monograph on Nigidius Figulus, also tends toward assuming that events of his day contaminated the *Brontoscopic Calendar.* For instance, she suggests that the prediction of July 14 must have meant Julius Caesar, whose birthday fell on July 12 or 13.[34] Yelnitsky (1971) did offer a closer analysis of the *Brontoscopic Calendar* as documentation of Etruscan society, but his work has received less notice because his publication in Russian was not so widely disseminated. Carmine Ampolo (1990–91) is one of the few scholars to take this into account

[30] Gundel and Gundel 1966:138.

[31] Heurgon 1964:227.

[32] Published intermittently as Suppl. vols. 52, 54, 56, 61, 65–67 to *Caesarodunum*.

[33] Guittard 2003:464.

[34] Della Casa 1962:33.

in his brief yet thorough survey of the literature. William Harris (1971) evinced skepticism, linking the *Brontoscopic Calendar* to the tone of the *Prophecy of Vegoia* and saying that the only clear reference to purely Etruscan social systems or historical events is the entry preserved at September 11, which alludes to a revolt of dependents against the nobility.[35] Subsequent studies of Etruscan social order and onomastics, however, show a much more variegated picture with an earlier start to such a stratified society.

Ambros Pfiffig raised the question of a "disconnect" between the complexity of Etruscan lightning analysis (which received a full volume, *Die Blitzlehre*, in Thulin's study) and the relative simplicity of the *Brontoscopic Calendar*.[36] According to Roman sources, Etruscan priests could differentiate numerous types of lightning bolts as sent by several gods for a variety of purposes, yet the Figulus document does not qualify any of the thunder protases with any conditions such as direction, type, and so on. I would suggest that determining such factors as the directional source and magnitude, for example, of thunder is even more subjective than for visible omens, like lightning bolts: one can note the location of an apparition in the sky by reference to landmarks, but identifying the source of a sound is difficult. And because one cannot provoke such omens, priests would presumably have to rely on reports of events from the laity, so a simple switch – thunder or no thunder? – is probably as accurate as one could expect. Pfiffig was quite right in pointing out, however, that our sources on the *etrusca disciplina*, albeit fragmentary, do not mention *libri tonitruales* ("books on thunder-divination") specifically in the corpus, but they do offer *libri rituales* and *libri acherontici*, works on rituals and the afterlife.[37]

Dora Liuzzi (1983), writing on Nigidius Figulus, included an Italian translation of the text of *De ostentis* 27–38 in her book, and a recent translation of the entire *De ostentis*, by Ilaria Domenici and Erika Maderna (2007), includes an up-to-date synopsis of previous scholarship. They emphasize the place of Lydus' work in the milieu of sixth-century AD Constantinople and the repercussions of such literature in later ages (cf. Turfa 2008A). Their Italian translation has made *De ostentis* more accessible for scholars, who should now be able to appreciate the treatise as a whole and get a perception of its intellectual impact on the scholarly readership of the Byzantine Empire.

Alfredo Valvo (1988), writing on the one ancient document that seems to parallel the *Brontoscopic Calendar*, the *Prophecy of Vegoia* (see Appendix A, Text 2), again viewed the Figulus text as another product of the first century BC, or perhaps a composite with remnants of a second-century text – once again, effectively divorcing this text from Etruscan ritual works. There has

[35] Harris 1971:23, 31–40.

[36] Pfiffig 1975:138.

[37] Cf. Valvo 1988:87; Thulin 1906–11.

always been a tendency to view the Vegoia text as a pastiche or creation of the Late Republican period – the Gracchan land reforms immediately come to mind, or the reforms attempted by M. Livius Drusus (ca. 91 BC). But again, in tandem with the *Brontoscopic Calendar*, Ampolo has noted the affinities of the cosmology of the Vegoia text with Near Eastern prototypes.[38]

In 1999, I made a preliminary English translation of the *Brontoscopic Calendar* available for the Sixth Langford Symposium, on the topic of Etruscan Religion, at Florida State University, chaired by Nancy de Grummond and Erika Simon; a revised version has been published in *Religion of the Etruscans*, which they edited (see Turfa 2006D). The version now offered has been slightly improved. Other recent publications have taken the *Brontoscopic Calendar* into the public sphere, with differing effects. For example, Kornmüller (2006) purports to furnish the "first complete manual of Etruscan divination" ("primo manuale completo di divinazione etrusca"; the *Brontoscopic* text is rendered at 89–131). I ceased checking the Web for references to this field when I found my abstract from the British Museum symposium, *Etruscans Now!* (2002, now Turfa 2007), quoted on *MysticWicks Online Pagan Community and Spiritual Sanctuary*![39]

Apart from Legrand's 1931 work on Nigidius Figulus and the short analytical notes by Piganiol and Heurgon, what follows is the first intensive analysis of the information embedded in the *Brontoscopic Calendar* text. Many other scholars, such as Weinstock, Rawson, Harris, Maas, and others, have commented on it as an aside to either the situation of Figulus' lifetime or in illustration of the character of scholarship in Lydus' Byzantium. Some Classicists and historians have doubted its authenticity, at least as a Late Republican or Etruscan document, and some who do accept a measure of reality for it tend to place its date of origin later than the Iron Age. They see the social references as applicable to fourth- or third-century Etruria or Late Republican Rome, because Rome by then was experiencing dependency on foreign slave labor, along with social unrest or downright slave rebellion. Others have identified Archaic conditions in the text, and I will demonstrate that the calendar's social and economic phenomena can all be found in eighth- to seventh-century BC Etruria.

Some early commentators, especially Bezold and Boll (1911), very astutely pointed out the similarities between the Etruscan *Brontoscopic Calendar* and genuine Mesopotamian texts (but few had yet been read at that time). Now that a wider selection of the Akkadian texts has been published, it has facilitated an analysis of the relationships between them and the *Brontoscopic Calendar*. In the continuing absence of an Etruscan-language brontoscopy, we can only go so far, but it will become apparent that the document we have was

[38] Ampolo 1990–91:190–92.

[39] The site is www.mysticwicks.com.

not produced in the same scholarly atmosphere as all the later, astrological brontoscopies, namely, those of Ptolemaic Egypt or possibly Hellenistic Syria. It must be older in its inception than the two similar works (an Aramaic papyrus from Qumran Cave 4 and the first-century BC publication by Fonteius Capito translated at *De ostentis* 39–41). It shows a completely different outlook and format, much closer to those of the cuneiform literature that was its ultimate inspiration.

The discussion that follows should serve as an introduction to the Iron Age world in which this remarkable, early, Etruscan brontoscopy was created and used. The next chapter discusses the intellectual and cultic background of Etruria, and Chapter 3 associates this with its natural, ecological background. Etruscan worshipers saw themselves as mere mortals in a large and menacing environment, and it just happened that, during the period when they came to be recording religious scriptures, that world was more menacing than usual, because of changes in climate and cosmic ray profiles. At the time when divination by thunder was receiving attention, cyclonic storms, and all their potential effects, were occurring more frequently to reinforce the calendar's predictions.

Thematic analyses follow in Part III for the Etruscan world during the Iron Age and subsequent period of political autonomy of the Etruscan cities. These have been compiled according to the categories of human life that receive the most attention in the predictions of the *Brontoscopic Calendar*: food and the environment, health and disease, and society, in which Etruscan culture differed most notably from that of other peoples. Archaeological evidence has greatly enhanced our picture of life in the villages and first cities of central Italy. Most of the concerns of the calendar, about food supplies, wars and urban unrest, health, disease, and prosperity or famine, would have been daily sources of anxiety for any residents of the ancient world, and so the descriptions of Iron Age society, food, and disease can serve as background for contemplating ancient life in general. The concern seen in the text for the world of women, the actions of slaves, and other aspects of the early urban systems of Italy may serve to illustrate some of the differences that made Etruscan culture seem so alien to its neighbors in Italy and the Mediterranean.

Chapter 8 deals with the inspirational background, so to speak, of the *Brontoscopic Calendar*, in considering the Near Eastern sources for certain forms of divination in Etruria and the depth of interaction and influence that may now be discerned in archaeological evidence of the Orientalizing period. Finally, in Chapter 9 an examination of the Hellenistic, Roman, and Byzantine successor documents to the Etruscan brontoscopic text will illustrate the ongoing interest in such forms of divination, through the Byzantine period and beyond. The later texts were rooted in the same inspirational sources that had served the Etruscan scriptures, but they were the work of Hellenistic and later scholars, and the Etruscan text had a different pedigree.

CHAPTER 2

ETRUSCAN RELIGION IN THE CLASSICAL WORLD

A ROMAN FASCINATION WITH ETRUSCAN RELIGION

The literature of *Etruscan* religion held a special cachet in the Roman world, probably ever since the time of Rome's founding. Golden and Silver Age authors such as Cicero and Seneca referred to Etruscan wisdom – and scriptures – as the source for the rites by which cities should be founded, for divination by hepatoscopy (examining the livers of sacrificial victims), for expiation of prodigies, and more. Aspiring Roman statesmen found that, in addition to military service, officiating in certain cults could enhance political success. In the first century BC, the time of Nigidius Figulus and Cicero – and the upheavals of the dying republic – Etruscan religion was visible in Rome in the form of the haruspices, specially costumed priests certified to practice liver divination. In fact, the interpretation of prodigies had been of some recent consequence during the troubled period of the Catilinarian conspiracy (cf. Cicero *In Catilinam* 3.9). The mingling of politics and religion came to a head with the crisis of 56 BC, when Etruscan haruspices were officially summoned to Rome to give opinions on an omen, a loud noise in the heavens above Latium: possibly the only extant account of a thunder-omen like those envisioned in the *Brontoscopic Calendar.* Cicero's speech *De haruspicum responso* (*On the Response of the Haruspices*) publicly replied to this event, hinting at the evils of perverting religious authority for political goals. The fact that Cicero's audience understood this – and that his enemy Clodius Pulcher could interpret the *same* event as *Cicero's* fault instead of his own – is proof of the commonality of traditions of divination throughout the Roman citizenry, and a confirmation of the abuses of politicians in general.[1]

[1] See Beard et al. 1998:vol. 1, 137–40; vol. 2, 175–76, no. 7.4a.

ETRUSCAN RELIGION AS A RECEIVED, SCRIPTURAL DOCTRINE

Our few literary sources on early Etruscan religion are all culled from Latin or Greek authors and leave much to be desired. Etruscan books were recorded on cloth, leather, or papyrus, and only one fragment has survived, the *Liber linteus* ("linen book") now in Zagreb – it may date to the period of Figulus, or more likely circa 200 BC.[2] Archaeological finds contribute to our knowledge of Etruscan cults, but we lack examples of the official *etrusca disciplina*. Romans saw this body of written "Etruscan training" as the definitive source for religious lore, and especially for rituals. This is a sobering lesson: the combination of flimsy, organic media for recording literature and the prejudicial preservation of culture when victorious rivals and enemies write the history of vanquished peoples have almost completely obscured our understanding of a once-extensive, famous and widely respected literature.

During the Late Bronze Age and Iron Age, a core of cultic ritual and belief had developed, as attested at the sites of future sanctuaries, in deposits interpreted as ceremonial or votive, including such offerings as discs cut from antlers, small, nonutilitarian valuables, and prototypes of vases and metal objects recognized, in later contexts, as offerings. Other products of formalized religion are metal hoards that represent boundary offerings at the edges of village territories and the elaborate funerary rituals of the Final Bronze Age and later.[3] Animal sacrifice in funerary contexts must also reflect some common religious beliefs, and burnt sacrifices probably also occurred in sacred precincts.[4] Early first-millennium BC environmental changes threatened the cultures of Iron Age Italy and stimulated the stratification of warrior-dominated societies. The new, formalized religious doctrine came at just the time that a burgeoning population, city life and its pressures, and the challenges of intensified foreign contact brought social strata and discord to Etruria. State cults would have served to enhance the compliance of the citizen body.

THE TRADITION OF TAGES

One unique feature of Etruscan religion as related by the Classical authors contrasts with Greek and Italic practice: Etruscan religious doctrine was based on received scripture. A prophet had revealed the words of the gods, which were formally recorded and preserved by priestly, aristocratic families. The mythological version of this story has the chieftain Tarchon, founder

[2] See van der Meer 2007.

[3] Damgard Andersen 1993; Bartoloni 2002:32–37; Bietti Sestieri et al. 1989–90; Maggiani 1997.

[4] Becker et al. 2009:104–9; Bonghi Jovino 2010:passim.

of Tarquinia, plowing a field when a small creature, who looked like a child with an old man's gray hair and teeth, popped up from the plow's furrow. This person, called Tages in Roman accounts, then dictated (some versions say sang, implying verse) the *etrusca disciplina*. Tarchon then summoned an audience from all Etruria and wrote down his words.[5] The Augustan historian Verrius Flaccus stated that Tages was the son of the god Genius, thus grandson of Jupiter (Festus *Gloss. Lat.* 492 Lindsay).[6]

Logic complicates any attempt to link this story with actual events or place, but the account establishes that the Etruscans believed they had a divinely revealed set of "scriptures." In contrast, Greek and Roman religion had only myth and literature, with the exception of the famous (Greek) Sibylline Books, obtained by King Tarquinius Priscus of Rome. In the years of distrust after Constantine Christianized the empire, pagan apologists made much of Etruscan religion to counter the claims of divine revelation in Christianity and Judaism.[7]

TAGES IN CICERO'S *DE DIVINATIONE*

Cicero is our early source for the story of the prophet Tages, at *De divinatione* 2.50–51, written probably around 44 BC. (See Appendix A, Text 1.) Cicero found the story "in the books of the Etruscans," supposedly written when Tages appeared. This is a problem for modern critics, because, with archaeological hindsight, it appears anachronistic: Lydus (*De ostentis* 2.6.B)[8] says that Tarchon, eponymous founder of Tarquinia, recorded the prophet's words in the early days of the city.[9] The process of urbanization was a continuum that began in the Protovillanovan period, when much infrastructure for the Etruscan cities (road networks, water supply, waste management, exchange circuits, and industries developing out of cottage crafts) was already established – before first contact with the Greek colonial world in the eighth century. At some point, an official ritual was performed that "founded" an Etruscan city – as in the story of the circuit of Romulus' Rome plowed by a bull and cow in a symbolic furrow.[10] Foundation rituals, proto-cities, and formalized religion all existed by the mid-eighth century – but writing did not appear until the late eighth century, surely later than the (proto)historical Tages event.

[5] See de Grummond 2006B:23–27; Turfa 2006B:80–81.

[6] The myth of a people, individual, or prophet sprung from "native" soil is universal, cf. Sita from the *Ramayana*, Hesiod's Age of Gold, the Theban Spartoi, the Athenian hero Erechtheus, and others, as is the child-priest or philosopher (notably the Romano-Celtic Ambrosius-Merlin and Welsh Gwion Bach). These would make an interesting comparative study, and I thank Stephanie Budin for pointing them out to me.

[7] Briquel 1997.

[8] Cf. Ovid *Metamorphoses* 15.552–59.

[9] See de Grummond 2006C:27ff.

[10] Varro *De lingua latina* 5.143; see Bartoloni 2007:26, for 7th-c. Veii evidence.

If we interpret the time of the Tages event as the point after which *we* are able to identify the physical trappings of a city in Etruria, namely, Tarquinia, then the creation of the cities occurred long before the advent of writing in Etruria, yet Tages' words were supposedly recorded on the spot. A controversial inscription from Latium (Gabii/Osteria dell'Osa; see Chapter 7) may be as early as the late ninth century, but alphabetic writing developed only during the eighth century, too late for mythical Tarchon and Tages (but see Chapter 7 for unusual, ninth-century potential evidence of a child-prophet.)

Cicero's other remark is intriguing: after the Tages event, as new things were learned, they were examined and incorporated into the literature (the *disciplina*), specifically *haruspicinae disciplina*, liver divination. If some notable event, later recalled as the apparition of Tages, did spark the first formation of the *etrusca disciplina*, then we may assume that almost immediately the oral "literature" was being redacted and grafted with fresh material, as Cicero says, "in keeping with the principles of Tages." This, the oldest extant account of the Tages legend, is couched in deprecating terms – for instance, at 2.38.80 Cicero calls the prophet *exaratum puerum*, "the boy who was ploughed up" (like an accident) by some "country bumpkin" (*bubulcus*). He attacks the Tages narrative with rhetorical techniques, challenging the story of a person plowed up from the ground, questioning where this Tages, if he was mortal, had learned the *disciplina*. Presumably the story was well known to his audience, whether they were skeptics or believers.

A mirror from a woman's tomb at Tuscania, engraved circa 300 BC, may depict either Tages' apparition or the initiation of an Etruscan youth as a haruspex (Fig. 1).[11] A youth, labeled *Pava Tarχies*, the sun behind his head and the pointed hat of a haruspex slung over his shoulder, stands barefoot on the earth to examine a sheep's liver. He is observed by Avl Tarχunus, presumably Tarchon, with walking staff and pointed hat, and Veltune (Vertumnus?) and other gods. The image must have held special meaning for its owner, and the period around 300 BC was a time of religious change or renewal, as attested by other developments, such as the sudden popularity of anatomical votives and representations of prophecy or ritual.

EXTANT ETRUSCAN RELIGIOUS TEXTS

The written word was of great importance to religious practice and to society in general in later Iron Age Italy. As Mary Beard (2004) emphasizes, Late Republican Rome knew of such things as the hymns of the Salii and the Arval Brethren, which were still being sung/chanted in their original, archaic language. By the first century BC, this was essentially unintelligible to audiences, even scholars outside the cults themselves: such information must have been

[11] Pallottino 1979; Wood 1980; Torelli 1988; de Grummond 2000, 2006a.

transmitted over the course of several centuries, in written form. Mere recitation, without reference to some written text would not have been sufficient to transmit and preserve such a confusing text. Could this have been recorded when state religion was being codified in the Regal period, even during the reforms of Numa? Michels illustrated the importance of written documents in the history of Cnaeus Flavius, son of a freedman, who rose from scribe to curule aedile and in 304 BC published the religious calendar of Rome, previously kept secret by the *principes*.[12] The commons had to depend upon the priests who kept the calendar divulging to them the dates when court proceedings were possible or forbidden, and when the special holidays (*feriae*) and religious observances were to occur.

The few surviving Etruscan religious texts all emphasize their cosmic authority by reference to the written word.[13] Three happen to be the closest surviving analogues to the genres of Etruscan scriptures noted by Roman authors, the *libri rituales, Acherontici*, Capua tile (Fig. 4). The so-called Capua tile (*Tabula Capuana*) is one "leaf" of an early calendar for performance of specific rituals (perhaps funeral cult) in the region of Capua; it retains the note (at line X.62), *]ulis ziχunce*, "[Can?]ulis has written,"[14] and was very likely copied from an older text.[15] Large slabs of terracotta were incised with minuscule text before firing and posted or stacked in some public place in or near the necropolis. Consultation of the document could not have been casual: either the plaque or its reader must be turned 180 degrees to read alternate lines.[16] Dates for March to October are preserved. Most lines are damaged, but names of months, gods (e.g., the Underworld Letham), and rituals may be discerned.

ZAGREB LIBER LINTEUS

The Zagreb *Liber linteus* (Fig. 5), now in the Zagreb Museum, is a fragmentary liturgical calendar written in black ink with red rubrics on a textile folded into book format.[17] In Egypt in the first century BC it had been torn into strips and used to wrap a woman's mummified body. The text (I.21) states the document's authority, *ziχri cn*, "let this be written." Funerary images show such linen books on the biers of priests and on a bedside chest in the Tomb of the Reliefs at Cerveteri (Roncalli 1980 and 1985). The cloth may have been woven as early

[12] Michels 1967:108–13; followed by Beard 2004:127–28.

[13] See Bonfante 2006.

[14] *ET* TC, ca. 470 BC; Cristofani 1995:85–88.

[15] Cristofani 1995:88, 122.

[16] See Cristofani 1995.

[17] *ET* LL; van der Meer 2007; Turfa 2008. The representation of *libri lintei* in funerary images, such as the bedside chest of the young girl, Ramtha Matunai (Tomba dei Rilievi, Cerveteri), and men's sarcophagi and urns at Cerveteri and Chiusi (see Roncalli 1985:23), poses the possibility that these individuals, too, were associated with brontoscopy or related forms of divination.

as the fourth century BC, but the book's script, associated with the region of Perugia, is dated circa 200–150 BC. The text preserves entries for dates from March and June to September for animal sacrifices, food offerings, and other ceremonies to honor several gods, including *Nethuns*, *Thesan*, *Tin*, *Vetis/Veive*, *Satr*, and *Uni* (Neptune, Dawn, Jupiter, Veiovis, Saturn, and Juno).

For "the sacred priesthood of the citadel and of the community,"[18] priests are instructed *haθrθi repinθi-c śacnicleri cilθl śpureri meθlumeri-c enaś śveleri-c* (I.7–8), "be benevolent and bow to the temples of the people, to the cities and districts and hearths."[19] An alternate translation of *haθrθi repinθi-c* is a blessing, "in peace and prosperity."[20] Other instructions are *raχθ tura nunθenθ cletram śrenχve tei faśei*, "prepare the incense, offer with the ceremonial vessel these breads", and *Celi huθiś zaθrumiś flerχva neθunsl śucri θezeri-c/ ctnam θesan fler veiveś θezeri etnam aisna* [...] *iχ huθiś zaθrumiś*, "In the month of Celi [September], on the twenty-fourth day the offerings to Neptune must be made and immolated. And the same morning the offering to Veiovis must be immolated, and furthermore the divine service, as on the twenty-fourth day."[21] Recognizable dates, written in words rather than numerals, are preserved from June 18 (*eslem zaθrumiś acale*) through September 24 (*celi huθiś zaθrumiś*).[22] Its presumed starting date in January fits the late date and Romanized milieu.

PERUGIA CIPPUS

The Perugia *cippus*, a stone boundary marker stating the terms of an authorized agreement between two families over land boundaries and tomb ownership, ends with *iχ ca ceχa ziχuχe*, "as this agreement has been written down."[23] It shows that as late as the second century BC, the power of writing and of posting a text was still respected. The *Prophecy of Vegoia*, preserved only in Latin, but believed by the Romans to have been an Etruscan creation, also affirmed the authority of the written word. Near Eastern traditions venerated boundary monuments, such as the Assyrian *kudurru* (cf. Fig. 14).

VOTIVE EPIGRAPHY: PYRGI PLAQUES

The most famous Etruscan cultic texts are the gold Pyrgi plaques, once nailed (ca. 500 BC) to a monument, perhaps the portal of a temple of Leukothea-Uni

[18] According to van der Meer 2007:51–54.

[19] Bonfante and Bonfante 2002:183–84.

[20] van der Meer 2007:62.

[21] van der Meer 2007; Rix 1997; Turfa 2008; for months, see Edlund-Berry 1992:331.

[22] See van der Meer 2007:28–31, for additional possible dates.

[23] *ET* Pe 8.4; Bonfante and Bonfante 2002:176–78. See Facchetti 2000:9–58. The Cortona Tablet, recording a mortgage agreement in bronze, also says *cen.zic.ziχuχe*, "this document has been caused to be written" (see Agostiniani and Nicosia 2000:87; Facchetti 2000:59–88.)

in the Caeretan seashore sanctuary at Pyrgi and ritually buried in the third century BC around the time of the Roman takeover. One is beautifully incised in the Phoenician language and another paraphrases it in Etruscan; the third is an additional dedication by the same ruler, Thefarie Velianas.[24] Called in the Phoenician-Punic version *MLK*, "king," Velianas declares his gratitude for help received from the goddess "Uni-Astarte" (Etruscan *unialastres* – in the Semitic version she is "the Lady Astarte."[25] Punic involvement in the affairs of Caere at this time was not suspected until these texts were discovered. They represent the most lavish version of the votive dedication, the rarely preserved offerings of a ruler.

THE ETRUSCAN BRONTOSCOPIC CALENDAR

The Etruscan version of the *Brontoscopic Calendar* that Figulus handled could have been monumental in form, such as a set of plaques in bronze or terracotta (like the Capua tile), or duplicated on *libri lintei*, such as the Zagreb *Liber linteus*, for use by individual priests. Its "manuscript" tradition probably began early in the seventh century BC. No city or sanctuary has been exclusively linked to it, but a likely candidate was Tarquinia, where Tages dictated the *etrusca disciplina*. According to Lydus (*De ostentis* 38), Figulus noted that the version he translated and published was appropriate to the region of Rome only. There is, however, no significant climatic or geographic difference between Rome and the major Etruscan cities as to environmental region or latitude, so this comment remains obscure, yet it is often cited as evidence that Figulus tampered with the original text. He must have been assuming credit (or blame, from an Etruscan point of view) for some alteration in details – any great change would have voided this text from its proclaimed pedigree in the ancient Tagetic line of scriptures.

Any translation is a sort of alteration. The original calendar must have been cryptic in style, and some Byzantinisms are evident in the Greek of Lydus' translation: for instance, use of the term "the queenly city," meaning Rome (at June 29), or οἵ δῆμοι, "common people," which in Lydus' Byzantium carried the more specific connotation of "factions." Hints of lingering Etruscan syntax are paralleled in Etruscan cultic documents like the Zagreb *Liber linteus*. Other expressions such as "winged ones" (τὰ πτηνά) are close to the phrases used in Archaic Italic texts, such as the Umbrian *Iguvine Tables*. Even the oldest surviving Etruscan texts, like the literary passages describing Etruscan religion, are centuries more recent than the supposed prophetic events: archaeology has filled in the gaps with votive deposits, sanctuary structures, and artistic representations of mythology and rituals. There is no perceptible alteration

[24] Tiberius Velianus; see *ET* Cr 4.4–5; Bonfante and Bonfante 2002:64–68.
[25] *LRBT L'ŠTRT*; see Schmitz 1995.

in character (apart from a gradual and predictable increase in cult activity by nonaristocrats) between the Iron Age and the first century BC except for fluctuations in wealth and foreign artistic influence.

EXTANT ETRUSCAN RELIGIOUS TEXTS (IN FIRST-CENTURY BC ROMAN LITERARY CONTEXTS)

The role of *literati* of Etruscan ethnicity in Late Republican Roman religion and politics is now recognized as extensive.[26] By 89 BC, many Etruscan cities and families had gained Roman citizenship, and their scions saw their own fortunes as tied to the Roman Senate, often punctuated by participation in the army and/or prominent cults.[27] Much has been written on the partisan status of Etruria or of specific Etruscan cities and families in the period of the Marian movement, the Sullan regime, and the days of Julius Caesar. Cicero's remarks show that several well-known senators and *equites* were caught up in the wrangling, usually to their detriment. Aulus Caecina, father of an author of religious topics, was one such, and Nigidius Figulus another. They were accompanied, tragically, by "all Etruria's noblest" when they variously made their stands for or against major political figures.[28] Many other Etruscans or persons from Etruscan cities likewise were known as proponents of Pompey and remained fanatically opposed to Caesar. The rise to power of Augustus, aided by the Etruscan Maecenas, finally brought a respite to the worst fighting.

Divination has always been a tool in political power plays, as demonstrated in the Near Eastern kingdoms and Imperial Rome, where fortune-telling could be banned if it threatened to affect a regime's hold on power. Many modern analysts see the *etrusca disciplina* as having been manipulated as propaganda for the political factions in Republican Rome or for social control.[29] In fact, of the Latin excerpts or paraphrases preserved in Roman sources, Rawson could say objectively that "there is certainly fair evidence for parts of the *disciplina* reflecting a narrowly upper-class outlook."[30] The documents best known or most often alluded to in first-century Rome were the Sibylline Books, and the *Prophecy of Vegoia*, a short text in circulation in Latin during the first century BC (see Appendix A). But other texts come equally into play.

In a passage in book 1 of the *De bello civili*, Lucan poetically describes a series of prodigies and portents that presaged the Civil War, and he provides an impression of the effect of divination on the Roman people. The gods filled earth, air, and sea with prodigies: shooting stars, comets, lightning; a silent

[26] Cf. Hall 1984 and 1996; Macfarlane 1996.

[27] Cf. Torelli 1969.

[28] See Rawson 1978/1991:299.

[29] Cf. Schultz 2006.

[30] Rawson 1978/1991:300.

thunderbolt struck Alba Longa, along with a lunar eclipse and a solar eclipse; Etna erupted, disrupting the Straits of Messina; thaws in the Alps and Atlas Mountains brought floods; statues wept; wild animals behaved strangely, and others spoke; women bore monsters; groans came from the ashes in urns; a Fury was seen; and the ghosts of Sulla and Marius appeared (1.525–83). Although thunder is not mentioned, noises in the heavens included the clangor of arms in spectral battles, the sound of trumpets, and shouting (1.569–70, 578–80).

The mystique of Etruscan divination (and, perhaps, the impetus for Figulus' translation) is portrayed when Lucan then tells of the decision to observe "old traditions" (*more vetusto*) and summon "seers" (*vates*) from Etruria (*De bello civili* 1.584–637). The most venerable, "Arruns," comes from the deserted walls of Lucca; his specialties are the quintessential Etruscan skills – interpreting thunderbolts, entrails, and the flight of birds. (The last is described poetically as *monitus errantis in aere pinnae*, 1.588, "the warning of a wing wandering through the air," perhaps similar to the calendar's τὰ πτηνά, "winged ones," June 15.) The main activity of Arruns, however, is expiation – he must tell the Romans how to get out of the troubles of which they have been warned, and the prognosis is not good. They must destroy any abnormal births, march in procession around the city *pomerium*, and perform lustration, ending with the sacrifice of a bull (which itself turns out to furnish terrible omens). Before the sacrifice, Arruns "collects the scattered fires of the thunderbolt and buries them in the earth" (*dispersos fulminis ignes colligit et terrae maesto cum murmure condit*, 1.606–7). Although this is certainly not the observation of thunder, the citation of lightning (*fulmen*) at such a point is interesting.

The "prophecies of Marcius," actually verse *carmina* (Livy 25.12.2–15; Macrobius *Saturnalia* 1.17.27–28), were another written source of revelation, which, along with the Vegoia and Sibylline texts, were locked in the crypt of the Palatine Temple of Apollo, by none other than Augustus.[31] An example of the Archaic Italic religion, Marcius was a *vates inlustris*, "a shadowy figure from Italy's prophetic past," according to Klingshirn, and the prophecies tagged with his name apparently had begun to circulate as early as the third century BC.[32] One Marcian prophecy, recorded by Livy (25.12.9–15) established the festival in honor of Apollo; the other (Livy 25.12.4–8) predicted the disaster at Cannae in 216 BC, warning that the fish, birds, and wild beasts that inhabited that area would devour the flesh of those killed, just as Jupiter had told Marcius.[33]

[31] Underground passages link it to the nearby houses of Augustus and Livia, yet more apparent appropriation of divination by the state.

[32] Klingshirn 2006:146; Servius *Ad Aen.* 6.72; Beard et al. 1998:vol. 2, 182–83, no. 7.5c.

[33] *Piscibus atque avibus ferisque quae incolunt terra iis fuat esca caro tua. Nam mihi ita Iuppiter fatus est* (Livy 25.12.6).

THE SIBYLLINE BOOKS

No one knows the full content of the Sibylline Books because they originally had been offered to Tarquinius Priscus, the Etruscan king of Rome, in the early sixth century: the tale of his rejecting the sibyl's offer to sell them is well known.[34] She burned three of the nine books, and offered him the rest at the same price; rejected again, she burned three more, and Tarquin was then persuaded by the augurs to pay full price for the remainder. The books were locked up, originally in a stone chest beneath the Temple of Jupiter Optimus Maximus (then in construction) on the Capitoline. Official consultations occurred during a famine in 496, and plagues in 293 and 204 BC (leading to introduction of the cults of Aesculapius and Cybele respectively). When the temple burned in 83 BC, the books were restored by reference to oracular literature stored elsewhere in Italy (76 BC). In 12 BC, Augustus had the collection deposited in the Temple of Apollo on the Palatine, in the custody of the *Decemviri* (later *Quindecemviri*) *sacris faciundis*. Fragments preserved in a literary account are in Greek hexameter with an acrostic first verse, like other Greek prophecies. Still, the "history" of the Sibylline Books clearly implicates the Etruscan king and the relationship of divination and state.

The *Book of Marvels* of Phlegon of Tralles, written in the second century AD, gives a sample, in poor preservation, of the Sibylline oracle later published by priests concerning the birth of a hermaphrodite in 125 BC: money offerings, sacrifices, and choral rituals are prescribed as expiation, with particular attention to the chthonic gods, and especially Demeter and Persephone.[35] Their stated oracular source, Greek verse composition, and system of expiatory prescriptions set the Sibylline Letters in a different genre from a text like the *Brontoscopic Calendar.*

PROPHECY OF VEGOIA (SEE APPENDIX A, TEXT 2)

The Vegoia prophecy, in contrast, was supposedly given to an Etruscan "Arruns Voltumnus" and, in Rawson's words, "it is pretty indubitably hostile to any kind of agrarian reform in Etruria: slaves, or masters in concert with the slaves, are not to move the *limites*" (boundary markers).[36] Like the Sibylline Books, the Vegoia document was ultimately deposited in the Palatine Apollo temple; supposedly it is there that Tarquitius Priscus transcribed or translated the book(s) of Vegoia from Etruscan into Latin in the first century BC.[37]

[34] Dionysios of Halikarnassos *Ant. Rom.* 4.62; Pliny *Nat. Hist.* 13.27.88; Lactantius *Inst.* 1.6.10–11; Servius *Ad Aen.* 6.72.

[35] Phlegon *Marvels* 10 = Hansen 1996:40–43, 126–137; Beard et al. 1998:vol. 2, 179–81, no. 7.5a; see MacBain 1982:135; Parke 1988; Potter 1990.

[36] Rawson 1978/1991:300–301.

[37] Cf. Turfa 2006B:81–82.

Mention of slaves and land has prompted most commentators to link it to the Gracchan reforms or perhaps the time of the Social War and the reformer tribune M. Livius Drusus (ca. 91 BC).[38] Other scholars, in contrast, see the Etruscan "Arruns" as indicative of an Archaic situation, or at least place its inception well before 91 BC: Turcan (1976), for instance, said that the travails of third-century Volsinii (revolution by freed slaves ended by Roman conquest) were implied. But merely to link this text to Late Republican Roman debates over land reform would be to claim that some influential Roman forged a religious document for purely political (and very indirect) purposes. In fact, all the features noted in the Vegoia document, such as slaves, storms, crops, and disease, are found in different context in the *Brontoscopic Calendar*, and there is good reason to believe that these conditions all pertained in Etruria by the early Archaic period.

Vegoia's prophecy is known only from a Latin text preserved in the archives of the Roman surveyors' guild, but older Etruscan boundary inscriptions attest circumstantially the weight of its teachings.[39] After a cosmological opening attributing the apportionment of land among mortals to Jupiter, the prophetess warns that human greed will lead men to transgress boundaries during the eighth *saeculum*. She enumerates punishments for those who disturb boundary markers, whether masters or their slaves: death, diseases, storms, destruction of crops, and civil unrest, concluding "be not false or double-tongued. Keep this teaching in your heart." Etruscan iconography of the nymph Vecu, the prototype for Latin "Vegoia," has been identified on hand mirrors and a gold ring of circa 300 BC: she is a winged female, *lasa vecuvi[a] or lasa vecu*, seen in the company of Tinia, Menrva, and thunderbolts (see Fig. 11).[40]

THE ETRUSCAN DOCTRINE OF SAECULA

One other point of doctrine in the *etrusca disciplina* is directly attested only through Latin literature. Noises in the heavens announced the *saecula*, in the Etruscan doctrine, of cyclic historical periods allotted by the gods for different cultures to flourish.[41] Of course, *Roman* authors had compelling reasons for retailing the history of *saecula*, because by the Etruscans' own reckoning, their domination of Italy was ended – *finem fore nominis Etrusci*, "the end of the Etruscan name." Censorinus (*De die natali* 17.5–6, citing Varro) is our source on the succession of cosmic *saecula*, which were heralded by portents: according to the Roman sources, ten *saecula* were allotted to the Etruscan people and twelve to the Romans. Certainly Sulla was not above

38 See Valvo 1988:103ff.; Harris 1971:202–12; for full references, see Schultz 2006:223.

39 *Corpus Agrimensorum* 1:350–1; Schultz 2006:222–23; Harris 1971:31–40; Valvo 1988; Pfiffig 1975:160–61; Heurgon 1959; Turcan 1976.

40 de Grummond 2000, 2002, and 2006C:30–33.

41 Turfa 2006B:82–83; see Valvo 1988:20 (23–53 for modern interpretation of the doctrine).

TABLE 2.1. *The* saecula *of the nomen Etruscum*

#	# Years	Years = BC	Criteria
1	100	968–868	Arbitrary: 1 century
2	100	868–768	Arbitrary: 1 century
3	100	768–668	Arbitrary: 1 century
4	100	668–568	7th c. sees new timekeeping methods Span: 100 years
5	123	568–445	Span: 123 years
6	119	445–326	Span: 119 years
7	114	326–207	Span: 119 years
8	129	207–88	Heavenly trumpet ends eighth *saeculum*: Plutarch *Sulla* 7.4ff.
9	44	88–44	Comet announces end of ninth *saeculum*: Servius *Ad Buc.* 9.46 – cites Augustus *Memoirs* bk. 2
10		44–?	

Source: Calculated after Pfiffig 1975:160–61.

manipulating the public announcement that a regime change (in his favor) had been proclaimed through a loud noise, a heavenly trumpeting, heard in the skies (Plutarch *Sulla* 7.6–9; Censorinus *De die natali* 17.6). Plutarch gives 88 BC as the end of this eighth *saeculum*. Servius at *Ad Bucolica* 9.46 says that 44 BC was the end of the ninth *saeculum*, announced by the appearance of a comet. If we follow one reckoning suggested by Pfiffig, the ten *saecula* may be listed as in Table 2.1.

Much hinges on the actual length of a *saeculum*. According to Censorinus (*De die natali* 17.2), *saeculum est spatium vitae humanae longissimum partu et morte definitum*, "the *saeculum* is the longest lifespan, between birth and death, of a man born in the first year of the new *saeculum*," so a century or less is a reasonable approximation. A century may seem extreme, but of the Etruscan *saecula*, 5 through 8 are actually longer. (Modern scholars often estimate the length of a generation as thirty years, or one-third of a century, the time allowed for a child to mature and produce offspring.) The birth date calculations recorded in family tombs at Tarquinia seem to corroborate a certain longevity among the Etruscan aristocracy.[42] Some scholars indicate that (in a similar magical logic to the phenomenon of a *ver sacrum* migration, which was composed of age-cohorts all born in the same year), a *saeculum* ended with the death of the oldest person born in the year of its inception (Censorinus *De die natali* 17.5–6, citing Varro). If so, the Etruscan

[42] The oldest attested Etruscan lifespan is that of the remarkable Larth Felsnas, who campaigned with Hannibal and died at the age of 106 in the 2nd c. BC (*ET* Ta 1.107, Bonfante and Bonfante 2002:176, no. 63; Turfa 2006B:83n105).

statesman M. Perperna, who died in 49 BC at ninety-eight years of age, could have furnished the end date for the ninth *saeculum*, and thus the omen of civil war.[43]

According to Censorinus, Varro said that the Etruscans only wrote down this doctrine during the period of the eighth *saeculum*: astonishingly late, in the third to second century BC, at least 500 years after written records began in Italy. The beginning point of *saecula* reckoning is curious. Who would know when the first *saeculum* began, unless it was signaled by Tages? But the end of the first *saeculum* falls very close to the mid-ninth century, a time when natural – and social – events had begun to change the face of central Italy, and material manifestations of (proto-)Etruscan cult were beginning to appear, as will be seen.

The recognition of a heavenly noise, in the cycle of 88 BC, as the delineator for a new *saeculum* presumably was a long-standing tradition, or there could have been no consensus about it – a hint perhaps of the antiquity of thunder-omens in Etruria. The Roman Senate consulted the haruspices for interpretation, and they pronounced it an ill-omen, predicting a class war about to begin – which certainly sounds like some of the apodoses of the *Brontoscopic Calendar*: it appears that it or similar texts were being consulted openly by Romans in the first century BC.

As for the final Etruscan *saeculum*, Octavian's memoirs show that when a comet was sighted in 43 BC, the haruspex Volcatius (note Latin spelling) announced that it was the omen for the progression from the ninth to the tenth *saeculum* of the Etruscans – he then dropped dead, as he had cautioned, stricken for revealing the secrets of the gods.[44] Comets were also said to figure in the *etrusca disciplina* and were covered in the "writings of Tarquitius Priscus" (Ammianus Marcellinus 25.2.7). Octavian took up the popular rumor that the comet was Caesar's soul; there would later be a host of portents or sacrificial *omina* during his Perusine War, and for many Etruscans the fall of Perusia must indeed have been read as the end of the *nomen Etruscum*.[45]

CICERO AND THE *DE HARUSPICUM RESPONSO* (*SEE* APPENDIX A, TEXT 3)

Cicero constitutes a complex source for any information on Etruscan religion: his own religious treatises were written circa 45–44 BC, thus after Figulus' death, but he probably owned copies of his friend's works. Cicero the friend respected the pious endeavors of his political comrades and sometime legal clients, Figulus and Caecina; yet Cicero the philosopher heaps scorn on the

[43] See Rawson 1978/1991:313.
[44] Haack 2006:135–37.
[45] See Haack 2006:137n6.

basic practices of divination, and on the legendary background of the Etruscan scriptures. Cicero the politician knew how to make the most of all possible attitudes.

The portent of 56 BC comes closest to what may have been a historical consultation of the Etruscan *Brontoscopic Calendar*, but we have only hints of actual phrases in Cicero's Latin speech delivered, embellished, and published for the occasion. A disturbing, thunderous, rumbling noise was heard in the heavens outside Rome (in the *Ager Latiniensis*); multiple reports led the Senate to summon Etruscan haruspices (this ought not to be rendered "soothsayers," however, as in so many translations) to interpret the portent. If this was perceived as a thunder-omen, the event throws into sharper relief the translation, at about this time, of the Etruscan *Brontoscopic Calendar* by Figulus. According to Beard and colleagues, Cicero's speeches, especially the *De haruspicum responso*, demonstrate "a striking consensus of religious ideology" among his audience, the citizens and intellectuals of Rome.[46] Rumors of the portent were causing concern among the populace, and Cicero delivered a speech intended to calm people but additionally to work up sentiment against his (and Caesar's) political enemy, Clodius Pulcher. Cicero argued that the misdeeds of Clodius caused the gods' displeasure, and expiation was needed if the state was to avert disaster. (For Americans who have seen more than one leader blame a major hurricane on civil rights reforms, this will seem familiar, but no more logical.)

De haruspicum responso quotes several lines of the official pronouncements of the haruspices, presumed to have been uttered in Latin, as given here by Cicero. The replies note which gods are to be appeased (*De har. resp.* 20), and they list crimes committed: religiously sanctioned public games polluted (21), the Egyptian envoys to Rome murdered (34), oaths neglected (36), and venerable sacrifices profaned (37). For Cicero, this meant Clodius' disruption of the Megalesian Games, his invasion of the women's festival for *Bona Dea*, and bribing of jurors in his trial for same. There is specific mention (40) of preventing power from falling into the hands of a single man, and in reply Clodius claimed this meant Cicero. The *Brontoscopic Calendar* actually does refer to a single man, demagogue, or lone ruler at several dates, and in quite terse language, putting it close in tone to the venerable, formal expressions of the haruspices.

Cicero inadvertently gives circumstantial confirmation of the authenticity of the *Brontoscopic Calendar* in referring to Etruscan books as couched in aristocratic terms that speak of the factions ranged against the *patres*, *principes*, or *optimates* as being *deteriores* and *repulsos* ("baser and rejected").[47] The language of Figulus' brontoscopic document (even filtered through his Latin

[46] Beard et al. 1998:138.

[47] *De har. resp.* 53; see Rawson 1978/1991:301.

translation and Lydus' Greek version) does indeed encompass "commoners," "lower classes," "powerful men," "rulers," and "slaves." A number of haruspices, including members of the *Ordo LX*, during the first century BC can be shown to be ethnically Etruscan, and thus they must have been cognizant of this text.[48] Short, authoritarian quotations, not necessarily in the original order of the public reply of the haruspices, were recorded in the manuscripts of Cicero's speech in capital letters. They give a pallid hint of reliance on the authority of ritual documents like the *Brontoscopic Calendar.* Although we cannot judge the accuracy of the recording (or Latin translation), it seems likely that these stilted passages are true to the original phrases uttered by the officials.

Paralleling the sound of thunder is the phrase *strepitus cum fremitu*, translated as "a rumbling and a noise," and later "a subterranean rumbling and an awful noise as of arms."[49] Section 40, although damaged in the manuscripts, has terms that sound very much like those of the *Brontoscopic Calendar*, referring to nobles, the state, autocrats, discord, and division. (Compare July 14, September 25.) Likewise, the terms of section 56, "let not honor be increased for the baser and the rejected,"[50] invite consideration of the lower orders of a stratified society, a situation just as possible by the late seventh century as it was in the mid-first (the *Brontoscopic Calendar* notes dissension among or troubles for the lower classes [Oct. 4, Nov. 3, 29] and commons [July 15, 23, Aug. 3, 21, Sept. 2, 19, Oct. 3, 19, 24, Nov. 9, Jan. 4, 19, Feb. 27, May 1, 8]).

Cicero notes in his conclusion (§ 62) that this was not the only portent provided by the gods – a terrible earthquake (*terrae motus horribilis*) was felt in the Picene territory. An earthquake appears often as a qualifier in Near Eastern omen literature that influenced the Etruscan *Brontoscopic Calendar*, and also in the later Hellenistic and Roman *brontologia*. Many times during the waning of the Republic, players on the political stage exploited divination, as in the tales of Sulla and his personal haruspex, C. Postumius.[51] Pompey, too, had a tame haruspex, and Julius Caesar's disdain of the sacred chickens and haruspicy in general was legendary. Another emblem is the scene of the haruspex Spurinna, an aristocratic Etruscan, warning Caesar of dire consequences when the bull sacrificed at the *Lupercalia* in 44 BC turned out to have no heart.[52]

"FOREIGN" RELIGION IN FIRST-CENTURY ROME

Cicero (*De divinatione* 1.3) seemed to compartmentalize types of divination and identified haruspicy and the interpretation of prodigies as Etruscan disciplines, in contrast (it seems) to the "Roman" augury by observation of

[48] Cf. Haack 2006:9; Turfa 2006C.
[49] Watts 1923:341.
[50] Watts 1923:391.
[51] Cf. Haack 2006:99–101.
[52] Haack 2006:110–12; see Rawson 1978/1991:304–12.

birds. Etruscan omen literature was undoubtedly one of the tools of the men in the fore of the lethally polarized Roman political and cultic arena. Rawson suggested that the impetus for the translation of genuine Etruscan texts into Latin during the first half of the first century BC was the perception by Roman scholars of "the decline of the Etruscan language and probably, with this and the loss of independence, of many Etruscan religious usages."[53]

Haruspices using charged, traditional language predicting class wars and tyrants in the mid-first century could not have counterfeited it only days or years earlier without being exposed – they had to cite literature familiar to other priests and recognized as long-standing tradition. Figulus (and his contemporaries) promoted the *etrusca disciplina* and would have expected colleagues in the government and priestly colleges to understand it as a "native" religious tradition just as venerable as the famous Greek cults, and also a sensible counterfoil to the lore of the East that was flooding into Rome, like exotic consumer goods, along with foreign educated slaves and the Chaldeans (astrologers).

Roman fears of corrupting foreign cults are evident as early as the Second Punic War, with the influx of dispossessed peasants into Rome, and in the state's crackdown on persons performing exotic rituals or circulating *libri vaticini*, "publications of itinerant prophets."[54] In other cases, foreign cults were banned, marginalized, or kept under surveillance in Late Republican Rome. It took a special situation and decree to import the cult of Cybele or Asklepios into a beleaguered Rome, and in the 40s the "crisis of the Bacchanalia" that had prompted Senate action (186 BC) was still remembered.[55] Such cults as Egyptian Isis and Serapis were relegated to the foreign quarters of ports and towns, according to Vitruvius (1.7.1), or banned by decree from Rome.[56] In fact, the *Chaldei* had actually been expelled from Rome in 139 BC.[57]

Chaldean doctrines professed Babylonian origins: links to an ancient and sophisticated land known for the authoritative powers of kings and priests. If Lydus was correct in his attributions for the other works embedded in *De ostentis*, then other influential Romans such as Fonteius Capito overtly embraced Eastern texts and propagated them in Latin versions of *brontoscopia, seismologia* (divination by earthquake), and other omen texts.

A colophon from the translator was a hallmark of Near Eastern divination texts, and similar expressions recur in the Etruscan cultic texts of the Capua tile and Zagreb *Liber linteus*. Nigidius Figulus does not advertise any Eastern origin for his *Brontoscopic Calendar* but says he translated it "word for word" from a complete Etruscan document. In fact, he probably

[53] Rawson 1985:299.

[54] See Klingshirn 2006:145–47.

[55] Harris 1971:109.

[56] During the 50s BC; Rawson 1985:299.

[57] See Beard et al. 1998:vol. 1, 114–66, on Late Republican religion.

did not know that the Etruscan original was based in part on Mesopotamian texts, because it had been formed centuries before his day. For Figulus, the Etruscan text had been transmitted over many generations from some original lore of the Etruscan Iron Age, when a supernatural prophet had appeared and pronounced them.

THE INTELLECTUAL BACKGROUND OF NIGIDIUS FIGULUS AND THE *BRONTOSCOPIC CALENDAR*: ADDITIONAL EPIGRAPHIC AND LITERARY SOURCES

Roman aristocrats with scholarly inclinations and a good education concerned themselves with a rather abstruse selection of literature on all sorts of topics, with philosophy and religion often at the forefront. The library of charred scrolls found in the Villa of the Papyri at Herculaneum, although not fully excavated or studied, offers an impression of what circulated in the intellectual milieu of Rome in the first century BC.[58] The obscure Epicurean philosopher Philodemus is also found there.[59]

Background for Etruscan divination literature may appear in the prodigy lists and in the preservation of the *Annales Maximi*, both prime features of Republican historiography.[60] The earliest preserved references may be to an eclipse of around 400 BC and events of the early Republican period. Romans believed that written documents were generated and used in Rome during the period of the kings, that is, the seventh to sixth centuries BC, thus touching on the period of creation of the Etruscan *Brontoscopic Calendar*.[61] Roman scholars believed that such texts as the "books of Numa" were discovered in papyrus scrolls unearthed in Rome in 181 BC (Livy 40.29.8; Pliny *Nat. Hist.* 13.84). Their contents remain unknown.

Fragments of lost omen lists were preserved by Julius Obsequens, an Imperial (3rd-c.? AD) author who epitomized the works of Livy and published a list of the portents linked to historical events.[62] Because these were culled from Livy, Iron Age data are sparse, but a steady stream of events begins in 190 BC, over 500 years after the founding of Rome.[63] Many of the preserved omens feature lightning strikes, tempting modern commentators to offer "scientific" interpretations. For instance, the shower of stones in Picenum in 186 BC, in which *ignes caelestes* ("airborne fires") singed bystanders' clothing, sounds

[58] Many of its scrolls were written or copied long before 79 AD; see Sider 2005.

[59] See Rawson 1985:301; Delattre 2006.

[60] Rawson 1971/1991:1.

[61] Bucher 1987–88:40–44.

[62] Pliny notes many of these prodigies in his second book; see Guittard 2005 for full references.

[63] See Schlesinger 1959.

like a meteor shower.[64] Many phenomena are similar to Near Eastern omens, such as rains of blood, growth of "wool" on trees, and monstrous births, but although fire or lightning are featured in several events, there is no account of a specific thunder-omen.[65]

CONCLUSION

Although it is clear that the intellectuals of first-century BC Rome certainly courted religious studies and the fruits of divination, there is little direct correlation between the known documents of that period and the details or outlook of the Etruscan *Brontoscopic Calendar*. It would appear that this text was already archaic and of limited circulation, but its stilted language and unusual omens resonate with some other divinatory documents of the time, such as the haruspices' response to the Senate inquiry of 56 BC. What little we know of the works of Publius Nigidius Figulus shows a highly educated and ethical scholar, whose personal, creative interests lay in the field of astrology and astronomy. The *Brontoscopic Calendar*, however, does not incorporate astrological elements, and it does not match up with the known astrological or brontoscopic literature of *his* day; it stands at the head of a long series of increasingly astrological *brontoscopia*. Figulus, it would seem, really did execute an accurate translation of a different document and refrained from injecting his own knowledge or interests into it. Likewise, there is no evidence of serious contamination of the text by John the Lydian when, six centuries later, he came to translate Figulus' Latin translation into Greek. Many additional circumstantial points support Lydus' honesty – his interest was avowedly antiquarian, and there is no tampering in his treatise to make the dates of omens match great events that had occurred in Constantinople. A few circumstantial points, such as the archaic language of the *De haruspicum responso*, the Sibylline Books, and prophecies of Marcius and of Vegoia, further support an early, Etruscan character for the document known as the *Brontoscopic Calendar*.

[64] Obsequens 3 = Schlesinger 1959:238–41.

[65] For basic background on religion and politics in 1st-c. Rome, see Hall 1996; Macfarlane 1996; Rawson 1985; Rawson 1978/1991; Harris 1971.

CHAPTER 3

AN OMINOUS TIME: THUNDER, LIGHTNING, WEATHER, AND DIVINATION

A FEW WORDS ON HOW WEATHER WORKS (IT IS NOT REALLY TINIA)

The immense power of electrical storms is obvious to all. It is no surprise that they were regarded as omens, or that many cultures should attribute their power to the personal and deliberate actions of gods. At any given time of day, up to 1,800 thunderstorms rage somewhere on earth; 100 times a second lightning bolts strike the ground or spark from cloud to cloud. Most observers will see streak lightning that travels in a line from a cloud to the ground, and forked lightning appears even more dramatic.[1] The sudden discharge of electrical energy in the air creates temperatures of thousands of degrees Fahrenheit – hotter than the sun's surface – but lasting only a second.[2] Air rent by a lightning bolt superheats briefly, and its expansion and contraction result in pulses of the sound that is thunder.[3]

Ever since Aristotle, scholars have tried to formulate explanations of how thunder occurs.[4] In fact, the current opinion, articulated by Hirn in 1888, is not so very far from the descriptions framed by Aristotle and Lucretius: thunder arises from the violent expansion of air/gas when the electric spark of lightning raises the temperature suddenly. Both Aristotle and Lucretius also noted that the sound traveled more slowly than the light generated.[5] Flashes are seen almost instantaneously, but thunder's sound travels approximately 1 km in three seconds (or 1 mile in five seconds), so its link to lightning remains evident.

Weather is affected by the level and changes in GCR – galactic cosmic rays (which also happen to stimulate the production of the radioisotope

[1] See Uman 1984:1–13.

[2] Pielou 2001:178–80.

[3] Schaefer and Day 1981:195–96.

[4] See *Meteorologica* 2.9–3.1 = 369a10–371b15, Lee 1952/1987:223–41.

[5] For a full history of research into thunder and lightning, see Uman 1984:181–201.

Carbon-14) – and also affect ionization and thus the electric field of the earth's atmosphere.[6] As negative and positive ions are formed, the negative particles migrate to the upper atmosphere, and positive ions are concentrated in lower regions of air and clouds: the electric field they support is approximately 100 volts per meter on a calm day. As a thunderstorm forms, there is more separation of negative and positive particles, and the strength of the electric field increases to tens of thousands of volts per meter. Lightning results when the field discharges.[7] The earth receives over one billion lightning strikes annually; many are cloud-to-cloud discharges heard but not seen from the ground. For Etruscan observers (and others), other phenomena also conceivably could have become sources of "thunder" noise, especially if heard on a clear day: radio noise from the stars, or volcanic or seismic activity, but stormy weather was always the most likely source.

Lightning discharges during thunderstorms are the most common source of atmospheric noise, which varies according to geographic location and time of year and day. Central Africa experiences the most intense activity. Audio frequencies, including thunder, blend into VLF (very low frequency sound) frequencies: these have been exhaustively monitored worldwide; their patterns of intensity are of vital importance for modern communications and military applications. The *CCIR Report* on VLF noise notes that it is particularly severe during the rainy season in certain parts of the world, especially equatorial Africa, the Caribbean, and northern India. The map of VLF noise intensities for summer afternoons shows high noise frequencies extending north into the Mediterranean basin and the Italian archipelago.[8] Thunderstorms and related thunder booms are frequent there today, especially in spring and fall, and were probably more frequent during the first millennium BC because of the heavier storm cycles occurring during the first half of the millennium.

Audio frequencies extend from approximately 20 to 20,000 hz (cycles per second), and the frequency is a factor of propagation due to the reflective characteristic of the (D layer) ionosphere as the high audio blends with VLF. Galactic or cosmic noise emanating from sunspot activity and several sources along the galactic plane can also be perceived at frequencies of 15 to 100,000 Mz. The atmospheric noise levels for Italy in summer, between noon and 4:00 P.M., when thunderstorms often occur, are about 65 decibels.[9] For comparison, 65 decibels is approximately the level of sound heard when people converse or

[6] See Carslaw et al. 2002.

[7] Current research is centered on the correlation between cosmic ray flux and cloud formation. For explanations and illustration, see the Web site of the Stanford Linear Accelerator Center, Menlo Park, CA, http://www.slac.stanford.edu/. For recent scholarship of cosmic rays and climate, see Svensmark 2000; Mavrakis and Lykoudis 2006, and references therein.

[8] Vincent 1985:34–1 to 34–5; 34–4, fig. 2.

[9] See *CCIR* 1963.

hear small appliances or office keyboards, for example – so thunder booms are likely to be heard by multiple individuals during a storm event.[10] For ancient divination, this means that, whether or not a haruspex or other priest heard thunder during a storm, it is likely that many people would have, and so there could be no doubt that an event had happened: the diviner could take the observers' word and proceed to interpretation. In contrast to lightning, however, it would have been too difficult to determine the exact direction or source of the sound.[11]

A CLIMATIC HISTORY OF IRON AGE ITALY

Arlene Miller Rosen (2007) has proposed climate change as a potentially "civilizing factor" in human societies, with case studies from the Levant of Neolithic hunter-gatherers, Early Bronze Age complex societies, and the Byzantine Empire coping with climate-induced food crises (in the Near East it is usually drought rather than cold, wet weather). She concludes that "the need to adapt to sometimes rapidly changing climates is one of the challenges that can stimulate technological, social, and political innovation."[12] Archaeological and historical evidence suggest that social change on the Italian peninsula might be correlated with climatic change, and this is especially so for the period in which the *Brontoscopic Calendar* emerged. Iron Age communities were growing, society was stratifying, and armed conflict was increasingly sought, just as religious institutions were being formalized and linked to the state. At the same time, a century of severe weather occurred. Omens delivered by thunder and lightning were bound to draw attention to cults that could interpret them, and frequent storms meant more scope for predictions.

An extensive literature on the subject of ancient climate is developing, because excavations and paleoecological studies of the Swiss and Italian lake villages and other settlements are increasing.[13] An abrupt change in cosmic rays and climate, the Sterno-Estrussia solar excursion, had been identified around 850 BC.[14] In Europe, this period of cool, wet weather raised lake levels, causing some villages to be abandoned. In the Balaeric Islands, for instance, drastic social changes were triggered by the climate, even though it would have been buffered there by a moderating island effect: the result was the insular, Talayotic culture that changed burial practices and began erecting large, stone tower-like structures (*talayots*). By the end of the ninth century, Menorca

[10] Telephones, automobiles, and lawn mowers are slightly louder, at 80 to 100 decibels; see Schomer 1985:40–8, fig. 1.

[11] I am indebted to Alwin G. Steinmayer for invaluable advice and references on this topic.

[12] Rosen 2007:172.

[13] Marzatico 2004; van Geel et al. 1996; Tinner et al. 2003.

[14] See van Geel and Berglund 2000; van Geel et al. 1998.

and Mallorca had become isolated from Mediterranean trade routes.[15] In the Balaerics, the response to relatively mild but unremitting climate pressure was to protect the islands from any influx of migrating mainlanders, even if that meant the loss of foreign trade.

In Italy, atmospheric and planetary conditions around 800 BC – known as the Hallstatt Minimum – had entered a cycle that would have generated more than the usual number of cyclonic storms, rain, and chilly weather, and presumably, more than their share of thunder. Failed harvests, competition for scarce resources, the disappearance of some wild animals, and appearance of others less friendly would have served to further stratify society, make land ownership more important, and foster wars and raids, engendering all the elements of social stress that are recounted in the *Brontoscopic Calendar.*

This environmental challenge recently has become evident because of studies in another field – the radiocarbon dating of archaeological contexts. Problems were discovered in the use of the radioisotope Carbon-14 to date organic materials (from wood and seeds to human bones) that typologically and stratigraphically must have been deposited in the late ninth to eighth centuries BC, but these materials fail to give the correct values of Carbon-14 isotope when tested. An event with planetary impact occurred around 850–800 BC that resulted in abrupt and noticeable climate change that was to continue for perhaps three centuries. Only after the crucial Archaic period do Carbon-14 dates become reliable again.

Paleoecological or proxy records, too, such as pollen samples and tree rings, seem to show a time of climatic change in the late ninth century, generating cold, shorter growing seasons, and less food in Europe and Italy. The tropics also experienced climate change at this time, becoming more arid, and in many regions all over the world, migration and cultural change ensued. Periods of low sunspot activity seem to follow a solar cycle of approximately 2,400 years. This creates an abrupt change to cooler, wetter weather, and with it (for modern observers) comes a much higher production of Carbon-14 in the atmosphere: "When solar activity is high, strong magnetic fields block out the high-energy galactic cosmic rays approaching the Earth and less ^{14}C is produced in the atmosphere. During periods of low solar activity the magnetic fields decrease resulting in higher ^{14}C production."[16] The period at issue is dated at 2750 to 2450 years BP (Dates determined by radiocarbon analysis are given in years "Before Present," which can then be converted into dates BC or AD.), and takes its name from the European Hallstatt Iron Age cultural phase named for the site in Austria (near the famous salt mines), where its artifact typology was initially studied. Thus the "Hallstatt Minimum/Hallstatt

[15] van Strydonck 2004.

[16] van Strydonck 2004: 248; cf. diagram of Magny 1995:52.

Plateau" (sometimes "Hallstatt Disaster") covers the crucial historical period 760–400 BC.[17]

Proxy records of the central Italian climate, especially those derived from pollen deposited in lakebed sediment cores, all show a change in climate/weather around 2800 years BP. At this time, around 800 BC, lake levels rose and weather was generally cooler and wetter, meaning more storms and floods across Etruria (and beyond, especially in transalpine Europe). The site of Poggiomarino in Campania has furnished invaluable evidence of early life in Italy and also of the condition of the peninsular environment (see Chapter 5).[18] A key feature of the settlement is a fine system of canals and water channels for drainage and water supply. The environmental crisis of 800 BC is seen in the process of severe silting up of the village's fine water system, until it no longer functioned by the eighth to seventh centuries BC.

Recent studies of Etruscan lakes furnish data on Iron Age climate, at Lago di Vico, Valle di Castiglione, and, most notably, Lago dell'Accesa south of the Colline Metallifere, where a mining and industrial town of the seventh and sixth centuries BC has been excavated, along with its tombs.[19] During the seventh to sixth centuries BC, torrential rainfalls were at times so violent they caused the collapse of houses.[20]

The response of the communities of Tyrrhenian Italy to the stormy, wet chill was quite different from the Balaerics: the eighth century saw the *opening up* of Etruria and central Italy to foreign exchange on many levels and with several cultures. Radiocarbon evidence deposited just before the Hallstatt Plateau, in Italy and also in the Phoenician colonial world, is now being analyzed. Latin communities at Satricum and Ficana flourished: towns of the transitional period from Latial IIB to III have been dated as early as 820 BC, at least by some experts.[21] The establishment of a Phoenician colony at Atlantic Huelva by the early ninth century gives reason to envision the long-distance ships of Tarshish dispatched by Solomon and Hiram I of Tyre, and radiocarbon dates of cattle bones at Carthage support the founding of that colony in the second half of the ninth century.[22] At the Iberian Phoenician colony at Toscanos, it appears that Levantine masonry structures, such as the tripartite pillared building, were transplanted to the "new world" of the western Mediterranean during the eighth century soon after their prototypes

[17] Speranza et al. 2003; Dergachev et al. 2004; for Italian radiocarbon dates, see Skeates 1994 and Turfa 2006F.

[18] Albore Livadie and Cicirelli 2003.

[19] See Drescher-Schneider et al. 2007, for full documentation.

[20] Camporeale and Giuntoli 2000:19; with potentially lethal complications for miners living in the midst of toxic metallurgical by-products; see Harrison et al. 2009.

[21] Dating: Nijboer et al. 1999/2000; Bietti Sestieri 2010:270–84.

[22] See Nijboer 2006A and 2006B; Docter et al. 2005.

had been designed in Palestine.[23] Along with the 800 BC date of the inscriptions of the so-called Nora stones in southern Sardinia,[24] these finds point to early and intensive interaction between the eastern Mediterranean and the Italian archipelago already in progress.

Albert Nijboer and his colleagues have formulated a list of ramifications for the study of Italian protohistory as affected by the recent breakthroughs in radiocarbon and related studies.[25] These include a longer, and earlier, period of "precolonial" contacts between Italians and groups from the eastern Mediterranean, "less abrupt socio-economic transitions in Italy during the eighth century BC," "lengthening of the proto-urban phase during the urbanization process of central Italy," "synchronous evolution of the Italian and Greek Orientalising periods, which currently are somewhat dissociated in time," and some revision of our understanding of the phenomenon of Greek colonization, as linked to the comments of Thucydides. One issue developing is the precise dating of the manufacture and deposition of the numerous valuable Near Eastern imports found in the so-called princely tombs of Etruria, Latium, and Campania: many of the tombs have been dated about 680–650 BC or slightly later, but some objects may have been heirlooms when they were buried. For instance, metal vessels found in the so-called Tomb of Midas (Gordion Tumulus MM) including a siren-cauldron and lion-head *situla* (cf. Fig. 16) have close parallels in objects from Etruria and Latium.[26] But fresh studies of the tree-ring evidence and the Greek pottery imports have led to a re-dating of the tomb: its logs were felled circa 740 BC, perhaps fifty years *before* the demise of Midas.[27] Some other Phoenician imports in Villanovan burials, such as a bronze bowl at Vetulonia (Poggio alla Guardia Tomb VII, ca. 700 BC?) also support this earlier dating and picture of more intensive interaction.[28] Nijboer and colleagues warn that the great value of such prestige goods means they probably circulated a long time before they were buried with rulers in the mid-seventh century.[29]

The Hallstatt Minimum coincided with the full-blown Villanovan phase (*Villanoviano Evoluto*) in Etruria, beginning about 780–770 BC, when intensive contact with foreigners, marked by imported Greek pottery and Levantine amulets, began. Processes that had been underway for generations would now come to fruition: the stratification of society and the unequal and apparently hereditary apportionment of wealth and land. Industries and skilled

[23] For example, at 9th-c. Tel Hadar; see Kochavi 1998; corrected dating in Kourou 2005:499–500; Botto 2005:592.

[24] Or second half of the 9th c.; Amadasi-Guzzo 1990:72–73.

[25] Nijboer et al. 1999/2000:174.

[26] Veii Casale del Fosso 871, Rocca di Papa, Regolini-Galassi, Bernardini, Barberini Tombs; see Babbi and Piergrossi 2005:308; Arietti 2005.

[27] See DeVries et al. 2003, DeVries 2005; Delpino 2005:395–96, 643; Turfa 2006F.

[28] Maggiani 1973.

[29] Nijboer et al. 1999/2000:163.

professions arose where there had previously been only home or seasonal production of pottery and metal goods; workers could enjoy unprecedented comfort and affluence, and warriors' services earned them great prestige. The Late Villanovan period, the second half of the eighth century, saw settlements on the site of future cities, elite warriors, wealthy women, and elaborate funerary rituals; it was crucial in determining the character of ancient Italy.[30]

The absolute chronology of these cultural phases is hotly contested and depends upon cross-referencing artifact typologies to the well-documented European chronology of radiocarbon and dendrochronological data, and, via imported valuables deposited in eighth-century Italian tombs, to the established king lists and chronologies of the Aegean and Near East. The next turning point, the Orientalizing period, is now considered to have begun about 740–730 BC.[31] It must have been then that the additional, foreign inspirational sources of the *Brontoscopic Calendar* were first acquired in Etruria, perhaps through the agency of long-term Levantine trading missions or even diplomatic exchange. By the early seventh century, circa 680 BC, the major cities of Etruria were in full swing, with the rule of law enacted by a small ruling group or perhaps a king, elaborate cult rituals and formal sanctuaries, regular armies or militias, and stratified societies with commoners, women, and other groups vying for power. It is at that point that the *Brontoscopic Calendar* must have been committed to writing (in Etruscan) in a formalized manner. The generation who accomplished this would be buried with their heirloom treasures in the so-called princely tombs of Caere, Praeneste, Veii, Marsiliana d'Albegna, and Verucchio.

Iron Age Italy seems to have been more dynamic in social, political, and technological evolution than has been previously acknowledged. Technological and social interactions with Europe flourished. If the Iron Age farm communities of northern and central Italy were formulating oral traditions around the seasonal year in the century prior to the advent of writing and concerted Near Eastern exchange, they would surely have been influenced by the steady battering of violent storms and thunder, and it would be understandable if they tended to take such fierce activity personally. When, in the next generation or so, foreigners showed up with a long tradition of celestial divination to share, the learning must have fallen on receptive ears.

A CLIMATIC HISTORIOGRAPHY OF IRON AGE ITALY

The Classical authors were not oblivious to weather or to its effects on history. Portentous climatic events were observed and expiated in Italy at least as early

[30] Cf. Docter et al. 2004; Nijboer et al. 1999/2000; Nijboer 2006.

[31] de Marinis and Gamberi 2004; Peroni and Vanzetti 2004; Ferranti 2004.

as the eighth century, according to the legend of Romulus. Three eclipses are associated with events in his life, marking his birth and the foundation of Rome – an eclipse recorded in the Near East for 763 BC could have been one of these events.[32] As with Rome, a period of intensive storms, vulcanism, or celestial phenomena would surely have served to enhance the credibility of and reliance on the *Brontoscopic Calendar.*

Grove and Rackham have noted changes in the climate of Italy between the Classical period and today; they cite a number of Classical authors who seem to indicate a perception that climate had altered between their day and the past.[33] Theophrastus, writing in the fourth century BC, said that the Cretans found that the summers were warmer than it was in his day, and winters more severe, terminating cultivation in some regions.[34] In Italy, first-century authors were aware of an earlier change. Saserna, the famous Late Republican Etruscan agricultural expert, claimed that Italy used to be colder (say, before the second century BC), claiming that olives and grapes could now be grown in areas that used to be too cold.[35] The early, cool period to which this referred was probably the Iron Age-Archaic period of the Hallstatt Minimum. Much later, catastrophic weather would be long remembered: unusual rains in 241 BC caused a devastating flood of the Tiber in Rome, according to Orosius (4.11), who wrote 700 years after the event. Such recollections are subjective of course, but of great significance to farmers, so Saserna's comment may preserve folk memory or myth from the period of the formation of the cities and of the *Brontoscopic Calendar.*[36]

STORMY ROMAN KINGSHIP?

Of particular interest is the correlation of radiocarbon data for the ninth-century cosmic ray event and ongoing Hallstatt Minimum with the meager genealogical evidence for the legendary rulers of Iron Age Italy. Dionysios of Halikarnassos (1.71) surveys the rulers of Alba and Latium from Aeneas to Romulus. He states that Alba was founded in 1151 BC, thirty years after the foundation of Lavinium; Numitor ruled 432 years after the fall of Troy (which forms the benchmark for this chronology) in 1184 BC. Romulus and Remus began their colony in the first year of the 7th Olympiad (751 BC). Dionysios tells us how many years each of Aeneas' successors ruled, from Silvius (29 years) to Amulius (42 years). Earlier within this dynasty were Tiberinus, killed in battle in 921 after only an eight-year reign, and Allodus, an evil ruler, who attempted to fake thunder and lightning as a means of control over his

[32] See Nizzo 2005.
[33] Grove and Rackham 2001:141–42; cf. Neumann 1985.
[34] *On Winds* 13; see Taub 2003:192n21.
[35] See Columella *De re rustica* 1.1.4–5; Speranza 1974:40, fr. 7; White 1970:21.
[36] Cf. Barber and Barber 2004 on the power of myth-making to record natural disasters.

people, but he was then killed *by lightning* when heavy rains flooded the Alban Lake, destroying his house. The demise of Allodus and the remarkable storms and floods would thus be dated at the end of his nineteen-year reign, 872 BC. Although the real year of 872 is too early for the cosmic event that triggered the Hallstatt Disaster, this falls close to the projected date of 850 BC. Is it possible that the original storytellers had in mind a period of genuine catastrophic events?

By the early empire, when Pliny the Elder wrote his *Naturalis Historia*, Etruria was still known for its violent thunderstorms, which often brought torrential rainfall, flooding, and damage to vital crops in the fields. Pliny said:

> *Vere autem et autumno crebriora fulmina, correptis in utroque tempore aestatis hiemisque causis; ... sunt in Italiae quoque partibus iis quae a septentrione descendunt ad teporem, qualis est urbis et Campaniae tractus, iuxta hieme et aestate fulgurate, quod non in alio situ evenit.*
>
> In the spring and autumn, however, lightning bolts are more frequent, for in each season their summer and winter causes are combined; ... also in Italy, in those regions that face from the north toward the warm air, such as the regions of Rome and Campania, there is lightning in both winter and summer, which does not happen in any other place. (Pliny *Nat. Hist.* 2.51.136 et passim)

The possibility of electrical storms or simply of thunder and lightning occurring at all times of the year, not just summer, would be an additional factor in the mystique and credibility of thunder-divination.

VOLCANOES AS A SOURCE OF "THUNDER" AND ITS WOES

The doctrine of *saecula* depended in part on the observation of some loud or heavenly noise to herald the end of a cycle; surviving accounts specify a trumpet-like noise for the event of 44 BC. Certainly there were always other sources for noise in the lower range that could have been perceived as thunder, and especially as the most ominous type, thunder heard on a clear day. Vulcanism, the activity of volcanoes, and related seismic activity produce noise from underground that could be interpreted as thunder. Volcanoes also frequently caused disruptions to climate and food production: some of the larger eruptions produced cooling events that lasted two to three years, because of the dust sent into the atmosphere. Just as the ice particles and finely divided dust in storm clouds are involved in the formation of lightning, so the dust clouds of active volcanoes become electrically charged, and lightning flashes (and thunder) occur within the volcanic plumes.[37]

[37] See Pielou 2001:178.

Volcanic eruptions can have a direct impact on health and thus might be implicated in some of the omens concerning health in the *Brontoscopic Calendar.* Irritant particles could spread, causing skin infections and respiratory problems, and some possible candidates occur in the *Brontoscopic* omens (see Chapter 6). The opportunity existed to forge a direct link between the predictions and a volcanic event. Major eruptions of Vesuvius, Vulcano, Etna, and other volcanoes occurred at least thirteen times between 1080 and 450 BC (and did not stop then). Six major events, marked by dust in the Greenland ice cores, are known to have occurred during the crucial years between 800 and 600 BC, at which point Vesuvius erupted with a VEI of level 3.[38] Such an event would have been noticed widely for its deadly effects; emissions from Etna in 1329 AD killed a wide swathe of birds and animals.[39] Volcanic eruptions are among the cataclysmic phenomena often recounted in the mythologies of all cultures.[40] The known volcanic events of the eighth and early seventh centuries (735 and 693 BC) could well have had repercussions on the formative years for the *Brontoscopic Calendar.*

A plinian eruption of Vesuvius occurred about 3780 years BP (thus ca. 1780 BC), as is amply attested by scientific findings in the region of Nola, some distance from Pompeii. The "Avellino event" included a violent fallout of pumice and ash, followed by an overwhelming pyroclastic surge that destroyed an Early Bronze Age village at Nola-Croce del Papa; mud surges (lahars) account for incredible preservation of huts with foodstuffs and organic belongings left behind, as dozens of adults and children, accompanied by flocks and herds, fled for their lives.[41] In other discoveries, an "interplinian" event, possibly the final phase of the Avellino event, is attested by layers of tephra that fell on lakes at least 120 km from Vesuvius.[42] An eruption so strong that it could rain stones far away could well have generated noise and darkness, even storms, to remain in the memory of subsequent generations, reinforcing their awe for the power of thunder and its link to catastrophe.

Tradition may have hinted at the past volcanism of southern Etruria: in the area of Bolsena is the Distretto Vulcanico dei Monti Vulsini; these volcanoes were active from 600,000 to 100,000 years ago.[43] An ancient observer might have deduced the power of past events around Lake Bolsena: emanations of gas (carbon dioxide and hydrogen sulfide) and hot springs with temperatures

[38] VEI is the Volcanic Explosivity Index, with VEI 5 the level of the rare, Plinian eruption of Vesuvius in 79 AD. Debiasi (2000) has plotted presumed eruptions of Etna in Sicily at 1500, 1470, 735, 693, 565, 475, 426, and 394 BC. For a good survey of volcanoes, including the Italian archipelago, see Scarth 1994, and Stothers and Rampino 1983.

[39] Simkin and Siebert 1994:165, 183–84.

[40] See Barber and Barber 2004.

[41] Mastrolorenzo et al. 2006; Albore Livadie 2002.

[42] Lago Grande di Monticchio and possibly beyond; see Wulf et al. 2004; Rolandi et al. 1998.

[43] Cioni 1995.

ranging to 40° C. The springs with a high calcareous content, forming travertine, were probably more active in antiquity than today.

Although there is no confirmed evidence of specific earthquakes in Italy prior to the sixth century BC, there is archaeological evidence of seismic activity around 550 BC.[44] At the settlement of San Giovenale (Borgo), houses standing since the late seventh century received structural damage when an earthquake opened cracks in bedrock and masonry.[45] Historical events of vulcanism were recorded by some ancient authors, but by the time Livy and Pliny wrote, Volsinii (Orvieto) was a ghost town anyway; its survivors had been deported to the settlement of Bolsena near the eponymous crater lake. The sources link Volsinii to lightning strikes and firestorms: Livy (27.23.3) cited a portent of *Volsiniis sanguine lacum manasse*, "at Volsinii, the lake was drenched with blood." Pliny (*Nat. Hist.* 2.53.139) described a past event that probably occurred before 263 BC: *Volsinii, oppidum Tuscorum opulentissimum, totum concrematum est fulmine*, "Volsinii, the richest of the Etruscan towns, was completely consumed by a lightning bolt." Other catastrophes at Volsinii included fires early in the first century BC. In his list of portents, mostly culled from Livy, Julius Obsequens (§ 50) lists a fire-setting bird killed in 95 BC; for 94 BC, he again names a firestorm there. Later writers must have been influenced by these Late Republican works; Tertullian (*Apologeticum* 40.8.10; *Ad Nationes* 1.9.7; *De Pallio* 2.4.47) linked the fate of Pompeii with that of *Volsinii etrusca*, and so the ominous quality of rumbling, whether in earth or sky, continued to carry dread.[46]

ANCIENT WRITTEN SOURCES ON THUNDER IN DIVINATION AND RELIGION

ETRUSCAN EPIGRAPHIC EVIDENCE FOR THUNDER-DIVINATION: TRUTNUT FRONTAC

Although a growing collection of archaeological material points to an Iron Age date for the formation of the *Brontoscopic Calendar*, the earliest extant literary or epigraphical references relevant to brontoscopic divination are relatively late (Hellenistic period/Late Republic). Earlier than any Latin authors are a small number of Etruscan funerary inscriptions and images that attest to the involvement of noble families in this form of divination. Some votive inscriptions also appear to refer to thunder, lightning, or specialized diviners, representing southern Etruria (Tarquinia and Orvieto), and also the Umbrian region. The paucity of references, however, is perplexing for such an important profession.

[44] Cf. Guidoboni et al. 1994; Blomé et al. 1996. I am grateful to Carl Nylander for sharing this article with me.

[45] Blomé et al. 1996.

[46] See Massa-Pairault 1999:80–83.

On the other hand, the calling of the haruspex (Etruscan *netśvis*), and the literature of *haruspicina*, in Etruscan *neθśra(c)*, has even fewer references, yet artistic representations and other evidence attest to their Archaic importance. Two funerary inscriptions unequivocally name haruspices. A Chiusine Hellenistic epitaph (*ET* Cl 1.1036) notes that Nae Cicu, son of Pethna, was a *netśvis*. One priest linked to various cults, Laris Pulena, wrote a haruspical treatise.[47] The prosopography of thunder-diviners rests upon the interpretation of the bilingual epitaph of the *netśvis*, Lars Cafates, the Latin of which leaves no doubt as to a link with lightning, and the Etruscan version apparently attests thunder as a separate specialty.

LARS CAFATES (PESARO). An Etruscan term for thunder and/or lightning may appear in the title of a priest buried, not in Etruria but at the Umbrian city of Pesaurum (Pesaro) during the last quarter of the first century BC.[48] (See Fig. 6.)

Latin:

L. Cafatius. L. f. Ste. haruspex fulguriator
(Lucius Cafatius L(uci) f(ilius) Ste(llatina tribu) haruspex fulguriator)
"Lucius Cafatius, son of Lucius, of the Stellatina tribe, haruspex, interpreter of thunderbolts"

Estrucan:

cafates. Lr. Lr. netśvis. trvtnvt. frontac.
(cafates l[a]r[th] l[a]r[is] netśvis trvtnvt frontac)
"Cafates Larth son of Laris, haruspex, priest, interpreter of thunderbolts"

Lars Cafates/Lucius Cafatius was both *netśvis* and *trutnut frontac*, but the three Etruscan words are rendered cryptically by two in the Latin, *haruspex fulguriator*. *Haruspex* is certainly a liver-divination priest, but to be technically precise, a *fulguriator* is the Roman augural official who interprets lightning. The word *frontac* is not a normal (or early) Etruscan spelling, because they did not use the vowel "o," which is written as a strange, looped character here. It sounds suspiciously like Greek βροντή, "thunder,"[49] but if so, it would imply either that Etruscans did not know thunder prior to contact with Greeks, which is patently impossible, that they did not perform the official activity of thunder-divination until after Archaic contact with Greeks, or that this is a newer office instituted under the Roman state. This is perplexing because there are no pre-Roman Greek parallels

[47] *ET* Ta 1.17; Bonfante and Bonfante 2002:149–51, no. 31; see Turfa 2006C.
[48] *ET* Um 1.7; Bonfante and Bonfante 2002:69; Haack 2005:cat. no. 18.
[49] Pfiffig 1969:279.

for a *Brontoscopic Calendar.* I suggest that by the first century BC, when families felt the need to record epitaphs in Latin as well as Etruscan, that the Hellenistic style of literary culture already prevailed, or it had become fashionable to use the terminology of Ptolemaic-Hellenistic astrology as adopted by Latin speakers. If so, then we do not know the original Etruscan words for this priesthood. If we are going to entertain a Greek derivation for *frontac*, then perhaps we also should also investigate Akkadian terms such as *birûtu* ("divination")[50] or *birqu* ("lightning, lightning bolt, thunderbolt").[51] There is also the possibility that the "c" ending the last term, *frontac*, is the Etruscan enclitic "c," meaning "and." If this is so, then we might translate *three* specialties in the Etruscan line, "*haruspex, fulguriator,* and thunder-diviner." The *fronta*-divination would then stand as a native Etruscan characterization of Cafates, not precisely paralleled in his Latin persona, just as the Roman voting tribe of Cafatius (the Stellatina) had no place in the Etruscan side of his life.[52]

It is possible that Cafates, with his part-Latin genealogy, was attached to some Italic cult that used different words, but it remains puzzling that, rare though thunder-divination may have been, few other Etruscans used similar terminology for their religious distinctions. Perhaps brontoscopy – in contrast to *disciplina fulguralis* – was an Etruscan specialty not practiced by Roman priests. This epitaph does establish thunder-divination during the first century BC: Cafates must have practiced his craft around mid-century, just the time that Figulus rendered the Etruscan document into Latin. If there was not already an established tradition and hierarchy for this doctrine, no one would be publicly taking up and preserving Figulus' translation.

ARNTH APRIES (TARQUINIA). One securely attested *trutnuθ* (?) was Arnth Apries, whose funerary *cippus* was erected at Tarquinia; it names him simply as a *trutnuth*, indicating a stand-alone profession or calling: *apries. ar. vθ. / trutnuθ* ("Arnth Apries son of Velthur, *trutnuth*").[53] Versions of the term *trutnuth* also appear in two Orvietan funerary portraits (4th-c. Tomba Golini II) and a votive inscription from southern Etruria, reinforcing our perception of *trutnuth* and *fronta(c)* as two separate offices.

[50] From *bārû*: *CAD*, vol. 2 "B," 268.

[51] *CAD*, vol. 2 "B," 258–59.

[52] Steinbauer (1999:194–96, B9) translates the three words "Eingeweideschauer, Blitzdeuter, aus F(e)r.nt(a)," giving *frontac* a geographical meaning ("from Ferentum"?) to parallel the Latin tribal designation, but this is not widely accepted. See Benelli 1994:13–15; Wallace 2008:152–53, no. 10.25. Emiliozzi restores the broken ends of the block to read "[L.Ca]fatius.L.f.Ste.haruspe[x.êt]/fulguriator" (A. Emiliozzi, "Epigrafia della 'Bilingue di Pesaro': Un'integrazione nel testo latino," in *Etruria e Italia Preromana: Studi in onore di Giovannangelo Camporeale* (Pisa and Rome: Fabrizio Serra, 2009), 1:371–72).

[53] Dated stylistically 4th to 2nd c. BC: *ET* Ta 1.174 = *CIE* 5487.

The early Hellenistic period, circa 300 BC, seems to have been a watershed in the development of Etruscan religious cults. This is reflected in several epigraphic and artistic works, such as the Tuscania mirror portraying Pava Tarchies and others (Fig. 1), that seem to depict newly developed social aspects of some Etruscan cults.[54] Where the later priest Lars Cafates might have been relegated to the Italic (Umbrian) fringes of Etruscan culture, Arnth Apries practiced at Tarquinii and was interred in the family tomb there. Briquel (2005) accepts *trutnuθ* as relating to a haruspex, considered a mainstream profession.

VERCNA-CNEZUS-LEINIE FAMILY (ORVIETO). The earliest attestation of the office of *trutnut* is perhaps a badly damaged funerary inscription from the Orvietan necropolis at Settecamini. The word *truθun*, believed to be related to *trutnuθ*, appears painted above one, perhaps two, images on the wall of the Tomba Golini II, used in the third quarter of the fourth century BC by the Vercna-Cnezus family.[55] The entrance wall has mirror images, possibly of the founder, driving a *biga* and wearing the *apex* headgear of a *haruspex* or *flamen*. So important is the outsized hat that it cuts into the border of the wall decoration. The inscription may preserve the priest's name, Lars Vercnas (*ET* Vs 7.27 = *CIE* 5108). Tomba Golini II was decorated with scenes of a procession featuring the mirror-image *bigae* and musicians, male and female family members reclining at a banquet, chthonic bearded snakes and an underworld demon, and, in upper corners of the walls, armed warriors, perhaps for gladiatorial games. (The Leinie branch of the family erected Tomba Golini I, with famous scenes of servants preparing food for the banquet attended by Hades and Persephone.)[56]

TRUTVECIE (?) (SOUTHERN ETRURIA). One possible mention of the *trutnut* office is a Hellenistic bronze votive figurine inscribed *trutvecie* (*ET* OA 3.5 = *CII* 2603 bis) in southern Etruscan spelling, and so perhaps also from the region of Orvieto.

> tite: alpnas: turce: aiseras: θuflθicla: trutvecie
> Perhaps: "Tite gladly gave to the gods to Thuflthas as a sacred action"

The term *aesir/aeseras* is a gloss noted by several Classical authors as the Etruscan term for "gods."[57] *Thuf* appears as a divine name in three cells of

[54] Cf. Turfa 2006B:62–64.

[55] *ET* Vs 7.6 = *CIE* 5082; see Steingräber 1986:279, no. 33.

[56] The epitaph there of brother Vel Lathite gives a noble genealogy and calls him a *zilaθ meχl rasnal* (*praetor Etruriae*), saying he "served as zilat at Chiusi of the Etruscan people" (*ET* Vs 1.179 = *CIE* 5093).

[57] See de Grummond 2006B:241n24; Bonfante and Bonfante 2002:186, with references.

the Piacenza liver, and *Thuflthas* is named on a few Hellenistic votive figurines; Nancy de Grummond has made the case for identifying him with the Roman god Favor.[58] The term *eiseras thuflthi[cla]* also appears on a Hellenistic votive figurine from Chiusi, where it is unequivocally labeled *cvera*, "offering."[59] The similarity of *trutvecie* to *trutnut* might mean that the "sacred action" was "expiation of a lightning strike."

THE ZAGREB *LIBER LINTEUS*. The root *truθ* appears possibly six times in the text of the Hellenistic Zagreb linen book, a calendar for religious rituals. The lines, following van der Meer's (2007) reading, are

V.17 – heczri. truθ. celi. epc. śuθce. citz. trinum
V.18 – hetrn. aclχn. ais. cemnac. truθt. raχś. rinuθ
X.12 – ścanin ras truθur tutimc an masnur
XI.2 – vacl. vinum. śantiśtś. celi. pen. trutum
XI.3 – θi. θapneśtś. trutanaśa. hanθin. celi
XI.6 – [murce θi.] nunθen. etnam. θi. truθ. etnam

Perhaps we should read some of the *trut* words, following Bonfante and Bonfante, as "libation, sacred action," rather than lightning or thunder.[60] Van der Meer interprets *truθ*- as an infinitive used as imperative, so "judge/interpret!"[61] Such a reading is not far from a definition of the *trutnvt* as a divination expert. At the least, Lars Cafates' epitaph remains unequivocal testimony to the practice of divination by electrical storms during the period when Nigidius Figulus was translating the *Brontoscopic Calendar*.

CLASSICAL TEXTS: GREEK AND LATIN WRITTEN SOURCES ON ETRUSCAN DIVINATION

Perhaps Figulus' publications engaged Romans' interest in Etruscan divination. Various ancient authors have made reference to thunder or lightning among the specialities of Etruscan divination (Ammianus Marcellinus 23.5.12). References to divination by thunder and lightning are found in major works of famous authors during the first century BC through first century AD, especially Cicero, Seneca, and Pliny the Elder. The pointed remark of Lucretius about "Etruscan scrolls" *versus* scientific explanation of thunderbolts, however, sets the tone of Roman recollection.

[58] de Grummond 2005 and 2006B:151.
[59] *ET* Cl 3.7 = *CIE* 2341, *TLE* 558; cf. Maggiani 1992:206.
[60] Bonfante and Bonfante 2002:219.
[61] van der Meer 2007:94–95, 137, 147.

One odd feature pervades the Roman literature: many terms and names bequeathed to us by Latin authors are indeed Latin – they do not preserve correct Etruscan spelling or pronunciation, and they thus cannot be exactly what Etruscan diviners said.[62] The names include that of the prophet Tages and extend to the monster/portent called Olta, who ravaged the fields of Orvieto before King Porsenna summoned a lightning bolt to dispatch him.[63] We can hardly belabor Pliny for a divergent spelling of a 500-year-old foreign name, but still, if Etruscan lore was truly central to the mystical power of Roman religion, why was it not more accurately replicated? Were Etruscan religious texts so seldom written that there was no template to hand for the Late Republican authors, or for Pliny? Or had the doctrines been adopted by Latin speakers so long before that even proper names had morphed into Latin by the first century BC? Other names and terms, too, were garbled. The nymph Vegoia must be *Lasa Vecu[via]*, according to engraved mirrors and other images (Fig. 11),[64] and the name of the mortal to whom she revealed the prophecy, Arruns Veltymnus, seems related to the name of the god rendered Veltune on the Tuscania mirror of 300 BC (Fig. 1). The Latin name of the "principal god of Etruria," Voltumna (as in *fanum Voltumnae*), must have sounded Etruscan to a Roman of the first century BC,[65] but it could not have been mistaken by an Etruscan speaker. Were these anomalies fostered by centuries of Roman familiarity, in which names and myths were so well used that they acquired Latin pronunciation?

Other terms, like *haruspicina*, may sound alien to Latin also, but to the best of our knowledge, they do not have cognates, even misspelled ones, in the Etruscan language: *haruspex* is registered *netśvis* in Etruscan epitaphs, and *haruspicina* is probably *neθśrac*. A movement began in the second century BC with Caecina, and flourished in the mid-first century, to translate the Etruscan treatises into Latin. It probably arose out of individuals' concerns that documents in Etruscan were not going to survive and that even people with intellectual curiosity were completely ignorant of their context. It must also indicate that the Etruscan language was never used in Roman cult.

DIODORUS SICULUS. The first-century Greek author Diodorus Siculus, in his survey of Italy (5.40), has provided a vignette of Etruscan culture – seen through alien eyes – that includes all the key features of Etruscan fame and the heritage it left to Rome. (See Appendix A, Text 4.) For Greek and Roman

[62] Turfa 2006B:80–84.

[63] Etruscan did not use a "g." For that matter, even Porsenna's name is spelled diversely and with an "o," with Pliny (*Nat. Hist.* 2.140) using "Porsina"; cf. de Grummond 2006B:23–27, 13–14.

[64] de Grummond 2006B:29–30.

[65] Varro *De lingua latina* 5.46; cf. de Grummond 2006B:41.

historians, Archaic Etruria was credited with many of the underpinnings of Roman social and political success, including lictors, the ivory folding stool of senators and magistrates, and the purple-bordered toga. Some disconcerting aspect of the Etruscan social system is expressed by Diodorus' comment that the servants were dressed better than was proper for the status of slaves (5.40.3).

It was especially in the study of nature and theology that Etruscans created the basis of cult and culture for the Romans, "the people who rule the world." Etruscans were preeminent in the art of divination by lightning bolts (κεραυνοσκοπία), and Diodorus noted that the Romans still used Etruscan diviners for this. It could not have been noteworthy if Etruscan priests had only acquired their thunder lore in the first century BC.

LUCRETIUS ON THUNDER. In examining divination by thunder, most scholars turn immediately to Seneca's theological comment, and to Lucretius for the "natural" component. In book 6 of *De rerum natura* (especially 6.96–422), Lucretius discusses thunderstorms critically, using the *etrusca disciplina* as a focal point for his arguments for atomic theory. He says that thunder arises from the collision of clouds and that it makes a noise like awnings flapping in a high wind. Clouds scraping past each other, winds cutting or rippling the surface of clouds, along with hail, ice particles, and lightning strikes contribute to thunder noise, heard only after we see lightning. (See Appendix A, Text 5.)

The oft-quoted lines of *De rerum natura* (6.379–86) point to the atomists' view and furnish circumstantial data on the recording of Etruscan divination texts: *Tyrrhena carmina*, "Etruscan hymns," thus verse, written retrograde in scrolls that had to be "unrolled backwards" (*retro volventem*). We might assume the scrolls were of papyrus, but skins are also an option, and the Zagreb *Liber linteus*, though it seems to have been folded, demonstrates that cloth also was used for religious documents. "Backwards" surely implies the reverse direction of normal Etruscan writing. Lucretius (6.388–422) offers his readers critiques of natural phenomena by reductio ad absurdum: If the gods themselves send thunderbolts, then why must they wait for a stormy day to do so? Why do they strike the innocent indiscriminately and not the guilty? Why strike the sea, the desert, or their own temples and statues? This skeptical attitude, so evident in Cicero, Lucretius, and Seneca, may be a clue to the lack of Roman interest in preserving Etruscan cultic texts. If the archives were not maintained during and immediately after the first century BC, it is less likely they would have survived.

CICERO'S *DE DIVINATIONE* AND SKEPTICISM. Cicero, the friend of Figulus, has left perplexing vitriol where we might have hoped for factual information, gleefully repeating the anecdote attributed to Cato: How can two diviners look at each other in the street without bursting out laughing at the hoax they are all perpetrating on the rest of society?

> Vetus autem illud Catonis admodum scitum est, qui mirari se aiebat quod non rideret haruspex haruspicem cum vidisset. Quota enim quaque res evenit praedicta ab istis?
>
> But that old quip of Cato was pretty sharp, when he said he was amazed that one haruspex did not burst out laughing when he saw another one. For how many events predicted by those guys actually come true?
>
> (De divinatione 2.24.51–52)

Cicero thus seems to dismiss the piety of generations of Italic and Etruscan worshippers. In *De divinatione* (2.42), maintaining his skeptical tone, he comments on augury in an imaginary dialogue with his brother Quintus. With omens delivered by lightning, all we can do is ask: Did it happen or not? And if so, how do we index or quantify the phenomenon? The one means, in Cicero's view, is by location. (See Appendix A, Text 6.)

"The Etruscans divided the heavens into sixteen parts" by quartering and quartering again, beginning with cardinal points; they would then note the sector where the lightning bolt had originated. Cicero also quotes a traditional response, *Iove tonante, fulgurante comitia populi habere nefas*, "with Jupiter thundering, sending lightning, to hold a council of the people is impious." He says that this restriction, recorded in the "augural annals," was probably made for political purposes,[66] thus implying that written commentaries, annals, and archives survived from primitive times. No doubt for rhetorical drama, Cicero seems unfamiliar with actual Etruscan literature, even though by 44 BC, Figulus' translation of the *Brontoscopic Calendar* had been circulated (and Figulus already deceased), and several other Late Republican authors known to Cicero, from Caecina to Tarquitius Priscus, had written on the *etrusca disciplina*.

If we assume that Cicero had read the works of his devout and respected friend, then we must assume that he deliberately ignored their import for the sake of intellectual debate. We cannot know what he would have said to Figulus when he was alive – or would they have agreed not to discuss a topic on which they apparently had almost no common ground? Susanne Rasmussen reminds us that "to Cicero, in practice, philosophy was an intellectual exercise and educational matter, whereas official Roman religion was a historical, social and political reality and necessity which had to be preserved for the welfare of *res publica*."[67] The fictional Cicero of *De divinatione* (2.43–45) goes on to say that of course they do not interpret literally the myth

[66] Translation of Falconer 1923/1996:417 (*in nostris commentariis scriptum*).

[67] Rasmussen 2000:20.

that Jupiter's thunderbolt (*fulmen*) was forged in the fires of Mount Etna by the Cyclopes. He points out the incongruity of Jupiter hurling multiple bolts if he only had the one famous one, and he goes on (2.44) to describe the Stoics' theory of electrical storms, which begins with heat convection creating global winds that break clouds; thunder and lightning result if this occurs with sufficient speed and force. This almost could be a poetic expression of how modern science describes electrical storms. Other authors also commented on the "mechanics" of electrical storms, such as Pliny reprising Stoic doctrine to explain the rarity of electrical storms in extreme summer and winter (*Nat. Hist.* 2.135–36). The milder weather of Italy, however, allowed for frequent storms. Cicero's diatribe is almost a denunciation of the brontoscopic texts: "no reasoning can teach what the thunderbolt (*fulmen*) means," *quid significet autem, nulla ratio docet* (*De divinatione* 2.45).

PLINY *NATURALIS HISTORIA*. Pliny noted the assorted meanings carried by different varieties of thunderbolt (*Nat. Hist.* 2.137–39). Some lightning strikes cause explosions but not fires, others destroy the contents of vessels but leave the exteriors unscathed, and so on. He notes as portents accounts of people being struck by lightning, and it is Pliny who gives us the reference to "Etruscan writings" (*Tuscorum litterae*), which must have had some doctrinal relation to the genre of the *Brontoscopic Calendar*, but they do not correspond to its text. They say that nine gods can send thunderbolts, of which there are eleven different types; Jupiter alone wields three different bolts. Jupiter can throw lightning bolts in the daytime, but at night they come from the god Summanus; the names of the others "are mostly forgotten." (In Near Eastern divination, phenomena are cataloged according to whether they fall in day or night, and some post-Figulus brontoscopies do note this condition.) Servius reprises the doctrine of types of lightning, but he states that there were twelve varieties (*Ad Aen.* 1.42).

When authors speak dismissively of Etruscans, they tend to lump them together as "Etruscans," "Tusci," or "Tyrrhenians," whereas in real life, individuals would have been known by their city ("the Tarquinians, the Chiusines"), so Pliny probably intended to evoke skepticism when he said, "Etruria thinks that sometimes they [lightning bolts] can erupt from the ground, these are called *infera* [underground]" (*Nat. Hist.* 2.138). Such oddities are thought to come not from the stars but from the atmosphere; star-born bolts strike in a slanting path, the earthly bolts in straight lines. Pliny says the portentous thunderbolt that destroyed Volsinii came from Mars, so perhaps war was somehow in the background of the catastrophe (2.139). We lack any date or other point of reference by which to analyze the event. This excursus (2.139) sounds similar to the Mesopotamian omens (like the *Šumma ālu*). The first lightning strikes that occur after a man sets up his own home foretell his fortune; "public lightning omens" (*publica*) have the potential to predict the next thirty years, and those observed at the foundation of a colony have even longer duration. His excursus on types of lightning and anecdotes of past portents precedes this at 2.137–38 (see Appendix A, Text 7).

Robin Lorsch Wildfang (2000) has examined several ancient texts, especially Pliny, Seneca, and Servius, that mention the destructive capabilities of lightning bolts, which variously stab, shatter, or burn the things that they strike, finding that burning is of a less serious nature, requiring *supplicatio* in expiation. The *fulmen quod discutit*, which "shatters" the object struck, is most dire, and it can only be expiated with the more elaborate and costly ritual of *obsecratio*. Clearly, ancient *philosophers* recognized a much greater variety and hierarchy of omens than is apparent from surviving *religious* texts.

Pliny knew that the Roman king Numa Pompilius, famous for his piety, was able to summon lightning bolts: Numa's traditional dates, 715–672 BC, fit the crucial period of formation of states and state cults,[68] seen in archaeological evidence at the Tarquinia Pian di Civita cult place and at princely tombs from Verucchio to Praeneste. This would be the ideal time for the formalization of religious texts. Lightning fell on the impious, too: Tullus Hostilius was blasted for trying to imitate Numa's rituals (cf. Livy 1.21.8). More interesting than the tale itself is Pliny's note that it was recorded in the *Annals* of the "reliable author" (*gravis auctor*), Lucius Piso: evidence for private/gentilicial archive keeping, such as could have served to record and preserve the *Brontoscopic Calendar*. Pliny also recalls the Roman institutions of the shrines of the three categories of Jupiters: the *Statores*, *Tonantes*, and *Feretrii Elicium* (Stayers, Thunderers, Receivers of Offerings [*spolia opima*]) (2.139). (See Appendix A, Text 8).

Pliny notes that the flash always precedes thunder; no one hit by lightning has ever seen or heard the bolt first (2.142). His commentary meshes with other authors when he states that flashes on the left side are good omens, because the sun rises on that side: this is appropriate if the observer, like the Etruscan augur, stands facing south.[69] The principle of quartering and requartering the circle of the earth or sky is elaborated by Pliny (2.143–44) (see Appendix A, Text 9). Cicero was familiar with this, and there is some archaeological evidence for the cordoning of a precinct or symbolic circle on the ground.[70] Mesopotamian divination, too, used this sort of binary configuration, right versus left, dark versus light, and so on.

Pliny states that the direction in which thunderbolts travel is very important, with the eastern portion of the sky being the best. There must be some lost reference behind his next statement that some people will only give or hear reports of lightning bolt omens if they are being told to a guest or a

[68] Plutarch *Numa*; Livy, bk. 1, passim.

[69] *Laeva prospera existimantur quoniam laeva parte mundi ortus est*; cf. van der Meer 1979.

[70] Cf. Torelli 1986.

parent. In this story, the parent as head of household might represent the cult of the noble family, but the presence of the guest (*hospiti*) is puzzling. It is from him that we learn that men killed by lightning may not be cremated, but must be buried in the earth (*hominem ita exanimatum cremari fas non est, condi terra religio tradidit*; 2.145). Lightning does not strike laurel bushes, deep caves, sea-calves, or eagles; but during the Caesarian war it struck buildings so often in the region between Terracina and the Lucus Feroniae that people stopped building towers there (2.146).

SENECA *QUAESTIONES NATURALES*. Gérard Capdeville (1989) has criticized Pliny's explanation of the nine kinds of thunderbolts and his confusion on the number of gods who wield them. He suggests that an apparent misreading by Pliny of an Etruscan term rendered in Latin as *Novensiles*[71] came to be perpetuated by Late Republican authors and commentators, beginning sometime after the Etruscan *libri* were first translated into Latin. Capdeville suggests that Seneca got a better version of the original information because he consulted the writings of Caecina, an undisputed Etruscan expert and practitioner of divination. Aulus Caecina, related to the Caicna family of Volterra and elsewhere, was probably the son of the defendant of Cicero's oration of 69 BC, and an acknowledged expert in the *etrusca disciplina*.[72] Capdeville maintains that there never were "nine Etruscan gods" who wielded thunderbolts (which would explain why Pliny was incapable of naming all but Jupiter and Summanus).[73] Instead, Jupiter is bound by two different divine councils in his decisions to hurl lightning: the twelve *Dii Consentes* and the *Dii Novensiles*, an epithet that does not mean nine in number, but remains obscure.

This controversy of Pliny's "nine gods" aside, Seneca's *Quaestiones Naturales* are a rich source for Etruscan cosmic beliefs. He must have researched his subject in secondary works on the Etruscan *disciplina*, and he would have had access to treatises like the writings of Aulus Caecina, to which he refers at *Quaestiones Naturales* 2.39.1. The details of types of lightning, the gods involved, and the meanings of such portents are discussed at 2.39, 2.41.1–2, 2.45.3, and 2.47–9. Seneca (*Nat. Quest.* 2.41.1–2) describes the three *manubiae*, different kinds of bolts, hurled by Jupiter, and this is nearly paraphrased by Festus (see Appendix A, Text 10).[74] As Capdeville has shown, this information almost certainly comes from a text of the *etrusca disciplina*.

Seneca tells us that the council voting on the second type of bolt comprises twelve gods; he also adds, to the information also eventually preserved

[71] Cf. Varro *Antiquitates* 3, fr. 55 Cardauns.

[72] He is also cited by Pliny (*Nat. Hist.* bk. 1, *index auctorum* to bk. 2; cf. Weinstock 1951 for contemporary authors.

[73] Capdeville 1989:1190.

[74] For Seneca *Quaest. Nat* 2.41.1–2; see de Grummond and Simon 2006:214.

by Festus, that the third, most destructive, category of *manubiae* is under the surveillance of the council of the gods who are *superiores* and *involuti*, which Capdeville (1989 and 1992) and most scholars read as a separate group, a partial subset of the twelve who guide the second *manubiae*. Many scholars have associated the council of twelve with the *Dii (Penates) Consentes et Complices* named by Arnobius (and Manilius) that are composed of six male and six female gods exercising oversight of Jupiter. Several authors[75] describe lightning bolts as if they could be tridents or forks of some sort, and this symbolic format may be an additional link between Italy and the Near East.

The most famous comment on Etruscan religion is probably Seneca's *Quaestiones Naturales* 2.32.2: the contrast between Etruscan and Roman religious philosophy (see Appendix A, Text 11). The Etruscan "disconnect" between natural phenomena and nature seems to have been foremost in Roman minds; for them, Greek natural philosophers cataloged phenomena precisely, but Etruscans took a more mantic approach. We may glean some impressions of Etruscan thunder literature from stray remarks (*Quest. Nat.* 2.48.2): *Quae inspicienda sint in fulgure, passim et vage dicunt, cum possint sic dividere quemadmodum ab Attalo philosopho, qui se huic disciplinae dediderat, divisa sunt*, "The Etruscans speak in general terms and vaguely about what really ought to be noted in lightning. They could have set up classifications as was done by the philosopher Attalus, who had devoted himself to this study."[76] The vagueness of the *Brontoscopic Calendar* protases, which merely note "if it thunders" or "if in any way it should thunder," should have been familiar to Seneca; they contrast markedly with the exceedingly specific daily predictions and relate to the state or the entire community. Presumably, Seneca only intended here to demonstrate that the *etrusca disciplina* had no relation either to empirical science or to Greek-style philosophical analysis of reality.

Lightning and thunder were sent by the ruler of the gods: *Idem Etruscis quoque visum est, et ideo fulmina mitti dixerunt a Iove quia sine illo nihil geritur*, "The same notion was seen by the Etruscans, and thus they said that lightning bolts were sent by Jupiter because without him nothing is accomplished" (*Quest. Nat.* 2.45.3). Seneca describes the three types of lightning allegedly discussed by Caecina: advising, confirming, and conditional (*Quest. Nat.* 2.39; see Appendix A, Text 12). As Seneca tendentiously reminds us, these are not kinds of lightning but kinds of interpretation (2.40.1). Etruscan doctrine differentiated perpetual, limited, and deferred lightning bolts/messages, and here we see a glimpse of divinatory information at *Quaestiones Naturales* 2.47 (see Appendix A, Text 13).[77]

[75] Following Ovid *Ibis* 369; *Metamorphoses* 2.848–50; Seneca *Herc. Oet.* 1994–96; Festus 480.10–12 L, s.v. *trisulcum* [*fulgur*].

[76] Translation of Corcoran 1971/1999:177.

[77] Cf. Pfiffig 1975:134–35.

THUNDER AND LIGHTNING IN CLASSICAL RELIGION

Historical accounts link the kings of Rome and Etruria with electrical storms and other celestial phenomena, for good or bad. The life of Romulus was correlated by ancient historians with an eclipse (or three), some documented by Greek historians and Assyrian annals.[78] Livy's account of the aftermath of Romulus' disappearance during an electrical storm notes that some witnesses suggested he was eliminated by the senators around him, but the official story was a sort of assumption by lightning bolt, with Romulus subsequently venerated as the god Quirinus.[79] Tullus Hostilius was killed by a lightning strike (640 BC) in response to his impious behavior during an epidemic that struck Rome. He tried to imitate Numa's expertise, but improperly, and Livy (1.31.8) says, *fulmine ictum cum domo conflagrasse*, "struck by the lightning bolt, he burned up with his house." (At August 23, thunder in the *Brontoscopic Calendar* "signifies that the lightning bolt shall fall, and warns of slaughter.") It appears that, in the first era of urbanism, the most direct of the meteorological omens were invariably intended for kings.

For the Greeks, death by Zeus' thunderbolt was meted out to Asklepios and Semele (Pindar *Olympian Ode* 2.25–26); the death of Erechtheus on Poseidon's trident is also sometimes associated with a sort of deification, because all these humans ultimately became divinized or heroized. The lightning bolt could also be punishment, as when Zeus used it to kill Kapaneus at the siege of Thebes: the body had to be cremated and interred on the spot, which was treated as sacred ground thereafter (Euripides *Suppliants* 934–35).

Roman religion required that lightning victims be buried on the spot (Pliny *Nat. H* 2.145), and Plutarch even implied that the unburied bodies of victims were simply fenced off and did not decompose.[80] The earth where a bolt struck became "sacred" (*sacer*), was cordoned off, and inaugurated with the sacrifice of a sheep.[81] A number of ritual graves/tombs are known marking the spot where a lightning bolt hit the earth, including one seen at Vulci.[82] Etruscans had a technique for expiation of lightning strikes to "bury the lightning bolt" (*fulmen condere*); the now-sacred spot was called *bidental*. Apuleius (*De deo Socr.* 7) said: *Tuscorum piacula, fulguritorum bidentalia*, "the expiatory offerings of the Etruscans, the *bidentalia* of those who were struck by lightning." As Pfiffig points out,[83] the term

[78] Nizzo 2005:645–46.

[79] Livy 1.16; cf. Ovid *Fasti* 2.475ff.; Plutarch *Romulus* 27.

[80] *Quaest. Conv.* 4.2.3.665; but contrast Seneca *Quaest. Nat.* 2.31.2.

[81] The earliest reference is Lucan's *De bello civili* 8.863–64: *in Clusium Tusco venerantur caespite fulmen*; see Lambrinoudakis and Petrounakos 2005 for references.

[82] All of Roman date; see Marcattili 2005; Pfiffig 1975:135–38; Quintilian *Declamationes Minores* 274.

[83] Pfiffig 1975:135–37; Thulin 1906:vol. 1, 93–98.

was so old that none of the Late Republican authors could properly explain it. Lucan (*De bello civili* 1.584–637) described the state's observance of ancient traditions of lustration and sacrifice at the behest of the Etruscan seer (*vates*) Arruns, who "collects the scattered fires of the thunderbolt and buries them in the earth," *dispersos fulminis ignes colligit et terrae maesto cum murmure condit* (1.606–7).

One explanation claimed the expiatory sacrifice was of a lamb "who had just cut two teeth," but this appeared highly dubious to the Scholiast on Horace, who noted that "those are wrong." Macrobius (*Saturnalia* 6.9.5) noted that Publius Nigidius, writing on haruspicy, used the term *bidentes*, a misspelling of *bidennes*, for all sacrificial animals that were two years old. The term remains obscure.[84] The monument itself was a form of expiation, inscribed *fulgur conditum*, sometimes in various acronyms (FCS, FDC = *Fulgur Conditum Summanium, Fulgur Dium Conditum*). Objects (or bodies) struck were to be left or buried where they lay, and denied standard funerals.[85] The *bidental* monument, like the *mundus* for chthonic offerings, had to have a portion below ground and some portion projecting above as the focus for offerings. Most *fulgura condita* are attested only by Latin inscriptions found out of context; at least eight were marked by small *sacella* or shrines inscribed with variations on *Fulgur Conditum* (Minturnae, Luni, Vulci, Pompeii – Casa dei Quattro Stili, Ostia, Rome – Praetorian Camp.) Another found near the Theater of Pompey held a bronze statue of Hercules that had indeed been struck by lightning.[86]

ETRUSCAN CELESTIAL DEITIES IN REPRESENTATIONAL (AND NONREPRESENTATIONAL) ART

In Etruria, of course, the thunderbolt was the prerogative of Tinia, whose name is derived from Etruscan *tin*, "day." Votive inscriptions and images attest him as a father god (occasionally addressed as *apa*, "father") and also as the head of the gods, the *aeser*.[87] He is well attested in both art and epigraphy, and a stylized lightning bolt is his chief attribute. Engraved mirrors and votive figurines show Tinia at rest, holding one or two bolts, as when he participates in the court on Mount Olympos (Fig. 9), the birth of Menrva, his wedding to Uni, or his amorous trysts.[88] On the side panel of the sixth-

[84] Ps.-Acr. *Hor. Ars.* 471; see Marcattili 2005:201.

[85] Scholia ad Perseus 2.27: *etiam homo fulmine percussus eodem loco sub bidentali humatur.* Festus (450 L.) says that the site had to be left untouched, in the open: *fulgur conditum* [...] *<ne>fas est integi, semper foramin<e ibi aper>to caelum patet.* Marcattili (2005) associates the sacrifice of young sheep (two years of age) with offerings made at this sacred site.

[86] Marcattili 2005:202, no. 4.

[87] de Grummond 2006B:53, emphasizes this link to the Norse gods and to Tinia's identity as the storm god.

[88] See de Grummond 2006B:53–70.

century BC parade chariot from Castel San Mariano (near Perugia, Fig. 7), Tinia, speaking to Hercle, casually holds, as if it were a spear, a thunderbolt with four curved, toothed blades.[89] On the other panel of this chariot, in combat with a giant, Tinia brandishes his usual short, flower-like *keraunos* – the two designs must correspond to different categories of *manubiae*. On the badly damaged Pyrgi Temple A *columen* plaque (ca. 460 BC), Tinia hurls a bolt (not preserved) at Capaneus in the battle of the Seven Against Thebes.[90] Sometimes he holds two (but apparently never three) different kinds of thunderbolts, as in a bronze mirror in the Museo Gregoriano Etrusco depicting the visit from Eos and Thetis (Fig. 8, ca. 470 BC). One is the commoner bolt that looks like a Greek *keraunos*, with one end forked and almost floral in design; the other is longer, a set of three wavy bolts that look like the prongs of back-to-back tridents: it is very similar to some Near Eastern representations (see Figs. 14, 15).[91] Many examples of the commoner type have a spear-like distal end. De Grummond suggests that the diversity in renderings makes "it impossible to construct a typology, quite unlike the situation in representations of the Hellenic Zeus, whose bolt is regularly symmetrical, usually with a flourish on each end. Thus the Etruscan cultic understanding of lightning as having a rich and widely varying repertory is quite consistent with the representations in art."[92] As with the enigmatic *Di Involuti*, such variety signals Etruscan beliefs with roots much earlier than the Orientalizing period. Menrva, too, wields the lightning bolt, with floral end(s), but at times, as on a third-century mirror in Berlin, it is as large and businesslike as a warrior's spear (Fig. 10).

In the early art of Etruria, the gods of prophecy, omen, and cataclysm are conspicuously absent, or at least they do not appear in humanized form. In the *Brontoscopic Calendar*, on July 9, "if in any way it should thunder, it foretells a *vision of the gods* [my italics] and the advancement of many good men"; yet if thunder falls on December 6, "men shall be visited with *visions of the faces of the gods*, they shall experience a bad outcome." Ambivalent attitudes to the sight of gods, or their countenances, seem to occur in Etruscan art also. Dynamic or anthropomorphic representations of celestial gods are rare before the end of the sixth century BC, long after the initial formalization of the *etrusca disciplina*. When gods do come to be depicted in Archaic Etruscan art, they seldom "show their faces" or appear full-face or gazing directly at viewers. Of the *Di Involuti*, however, the celestial gods who are shrouded in clouds, there is never anything to see. The "faces" of the elemental gods – their true, powerful natures – were not for mortal view.

[89] See Feruglio 1991:132–33, fig. 1.5.

[90] Simon 2006; Ridgway 1990.

[91] Pfiffig 1975:131–33, figs. 54–57.

[92] de Grummond 2006B:54.

The original elemental forces still held sway in divination. Among the twenty-eight augural gods named on the Piacenza liver, less than half (thirteen) are the recipients of conventional cult, and only twelve have been subjects for representational art.[93] Of these augural gods, only Tinia (Jupiter) and Usil (Sun) appear in art as early as the seventh century BC. Etruscan artists rarely depicted heavenly bodies, clouds, or lightning, apart from the scepter-like thunderbolt of Tinia/Zeus (cf. Fig. 18, an Iron Age bronze belt). Thunderbolts must have been accounted for in prehistoric belief; yet they are absent until the later seventh century. A stylized bolt is depicted in the first preserved Etruscan image (derived from Greek scenes) of the birth of Menrva on a Caeretan white-on-red cinerary pyxis now in the Louvre.[94] Outside the realm of narrative, however, there are earlier representations, in the form of pronged metal objects used as funerary markers and perhaps processional symbols, resembling insignia used in the Near East for three millennia.

Images of celestial gods were confined to a very few personalities and did not begin until after Etruscan art had embraced Greek narratives. The earliest identified image of the sun god – Usil – already imitates human form. Although cited in Etruscan personal names by 660 BC,[95] Usil is first depicted on a late sixth-century mirror now in Paris, and on a succession of later mirrors in various guises.[96] Tiur (or Tiv), Moon, whose name is used for the Etruscan term "month," seems almost never to appear in human guise.[97] She is denoted by the crescent symbol, as in the heraldic device of the Tiuza family of Chiusi, whose third-century tomb at Tassinaia is painted with crescent moons, and by a shield marked with a lunar phase.[98] Thesan, Dawn, although invoked in girls' names since the seventh century (e.g., Thesathei on the Tragliatella *oinochoe*, Fig. 20), is not *seen* until the end of the Archaic period. She continues on vases of the fourth to third centuries, where she tends to appear in a window with her team of horses. We occasionally see Thesan's face, as on a mirror in Essen,[99] not frontally, but in three-quarter view, but she is actually a personification, not an elemental in the same way as the sun and moon, and she is not cited in divinatory texts.[100] It does appear that the elemental Etruscan gods, who must have been venerated long before contact with Near Eastern divination or Greek mythology, were radically different in nature and in (non)appearance.

[93] See de Grummond 2006B:53; van der Meer 1987.

[94] See Martelli et al. 1987:95, 266–68, no. 43; cf. Pfiffig 1975:127–33 for additional representations.

[95] van der Meer 1987:136–40; de Grummond 2006B:132–33.

[96] See Tirelli 1981, and Ridgway 1990.

[97] van der Meer 1987:133–35; but cf. de Grummond 2006B:133.

[98] Pallottino 1952.

[99] de Grummond 2006B:108 fig. V.39; *CSE* Bundesrepublik Deutschland 1.13a.

[100] See de Grummond 2006B:106–11.

ETRUSCAN TRIDENTS, BIDENTALIA, *AND NEAR EASTERN EQUIVALENTS*

ETRUSCAN "TRIDENT" SYMBOLS. A number of unusual metal implements found in early Etruscan tombs may be symbolic tokens of the deceased man's honored position in the priesthood of his city. Some look like tridents, and others have only two prongs. Roman usage may have termed such objects *bidental*, but this is not confirmed by inscriptions. We are unsure of the Etruscan term for these items, and, as with other symbols of state and cult (*lituus, triumphus, sella curulis, fasces*), we must lapse into using the Latin terms. The implements have variously been identified as meat hooks or fishing tridents, but several recent, intensive studies show that they are too fragile for such purposes. Their highly decorative appearance further reflects their symbolic display, and close analogies with Anatolian and Mesopotamian cultic objects support their interpretation as images of the thunderbolt.

One of the earliest, dated to the second or third quarter of the eighth century BC, was excavated in 1878 in a tomb of the Golasecca culture of northern Italy, in Vigna di Mezzo at Rondineto near Como.[101] The tomb was made for a Villanovan chariot owner and warrior. The forked implement has a twisted shaft with an elliptical eyelet in the middle, two long, thin prongs, and an ornamental circular end with bronze pendants.[102] Colonna associated the eyelet shaft with ritual objects carried in the civic processions that continued to be depicted in the funerary art of fourth- to third-century Etruria, beginning with the Tomb of the Reliefs at Cerveteri.[103]

The most spectacular Etruscan trident is a fine, early seventh-century cast-bronze implement excavated at Vetulonia in the Tomba del Tridente (Fig. 13).[104] The bronze is 89.5 cm long and would have been fitted onto a wooden shaft. It is elaborately constructed of two parts, shaft and prongs, fastened with a pin and chain; the mechanism is decorated with a pattern of circles. (The purpose of the two-part breakdown is not clear.) Many authors emphasize the symbolism of fishing tridents and the princely owner's "rule of the sea," but it is too fragile to be anything but symbolic. Massa-Pairault has identified a black-figured vase of the Gruppo di Orvieto that shows a ceremony in progress with a curved, trident-like thunderbolt standing upright in the ground; the men may be *fulguriatores* or *sacerdotes bidentalium*.[105]

An iron trident was excavated in Pisa, in the Via San Jacopo necropolis.[106] A wooden chamber tomb held offerings of the end of the eighth to the

[101] See De Marinis and Gambari 2005:208; De Marinis 1988:178, figs. 146–49.
[102] Gambari and Colonna 1988:159n165, 159, fig. 21; Como 1986:63–64.
[103] Cf. Blanck and Proietti 1986:30–32.
[104] Pagnini 2000; Cygielman and Pagnini 2006:72–74, pl. VII,f,I; Sciacca 2004:271, figs. 2–3.
[105] Massa-Pairault 1999:82–83, 96, fig. 1.
[106] Bruni 1998:106–7, pl. 28; Sorrentino 2004.

beginning of the seventh century BC, but no body, beneath a tumulus 30 m in diameter. In the trench and mound were animal bones perhaps from sacrifices: sheep, cattle, pig, cow, dog, tortoise. Vases and offerings, including a fragmentary gold earring, were recovered. Above the chamber stood a special base that held an upright "trident" with a deliberately broken shaft. This was buried in turn and covered by a large altar made of stone blocks, beside which was placed a jar containing remnants of burnt wood (oak), nails, and gold wire, interpreted by the excavator as remains of a pyre on which a statue had been cremated. The ritual here must have been elaborate, and violent, because the stone altar atop the mound had been broken up with hammers. Almost immediately afterward, the tumulus was erected, and burials began to be placed in it. This tomb has been interpreted as a cenotaph, with the buried trident a symbol of the prince's seafaring aspect or, according to others, of his priestly office and lightning-reading skills.

A similar object, possibly a trident of iron, was discovered in the seventh-century Circolo della Fibula (Tomb XLI) at Marsiliana d'Albegna.[107] This tomb was one of the larger, richer structures, with a travertine-paved rectangular platform on which were found several metal vessels and implements, including a grill and tripod. On one corner of the pavement were four iron spear(point)s, and the trident lay across the center of the platform, near two fine bronze bracelets. Near the center of the stone-ringed Circolo was the actual burial, in a wood-paneled rectangular trench, along with a chariot, shield, tripods, bronze and ceramic vessels, and an iron-bladed dagger. Several famous art objects were also found, including an ivory box lid carved with Orientalizing animals, an ivory nude goddess figurine, a sheet-bronze bust with sphere for a head, and several elaborate pieces of jewelry in gold and silver.[108] It appears as if the trident, spears, and vessels were displayed in some ceremony performed on the paved rectangle beside the burial trench. The three-pronged iron "trident" was made of two twisted rods, the shorter one curved and fastened across the longer to make the outer prongs slightly shorter than the central one. All three tips are finished with spherical ends, so it is evident that this object was not for spearing fish or grilling meat. The goods characterize a male ruler, and the separately placed trident can only be of ritual significance, circa 680–650 BC. Was the prince of Tomb XLI, like the ruler of Verucchio's nearly contemporary Tomba del Trono, also a warrior-priest, one with special talent for lightning?

The lack of human remains in the Pisa cenotaph and the Vetulonia tomb leaves one to wonder whether the burial was in lieu of a man who died at sea, or instead reflected the aftermath of a lightning strike or other omen. The tomb at Vetulonia certainly commemorated someone of high rank whose

[107] Pagnini 2000; Minto 1921:81–92, trident pl. XLVIII, 2.
[108] Minto 1921:pls. XVI, XLIII, XXXVII, XI.

family offered a wide array of goods, including at least three chariots, and exotic imports from Bologna, Sardinia, North Syria, and the Levant.[109] If the fragile trident of Marsiliana d'Albegna is indeed a lightning symbol rendered in the preferred metal, iron, then it marks the earliest material evidence of this cult in Etruria, in roughly the same generation that saw the burgeoning of cities and ruler-established cults.

If this interpretation is correct, then, although no exemplar of the *Brontoscopic Calendar* can be traced so far back, we do have evidence that thunder and lightning were incorporated in the religious activities of the ruling classes of several Etruscan cities, from northern Italy (Como) to Etruscan Pisa and Vetulonia, during the Iron Age. At this time, other regalia and traditions of the urban state were also being deposited at major centers. *Bidentalia* continued to be ritually buried through the Roman period. Pietrangeli (1949–51) suggested that many commemorated historical lightning events, such as thunderbolts that hit statues.

STONES COMMEMORATING LIGHTNING STRIKES? Roman monuments inscribed *fulgur conditum* marked where objects or persons were blasted by lightning; Etruscan religion may have included an additional form of marker. Pfiffig suggested that symbolic stones (*Steinkeile*) were buried in the place of objects that had been struck by lightning and completely consumed.[110] A serpentine, ovoid river stone, approximately 40 cm long, found in 1977 at Orvieto (Fontana del Leone) is one of several such stones from the territory of Volsinii believed to represent lightning. On its dark surface are veins of white mineral, and a shallow, longitudinal relief is carved in the form of a thunderbolt, rendered as a double-headed arrow. A groove around the narrower end of the stone probably served to attach it upright, like an Orvietan *cippus*, to a separate base. The hard stone was quarried on Elba or Monte Argentario on the Tyrrhenian coast. Other examples are ovoid or smoothed rectangular/oblong in shape, always in hard stone, such as basalt. The earliest dated pieces are a pair found in Tomb 7 of the Orvietan necropolis of Le Conce, dated to the end of the sixth century BC.

The use of such markers continued for centuries (perhaps older stones were recut), and later examples carry Latin inscriptions. In funerary contexts, some were incised with the name of the deceased, like canonical Volsinian *cippi*. Others have been found in Lake Bolsena and at the Orvietan Belvedere Temple (in a stratum of charcoal and ashes). A fragment from Bolsena has been associated with the site of a chthonic cult of the sixth or fifth century BC. The stone from Orvieto – Fontana di Leone may have been displayed in the sanctuary within the Cannicella necropolis; the Belvedere and Bolsena sanctuaries also

[109] Sciacca 2004:269, and notes 5–7; Camporeale 2000.
[110] Pfiffig 1975:135.

held examples.[111] At least five had been discovered in the Campo della Fiera sanctuary at Orvieto, and Colonna has suggested associations with the "dark" god Śuri (Apollo Sourios) as venerated at Pyrgi (and Mount Soracte).[112] Luigi Donati (2005A) identified others venerated as late as the Renaissance, when they were set into the masonry of the abbey at Sant'Antimo, in the region of Siena (others are known in Arezzo and elsewhere). The surviving stone from Sant'Antimo, now in the Montalcino museum, weighs approximately 31 kg, and measured 37 x 29 cm (its lost twin was 39 x 25 cm). Donati pointed out the resemblance of the carved reliefs to the iconography of thunderbolts on Etruscan mirrors of the fourth to third centuries BC. Some scholars have suggested links to Levantine cults that used baetyls or standing stones (*masseboth*), or to cults of Zeus Keraunos.

NEAR EASTERN LIGHTNING SYMBOLS. Virtually all Near Eastern tridents were found in sanctuaries or tombs of affluent, powerful families. G. Rossoni has emphasized that many now lacking provenance could have been looted from sanctuaries:[113] a gold model trident in wavy outline was preserved in the temple of Anu-Adad at Ashur.[114] Ferdinando Sciacca (2004) emphasizes the conclusions of Rossoni's (1997) thorough analysis of Near Eastern metal tridents: they are too fragile to have been actual tools or weapons. They range from third-millennium Mesopotamia (Ur) through the seventh century BC, and from Anatolia and Iran (Kültepe, Gordion, Bogazkhöy, Urartu, Tepe Sialkh), to Egypt (Tell Defenneh), and Syria-Palestine (Ugarit, Byblos, Akko, Gezer, Beth Shean, Lachish), and also Hala Sultan Tekke in Cyprus;[115] they also appeared in early first-millennium India. Further, the association of a trident with Poseidon/Neptune has been shown to be a late icon, developed from the sea god's control of storms, not from spearfishing. Near Eastern (especially Sargonid-dynasty Assyrian) and Iron Age Greek iconography place tridents and similar two-pronged items in the hands of storm gods like Assyrian Adad (whose voice is thunder) and Hittite Teshub, in the preserve of royal worshippers. For nearly two millennia, the storm god's iconographic identity included the thunderbolt. The Sumerian god Iškur (Assyrian Adad) is attested since the Early Dynastic period and often was depicted brandishing a thunderbolt like a two-pronged fork or a wavy, triple-bolt not unlike the *keraunos* of Greek Zeus.[116] The god is also symbolized by the thunderbolt alone, as in Assyrian *kudurrus*, boundary markers that depict a varying set of patron gods by their upright scepter or

[111] See Tamburini 1991, with references.
[112] Colonna 2007:117–20, pls. 23–24.
[113] Rossoni 1997:583.
[114] Sciacca 2004:273, fig. 7.
[115] Cf. Makkay 1983.
[116] See Black and Green 1992:110–11, fig. 89.

symbolic staff (see Fig. 14).[117] Ashurnasirpal (883–859 BC) on a stele from his palace at Kalḫu (Nimrud), stands pointing to symbols of his patron gods, among which is a forked bolt of Adad, imitated in the necklace the king wears.[118] Under the Neo-Assyrian kings, the god Ninurta wielded a pair of triple-pronged, double-sided thunderbolts, as in a relief from his temple at Nimrud (Fig. 15).[119]

In Near Eastern seal impressions of the third through first millennium, huge tridents stand upright in the ground, tines pointed skyward, while kingly personages perform acts of worship. The triple-bolt characteristic of storm-lightning is a fitting symbol for gods with power over the cosmos,[120] likewise attested in modern photographs of lightning strikes.[121] A text of Tiglath-Pileser I says, "I made bronze lightning bolts (and) I placed these lightning bolts therein."[122] A tablet excavated at Ashur surveys constellations in the northern sky, and in describing the "Smaller Twins" (part of Gemini and Canis Minor) states, "the second figure carries a lightning bolt (*bir-qa*) in its right hand."[123] Some Near Eastern tridents were made of iron, the earliest from a tenth-century tomb at Lachish. An iron trident was displayed at the entrance to a chapel room in Hasanlu destroyed in the ninth century BC.[124] A number of iron tridents, perhaps associated with the Anatolian god Teshub, were found at Urartian Toprakkale and Karmir Blur. Might the use of iron be a reference to the "iron of heaven" (meteoric iron/steel) that was used occasionally in Egypt and the Near East?[125]

Tridents were understood across the *eastern* Mediterranean as symbols of the king's place in the universe, as one who communes with the divine.[126] An Urartian seal depicts a king, accompanied by parasol-bearer and lion,

[117] Black and Green 1992:17, fig. 10; see also a stele of Nebuchadnezzar I (1125–1104 BC) at Sippar, 113, fig. 90.

[118] Black and Green 1992:31 fig. 21; another forked symbol is at 118, fig. 96.

[119] Black and Green 1992:142–43, fig. 117.

[120] See Rossoni 1997.

[121] Compare the forked emblem worn by Ashurnasirpal on a stele from his palace at Nimrud (Black and Green 1992:31, fig. 21) with a photo of lightning taken near Lugano, Switzerland (Uman 1984:41, fig. 2.12).

[122] See *CAD*, vol. 2 "B," 259 *birqu* 2.

[123] VAT 9428 line 11 obverse; Weidner 1927:76, 79.

[124] Research in progress on Hasanlu; lecture by Stephan Kroll, University of Pennsylvania Museum, January 20, 2010.

[125] Egyptian documents of the second millennium refer to *bia' n pet*, "iron of heaven," first in texts of the 18th Dynasty (see Wainwright 1932; cf. Lorimer 1950:118.) One way of writing "marvel, astonishment" in hieroglyphs was the sign *bia'* (a block of metal or a meteorite) on a sledge, with a determinative sign of a star. Contemporary Hittite texts refer to "black iron of heaven" (*nepišval*, see Pleiner 2000:8–11, for more). The terminology of the various cultures demonstrates that scholars, rulers, traders, and literate people in general understood that meteorites were of celestial origin, and artistic representations of thunder/bolts may have been influenced in the choice of iron by such perceptions; see Bjorkman 1973.

[126] Rossoni 1997:581–83.

approaching an upright trident.[127] A Neo-Assyrian seal shows a royal horseman hunting wild goats in the presence of an upright trident,[128] while a large trident is carried, tines up, on the shoulder of a tribute bearer, in a relief on the stool of the throne of Shalmaneser III at Nimrud.[129]

It is possible that the simple pronged design was spontaneously generated in two different parts of the world – but the close link to divine-assisted kingship in the Near East may be a clue that this is another case of a direct transmission of a religious icon along with more immediate, and useful, Near Eastern learning and manners. A Late Bronze Age parallel may suggest some means of transfer of such information. Although a small, barbed bronze trident on the Uluburun shipwreck is usually classified as fishing equipment, Rossoni found that it fit into his typological categories of symbolic/priestly equipment.[130] And the Canaanite ship, wrecked circa 1300 BC off the Anatolian coast on a voyage that carried royal goods from Egypt, Cyprus, the Levant, and even Bulgaria and Italy, was part of a Mediterranean cycle of diplomatic travel and exchange that by the Late Bronze Age sometimes included Sardinia, Sicily, and southern Italy.[131]

Persons of rank accompanied the ships that sailed west and demonstrably taught Etruria the complex ways of the East – to banquet and entertain in a way that imitated the kings of the great states.[132] The ritual use and display of parasols, fans, chariots, thrones, purple clothing, ivory objects, and other goods from the Near East are merely the surviving traces of a much larger phenomenon of close association and sharing of religious teachings. The manufacture and ritual burial of symbolic artifacts like tridents by the end of the eighth and early seventh centuries in Etruria implies that the adoption of rituals actually began in the preceding generation, during the middle or second half of the eighth century.

CONCLUSION

A variety of causes could produce sound in the heavens, perceived as thunder, for Etruscans of the first millennium as for modern observers. Around 800 BC and for at least two centuries thereafter, thunderstorms, harsh, cold, and wet weather would have been more frequently experienced in Italy, due to the cosmic ray phenomenon commonly termed the Hallstatt Disaster. Evidence for prehistoric vulcanism was familiar to the Classical authors, and we are

[127] Rossoni 1997:590, fig. 5.2; Sciacca 2004:274, fig. 8.

[128] Rossoni 1997:590, fig. 5.3.

[129] Sciacca 2004:277, fig. 14.

[130] Twenty-four centimeters long with traces of a wooden shaft (Rossoni 1997); the ship did hold other equipment, such as fishnets and a fishing spear; see Pulak 1998:210–11 fig. 23.

[131] Including a sword of Late Bronze Italian Thapsos-Pertosa type; Pulak 2001:45–46, fig. 5.

[132] See Magness 2001.

beginning to recognize the archaeological evidence of events of the Archaic period. All these phenomena would have enhanced, in the minds of ancient worshipers, the link between calamities and "thunder," and we know, from the comments of Cicero and Seneca, among others, that the *etrusca disciplina* paid especial attention to these phenomena. Some humans are better able to discern low-frequency noise that would be equated with thunder. Some priests, possibly the haruspices themselves, were skilled and trained in interpreting such things and would have been able to preserve and share texts of divination. Whether the rare finds of symbolic tridents and bidents commemorate the tombs of such experts is still uncertain. That the time of formation of the Etruscan *Brontoscopic Calendar* – the Iron Age or early Orientalizing period – must have been a time of distinct and troubling portents is indisputable.

PART II

THE *BRONTOSCOPIC CALENDAR*: GREEK TEXT AND ENGLISH TRANSLATION

DIARIUM TONITRUALE

JOHANNES LYDUS, DE OSTENTIS *27–38*

ΕΦΗΜΕΡΟΣ ΒΡΟΝΤΟΣΚΟΠΙΑ ΤΟΠΙΚΗ ΠΡΟΣ ΤΗΝ ΣΕΛΗΝΗΝ ΚΑΤΑ ΤΟΝ ΡΩΜΑΙΟΝ ΦΙΓΟΥΛΟΝ ΕΚ ΤΩΝ ΤΑΓΗΤΟΣ ΚΑΘ' ΕΡΜΗΝΕΙΑΝ ΠΡΟΣ ΛΕΞΙΝ

Εἰ ἐπὶ πάσαις <ταῖς> τῆς διοσημείας παραδόσεσι τὴν σελήνην φαίνονται λαβόντες οἱ ἀρχαῖοι (ὑπ' αὐτὴν γὰρ τά τε βροντῶν τά τε κεραυνῶν ἐκδέδοται σημεῖα), καλῶς ἄν τις ἄρα καὶ τὸν σελήνης οἶκον ἐπιλέξαιτο, ὥστε ἀπὸ τοῦ καρκίνου καὶ ἐνταῦθα τὴν ἐφήμερον ἀπὸ τῆς σεληνιακῆς νουμηνίας κατὰ τοὺς σεληνιακοὺς μῆνας ληψόμεθα τῶν βροντῶν ἐπίσκεψιν· ἐξ ἧς τὰς τοπικάς, ἐφ' ὧν ῥήγνυνται χωρίων αἱ βρονταί, παρατηρήσεις οἱ Θοῦσκοι παρέδοσαν.

ΜΗΝΙ ΙΟΥΝΙΩ

Σελ. α. ἐὰν βροντήσῃ, τῶν καρπῶν εὐφορία ἔσται, ἐξῃρημένων κριθῶν· νοσήματα δὲ ἐπισφαλῆ τοῖς σώμασιν ἐπισκήψει.

β. ἐὰν βροντήσῃ, ταῖς μὲς ὠδινούσαις ἀπαλλαγὴ μᾶλλον ῥᾳδία, τοῖς δὲ θρέμμασι φθορά, ἰχθύων γε μὴν ἀφθονία ἔσται.

γ. ἐὰν βροντήσῃ, καύσων ἔσται ξηρότατος, ὥστε μὴ τοὺς ξηροὺς μόνους ἀλλὰ καὶ τοὺς ὑγροὺς καρποὺς διαφρυγέντας ἀπακαυθῆναι.

δ. ἐὰν βροντήσῃ, νεφελώδης καὶ ὑετώδης ὁ ἀὴρ ἔσται, ὡς ἐκ σηπτικῆς ὑγρότητος φθαρῆναι τοὺς καρπούς.

ε. ἐὰν βροντήσῃ, ἀπαίσιον τοῖς ἀγροῖς· οἱ δὲ χωρίοις ἢ πολίχναις ἐφεστῶτες ταραχθήσονται.

ϛ. ἐὰν βροντήσῃ, ἀκμάσασι τοῖς καρποῖ ς ἐντεχθήσεταίτι θηρίον τὸ βλάπτον αὐτούς.

ζ. ἐὰν βροντήσῃ, νοσήματα μὲν ἐνσκήψει, ἀλλ' οὐ πολλοὶ ἐξ αὐτῶν τεθνήξονται· καὶ οἱ μὲν ξηροὶ καρποὶ ἐπιτεύξονται, οἱ δὲ ὑγροὶ ξηρανθήσονται.

η. ἐὰν βροντήσῃ, ἐπομβρίαν καὶ φθορὰυ σίτοη δηλοῖ.

θ. ἐὰν βροντήσῃ, φθορὰ τοῖς θρέμμασιν ἔσται ἐξ ἐπιδρομῆς λύκων.
ι. ἐὰν βροντήσῃ, θάνατος μὲν ἔσται συχνός, εὐθηνία. δὲ ὅμως.
ια. ἐὰν βροντήσῃ, καύματα μὲν ἀβλαβῆ, τὰ δὲ πολιτικὰ ἐν εὐφροσύνῃ ἔσται.
ιβ. ἐὰν βροντήσῃ, ὁμοίως ὡς ἐπὶ τῇ πρὸ ταύτης.
ιγ. ἐὰν βροντήσῃ, δυνάστου πτῶσιν ἀπειλεῖ.
ιδ. ἐὰν βροντήσῃ, καυσῶδες μὲν τὸ περιέχον ἔσται, εὐφορία δὲ τῶν καρπῶν καὶ εὔροια τῶν ποταμίων οὐχ ἥκιστα ἰχθύων. τὰ σώματά γε μὴν ἐξασθενήσει.
ιε. ἐὰν βροντήσῃ, τὰ μὲν πτηνὰ λυμανθήσεται τῷ θέρει, οἱ δὲ ἰχθύες φθαρήσονται.
ις. ἐὰν βροντήσῃ, οὐκ ἐλάττωσιν μόνον τῶν ἐπιτηδείων ἀλλὰ καὶ πόλεμον ἀπειλεῖ, ἀνὴρ δέ τις εὐτυχὴς ἀφανισθήσεται.
ιζ. ἐὰν βροντήσῃ, καύματα ἔσται καὶ μυῶν καὶ ἀσφαλάκων καὶ ἀκρίδων φορά· εὐθηνίαν δὲ ὁμῶς καὶ φόνους τῷ δήμῳ φέρει.
ιη. ἐὰν βροντήσῃ, φθορὰν τῶν καρπῶν ἀπειλεῖ.
ιθ. ἐὰν βροντήσῃ, τὰ λυμαινόμενα τοῖς καρποῖς θηρία φθαρήσεται.
κ. ἐὰν βροντήσῃ, διχόνοιαν ἀπειλεῖ τῷ δήμῳ.
κα. ἐὰν βροντήσῃ, ἐλάττωσιν μὲν οἴνου, ἐπίδοσιν δὲ τῶν ἄλλων καρπῶν δηλοῖ καὶ ἀφθονίαν ἰχθύων.
κβ. ἐὰν βροντήσῃ, ἐπίφθορον ἔσται τὸ καῦμα.
κγ. ἐὰν βροντήσῃ, εὐφροσύνην καὶ κακῶν ἀπαλλαγὴν καὶ νόσων ἀφανισμὸν δηλοῖ.
κδ. ἐὰν βροντήσῃ, εὐθηνίαν δηλοῖ.
κε. ἐὰν βροντήσῃ, πόλεμοι καὶ μυρία ἔσονται κακά.
κς. ἐὰν βροντήσῃ, ὁ χειμὼν τοῖς καρποῖς ἐπιβλαβὴς ἔσται.
κζ. ἐὰν βροντήσῃ, κίνδυνος στρατιωτικὸς τοῖς κρατοῦσιν ἔσται.
κη. ἐὰν βροντήσῃ, εὐετηρία ἔσται τῶν καρπῶν.
κθ. ἐὰν βροντήσῃ, τὰ τῆς βασιλίδος πόλεως ἔσται κρείττονα.
λ. ἐὰν βροντήσῃ, πρὸς βραχὺ θάνατος ἔσται συχνός.

ΙΟΥΛΙΟΣ

α. Ἐπὶ τῆς σεληνιακῆς νουμηνίας ἐὰν βροντήσῃ, εὐθηνία μὲν ἔσται, τῶν δὲ θρεμμάτων πτῶσις.
β. ἐὰν βροντήσῃ, καλὸν τῷ φθινοπώρῳ.
γ. ἐὰν βροντήσῃ, χειμῶνα βαρὺν σημαίωει.
δ. ἐὰν βροντήσῃ, ἀέριοι ἔσονται ταραχαί, ὡς ἐξ αὐτῶν σπάνιν γενέσθαι.
ε. ἐὰν βροντήσῃ, εὐθηνία μὲν σίτου, ἄρχοντος δὲ ἀγαθοῦ πτῶσις ἔσται.
ϛ. ἐὰν βροντήσῃ, νόσους θανατηφόρους ταῖς δουλικαῖς τύχαις ἀπειλεῖ.
ζ. ἐὰν βροντήσῃ, κατομβρία ἔσται βλαβερὰ τοῖς σπορίμοις.
η. ἐὰν βροντήσῃ, εἰρήνην μὲν τοῖς κοινοῖς, ὄλεθρον δὲ τοῖς βοσκήμασι καὶ βῆχα ξηρὸν ἐνσκῆψαι δηλοῖ.
θ. ἐὰν βροντήσῃ, ἐποψίαν θεῶν καὶ ἀγαθῶν πολλῶν πρόσοδον δηλοῖ.

ι. ἐὰν βροντήσῃ, σωτηριώδη ἔσται τὰ ποτάμια ὕδατα.
ια. ἐὰν βροντήσῃ, καῦμα σημαίνει καὶ ὄμβρον βίαιον καὶ σπάνιν σίτου.
ιβ. ἐὰν βροντήσῃ, ψῦχος τῷ θέρει ἔσται ἀδόκητον, δι᾽ οὗ φθαρήσεται τὰ ἐπιτήδεια.
ιγ. ἐὰν βροντήσῃ, ἑρπετὰ φανεῖται βλαβερώτατα.
ιδ. ἐὰν βροντήσῃ, εἰς ἕνα τὴν πάντων δύναμιν ἐλθεῖν φράζει· οὗτος δὲ ἔσται τοῖς πράγμασιν ἀδικώτατος.
ιε. ἐὰν βροντήσῃ, διχόνοια ἔσται τοῦ δήμου καὶ τοῦ σίτου ἔνδεια.
ις. ἐὰν βροντήσῃ, πόλεμον ὁ τῆς ἀνατολῆς βασιλεὺς ἐκδεχέσθω ...[...] καὶ νόσον ἀπὸ ξηροῦ καύματος.
ιζ. ἐὰν βροντήσῃ, διαδοχὴν μεγάλου ἄρχοντος δηλοῖ.
ιη. ἐὰν βροντήσῃ, ἔνδειαν καρπῶν ἐξ ἐπομβρίας ἀπειλεῖ.
ιθ. ἐὰν βροντήσῃ, πόλεμον δηλοῖ καὶ ἀπώλειαν δυνατῶν· ἀφθονία δὲ ἔσται τῶν ξηρῶν καρπῶν.
κ. ἐὰν βροντήσῃ, αὐχμὸν νοσώδη ἀπειλεῖ.
κα. ἐὰν βροντήσῃ, διχόνοια ἔσται τοῖς ὑπηκόοις, ἀλλ᾽ οὐκ εἰς μακράν.
κβ. ἐὰν βροντήσῃ, ἀγαθὰ μὲν τοῖς κοινοῖς πράγμασι, νόσους δὲ τοῖς σώμασι περὶ τὴν κεφαλὴν σημαίνει.
κγ. ἐὰν βροντήσῃ, ἡ διχόνοια τοῦ δήμου παυθήσεται.
κδ. ἐὰν βροντήσῃ, δυνατοῦ ἀνθρώπου δηλοῖ δυνατὸν ἀτύχημα.
κε. ἐὰν βροντήσῃ, τῇ νεολαίᾳ κάκωσις καὶ τοῖς καρποῖς μετ᾽ αὐτῆς, νοσώδης δὲ ὁ καιρὸς ἔσται.
κς. ἐὰν βροντήσῃ, ἀπὸ μεγάλης εὐθηνίας ἔνδεια ἔσται.
κζ. ἐὰν βροντήσῃ, ἐκβρασμὸν τοῖς σώμασιν ἀπειλεῖ.
κη. ἐὰν βροντήσῃ, ἔνδεια ὑδάτων καὶ ὄχλος ἑρπετῶν ἐπιβλαβῶν ἔσται.
κθ. ἐὰν βροντήσῃ, εὐετηρίαν δηλοῖ.
λ. ἐὰν βροντήσῃ, οἱ ἄνθρωποι ἐκ ποινηλασίας ἐπὶ τὰ χείριστα τῶν πταισμάτων ὀλισθήσουσιν.

ΑΥΓΟΥΣΤΟΣ

α. Ἐὰν βροντήσῃ, καὶ τὰ τῆς πολιτείας ἠρέμα πως καλλίονα καὶ εὐθηνία ἔσται.
β. ἐὰν βροντήσῃ, νόσους ἅμα καὶ ἔνδειαν τῶν ἐπιτηδείων ἀπειλεῖ.
γ. ἐὰν βροντήσῃ, δίκας καὶ λέσχας τῷ δήμῳ ἀπειλεῖ.
δ. ἐὰν βροντήσῃ, ἔνδεια τροφῶν λογικοῖς ἅμα καὶ ἀλόγοις ἔσται.
ε. ἐὰν βροντήσῃ, τὰς γυναῖκας συνετωτέρας δηλοῖ.
ς. ἐὰν βροντήσῃ, μέλιτος μὲν ἀφθονία ἔσται, ὕδατος δὲ καὶ τῶν λοιπῶν τροφίμων λεῖψις.
ζ. ἐὰν βροντήσῃ, ἀνέμους τραχεῖς καὶ νόσους ὁμοῦ σημαίνει.
η. ἐὰν βροντήσῃ, νόσον ἀκίνδυνον τοῖς τετραπόδοις ἀπειλεῖ.
θ. ἐὰν βροντήσῃ, ὑγίειαν μὲν ἀνθρώποις ὡς ἐκ τοῦ πλείονος ἐπαγγέλλεται.

ι. ἐὰν βροντήσῃ, λύπας καὶ μοχθηρίας τῷ πλήθει ἀπειλαῖ.
ια. ἐὰν βροντήσῃ, εὐετηρία μὲν ἔσται, πτῶσις δὲ ἑρπετῶν καὶ βλάβη τοῖς ἀνθρώποις.
ιβ. ἐὰν βροντήσῃ, χόρτου καὶ βαλάνου ἀφθονία ἔσται, τῇ δὲ πρώτῃ ἡλικίᾳ κάκωσις.
ιγ. ἐὰν βροντήσῃ, ὄλεθρος ἔσται τοῖς σώμασι λογικῶν τε καὶ ἀλόγων.
ιδ. ἐὰν βροντήσῃ, πόλεμον μὲν τοῖς κοινοῖς, ἀφθονίαν δὲ τοῖς καρποῖς δηλοῖ.
ιε. ἐὰν βροντήσῃ, ἐπὶ τὸ χεῖρον τὰ πράγματα.
ις. ἐὰν βροντήσῃ, εἰρήνην βαθεῖαν ἐπαγγέλλεται.
ιζ. ἐὰν βροντήσῃ, οἱ χείρους τῶν ἀνθρώπων στυγνάσουσιν.
ιη. ἐὰν βροντήσῃ, πόλεμον ἐμφύλιον ἀπειλεῖ.
ιθ. ἐὰν βροντήσῃ, φόνους αἱ γυναῖκες καὶ τὸ δουλικὸν τολμήσει.
κ. ἐὰν βροντήσῃ, ὄλεθρον βοῶν καὶ ταραχὰς τοῖς πράγμασιν ἀπειλεῖ.
κα. ἐὰν βροντήσῃ, εὐθηνίαν ἅμα καὶ διχόνοιαν τῷ δήμῳ δηλοῖ.
κβ. ἐὰν βροντήσῃ, καλὰ μετρίως τὰ πράγματα ἀνὰ πάντα τὸν ἐνιαυτόν.
κγ. ἐὰν βροντήσῃ, κεραυνὸν πεσεῖσθαι δηλοῖ καὶ φόνους ἀπειλεῖ.
κδ. ἐὰν βροντήσῃ, ἀπώλειαν εὐγενῶν νέων ἀπειλεῖ.
κε. ἐὰν βροντήσῃ, χειμῶνα καὶ σπάνιν ὀπωρῶν ἔσεσθαι προλέγει.
κς. ἐὰν βροντήσῃ, πόλεμον σημαίνει.
κζ. ἐὰν βροντήσῃ, πολέμους ἅμα καὶ δόλους ἀπειλεῖ.
κη. ἐὰν βροντήσῃ, εὐθηνίαν μὲν τῶν καρπῶν, φθορὰν δὲ τῶν βοῶν σημαίνει.
κθ. ἐὰν βροντήσῃ, οὐδεμίαν ἐναλλαγὴν σημαίνει.
λ. ἐὰν βροντήσῃ, νόσους τῇ πόλει, ἐφ’ ἧς ῥαγήσεται, ἀπειλεῖ.

ΣΕΠΤΕΜΒΡΙΟΣ

α. Ἐὰν βροντήσῃ, εὐετηρίαν ἅμα καὶ εὐφροσύνην δηλοῖ.
β. ἐὰν βροντήσῃ, διχόνοια τῷ δήμῳ ἔσται.
γ. ἐὰν βροντήσῃ, κατομβρίαν καὶ πόλεμον δηλοῖ.
δ. ἐὰν βροντήσῃ, δυνατοῦ πτῶσιν παὶ πολέμου παρασκευὴν δηλοῖ.
ε. ἐὰν βροντήσῃ, κριθῆς μὲν ἀφθονίαν, σίτου δὲ ἐλάττωσιν δηλοῖ.
ς. ἐὰν βροντήσῃ, δύναμις ἔσται ταῖς γυναιξὶ κρείττων ἤ κατ’ αὐτάς.
ζ. ἐὰν βροντήσῃ, νόσον καὶ πτῶσιν τῷ δουλικῷ ἐξ αὐτῆς ἀπειλεῖ.
η. ἐὰν βροντήσῃ, τοὺς μάλιστα δυνατοὺς τοῦ πολιτεύματος σκολιὰ ἐννοεῖν, ἀποτεύξεσθαι δὲ τῶν νοουμένων κατηγορεῖ.
θ. ἐὰν βροντήσῃ, ἄνεμον πνεῦσαι ἐπίνοσον ἀπειλεῖ.
ι. ἐὰν βροντήσῃ, ἔρις ἔσται τῷ τόπῳ ἐφ’ οὗ ῥαγήσεται πρὸς ἕτερον οὐκ ἄλογος.
ια. ἐὰν βροντήσῃ, οἱ ὑπεξούσιοι τῶν εὐγενῶν σκέψονταί τι καινὸν ἐν τοῖς κοινοῖς.

ιβ. ἐὰν βροντήσῃ, ἔπομβρον τὸν τοῦ ἀμητοῦ καιρὸν καὶ λιμὸν ἔσεσθαι λέγει.
ιγ. ἐὰν βροντήσῃ, λιμὸν βαρὺν ἀπειλεῖ.
ιδ. ἐὰν βροντήσῃ, νόσους ἀπειλεῖ.
ιε. ἐὰν βροντήσῃ, κατομβρίαν μὲν δηλοῖ, εὐθηνίαν δὲ ὅμως.
ις. ἐὰν βροντήσῃ, ὁ σπόρος πολὺς, ἄκαρπος δὲ δηλοῦται.
ιζ. εἰ βροντήσῃ, ἔνδειαν τῶν ἀναγκαίων ἀπειλεῖ.
ιη. εἰ βροντήσῃ, λιμὸν ἅμα καὶ πολέμους σημαίνει.
ιθ. εἰ βροντήσῃ, ὁ μὲν δεωδρώδης καρπὸς ἐπιτεύξεται, νόσοι δὲ καὶ στάσεις δημοτικαὶ ἔσονται.
κ. εἰ βροντήσῃ, ἀπώλειαν κλεινοῦ ἀνθρώπου καὶ πόλεμον ἀπειλεῖ.
κα. εἰ βροντήσῃ, λύπας καὶ ζημίας τῷ δήμῳ ἀπειλεῖ.
κβ. εἰ βροντήσῃ, εὐθηνίαν μέν, χειμῶνα δὲ βαρὺν καὶ ὑγρὸν σημαίνει.
κγ. εἰ βροντήσῃ, τὸν χειμῶνα τοῦ ἔτους ἐνδεᾶ προλέγει.
κδ. εἰ βροντήσῃ, λειψυδρίαν ἀπειλεῖ. τῶν δὲ ἀκροδρύων εὐφορία ἔσται, περὶ τὸ φθινόπωρον θυέλλαις διαφθαρήσεται.
κε. εἰ βροντήσῃ, ἐκ διχονοίας τοῦ πολιτεύματος τύραννος ἀναστήσεται, καὶ αὐτὸς θὲν ἀπολεῖται, ζημίαις δὲ ἀφορήτοις οἱ δυνατοὶ ὑποστήσονται.
κς. εἰ βροντήσῃ, ὁ κακὸς δυνάστης βουλῇ θεοῦ πεσεῖται.
κζ. εἰ βροντήσῃ, δυνατοὶ πρὸς ἑαυτοὺς διενεχθήσονται καὶ ἀλλήλους διαβολοῦσιν.
κη. εἰ βροντήσῃ, σημεῖα ἔσται πραγμάτων μεγάλων μηνυτικά. εὐλαβητέον δὲ μὴ καὶ πῦρ ἐπομβρήσῃ ἐνιαχοῦ.
κθ. εἰ βροντήσῃ, αὐχμὸν ἐπιβλαβῆ ἀπειλεῖ.
λ. εἰ βροντήσῃ, ἀπὸ χειρόνων ἐπὶ τὰ κρείττω τὰ κοινά.

ΟΚΤΩΒΡΙΟΣ

α. Εἰ βροντήσῃ, κακὸν τύραννον τοῖς πράγμασιν ἀπειλεῖ.
β. εἰ βροντήσῃ, εὐθηνία μὲν ἔσται, φθορὰ δὲ τῶν χερσαίων μυῶν.
γ. εἰ βροντήσῃ, θυέλλας καὶ ταραχὰς δηλοῖ, δι᾽ ὧν τὰ μὲν δένδρα φθαρήσεται· μεγάλων δὲ ζαλῶν ἔσται τοῖς κοινοῖς μήνυμα.
δ. εἰ βροντήσῃ, οἱ ἐλάττους <τὰ> τῶν κρειττόνων καθέξουσιν, ὑγιεινὸν δὲ ἔσται τὸ τοῦ ἀέρος κατάστημα.
ε. εἰ βροντήσῃ, πάντων ἐπιδόσεις τῶν ἀναγκαίων, ἐξῃρημένου σίτου.
ς. εἰ βροντήσῃ, ἡ ὄψις μὲν εὐθηνίας, ἀμητὸς δὲ ἐλάττων, καὶ τὸ φθινόπωρον ἐγγὺς ἄκαρπον.
ζ. εἰ βροντήσῃ, ὄσπρια μὲν ἄφθονα, οἶνος δὲ ἐλάττων ἔσται.
η. εἰ βροντήσῃ, σεισμὸν μετὰ μυκήματος προσδοκητέον.
θ. εἰ βροντήσῃ, ὄλεθρον θηρίοις ἀπειλεῖ.
ι. εἰ βροντήσῃ, ἀνδρὸς ἐπαινουμένου πτῶσιν δηλοῖ.
ια. εἰ βροντήσῃ, ἀλληλανεμίαν ἐπισφαλῆ ταῖς βοτάναις δηλοῖ.

ιβ. εἰ βροντήσῃ, εὐθηνία μὲν ἔσται, σκηπτοὶ δὲ ἐκθλιβήσονται.
ιγ. εἰ βροντήσῃ, καλὰ τὰ συναλλάγματα, καὶ εὐθηνία πρὸς αὐτοῖς· ὁ δὲ ἐπὶ τῇ πολιτείᾳ βαρὺς οὐκ ἐπὶ πλέον ἰσχύσει.
ιδ. εἰ βροντήσῃ, πόλεμον καὶ φθορὰν βοσκημάτων ἀπειλεῖ.
ιε. εἰ βροντήσῃ, σπάνις ἔσται πνεύματος ξηροῦ καὶ καυστικοῦ ἐμπίπτοντος τοῖς καρποῖς.
ις. εἰ βροντήσῃ, ἐξασθενήσουσιν οὕτως ἄνθρωποι, ὡς ἄγνωστοι εἶναι δοκεῖν.
ιζ. εἰ βροντήσῃ, πλουσίου ἀνδρὸς καὶ εὐγενῶν εὐτυχήματα.
ιη. εἰ βροντήσῃ, ἐπείσακτον εὐθηνίαν δηλοῖ.
ιθ. εἰ βροντήσῃ, δυνάστου πτῶσιν ἤ βασιλέως ἐκβολὴν ἀπειλεῖ, διχόνοιάν τε τῷ δήμῳ καὶ ἀφθονίαν δηλοῖ.
κ. εἰ βροντήσῃ, ἕλκη ξένα ἀπειλεῖ, τῷ δὲ πλήθει ἐκ διχονοίας ἐσχάτην ἀτυχίαν.
κα. εἰ βροντήσῃ, νόσοι βηχώδεις ἔσονται καὶ στηθῶν κακώσεις.
κβ. εἰ βροντήσῃ, μοχθηρίας τῷ δήμῳ καὶ νόσους ποικίλας ἀπειλεῖ.
κγ. εἰ βροντήσῃ, ὁ δῆμος εὐφρανθήσεται παραδόξως.
κδ. εἰ βροντήσῃ, ἐκ διχονοίας τῶν κρατούντων ὁ δῆμος περιέσται.
κε. εἰ βροντήσῃ, βαρεῖα ἔσται ἀπὸ τῶν κακῶν δειλία.
κς. εἰ βροντήσῃ, αὐξηθήσεται μὲν τὰ θηρία, λιμώξει δὲ ὅυως.
κζ. εἰ βροντήσῃ, ἐπομβρίαν δηλοῖ.
κη. εἰ βροντήσῃ, ἔνδεια ἔσται τῶν ἀναγκαίων.
κθ. εἰ βροντήσῃ, ἐπίνοσος ὁ ἐνιαυτός.
λ. εἰ βροντήσῃ, οὐκ εὐθηνίαν μόνον ἀλλὰ καὶ ἐλάττωσιν τῶν πολεμίων καὶ εὐφροσύνην τοῖς πράγμασι δηλοῖ.

ΝΟΕΜΒΡΙΟΣ

α. Εἰ βροντήσῃ, διχόνοιαν δηλοῖ τῇ πόλει.
β. εἰ βροντήσῃ, εὐθηνίαν προλέγει.
γ. εἰ βροντήσῃ, πράγματα ἀναφύεται, δι᾽ ὧν οἱ ἐλάττους περιέσονται τῶν μειζόνων.
δ. εἰ βροντήσῃ, κρείττων ὁ σῖτος.
ε. εἰ βροντήσῃ, ζάλην τοῖς πράγμασι δηλοῖ, αἰσθητοῖς τε καὶ ἀλόγοις νόσον.
ς. εἰ βροντήσῃ, σκώληκες τῷ σίτῳ λυμανοῦνται.
ζ. εἰ βροντήσῃ, τοῖς ὑπὸ τὴν δύσιν ἀνθρώποις τε καὶ ἀλόγοις νόσοι.
η. εἰ βροντήσῃ, ἀδδηφαγίας ἔσεσθαι δεῖ διὰ νόσους ἀπειλουμένας.
θ. εἰ βροντήσῃ, δημοτῶν ἀνασκολοπισμοὶ ἔσονται, ἀλλὰ μὴν καὶ ἀφθονία τῶν ἐπιτηδείων.
ι. εἰ βροντήσῃ, πέρας ἔχει τὰ κακῶς βουλευθέντα τοῖς κρατοῦσιν, ἄνεμος δὲ καυσώδης τοῖς δένδροις λυμαίνεται.

ια. εἰ βροντήσῃ, εὐχαριστείτωσαν ἄνθρωποι τῷ θεῷ·ἄνεμος γὰρ ἐξ ἀνατολῶν ῥεύσει.
ιβ. εἰ βροντήσῃ, ἐνύπνια τοῖς ἀνθρώποις πλείονα δόξει.
ιγ. εἰ βροντήσῃ, ἐπικερδῆ μὲν τὸν καιρὸν, λοιμικὸν δὲ ἀπειλεῖ, ἕλμισι δὲ ἐντοσθίοις κακοῖ τὰ σώματα.
ιδ. εἰ βροντήσῃ, ἑρπετὰ τοῖς ἀνθρώποις ἠρέμα πως λυμαίνεται.
ιε. εἰ βροντήσῃ, ὁ μὲν ιχθὺς ἀφθονώτατος, λοιμώξει δὲ τὰ ἔνυδρα θηρία, καὶ τὰ κοιωὰ δῆθεν καλλίονα.
ις. εἰ βροντήσῃ, ἀκρίδες καὶ μυῶν ἀρουραίων γένεσις, τῷ δὲ βασιλεῖ κίνδυνος, καὶ σίτου ἀφθονία ἔσται.
ιζ. εἰ βροντήσῃ, νομὴν ἄφθονον τοῖς κτήνεσι δηλοῖ.
ιη. εἰ βροντήσῃ, πόλεμον δηλοῖ καὶ τοῖς ἀστειοτέροις λύπας.
ιθ. εἰ βροντήσῃ, γυναικῶν εὐπραγίας.
κ. εἰ βροντήσῃ, λιμὸν ἀλλ' οὐκ εἰς μακρὰν δηλοῖ.
κα. εἰ βροντήσῃ, οἱ μύες φθαρήσονται, εὐθηνία δὲ οὐ σίτου μόνον ἀλλὰ καὶ νομῆς, καὶ ἰχθύων πλῆθος.
κβ. εἰ βροντήσῃ, εὐετηρίαν δηλοῖ.
κγ. εἰ βροντήσῃ, ἄνεμος νοσώδης πνεύσει.
κδ. εἰ βροντήσῃ, φρούριον τῇ πολιτείᾳ χρηστὸν ὑπὸ πολεμίοις τελέσει.
κε. εἰ βροντήσῃ, πόλεμος ἐπικίνδυνος ἔσται· καὶ ἄνεμος νοσώδης πνεύσει.
κς. εἰ βροντήσῃ, πόλεμον ἐμφύλιον καὶ πτῶσιν πολλῶν δηλοῖ· ὄμβροι τε ἔσονται λοιμώδεις.
κζ. εἰ βροντήσῃ, τὰ αὐτὰ ἀπειλεῖ.
κη. εἰ βροντήσῃ, τῆς συγκλήτου πολλοὶ ἀθυμίᾳ φθαρήσονται.
κθ. εἰ βροντήσῃ, οἱ μὲν χείρους κάλλιον πράξουσιν, αἱ δὲ ἐσόμεναι ὀπῶραι φθαρήσονται.
λ. εἰ βροντήσῃ, θεοφιλέστερον οἱ ἄνθρωποι ζήσονται· καὶ εἰκότως σύμμετρα τὰ κακά.

ΔΕΚΕΜΒΡΙΟΣ

α. Εἰ βροντήσῃ, εὐετηρίαν μεθ' ὁμονοίας δηλοῖ.
β. εἰ βροντήσῃ, ἀφθονία ἰχδύων καὶ καρπῶν διαφερόντως.
γ. εἰ βροντήσῃ, τοῖς θρέμμασιν ἄνθρωποι καταχρήσονται δι' ἔνδειαν ἰχθύων.
δ. εἰ βροντήσῃ, βαρὺς ἔσται χειμών, εὐθηνία δὲ ὅμως.
ε. εἰ βροντήσῃ, νόσους ψωρώδεις ἀπειλεῖ.
ς. εἰ βροντήσῃ, αἰσίους ὀνείρους οἱ ἄνθρωποι ἐνυπνιασθήσονται, οἵ κακὸν πέρας ἕξουσιν.
ζ. εἰ βροντήσῃ, τὸ αὐτὸ δηλοῖ πᾶσιν.
η. εἰ βροντήσῃ, νότος ἐπικρατήσει· ἐξ οὗ εὐθηνία μὲν καρπῶν ἔσται, ὄλεθρος δὲ θρεμμάτων.

θ. εἰ βροντήσῃ, κλεινοῦ ἀνδρὸς πτῶσις ἔσται.
ι. εἰ βροντήσῃ, φθορὰν ἀνθρώποις ἐκ νοσημάτων ἀπειλεῖ, οἱ δὲ ἰχθύες ἐπιδώσουσιν.
ια. εἰ βροντήσῃ, καυσώδης ἡ θερινὴ ἔσται τροπή, καὶ ἀφθονία ἐπείσακτος.
ιβ. εἰ βροντήσῃ, ἐκ γαστρορροίας νόσους ἀπειλεῖ.
ιγ. εἰ βροντήσῃ, εὐθηνίαν, νόσους δὲ ἀπειλεῖ.
ιδ. εἰ βροντήσῃ, ἐμφύλιον ἅμα πόλεμον καὶ εὐθηνίαν δηλοῖ.
ιε. εἰ βροντήσῃ, πολλοὶ ἐπὶ πόλεμον πορεύσονται, ὀλίγοι δὲ ἀναστρέψουσιν.
ις. εἰ βροντήσῃ, καινοπρεπῆ πράγματα ἐπὶ τῆς πολιτείας.
ιζ. εἰ βροντήσῃ, ἀκρίδας φύσεσθαι λεπτὰς ἀπειλεῖ, εὐθηνίαν δὲ ὅμως.
ιη. εἰ βροντήσῃ, πόλεμος ἔσται βαρύς.
ιθ. εἰ βροντήσῃ, ἐπίτασιν τοῦ πολέμου ἀπειλεῖ.
κ. εἰ βροντήσῃ, ἔνδειαν τῶν ἀναγκαίων λέγει.
κα. εἰ βροντήσῃ, ἄνεμον θερμὸν καὶ νοσοποιὸν ῥεῦσαι ἀπειλεῖ.
κβ. εἰ βροντήσῃ, καυσῶδες ἔσται τὸ θέρος καὶ πολύκαρπον.
κγ. εἰ βροντήσῃ, ἀνθρώποις νόσον δηλοῖ, ἀλλ' ἀκίνδυνον.
κδ. εἰ βροντήσῃ, ἐμφυλίους πολέμους τῇ πόλει καὶ ὄλεθρον τοῖς ὑλαίοις θηρίοις ἀπειλεῖ.
κε. εἰ βροντήσῃ, ἐκστρατείας κίνησις ἐπὶ πόλεμον, ἀλλ' εὐτυχήσει.
κς. εἰ βροντήσῃ, νόσους τοῖς θεράπουσιν ἀπειλεῖ.
κζ. εἰ βροντήσῃ, ὁ βασιλεὺς πολλοὺς ὠφελήσει.
κη. εἰ βροντήσῃ, ἀκρίδων γένεσις.
κθ. εἰ βροντήσῃ, ἰσχνότητα τοῖς σώμασιν ὑγιεινὴν δηλοῖ.
λ. εἰ βροντήσῃ, ἀνταρσίαν κατὰ τῆς βασιλείας δηλοῖ καὶ εἰκότως πόλεμον.

ΙΑΝΟΥΑΡΙΟΣ

α. Εἰ βροντήσῃ, ἐλαφρὸς ἄνεμος ῥεύσει, ἀλλ' ἀκίνδυνος.
β. εἰ βροντήσῃ, ἀδόκητος ἔσται πόλεμος.
γ. εἰ βροντήσῃ, μετὰ νίκην ἧτταν τοῖς ἐν πολέμῳ φράζει· εὐθηνία δὲ ἔσται.
δ. εἰ βροντήσῃ, ὁμονοήσει ὁ δῆμος πρὸς εἰρήνην.
ε. εἰ βροντήσῃ, ὑγίειαν τοῖς κτήνεσι σημαίνει.
ς. εἰ βροντήσῃ, βηχώδη νόσον ἀπειλεῖ, ἀφθονίαν δὲ ἰχθύων καὶ καρπῶν δηλοῖ.
ζ. εἰ βροντήσῃ, δουλομαχία ἔσται καὶ νόσος συχνή.
η. εἰ βροντήσῃ, ὁ δυνάστης τῆς πολιτείας πρὸς τοῦ δήμου κινδυνεύσει.
θ. εἰ βροντήσῃ, ὁ τῆς ἀνατολῆς βασιλεὺς κινδυνεύσει.
ι. εἰ βροντήσῃ, ἀνέμου κίνησιν σφοδρὰν καὶ σίτου μὲν εὐφορίαν τῶν δ' ἄλλων καρπῶν ἀφορίαν δηλοῖ.
ια. εἰ βροντήσῃ, λιμὸν σημαίνει καὶ μέχρις ἀλόγων.
ιβ. εἰ βροντήσῃ, οἱ ἄνθρωποι τὰς ὄψεις νοσήσουσιν, ἔσται δὲ πολὺς χόρτος καὶ ἰχθύων ἐπίδοσις.

ιγ. εἰ βροντήσῃ, νόσους ἀπειλεῖ.
ιδ. εἰ βροντήσῃ, σπάνιν καὶ μυῶν γένεσιν καὶ ὄλεθρον τετραπόδων ἀπειλεῖ.
ιε. εἰ βροντήσῃ, στάσιν δουλικὴν καὶ τιμωρίαν αὐτοῖς καὶ εὐθηνίαν καρπῶν.
ις. εἰ βροντήσῃ, ὁ δῆμος ὑπὸ τοῦ βασιλέως ταραχθήσεται.
ιζ. εἰ βροντήσῃ, νόσους ἀκινδύνους ἀπειλεῖ.
ιη. εἰ βροντήσῃ, πράγματα ἐκδειματοῦντα τὸν δῆμον ἀναστήσεται.
ιθ. εἰ βροντήσῃ, καὶ ὁ βασιλεὺς νικήσει καὶ ὁ δῆμος τὴν κρείττονα ἕξει τάξιν.
κ. εἰ βροντήσῃ, εὐθηνία μὲν ἐπείσακτος ἔσται, βηχώδης δὲ νόσος ἐνοχλήσει τοῖς σώμασιν.
κα. εἰ βροντήσῃ, ὁ βασιλεὺς πολλοῖς ἐπιβουλεύσας τέλος ἐπιβουλεύεται.
κβ. εἰ βροντήσῃ, εὐθηνία μὲν ἔσται, μυῶν δὲ καὶ ἐλάφων πλῆθος.
κγ. εἰ βροντήσῃ, εὐταξίαν τῇ πόλει δηλοῖ.
κδ. εἰ βροντήσῃ, νόσον μετὰ ἐνδείας δηλοῖ.
κε. εἰ βροντήσῃ, δουλομαχία ἔσται.
κς. εἰ βροντήσῃ, πολλοὶ πρὸς τοῦ κρατοῦντος ἀναιρεθήσονται, τέλος δὲ καὶ αὐτός.
κζ. εἰ βροντήσῃ, νόσους ἀκινδύνους δηλοῖ.
κη. εἰ βροντήσῃ, οἱ μὲν ἐνάλιοι ἰχθύες ἐπιδώσουσι, τὰ δὲ θρέμματα φθαρήσεται.
κθ. εἰ βροντήσῃ, λοιμικὸν καὶ νοσῶδες ἐπὶ πᾶσι τὸ τοῦ ἀέρος κατάστημα.
λ. εἰ βροντήσῃ, θάνατον συχνὸν ἀπειλεῖ.

ΦΕΒΡΟΥΑΡΙΟΣ

α. Εἰ βροντήσῃ, πόλεμον καὶ πτῶσιν ἀνδρῶν εὐπόρων ἀπειλεῖ.
β. εἰ βροντήσῃ, ὁ μὲν σῖτος ἐλάττων, ἡ δὲ κριθὴ κρείττων, καὶ θηρίων μὲν αὔξησις, φθίσις δὲ ἀνθρώπων ἔσται.
γ. εἰ βροντήσῃ, ἐμφύλοις ἔσται στάσις.
δ. εἰ βροντήσῃ, οἱ ἄνθρωποι οὐ προσώποις μόνον ἀλλὰ καὶ διανοίαις αὐταῖς ταραχθήσονται.
ε. εἰ βροντήσῃ, ἄμητος πολὺς καὶ ἀνθρώπων ἀπώλεια ἔσται.
ς. εἰ βροντήσῃ, φθορὰ τῶν ξηρῶν καρπῶν, καὶ διαφερόντως τῶν κριθῶν.
ζ. εἰ βροντήσῃ, φθορὰν οὐκ εἰς μακρὰν ἀνθρώποις ἀπειλεῖ.
η. εἰ βροντήσῃ, πρᾶγμα μέγιστον ἀναφύσεται τῇ πολιτείᾳ, καὶ οἱ μὲν ἰχθύες ἐπιδώσουσι, τὰ δὲ θηρία φθαρήσεται.
θ. εἰ βροντήσῃ, ἐλάττων ἡ κριθή.
ι. εἰ βροντήσῃ, τὰ θηρία τοῖς ἀνθρώποις λυμανεῖται.
ια. εἰ βροντήσῃ, εὐτοκία γυναικῶν.
ιβ. εἰ βροντήσῃ, θάνατον συχνὸν ἀπειλεῖ καὶ ἀνέμους ἀήθεις.
ιγ. εἰ βροντήσῃ, εὐθηνία μὲν ἔσται, στάσις δὲ ὅμως πολιτική.
ιδ. εἰ βροντήσῃ, ἀποβολὴν τέκνων καὶ ἔφοδον ἑρπετῶν ἐπιβλαβῆ ἀπειλεῖ.
ιε. εἰ βροντήσῃ, λοιμικὸς ἀὴρ ἔσται, θηρίων τε καὶ μυῶν γένεσος.

ις. εἰ βροντήσῃ, τῷ μὲν δήμῳ αἴσιον, τοῖς δὲ δυνατοῖς ἐκ διχονοίας κακόν.
ιζ. εἰ βροντήσῃ, θέρος ἔσται ἀφθονώτατον.
ιη. εἰ βροντήσῃ, ἄνεμον βαρὺν καὶ φλυκταινῶν ἀνάστημα τοῖς σώμασιν ἀπειλεῖ.
ιθ. εἰ βροντήσῃ, ἑρπετῶν καὶ ἑλμίνθων πρὸς αὐτοῖς ὄχλος ἔσται.
κ. εἰ βροντήσῃ, καλοὺς ἀέρας δηλοῖ.
κα. εἰ βροντήσῃ, εὐθηνίαν δηλοῖ.
κβ. εἰ βροντήσῃ, ἐπίνοσος ὁ ἀήρ, ἀλλ' οὐ φθαρτικὸς ἔσται.
κγ. εἰ βροντήσῃ, ἀνθρώποις μὲν δυσμορφίαν, ὀρνέοις δὲ φθορὰν ἀπειλεῖ.
κδ. εἰ βροντήσῃ, ὑγίειαν μὲν ἀνθρώποις, ἰχθύσι δὲ καὶ ἑρπετοῖς ὄλεθρον ἀπειλεῖ.
κε. εἰ βροντήσῃ, τοῖς τρυφῶσιν ἐναντίον· πόλεμοι γὰρ ἔσονται, καὶ ζάλη βαρεῖα.
κς. εἰ βροντήσῃ, αὐχμὸν καὶ λειψυδρίαν καὶ ψώρωσιν τοῖς σώμασιν ἀπειλεῖ.
κζ. εἰ βροντήσῃ, στάσιν τῷ δήμῳ δηλοῖ.
κη. εἰ βροντήσῃ, εὐθηνίαν μέν, ἄνεμον δὲ νοσοποιὸν ῥεῦσαι μαντεύεται.
κθ. εἰ βροντήσῃ, πόλεμον καὶ εὐθηνίαν δηλοῖ.
λ. εἰ βροντήσῃ, ἀγαθὰ σημαίνει μετὰ μακρᾶς τινος τοῦ δήμου διχοστασίας.

ΜΑΡΤΙΟΣ

α. Εἰ βροντήσῃ, ἅπαν τὸ ἔτος ἀψιμαχίαι καὶ διχόνοιαι ἔσονται.
β. εἰ βροντήσῃ, παύσεται τὰ προαπειλούμενα.
γ. εἰ βροντήσῃ, κακοδαιμονία μετὰ ἐνδείας τοῖς πράγμασιν.
δ. εἰ βροντήσῃ, εὐθηνία ἀφθονωτάτη ἔσται.
ε. εἰ βροντήσῃ, εὐήλιον τὸ ἔαρ καὶ εὔκαρπον τὸ θέρος ἔσται.
ς. εἰ βροντήσῃ, ὅμοια τοῖς πρὸ ταύτης.
ζ. εἰ βροντήσῃ, ἄνεμος βαρὺς ἀναστήσεται, ὁ δὲ κρατῶν τὰ πράγματα κινήσει.
η. εἰ βροντήσῃ, βροχὰς σημαίνει.
θ. εἰ βροντήσῃ, φθορὰν μὲν ἀνθρώπων, γένεσιν δὲ θηρίων ἀπειλεῖ.
ι. εἰ βροντήσῃ, φθορὰ τοῖς τετραπόδοις.
ια. εἰ βροντήσῃ, κατομβρίαν καὶ ἀκρίδων γένεσιν δηλοῖ.
ιβ. εἰ βροντήσῃ, δυνατὸς τοῦ πολιτεύματος ἢ στρατηγὸς κινδυνεύσει, διὸ μάχαι συρραγήσονται, καὶ τὰ θηρία τοῖς ἀνθρώποις ἐπέλθῃ.
ιγ. εἰ βροντήσῃ, εὐθηνία ἔσται, τὰ δὲ θηρία φθαρήσεται, καὶ οἱ ἰχθύες ἐπιδώσουσι· καὶ ἑρπετὰ τοῖς οἰκήμασιν ἐνοχλήσει, οὐ μὴν βλάψει.
ιδ. εἰ βροντήσῃ, εὐθηνίαν μὲν σημαίνει, θάνατον δὲ ἀνθρώπων καὶ γένεσιν θηρίων ἀπειλεῖ.
ιε. εἰ βροντήσῃ, καύματα σημαίνει καὶ λειψυδρίαν καὶ μυῶν ὄχλον καὶ ἰχθύων πολύν.
ις. εἰ βροντήσῃ, ὑγιεινὸν μὲν τὸ ἔτος, ἀλλ' ἐνδεὲς τῶν ἐπιτηδείων.

ιζ. εἰ βροντήσῃ, πρᾶγμα παράδοξον τῷ δήμῳ συμβήσεται, συχνὴ δὲ φθορὰ ἀνθρώποις τε καὶ θηρίοις τετράποσιν.
ιη. εἰ βροντήσῃ, κατομβρίαν καὶ νόσον καὶ ἀκρίδων γένεσιν καὶ ἐγγὺς ἀκαρπίαν δηλοῖ.
ιθ. εἰ βροντήσῃ, αὐχμὸς ξηρότατος καὶ φθοροποιός.
κ. εἰ βροντήσῃ, εὐπορώτερον οἱ ἄνθρωποι ἅμα καὶ ἀφθονώτερον ζήσονται.
κα. εἰ βροντήσῃ, εὐθηνίαν μετὰ πολέμους καὶ φθορώδεις αὐχμοὺς δηλοῖ.
κβ. εἰ βροντήσῃ, φθορὰν μὲν ὀρνέων, ἐπίδοσιν δὲ τῶν ἐπιτηδείων.
κγ. εἰ βροντήσῃ, στάσεις δηλοῖ.
κδ. εἰ βροντήσῃ, εὐθηνίαν σημαίνει.
κε. εἰ βροντήσῃ, καινὰ πράγματα τῷ δήμῳ ἀνα φύεται.
κς. εἰ βροντήσῃ, κτῆσιν ἐπεισάκτων ἀνδραπόδων δηλοῖ.
κζ. εἰ βροντήσῃ, ἐπείσακτον εὐθηνίαν δηλοῖ.
κη. εἰ βροντήσῃ, ἀφθονία ἰχθύων θαλαττίων ἔσται.
κθ. εἰ βροντήσῃ, αἱ γυναῖκες τῆς κρείττονος δόξης ἀνθέξονται.
λ. εἰ βροντήσῃ, δυνατός τις τῆς βασιλείας ἐγκρατὴς ἔσται, δι᾽ οὗ εὐφροσύνη.

ΑΠΡΙΛΙΟΣ

α. Εἰ βροντήσῃ, ἐμφύλιον στάσιν καὶ ἀποπτώσεις οὐσιῶν ἀπειλεῖ.
β. εἰ βροντήσῃ, δίκης σημεῖον, ἐσθλοῖς ἐσθλὰ φερούσης καὶ φαύλοις φαῦλα.
γ. εἰ βροντήσῃ, κέρδη ἐξ ἐπεισάκτου εὐθηνίας δηλοῖ.
δ. εἰ βροντήσῃ, ὀργὴν τῶν κρειττόνων ἀπειλεῖ τοῖς ἀξίοις.
ε. εἰ βροντήσῃ, αὐχμὸν μὲν τῷ ἦρι, ὑγιεινὸν δὲ τὸν ἐνιαυτὸν σημαίνει.
ϛ. εἰ βροντήσῃ, πόλεμοι ἐμφύλιοι ἀναστήσονται.
ζ. εἰ βροντήσῃ, ἀγαθὰ πάντα καὶ ἄφθονον εὐετηρίαν δηλοῖ.
η. εἰ βροντήσῃ, κατομβρίαν δηλοῖ ἐπίνοσον.
θ. εἰ βροντήσῃ, νίκην τῇ βασιλείᾳ δηλοῖ καὶ τοῖς δυνατοῖς εὐφροσύνην.
ι. εἰ βροντήσῃ, ἀνδρῶν ἀγαθῶν προκοπαὶ ἔσονται.
ια. εἰ βροντήσῃ, τὰ αὐτὰ σημαίνει.
ιβ. εἰ βροντήσῃ, βροχὰς καὶ εὐθηνίας καὶ φθορὰν ἰχθύων δηλοῖ.
ιγ. εἰ βροντήσῃ, ἀνθρώποις καὶ κτήνεσι φθορὰν ἀπειλεῖ.
ιδ. εἰ βροντήσῃ, ὑγίειαν καὶ εὐθηνίαν δηλοῖ.
ιε. εἰ βροντήσῃ, λοιμὸν σημαίνει.
ις. εἰ βροντήσῃ, εὐθηνίαν μέν, μυῶν δὲ ἀρουραίων γένεσιν δηλοῖ.
ιζ. εἰ βροντήσῃ, ἀφθονίαν τῶν ἐπιτηδείων σημαίνει.
ιη. εἰ βροντήσῃ, διχόνοιαν καὶ ἀνδρῶν ἀστοχίαν σημαίνει.
ιθ. εἰ βροντήσῃ, δυνατὸς ἀνὴρ τῆς πολιτείας οὐσίας ἅμα καὶ δόξης ἀφαιρεθήσεται.

κ. εἰ βροντήσῃ, θεοχολωσίαν σημαίνει.
κα. εἰ βροντήσῃ, τοῖς μὲν καρποῖς δεξιόν, τῇ δὲ πολιτείᾳ πόλεμον σημαίνει.
κβ. εἰ βροντήσῃ, φθορὰ τῶν μυῶν ἔσται.
κγ. εἰ βροντήσῃ, βροχὴν ὠφέλιμον τοῖς σπορίμοις δηλοῖ.
κδ. εἰ βροντήσῃ, διχόνοια ἔσται τῶν δυνατῶν, τὰ δὲ σκέμματα αὐτῶν ἀναφανήσεται.
κε. εἰ βροντήσῃ, εἰρήνη ἀνὰ πάντα τὸν ἐνιαυτόν.
κς. εἰ βροντήσῃ, ἐλπίδα καρπῶν μεγάλην καὶ ἀμήτων σπάνιν δηλοῖ.
κζ. εἰ βροντήσῃ, διοσημεία παράδοξος ὀφθήσεται.
κη. εἰ βροντήσῃ, ὅπλοις ὁ δῆμος ἀναζώσεται.
κθ. εἰ βροντήσῃ, ἐπικρατήσει ὁ ζέφυρος.
λ. εἰ βροντήσῃ, πραγμάτων καλῶν ἐπομβρία.

ΜΑΙΟΣ

α. Εἰ βροντήσῃ, φυγὴν τῷ δήμῳ καὶ ἀτιμίαν δηλοῖ.
β. εἰ βροντήσῃ, λιμὸν ἀπειλεῖ.
γ. εἰ βροντήσῃ, ἐπείσακτον εὐθηνίαν δηλοῖ.
δ. εἰ βροντήσῃ, εὔκρατος ἀήρ, καὶ οἱ καρποὶ εὐθηνοὶ ἔσονται.
ε. εἰ βροντήσῃ, ἐναλλαγὴ τῶν λυπηρῶν τοῖς πρά γμασιν ἔσται, καὶ πλείων ὁ σῖτος ἤπερ ἡ κριθή· τὰ δὲ ὄσπρια φθαρήσεται.
ς. εἰ βροντήσῃ, θᾶττον ἀκμάσαι τοὺς καρποὺς καὶ φθαρῆναι δηλοῖ.
ζ. εἰ βροντήσῃ, ὀρνέων καὶ ἰχθύων ἀφθονία ἔσται.
η. εἰ βροντήσῃ, ἀπαίσιον τῷ δήμῳ.
θ. εἰ βροντήσῃ, λοιμὸν σημαίνει, ἀλλὰ μετρίως ἐπικίνδυνον.
ι. εἰ βροντήσῃ, ταραχὰς κατομβρίας καὶ ὑπερχύσεις ποταμῶν ἐπιβλαβεῖς δηλοῖ, σαυρῶν τε καὶ ἑρπετῶν πλῆθος.
ια. εἰ βροντήσῃ, εὐθηνίαν κατά τε τὴν γῆν κατά τε τὴν θάλασσαν ἐλπιστέον.
ιβ. εἰ βροντήσῃ, φθορὰ ἰχθύων ἔσται.
ιγ. εἰ βροντήσῃ, ἐπίδοσιν ποταμίων ὑδάτων δηλοῖ, νόσους δὲ τοῖς ἀνθρώποις.
ιδ. εἰ βροντήσῃ, ἀνατολικὸς ἔσται πόλεμος καὶ φθορὰ πολλή.
ιε. εἰ βροντήσῃ, εὐθηνίαν δηλοῖ.
ις. εἰ βροντήσῃ, εὔχεσθαι δεῖ διὰ τὰ ἀπειλούμενα.
ιζ. εἰ βροντήσῃ, ὑετὸν σημαίνει.
ιη. εἰ βροντήσῃ, στάσιν καὶ ἐξ αὐτῆς πόλεμον καὶ ἔνδειαν τῶν ἐπιτηδείων.
ιθ. εἰ βροντήσῃ, εὐνοίᾳ τοῦ δήμου ἀνήρ τις εἰς ἄκρον εὐδαιμονίας ἀρθήσεται.
κ. εἰ βροντήσῃ, τοῖς μὲν περὶ τὴν ἀνατολὴν εὐθηνία, τοῖς δὲ ἐπὶ δύσιν οὐχ οὕτως.
κα. εἰ βροντήσῃ, εὐχῶν δεῖ διὰ τὰ ἀπειλούμενα.
κβ. εἰ βροντήσῃ, κατομβρίαν καὶ φθορὰν τῶν θαλαττίων ἰχθύων σημαίνει.

κγ. εἰ βροντήσῃ, εὐομβρίαν καρποφόρον δηλοῖ.
κδ. εἰ βροντήσῃ, μεγάλα κακά, ὡς ἐξ ἀθυμίας λειποθυμῆσαι τοὺς ὑπηκόους.
κε. εἰ βροντήσῃ, ἀνάπαυλαν καὶ ὕφεσιν τῶν κακῶν ἐλπιστέον.
κς. εἰ βροντήσῃ, καλὸν τοῖς περὶ γεωργίαν ἔχουσιν.
κζ. εἰ βροντήσῃ, διοσημεία ἔσται καὶ κομήτης ἐξαφθήσεται.
κη. εἰ βροντήσῃ, ὡσαύτως ἔσται.
κθ. εἰ βροντήσῃ, πόλεμον ἀρκτῷον σημαίνει, ἀλλ' ἀκίνδυνον τοῖς πράγμασιν.
λ. εἰ βροντήσῃ, τὰ φυτὰ τῷ ἀνέμῳ συντριβήσεται.

Ταύτην τὴν ἐφήμερον βροντοσκοπίαν ὁ Νιγίδιος οὐ καθολικὴν ἀλλὰ μόνης εἶναι τῆς Ῥώμης ἔκρινεν.

TRANSLATION OF THE ETRUSCAN *BRONTOSCOPIC CALENDAR*

Although I have made slight adjustments in my English translation of the text, the Greek version is essentially identical to that published as "Appendix A: The Etruscan Brontoscopic Calendar," in Nancy Thomson de Grummond and Erika Simon, eds., *The Religion of the Etruscans* (Austin: University of Texas Press, 2006), 173–90 (Greek text, 175–82). The translation has been made from the text published in 1837 by I. Bekker, *Ioannes Lydus* (*Corpus Scriptorum Historiae Byzantinae, editio emendatior et copiosior, consilio B. G. Nieburhrii C.F., instituta, auctoritate Academiae Litterarum Regiae Borussicae continuata*) (Bonn: Ed. Weber), 306–31. For additional background, see Domenici and Maderna 2007; for their new Italian translation, see 81–98.

In addition to the *The Oxford Greek Dictionary* (Liddell & Scott, 9th ed.), I have relied on Sophocles (1887) for special insights into Byzantine terminology. Note that the numbers of days of the month are given in the Greek text in the Greek alphabetic system (see Richardson 2004:13–14).

ENGLISH TRANSLATION

BRONTOSCOPIC CALENDAR

ARRANGEMENT ACCORDING TO THE LUNAR MONTH

BY THE ROMAN FIGULUS

FROM THE SAYINGS OF TAGES

ACCOUNT TRANSLATED WORD FOR WORD

Supposing that publicly, in all augural teaching, the ancients assumed the moon to be a reference point (for under this heading they classified both thunder and lightning signs), one likewise may correctly select the phase of the moon as a factor for reckoning, so that, beginning with Cancer, we shall make observations of thunder day by day, beginning with the first day of the lunar month, and following lunar months.[1] From this [study] the Etruscans transmitted observations localized according to the regions that are struck from the sky by thunder.

IN THE MONTH OF JUNE

1. Full Moon.[2] If in any way it should thunder, there will be an abundance of fruits, with the exception of barley; but dangerous diseases will be inflicted upon bodies.
2. If in any way it should thunder, women in labor will have an easy delivery, but there will be abortion of cattle, yet there will be an abundance of fish.
3. If in any way it should thunder, there will be a scorching and drying wind, such that not only grains but even the soft fruits will be parched through and through and shrivel up.
4. If in any way it should thunder, the air will be cloudy and rainy, so that out of a moldy dampness the crops will rot.
5. If in any way it should thunder, ill-omened for the countryside. Those responsible for villages or towns will be thrown into a state of disorder.
6. If in any way it should thunder, just as the crops are maturing, some sort of wild pest that has sunk deep into them will waste them.
7. If in any way it should thunder, diseases will infect [men], but not many shall die. And although the cereal crops shall be successful, the soft fruits shall dry up.
8. If in any way it should thunder, it indicates wet weather, and ruin of the grain.
9. If in any way it should thunder, there will be a loss of flocks through being overrun by wolves.
10. If in any way it should thunder, there will be frequent death, yet prosperity.

[1] The text as preserved by Lydus includes the note that "beginning with Cancer, we shall make observations of thunder day by day." This is essentially the only allusion to the zodiac – in marked contrast to all the subsequent brontoscopic works, all of which are later in date than the Figulus document. The issue cannot be finally resolved without an earlier Etruscan text, but because this comment occurs at the beginning, and not in the body of the precisely formatted calendar text, I suggest that it may derive from either editing by Figulus or his translation of an Etruscan term for some event or period like equinox or solstice. Had it appeared more than once, or been integrated into the actual predictions, this would place the text closer to the astrologically influenced treatises of Ptolemaic and later scholars.

[2] In Byzantine Greek, σελήνη can mean "Monday" (cf. Dio Cassius 37.18); here it would have meant "start with the first full moon in Cancer." The phrase νουμηνία κατὰ σελήνην, which means "by the lunar month," is not used here.

11. If in any way it should thunder, there will be days of heat, burning but harmless; there will be glad festivities in political affairs.
12. If in any way it should thunder, the same thing as on the preceding day.
13. If in any way it should thunder, it announces the fall of a ruler.
14. If in any way it should thunder, the atmosphere shall be burning hot, but there will be abundant harvest and good flow, not the poorest, of the river fish. Bodies, nevertheless, shall be utterly weak.
15. If in any way it should thunder, the winged creatures shall be injured during the summer, and also the fish shall perish.
16. If in any way it should thunder, it threatens not only dearth of the necessities of life, but also war, and a prosperous man shall disappear from public life.
17. If in any way it should thunder, there shall be days of burning heat and destruction by mice, blind-mice,[3] and locusts. Still, it brings abundance and at the same time murders to the people.
18. If in any way it should thunder, it threatens destruction to the crops.[4]
19. If in any way it should thunder, pests destructive to the crops shall perish.
20. If in any way it should thunder, it threatens discord for the community.
21. If in any way it should thunder, it means there will be a dearth of wine, but an increase in the other crops, and an abundance of fish.
22. If in any way it should thunder, the hot weather will be especially ruinous.
23. If in any way it should thunder, it announces good cheer, a putting aside of ills, and an end to disease.
24. If in any way it should thunder, it announces plenty.
25. If in any way it should thunder, there will be wars and countless ills.
26. If in any way it should thunder, the winter will be especially harmful to the crops.
27. If in any way it should thunder, there is danger from the army for the men in power.
28. If in any way it should thunder, there will be a good harvest for the crops.
29. If in any way it should thunder, the affairs of the queenly city[5] will be improved.
30. If in any way it should thunder, in a short time there shall be frequent death.

[3] Field mice that proliferate at harvest time.

[4] "Crops" translates καρπῶν – "of fruits" or "soft fruits." It may have been applied generically throughout, except where distinction is made between these and ξύλινα καρπά – "hard tree fruits," i.e., nuts, or ἀκρόδρυα, "hard-shelled fruits" (also "fruits that grow in the tops of trees").

[5] "Queenly" = royal city, said only of Rome or Constantinople. Either this had been adjusted by Figulus to mean "Rome," or the original version indicated the city of the calendar's origin, perhaps Tarquinia.

JULY

1. Upon the new moon, if in any way it should thunder, there shall be plenty, yet there shall be ruin[6] of the flocks.
2. If in any way it should thunder, the late autumn will be good.
3. If in any way it should thunder, it signals a heavy winter.
4. If in any way it should thunder, the airs will be turbulent, so that of them will be born scarcity.
5. If in any way it should thunder, there will be an abundance of grain, yet it is the downfall of a virtuous ruler.
6. If in any way it should thunder, it threatens death-bearing diseases to the fortunes of slaves.
7. If in any way it should thunder, there will be rains harmful to the grain fields.
8. If in any way it should thunder, it signifies peace for the community, but ruin for the cattle herds, and a dry cough shall infect.
9. If in any way it should thunder, it foretells a vision of the gods and the advancement of many good men.
10. If in any way it should thunder, there will be lifesaving river waters.
11. If in any way it should thunder, it signals hot weather and stormy rain and a scarcity of grain.
12. If in any way it should thunder, there will be unexpected cold in the summer, because of which the necessities of life will be spoiled.
13. If in any way it should thunder, there will appear the most poisonous reptiles.
14. If in any way it should thunder, it shows one man will come to power over many. But this man is most unjust in state affairs.
15. If in any way it should thunder, there will be dissension among the common people and a scarcity of grain.
16. If in any way it should thunder, the king of the East … will be overcome [by?] war …[7] and disease will be received from dry hot weather.
17. If in any way it should thunder, it signifies the succession of a great ruler.
18. If in any way it should thunder, it threatens a dearth of crops due to rainy weather.
19. If in any way it should thunder, it signifies war and the destruction of the powerful. On the other hand, there will be a plenty of cereals.
20. If in any way it should thunder, it threatens an unhealthy drought.
21. If in any way it should thunder, there will be disagreement among the subjects, but not for long.

[6] "Ruin" or "a falling" (πτῶσις) of the flocks: the connotation of falling down, or away, might have described a particular disease condition.

[7] Damaged manuscript.

22. If in any way it should thunder, it signals good things for the affairs of state, but for the bodies, diseases around the head.
23. If in any way it should thunder, the dissension of the common people will come to an end.
24. If in any way it should thunder, it shows the possible misfortune of a powerful man.
25. If in any way it should thunder, it will go badly for a band of youth and also for the crops along with them. It will be a disease-bearing time.
26. If in any way it should thunder, after great plenty there will be famine.
27. If in any way it should thunder, it threatens subcutaneous eruptions to [men's] bodies.
28. If in any way it should thunder, there will be a dearth of water and a plague of poisonous reptiles.
29. If in any way it should thunder, it signifies a good harvest.
30. If in any way it should thunder, men bent on vengeance shall slip into the worst kind of treachery.

AUGUST

1. If in any way it should thunder, the affairs of the state will be slightly better, and there will be plenty.
2. If in any way it should thunder, it threatens both diseases and at the same time a dearth of the necessities of life.
3. If in any way it should thunder, it threatens both [public] trials and debates among the common people.
4. If in any way it should thunder, there will be a dearth of foodstuffs for both humans and dumb animals.
5. If in any way it should thunder, it signifies that the women are the more sagacious.
6. If in any way it should thunder, there will be an abundance of honey, yet a lack of both water and the other foodstuffs.
7. If in any way it should thunder, it signals harsh winds and diseases at the same time.
8. If in any way it should thunder, it threatens harmless disease to the four-footed.
9. If in any way it should thunder, it proclaims good health for men for a full year.
10. If in any way it should thunder, it threatens bodily pains and wretchedness for the greater part of the people.
11. If in any way it should thunder, there will be a good harvest, yet the downfall of reptiles and harm to men.
12. If in any way it should thunder, there will be an abundance of cattle fodder and of acorns, but in the first ripening season, it will go badly.

13. If in any way it should thunder, there will be plague upon the bodies of both humans and dumb animals.
14. If in any way it should thunder, it signals war for all the people, yet an abundance of crops.
15. If in any way it should thunder, affairs will change for the worse.
16. If in any way it should thunder, it promises a deep peace.
17. If in any way it should thunder, the men of lowly degree shall be gloomy.
18. If in any way it should thunder, it threatens civil war.
19. If in any way it should thunder, the women and the servile class will dare to undertake murders.
20. If in any way it should thunder, it threatens a plague on the cattle and disorder in the affairs of state.
21. If in any way it should thunder, it threatens at once prosperity and discord among the commons.
22. If in any way it should thunder, affairs will be moderately good for an entire year.
23. If in any way it should thunder, it signifies that the lightning bolt shall fall, and warns of slaughter.
24. If in any way it should thunder, it threatens the loss of wellborn youths.
25. If in any way it should thunder, it foretells that during a stormy winter there will be a scarcity of soft fruits.
26. If in any way it should thunder, it signals war.
27. If in any way it should thunder, it threatens at once wars and treachery.
28. If in any way it should thunder, it signals both an abundance of crops and a loss by death of cattle.
29. If in any way it should thunder, it signals no sort of reversal.
30. If in any way it should thunder, it threatens diseases in the city over which it [the thunder] is cast down.

SEPTEMBER

1. If in any way it should thunder, it signifies both a good harvest and good cheer.
2. If in any way it should thunder, there will be discord among the common people.
3. If in any way it should thunder, it signifies heavy rains and war.
4. If in any way it should thunder, it signifies the downfall of a powerful man and preparation for war.
5. If in any way it should thunder, it signifies an abundance of barley but a decrease in wheat.
6. If in any way it should thunder, there shall be power among the women greater than [what is] appropriate to their nature.

7. If in any way it should thunder, it threatens a disease and out of it, a disaster for the servile class.
8. If in any way it should thunder, it indicates that those especially powerful will consider crooked dealings in government, but they will not achieve their aims.[8]
9. If in any way it should thunder, it threatens that a disease-bearing wind will blow.
10. If in any way it should thunder, there will be strife in the area in which the thunder is let loose; for another place [it is] not inapplicable.
11. If in any way it should thunder, the underlings of the wellborn will foment revolution in the state.
12. If in any way it should thunder, it says that the time of harvest shall be very rainy and there shall be famine.
13. If in any way it should thunder, it threatens grave famine.
14. If in any way it should thunder, it threatens diseases.
15. If in any way it should thunder, it signifies a wet spell, but at the same time, prosperity.
16. If in any way it should thunder, it is made known that there will be good sprouting, but [the plants will be] fruitless.
17. If it thunders, it threatens a lack of the necessities.
18. If it thunders, it signals both famine and wars.
19. If it thunders, the fruits of the trees will be successful, but there will be diseases and sedition among the commons.
20. If it thunders, it threatens the destruction of a famous man and war.
21. If it thunders, it threatens ills and losses for the people.
22. If it thunders, it signals prosperity yet a heavy and wet winter.
23. If it thunders, it foretells a time of need during the winter of the year.
24. If it thunders, it threatens a drought. There will be an abundant harvest of the hard-shelled fruits;[9] around late autumn though, they will be destroyed by storms.
25. If it thunders, out of civil unrest a tyrant shall be raised up, and he will be undone, but the powerful will be destroyed utterly with insufferable penalties.
26. If it thunders, a corrupt ruler will be felled by divine decision.
27. If it thunders, powerful men will work hatred toward themselves, and shall take sides against each other.
28. If it thunders, there will be signs revealing great things. Beware lest it pour rain upon the fire of joyful elation.
29. If it thunders, it threatens a severe drought.
30. If it thunders, affairs of state [shall change] from worse to better.

[8] Text not clear here.

[9] Perhaps nut-trees.

OCTOBER

1. If it thunders, it threatens a corrupt tyrant over the affairs of state.
2. If it thunders, there will be prosperity but the destruction of the mice of dry land.
3. If it thunders, it signifies hurricanes and disturbances by which the trees will be overturned; there will be a great disruption in the affairs of common people.
4. If it thunders, the lower classes will have the upper hand over their betters, and the mildness of the air will be healthy.
5. If it thunders, there will be a surplus of all the necessities excepting grain.
6. If it thunders, appearance of future abundance, yet harvest will be less plentiful and autumn practically empty of fruit.
7. If it thunders, pulses will be plentiful but wine less.
8. If it thunders, an earthquake with roaring is to be expected.
9. If it thunders, it threatens destruction to wild beasts.
10. If it thunders, it signifies the downfall of a praiseworthy man.
11. If it thunders, it signifies a strange sort of wind will be of service to the pastures.
12. If it thunders, there will be prosperity, but wind squalls will oppress.
13. If it thunders, commerce[10] will be good, and prosperity in addition. He who controls the government with heavy hand will not be strong for very long.
14. If it thunders, it threatens war and the loss of flocks to death.
15. If it thunders, there will be scarcity from a dry and searing wind falling upon the crops.
16. If it thunders, men will be weakened in such a manner that they will seem to be unrecognizable.
17. If it thunders, good fortune for a rich man and for men wellborn.
18. If it thunders, it signifies a plentiful grain supply brought in from foreign lands.
19. If it thunders, it warns the downfall of a ruler or the overthrow of a king, but it warns both discord among the common people and abundance.
20. If it thunders, it warns there will be a festering wound, and for the many, extreme suffering out of the discord.
21. If it thunders, there will be coughing sicknesses and oppression of the heart.
22. If it thunders, it threatens for the people, bad conditions and spotted diseases.
23. If it thunders, the people will be of marvelously good cheer.

[10] Literally, "contracts" or "business dealings" in Greek usage.

24. If it thunders, out of the discord of those in power, the common people will oppress [others].
25. If it thunders, there will be heavy misery resulting from misfortunes.
26. If it thunders, there will be an increase of animals, but at the same time they will suffer thirst.
27. If it thunders, it signifies heavy rains.
28. If it thunders, there will be a dearth of the necessities.
29. If it thunders, a year of serious disease.
30. If it thunders, it signifies not merely prosperity, but even fewer enemies, and good cheer for the state.

NOVEMBER

1. If it thunders, it signifies discord for the city.
2. If it thunders, it foretells prosperity.
3. If it thunders, situations will pertain through which the lower classes will oppress [their] betters.
4. If it thunders, grain will be better.
5. If it thunders, it signifies storm for the state, and disease for humans and dumb animals alike.
6. If it thunders, borers will ruin the grain.
7. If it thunders, for those who are in the west, both humans and dumb beasts, diseases.
8. If it thunders, it says gluttony shall come about from menacing diseases.
9. If it thunders, the common people will be led into misery, but an abundance of daily provisions.
10. If it thunders, for those in power, it makes an end to their perverted plans. A parching wind will wrack the trees.
11. If it thunders, men shall give blessings to the god, for the wind shall blow out of the east.
12. If it thunders, it indicates insomnia for some time for men.
13. If it thunders, a wealthy yet sickly period threatens, tormenting bodies with internal worms.
14. If it thunders, poisonous snakes shall somehow be gently undone by the men.
15. If it thunders, the fish especially plentiful, but it shall plague the water-bound beasts. The commonwealth rather better.
16. If it thunders, the creation of locusts and field-voles, to the king, danger, and there will be an abundance of grain.
17. If it thunders, it signifies plentiful fodder for the flocks.
18. If it thunders, it signifies war and woes for city folk.
19. If it thunders, welfare of women.
20. If it thunders, it signifies famine not of long duration.

21. If it thunders, the mice shall perish; an abundance not merely of grain but also of pasturage, and a plenty of fish.
22. If it thunders, it signifies a year of well-being.
23. If it thunders, disease-bearing wind will blow.
24. If it thunders, the watch post shall complete for the state good service against enemy tricks.
25. If it thunders, there will be a very dangerous war.
26. If it thunders, it signifies a civil war and the death of many.
27. If it thunders, it threatens the same.
28. If it thunders, many of the councilmen of the wealthier rank shall be ruined utterly by cowardice.
29. If it thunders, the lower classes will do better, but the hoped-for fruit harvest shall be destroyed.
30. If it thunders, the mortals shall live in a condition more favored by the gods. Naturally, evils [will come] in due proportion.

DECEMBER

1. If it thunders, it signifies a year of well-being according to concord.
2. If it thunders, a plenty of fish and especially of fruits.
3. If it thunders, men will excessively consume their flocks because of a dearth of fish.
4. If it thunders, winter will be heavy, yet [there will be] abundance also.
5. If it thunders, it threatens mangy diseases.
6. If it thunders, the men shall be visited with visions of the faces of the gods, they shall experience a bad outcome.
7. If it thunders, it signifies the same for all.
8. If it thunders, virulent disease; out of it, though, will be an abundance of crops, but a plague on the flocks.
9. If it thunders, there will be the downfall of a famous man.
10. If it thunders, it threatens slaughter for men from diseases, but the fish shall be abundant.
11. If it thunders, heat-bearing shall be the summer season, and plenty imported from foreign lands.
12. If it thunders, it threatens diseases from diarrhea.
13. If it thunders, plenty, yet diseases it threatens.
14. If it thunders, it signifies at the same time civil war and abundance.
15. If it thunders, many will set out for war but few shall return.
16. If it thunders, newfangled affairs for the state.
17. If it thunders, it threatens that small locusts shall be born, yet there will still be plenty.
18. If it thunders, there shall be a heavy war.
19. If it thunders, it threatens prolongation of war.

20. If it thunders, it tells a lack of the necessities.
21. If it thunders, it threatens a hot and disease-making wind will blow.
22. If it thunders, the summer will be hot but plentiful in crops.
23. If it thunders, it signifies a disease for men but a harmless one.
24. If it thunders, it threatens civil wars for the city, and a plague on the beasts of the woods.
25. If it thunders, a movement of troops to war, but it will turn out well.
26. If it thunders, it threatens diseases for the slaves.
27. If it thunders, the king will help many.
28. If it thunders, the hatching of locusts.
29. If it thunders, it signifies the most healthful leanness for the bodies.
30. If it thunders, it signifies a rebellion against the kingdom and, reasonably, war.

JANUARY

1. If it thunders, a fast wind will blow, but not dangerous.
2. If it thunders, there will be unlooked-for war.
3. If it thunders, it shows after victory, loss for those in the war. Still, there will be plenty.
4. If it thunders, the common people will agree to make peace.
5. If it thunders, it signals health for the flocks.
6. If it thunders, it threatens a coughing sickness, but signifies an abundance of fish and of fruits.
7. If it thunders, there will be a slave revolt and recurring illness.
8. If it thunders, the ruler of the state shall be in danger from the people.
9. If it thunders, the king of the East shall be in danger.
10. If it thunders, it signifies rapid movement of wind, and a plenty of grain, but a dearth of other crops.
11. If it thunders, it signals famine [reaching] just up to dumb animals.
12. If it thunders, men shall be damaged in their faces, but there will be much fodder [for horses/cattle], and a plenty of fish.
13. If it thunders, it threatens diseases.
14. If it thunders, it threatens need, and the birth of mice, and the destruction of four-footed creatures.
15. If it thunders, servile revolt, and punishment for them, and abundance of crops.
16. If it thunders, the people shall be oppressed by the king.
17. If it thunders, it threatens nondangerous diseases.
18. If it thunders, affairs circulating abroad shall make the people rise up.
19. If it thunders, when the king will have victory, then the common people will have the upper hand/stronger position.

20. If it thunders, there will be abundance of imported goods, but a coughing disease will afflict bodies.
21. If it thunders, the king hated by many shall be the object of a final plot.
22. If it thunders, there will be plenty, but also there will be an abundance of mice and of deer.
23. If it thunders, it signifies good order for the city.
24. If it thunders, it signifies disease following want.
25. If it thunders, there will be unrest among the slaves.
26. If it thunders, many shall be cut down by a man in power, but in the end he himself [will be killed].
27. If it thunders, it signifies nonthreatening diseases.
28. If it thunders, the fish of the sea shall be plentiful, but yet the flocks will be ruined by death.
29. If it thunders, the condition of the air oppressive, and disease-bearing for all.
30. If it thunders, it threatens plentiful death.

FEBRUARY

1. If it thunders, it threatens war and the ruin of wealthy men.
2. If it thunders, wheat in less supply, but barley better, and an increase in livestock, but there will be a wasting away of humans.
3. If it thunders, there will be civil unrest.
4. If it thunders, men shall be troubled not only in visage but also in their very minds.
5. If it thunders, there will be a large harvest, a destruction for men.
6. If it thunders, destruction of grain supplies and especially barley.
7. If it thunders, it threatens destruction though not for long to humans.
8. If it thunders, the greatest affair will inflame the state, and also fish will increase and yet dangerous wild beasts shall perish.
9. If it thunders, worse the barley.
10. If it thunders, the wild beasts shall undo the humans.
11. If it thunders, good deliveries [in childbirth] for women.
12. If it thunders, it threatens frequent death and unseasonable winds.
13. If it thunders, there will be plenty, yet at the same time, political unrest.
14. If it thunders, it threatens loss of progeny, and an onslaught of poisonous reptiles.
15. If it thunders, the air shall carry plague, creation of both wild beasts and mice.
16. If it thunders, to the people, auspicious, but of the powerful ones, bad [will come] out of discord.
17. If it thunders, summer will be most fruitful.

18. If it thunders, it threatens a heavy wind and eruption of pustules on bodies.
19. If it thunders, there will be a throng of reptiles, and in addition, of worms.
20. If it thunders, it signifies fine breezes.
21. If it thunders, it signifies abundance.
22. If it thunders, the air will be disease-carrying but not lethal.
23. If it thunders, it threatens deformity for men, but destruction for birds.
24. If it thunders, it threatens good health for men, but destruction for both fish and reptiles.
25. If it thunders, to those living luxuriously, a reversal. There will be wars, and a heavy storm.
26. If it thunders, it threatens hot weather, and a lack of water, and scabs on bodies.
27. If it thunders, it signifies unrest among the commons.
28. If it thunders, it prophesies abundance, yet at the same time, a disease-giving wind will blow.
29. If it thunders, it signifies war and abundance.
30. If it thunders, it signifies good things with long duration after great divisions of the people.

MARCH

1. If it thunders, for the entire year there will be strife and disagreements.
2. If it thunders, it shall end the threatening affairs.
3. If it thunders, for the state, discord following famine.
4. If it thunders, there will be boundless prosperity.
5. If it thunders, the spring will be sunny and the summer fruitful.
6. If it thunders, the same and even better.
7. If it thunders, a heavy wind will arise, which shall move the affairs of powerful men.
8. If it thunders, it signals rains.
9. If it thunders, it threatens ruin of man and creation of wild beasts.
10. If it thunders, destruction to the four-footed.
11. If it thunders, it signifies heavy rain and the creation of locusts.
12. If it thunders, a powerful man in politics or a general is endangered; on his behalf, battles will be waged, and the wild beasts shall fall upon man.
13. If it thunders, there will be plenty but the wild beasts shall be destroyed and the fish shall increase; and reptiles will trouble habitations but will not be harmful.
14. If it thunders, it signals prosperity but threatens a death of men and birth of wild beasts.

15. If it thunders, it signals hot spells and drought and a great throng of mice and fish.
16. If it thunders, healthful the year, yet lacking in necessities.
17. If it thunders, something unexpected will befall the people; ruin upon ruin for men and four-footed beasts.
18. If it thunders, it signifies a period of severe rain, and disease, and the birth of locusts, barrenness [of crops] near at hand.
19. If it thunders, a very dry summer and destructive.
20. If it thunders, man will live with better behavior at the same time as more prosperously.
21. If it thunders, it signifies prosperity after wars and hot spells causing destruction.
22. If it thunders, destruction of birds, but a plenty of daily supplies.
23. If it thunders, it signifies discord.
24. If it thunders, it signifies prosperity.
25. If it thunders, new affairs are given birth among the people.
26. If it thunders, it announces [the] acquisition of imported slaves.[11]
27. If it thunders, it signifies abundance imported from abroad.
28. If it thunders, there will be a plenty of marine fish.
29. If it thunders, the women shall obtain the better reputation.
30. If it thunders, there will be some powerful, self-possessed man of the kingdom, through whom [will come] good cheer.

APRIL

1. If it thunders, it threatens civil discord and the downfalls of fortunes.
2. If it thunders, sign of justice, bearing prosperity to good men, and paltry things to evil men.
3. If it thunders, it signifies profit out of a grain supply brought from abroad.
4. If it thunders, anger it threatens of those more powerful against the upright.
5. If it thunders, it signals a hot summer early [in the season] but a healthful year.
6. If it thunders, civil wars will arise.
7. If it thunders, it signifies all good things and a prosperous season.
8. If it thunders, it signifies heavy rains bearing disease.
9. If it thunders, it signifies victory for the kingdom and good cheer for the powerful ones.

[11] The connotation is of foreign prisoners of war sold as slaves, at least in the languages of Lydus and Figulus. Slaves with non-Italian names are recorded as early as the 7th c. in Etruria (see Chapter 7).

10. If it thunders, of upright men there will be advances.
11. If it thunders, it signals the same things.
12. If it thunders, rains and prosperity and ruin of fish it signifies.
13. If it thunders, for men and for cattle destruction it threatens.
14. If it thunders, good health and prosperity it signifies.
15. If it thunders, it signals a plague.
16. If it thunders, it signifies abundance but at the same time the birth of field-voles.
17. If it thunders, it signals a plenty of daily supplies.
18. If it thunders, it signals discord and thoughtlessness of men.
19. If it thunders, a powerful man in the state shall be deprived at once of both reputation and property.
20. If it thunders, it signals divine anger.
21. If it thunders, it signifies good fortune for the crops, yet war for the state.
22. If it thunders, it will be the destruction of the flies.
23. If it thunders, it signifies a rain helpful for the sprouting time.
24. If it thunders, there will be discord among those in power, but their plans will be exposed.
25. If it thunders, peace during the entire year.
26. If it thunders, it signifies great hope of fruits and scarcity of harvests.
27. If it thunders, omens from the sky incredibly shall be revealed.
28. If it thunders, by shields the people shall be saved.
29. If it thunders, a zephyrus will prevail.
30. If it thunders, a shower of good things.

MAY

1. If it thunders, it signifies flight for the common people and loss of honor.
2. If it thunders, it threatens need.
3. If it thunders, it signifies abundance imported from abroad.
4. If it thunders, the air will be mild and the crops will be plentiful.
5. If it thunders, there will be an interchange of hardships in political affairs, and wheat more plentiful than barley. The pulses, however, will be ruined.
6. If it thunders, it signifies that crops will ripen in haste and will be ruined.
7. If it thunders, there will be abundance of birds and fish.
8. If it thunders, ill-omened for the common people.
9. If it thunders, it signals plague, but not exceptionally life-threatening.
10. If it thunders, it announces storms, heavy rain, heavy floods of the rivers, a throng of lizards and of reptiles.
11. If it thunders, abundance to be hoped for both on land and sea.
12. If it thunders, there will be destruction of fish.

13. If it thunders, it signals an increase in river waters, but diseases for men.
14. If it thunders, there will be eastern war and great want.
15. If it thunders, it signifies abundance.
16. If it thunders, atonement must be made on account of terrible news.
17. If it thunders, it signifies rainy weather.
18. If it thunders, discord and out of it war and a lack of daily supplies.
19. If it thunders, through good will of the people, some man shall be exalted to the height of good fortune.
20. If it thunders, for those in the east, prosperity, but for those in the west, not the same.
21. If it thunders, atonement must be made on account of terrible news.
22. If it thunders, it signals heavy rains and destruction of marine fish.
23. If it thunders, it signifies a good and fruitful rain.
24. If it thunders, great evils such that those hearkening [to them] shall pass away from grief.
25. If it thunders, a hoped-for resting place and slackening of evils.
26. If it thunders, good for those working upon the tilled land.
27. If it thunders, there shall be prodigies, and a comet shall shine forth.
28. If it thunders, it shall be the same.
29. If it thunders, it signals northern war, but not dangerous for commerce.
30. If it thunders, the sprouting crops will be chilled by the winds.

This brontoscopic almanac Nigidius claimed was not universal, but was only for Rome.

PART III

THEMATIC ANALYSIS OF THE ETRUSCAN *BRONTOSCOPIC CALENDAR*

CHAPTER 4

ANALYSIS OF THE *BRONTOSCOPIC CALENDAR*

Topics of the calendar's predictions are discussed in detail in Chapters 5–7, and background and samples of Mesopotamian literature are treated in Chapter 8. This chapter considers the identity of the *Brontoscopic Calendar* as an Etruscan religious text and its place in the intellectual milieu of first-century BC Roman Italy.

FORMAL CHARACTERISTICS: THE CALENDAR

The most striking aspect of the document attributed to Figulus is its distinctive format, as a calendar covering an entire year of twelve months, each composed of an arbitrary thirty days. It begins with the new moon "when the sun is in Cancer" (the only reference to the zodiac), thus June, around the time of the summer solstice. As discussed in Chapter 8, this corresponds to the Mesopotamian Standard Calendar established by the first millennium BC, and in full force, at least for civic affairs, during the Late Assyrian period, and, in many areas, for some time thereafter.[1] A formal calendar of some sort was probably instituted in the Etruscan cities at the time of their foundation or legal consolidation by the end of the eighth century BC, because government affairs, judicial procedures, and markets needed such a framework for regular business. In the Greek cities, cultic calendars parted ways with the civic format,[2] but in the Near East, where religion and the kingdom were inextricably intertwined, the Standard Calendar made sense.

The *Brontoscopic Calendar* specifies that it has been "translated word for word" from the Etruscan original, but such background rarely appears in any of the Hellenistic or Roman *brontoscopia*.[3] Statements of authority

[1] Cohen 1993:297–342; see Brack-Bernsen 2007 for astronomical and epigraphic descriptions.

[2] See Hannah 2005:16–82

[3] But see the earliest lengthy text, attributed to Fonteius, at Appendix C, Text 1.

were incorporated in Etruscan documents, such as the cultic calendars of the Capua tile and Zagreb *Liber linteus*. Clearly, this written authority was of some importance – to Figulus at least – to establish the fidelity of the text to some older original. Further, the stipulation that this document was "valid only for Rome" is not mirrored in any of the other brontoscopic works – it seems that Figulus was asserting both the basic validity and antiquity of the source text and the fact that his Latin version had been adjusted so that it could only apply to Rome. But there is no natural condition that is met by Rome, in terms of climate, latitude, terrain, ecology, agriculture, and so on, that could not equally be applied to adjacent Etruscan (or Latin or Faliscan) communities. Perhaps the distinction was intended as cultural or linguistic, because cultures were determining factors in the doctrine of *saecula* (at least, according to the Roman authors who claimed that Rome's cycle of *saecula* would be of longer duration than the Etruscan era, 12 instead of 10 phases).

One might speculate on the possibility that some early documents of the *etrusca disciplina* were written in Etruscan Rome. The obvious atmosphere is that of the "great Rome of the Tarquins," which began around 615 BC with Tarquinius Priscus, when so much architecture, art, and technology developed according to Etruscan design. The woman who helped to usher in that period, Tanaquil, was described as a seer in her own right, interpreting portents as her husband embarked upon his career at Rome (*perita ... caelestium prodigiorum*, "skilled in heavenly portents," Livy 1.34.3–10). And the events of that time were indeed recorded as part of Etruscan history, to judge from the historical slaughter/rescue scene in the François Tomb of fourth-century Vulci.[4] But there were Etruscans in Ruma in the eighth century, too, according to the accounts of "Celer," who were responsible for the death of Remus (Plutarch *Romulus* 10.2).[5] Early inscriptions already show ethnic diversity among the residents of several cities, so an active Etruscan cult practicing divination in Rome is surely not out of the question.[6]

FORMAT

Many analysts[7] have seen in the *Brontoscopic Calendar*'s format of twelve months the Roman twelve-month year, often interpreted as the Julian calendar. The correlation is not exact, however, because all months are an arbitrary thirty days, and it begins in June – not the March of the Roman civil year, or the January of the Julian system. This is, however, a perfect match for the

[4] François Tomb: Moretti Sgubini 2004; Torelli 2009:175–76, 234–37.

[5] Wiseman 1995:9–10.

[6] Cf. Cristofani 1996.

[7] Even Ampolo 1990–91:190–92.

already ancient Mesopotamian Standard Calendar.[8] The most telling feature is the total absence of the Roman system of days, using Kalends, Nones, and Ides, even though ancient authorities gave Etruria credit for such concepts as "Ides." Several successors to the *Brontoscopic Calendar* do employ the Roman reckoning system.

ARCHAIC CALENDARS: VARIATIONS

We know Archaic versions of the Roman calendar from their later Republican successors, the *Fasti*, large-scale public calendars that were posted on walls so that all citizens could refer to them.[9] It appears that the earliest formal Roman calendar had only ten months and required a lot of tinkering to keep it synchronized with the natural year, but the evidence for the Archaic period is understandably very sparse.[10] There is no evidence whatever for an *Etruscan* Archaic calendar of ten months, unless the fragmentary later texts (5th and 2nd c. BC) of the Capua tile and Zagreb *Liber linteus* are understood as being nearly complete and thus close to their original lengths and sizes: if so, they *may* have terminated at ten months.[11] By the time of the earliest preserved monumental *Fasti* (the 1st c. *Antiates Maiores*), we already see twelve months.

The first Roman calendar "reform" was attributed to Numa Pompilius (715–673 BC), who instituted twelve lunar months of thirty days with intercalary days every other year. Numa determined which days would be "lawful" or "unlawful" (*dies fasti* or *nefasti*) for public business, so the twelve-month civil year was already in operation by the time of the first preserved epigraphical evidence. This event is noted by Livy (1.19.6–1.20.7),[12] who goes on to describe Numa's founding of cults and the offices of *flamines*, Vestal Virgins, the Salii, and the *pontifex*, all significant personages in the life of the city and state. The legend of Numa's establishment of religious traditions as support for the state occurs at just the time that Etruscan cities implemented exactly these sorts of tactics of urbanization.

The monumental Roman calendars list the days in a column for each month with markings to indicate workdays, festivals, and days forbidden for courts or government meetings. The *Fasti Antiates Maiores* were painted in red and black on a plastered wall in the Roman colony of Antium (Anzio). Terms are

[8] Although see Piganiol 1951:83; see Wachsmuth 1897:xxxvii, for other possible motives for beginning in June.

[9] See Michels 1967; Hannah 2005:98–130, 102–6, for *Fasti Antiates Maiores*. For the scientific background to calendars and a wealth of reference material for calculating the synchronism of dates in other calendars, see Dershowitz and Reingold 2008.

[10] See Michels 1967:16–30; York 1986; and Pedroni 1998.

[11] See van der Meer 2007:29.

[12] See Beard et al., 1998:vol. 2:4–5, no. 1.2.

condensed to letters and acronyms, and the year begins (reading left to right like Latin texts) at January, with market days indicated every ninth day.[13] This sort of calendar is not especially relevant to the *Brontoscopic Calendar*, but it serves to contrast Roman systems with Etruscan by its January start; it also shows that the twelve-month system was commonplace in Italy before the Julian reforms.

Later (Late Ptolemaic and Roman Imperial) brontoscopies also adhere to a calendrical format, but none show the same system as the *Brontoscopic Calendar*. The *Tonitruale* attributed to the Late Republican Fonteius Capito (Appendix C, Text 1) incorporates the zodiacal periods and begins in the sign of Capricorn; it also distinguishes between day and night, as in other astrological texts. The Augustan-period work by Claudius Tuscus, although it claims to be "from the sacred writings of the Etruscans, word for word," and also cites several Greek scholars whose works are supposed to be similar, actually has an even later, more Roman format: it begins in January and notes days according to the Roman system of Kalends, Nones, and Ides. All subsequent brontoscopies follow one or another system of zodiacal counting and usually begin either in Aries, October, or January; this is the same situation seen in regard to the other divinatory or astrological documents of the Late Hellenistic and Roman periods, such as the *seismologia* preserved in *De ostentis* and *CCAG*. (For details, see Table 9.1.) The clear format of the Figulus text, with an entry for all thirty days, and virtually no zodiacal contamination, is unique; and, as discussed in Chapter 8, neither did its original author copy directly from the Mesopotamian hemerologies and menologies.

Mark Cohen has shown that even the Mesopotamian Standard Calendar derived its month names from multiple cultures, supplementing old Sumerian terms with words from Ugaritic, Old Persian, and other sources – calendars, used every day, are constantly being adapted and adjusted, and the form inherent in the *Brontoscopic Calendar* should not surprise us.[14]

THE ETRUSCAN CALENDAR: AVAILABLE INFORMATION

The Etruscan calendar is partially known, thanks to the preservation of two related documents, the Capua tile and the Zagreb *Liber linteus*; our meager information on the basic calendar has been gathered by Ingrid Edlund-Berry (1992). In contrast to many other systems, the Etruscan day (*tin*, from the god Tin/Tinia) began at noon (*usil/uslane*, from Usil the sun god). For legal

[13] See Michels 1967 foldout facsimile; Beard et al. 1998:vol. 2, 61–66.

[14] Cohen 1993:302–5. On time reckoning in Greece and Rome, see Hannah 2009, who unfortunately makes no reference to Etruscan systems. Could the citation of Cancer at the very beginning have been a slip of Figulus the astrologer?

purposes, this may have been very useful, because noon, with the sun directly overhead, is easily observed, but other times of day may be subjectively perceived (how dark or light is it, etc.?), and citizens might not reach a consensus easily. In contrast, the Babylonian system, known since the third millennium BC, and well documented in astronomical texts, began the day at dusk, and its twelve lunar months started in spring when the new moon was sighted. The postexilic Hebrew calendar was similar and began with the spring equinox (*Nisan*). The Egyptian calendar, known since the third millennium, followed the so-called Sothic cycle and also had twelve months of thirty days, with five unlucky days added; its twenty-four-hour days began with sunrise. The Roman day originally was reckoned from sunrise, as it was in Western Europe before the invention of clocks in the fourteenth century; but later the hours were counted beginning at midnight, as in the modern, Western day. Curiously, the Roman feast of "appeasing the dead" (*placandis Manibus*), the *Parentalia* (February 13 to 21 or 22), officially began at the sixth hour – noon – of the Ides, according to Lydus.[15] This has prompted some scholars to posit for its origin some Etruscan feast in honor of the dead. If only we had so much hard information.

In Etruscan "month" is *tiu/tiur*, derived from Tiv or Tiur, the moon goddess; and *thesan*, "morning," is named for the goddess Thesan, Dawn. Religious texts such as the Magliano plaque, a lead sheet perhaps in the form of a sheep's liver, with a spiral inscription, record cultic rituals to be performed *tuthiu avils*, "every year," or *tuthi tiu*, "monthly."[16] The Zagreb text (*ET* LL 3.21–23) details rituals to be performed "on the proper day, month, year" (*eθrse tinśi tiurim avilś*) on behalf of the state and city.[17]

We only know the names of perhaps eight Etruscan months, some from epigraphic finds, such as the Capua tile, and others only from glosses noted by Classical authors: *acale*, *paralse/tranius*, *thucte*, *celi*, and *chosfer* are June, July, August, September, October; *velcitanus*, *apirase/cabreas*, *anpilie* are March, April, May (mere glosses are underlined, others are attested epigraphically). September, *celi*, is named for Cel, the goddess Earth. A link has been suggested between *thu*, "one," and *thucte* (August?), and between *cezp-*, "eight," and *chosfer* (October), but we simply do not have enough information on the names or date of the origin of these names. Some scholars interpret a month called *Mtn/Masan* as cited in the second Etruscan gold Pyrgi tablet of circa 500 BC (E2 = *ET* Cr 4.5), but van der Meer reads this differently, and we still have no indication of the season in which it would fall.[18] The Phoenician Pyrgi text also records the month *Krr*, in the line "year 3 in his reign in the month Karar on

[15] *De mensibus* 4.29; cf. Scullard 1981:74–76; Fowler 1899/1969:306–10.

[16] *ET* AV 4.1, ca. 450 BC; van der Meer 2007:172–73, no. 4.

[17] *ET* LL 3.21–22, 8.15; see van der Meer 2007:87; Bonfante and Bonfante 2002:183.

[18] van der Meer 2007:30, 117, 172, no. 2.

TABLE 4.1. *Etruscan months compared with Babylonian and Phoenician calendars*

	Etruscan	Modern (Roman)	Babylonian	Phoenician	Pyrgi plaque E1
1	acale	June	Nisanu	ZBH ŠMŠ	tulerase
2	paralse or tranius	July	Ayura	KRR	alsase
3	thucte	August	Simanu	<SH>? MP'LPNY?	
4	Celi	September	Du'uzu	MP'	
5	chosfer	October	Abu	'TNM	
6		November	Ululu	BL	
7		December	Tashritu	MRP'(M)	
8		January	Arakhsamnu	[PG]RM	
9		February	Kislimu	P'LT	
10	velcitanus	March	Tebetu	HYR	
11	apirase or cabreas	April	Shabatu	ZYB	
12	anpilie	May	Adaru	MTN	
13		intercalary		MP'LPNY?	

Sources: See Edlund-Berry 1992; Steiglitz 1998; Richards 1998:148, table 10.1.

the day of the god's burial (*zbh šmš*)."[19] There would have been an intercalary period to even up the number of days, but we have no data on the details.

If the name *chosfer* is correctly attributed to October, then the month sequence has obviously been nudged back by three months, a full season, presumably to match the Roman-Julian calendar system, and we are looking at a late, thus Roman, version of the Etruscan calendar.[20] It may be that the original names for months in the Etruscan *Brontoscopic Calendar* were different and have not been preserved. (The possible month *thucte* is even less well understood). But even the Mesopotamian Standard Calendar evolved in response to new situations of diversity, like the advent of other ethnic groups to the theater of Near Eastern politics and society.

CONTENT: CONTRASTS WITH NEAR EASTERN SOURCES AND WITH LATER DOCUMENTS

Although other commentators often stopped after noting the similarities to first-century Rome in the social hierarchy, rebellions, and so on predicted by

[19] See Schmitz 1995:564, 571.

[20] But even Etruscan numbers are still debated: Wallace (2008:54–55) reads *cezp*- as "seven" and *semφ*- as "eight."

the *Brontoscopic* text, Ampolo (1990–91) took its Etruscan identity a step further, analyzing select, specific predictions for their affinities to Etruscan social conditions, in contrast to the simpler events described in the Mesopotamian omen literature. Noting how greatly the Figulan document departed from Mesopotamian texts, especially in the social realm, he questioned the date at which these passages were edited and whether the procedure had to be attributed to Figulus himself, but he did not try to explicate the mechanisms by which these allusions became embedded in the calendar.

The variety of the social and political events described in the *Brontoscopic Calendar* does greatly exceed that of any of its counterparts in the Mesopotamian libraries. The *Brontoscopic Calendar* encompasses "revolts of the common people" (στάσεις δημοτικαί, September 19) and sedition or general rebellion (στάσεις, March 23), which do not specify a revolt against a single ruler, but imply a deeper-seated condition of social unrest. As Ampolo noted, this *stasis* is an internal, social phenomenon, not a simple political act of opposition to a particular ruler, as would have been appropriate to the kingdoms of Mesopotamia.[21] Another Etruscan social development is διχονοία, "discord," that would have been generated by friction between different social classes.

Ampolo also noted the similarities to the Babylonian method of using the cardinal points to divide up regions dominated by enemies or allies (such as the West, *Amurru*).[22] Further, references in Mesopotamian texts to the king of the East resonate with the figure who appears in the *Brontoscopic Calendar* at July 16 and January 9. The Near Eastern omens certainly influenced the Etruscan *Brontoscopic Calendar*, but in reality, such a subject could also have been perfectly applicable to the situation seen by the rulers of Iron Age and Archaic Etruria. Because the brontoscopic document does not name a specific culture or city, the king may be understood as ruling some settlement to the east of the one using the calendar. Many cities, including even Volsinii or Chiusi, and the territories of the Umbrians and other ethnic groups, could have been interpreted as lying to the east of Rome, Caere, or Tarquinia. Chiusi, Orvieto, and even Veii might be considered east of Vulci, Caere, or Tarquinia, and Cortona, Perugia, and Arezzo lie even farther to the east. The East might symbolize the Apennines, the prehistoric homeland of so many peoples, viewed in the Late Iron Age as the refuge of primitives.

ARGUMENTS FOR ROMAN INFLUENCES

Some critics have noted supposed points of contact between the *Brontoscopic Calendar* and Roman historical events; if this Romanization were proven to correlate exactly with specific events and to have permeated the text, then its

[21] Ampolo 1990–91:189–90.

[22] Ampolo 1990–91:188–89.

authenticity as an Etruscan document would be severely compromised. The usual points are presumed allusions to the personalities and strife between Caesar and Pompey, for instance, as posited by Weinstock, at July 14 and 19.[23] But such predictions could equally refer to Sulla, say, and his life was linked with several historically recorded omens (see Plutarch *Sulla* 5.11.6; 11–13.37.2). The haruspex Postumius, attested only in literary sources, seems to have been his private adviser.[24] Neither the exact dates nor the character of the predictions, however, warrant linking any apodoses with Roman political events.[25] The period encompassing the Ides of March, for instance, does note "something unexpected" and also "ruin for men" (not on the Ides, however, but on March 17). On the 12th it predicts trouble for a powerful man in politics – but this is not the Ides, and this category of prediction can be found at several other dates (see Chapter 7). Further, there are other, frequently cited apodoses in the set, and nothing especially unusual; and if Caesar is the intended "powerful man" at September 12, then who is the "king" who appears, sometimes good, sometimes evil, at eight *different* dates? Would a single line have been interpolated and glaring discrepancies left in several other places? And if there is a reference to the fall of a popular politician, is it not possible that the date – or the remembrance of the date – of a political action was influenced by preexisting, genuine Etruscan hemerological texts? Would Spurinna's famous prediction have been recalled if it had been incorrect, and could such a haruspex have been referring to documents of the *etrusca disciplina*?[26]

A preoccupation with linking lightning to autocratic or tyrannical power may be seen in other divinatory literature, but the actual *Libri fulgurales* no longer exist: its category of *fulmina regalia* may be a link to rulers' power. Seneca (*Quaest. Nat.* 2.49.2) cites a definition given by Caecina: *regalia, cum forum tangitur vel comitium vel principalia urbis liberae loca, quorum significatio regnum civitati minatur*, "royal bolts strike the forum or senate-house or seat of government of a free city, their meaning is that the state is threatened by monarchy."[27] Servius offers an interpretation of lightning bolts that strike a *princeps civitatis* ("leader of a state") or a king.[28] The divine thunderbolt and its various purposes are separate issues but again illustrate an association between celestial events and the well-being of a state or community.[29]

Other reasoning has at times been adduced for denying the Etruscan, early origin of the *Brontoscopic Calendar*. For instance, Della Casa quoted the ritual response recorded by Cicero (*De divinatione* 2.42.14), which he said he culled

[23] Weinstock 1951:140.
[24] See Haack 2006:99–101, no. 75.
[25] So, too, Ampolo 1990–91:190–92.
[26] Cf. Haack 2006:110–12, no. 88.
[27] See Thulin 1906–1909:vol.1, 70–71.
[28] Servius *Ad Aen.* 2.649, cited by Piganiol 1951:83.
[29] See Simon 2006.

in nostris commentaris: *Iove tonante, fulgurante, comitia populi habere nefas*, which she compares to the prediction at August 3, "it threatens to the people trials and public slanders."[30] But the Latin Cicero was quoting actually says, "with Jupiter thundering, sending lightning, to hold a council of the people is impious" – something quite different! It appears that Rome, by the time of Cicero, and sometime earlier, to judge from the archaic language of the quoted line, already used a different kind of thunder-related documentation for civic affairs.

POINTS OF CONTACT WITH NEAR EASTERN PROTOTYPES

In the text of the *Brontoscopic Calendar*, the apodoses are always much closer in spirit, and in detail, to the apodoses of Mesopotamian divination texts than they are to the predictions of the later (Ptolemaic-Roman-Late Antique-Byzantine) *brontoscopia*. Whatever the time of the creation of the Etruscan *Brontoscopic Calendar*, its crafters were most definitely not consulting the Hellenistic or Roman-era documents, but they somehow had access to genuine Mesopotamian literature.

As demonstrated in Chapter 8, many of the apodoses used (often repeatedly) in the Etruscan *Brontoscopic Calendar* were formed under the influence of Mesopotamian omen literature, from several radically different sources, covering celestial, teratological, and other phenomena. Some odd, closer, similarities remain also – for example, the mingling of a prediction about the ruler with warnings about crops or other basics of life. In the Babylonian texts, we occasionally see apodoses like that of a *Babylonian Planetary Omen*: "In that year the king of Akkad will die, the crops of the land will prosper."[31]

APODOSES IN THE ETRUSCAN *BRONTOSCOPIC CALENDAR* DERIVED FROM MESOPOTAMIAN TEXTS

A significant proportion of all the predictions in the Figulus document may be traced to antecedents that are also represented in Mesopotamian divination literature, including tablet texts that are found in the libraries of Nineveh, Nimrud, and Babylon. Of course, the several iterations of translation, and our loss of the Etruscan and Latin originals, precludes a word-for-word or line-for-line comparison, but if we allow for slightly different wording, as much as three-quarters of the omens may be traced to Mesopotamian influence. (The

[30] Della Casa 1962:129.
[31] Reiner and Pingree 1981:Text III, line 13b; Chadwick 1989:187; Thompson 1900:195, r.6.

main discrepancies, those predictions invented to fit the society and climate of Etruria, are discussed in the next three chapters.)

Many aspects of the text show the influence of Akkadian and Babylonian divination, in outlook and meaning if not in exact terminology and syntax. This includes the use of thunder as a determinant (though it is nowhere else the exclusive trigger for a whole year), and the combination of two or three predictions for a single date. In the Mesopotamian prototypes, a list of calamity options may be joined to a single omen: "in the entire land there will be pestilence, epidemic among cattle, or famine" (*Enūma Anu Enlil* 51), or two seemingly unrelated events may be predicted: "there will be hostilities in the land, the harvest will prosper" (*Enūma Anu Enlil* 63). And some things all too often go hand in hand: "defeat of an army, an epidemic will occur" (*Enūma Anu Enlil* 24[25]). As in the *Brontoscopic Calendar*, the Mesopotamian texts often link barley or wheat "futures" with some other phenomenon: "there will be barley in the land and animals will 'fall,' a great army will be captured" (*Enūma Anu Enlil* 63). Many Mesopotamian apodoses sound familiar: "epidemic in the land" (*Enūma Anu Enlil* 16), "pregnant women will miscarry" (*Enūma Anu Enlil* 22), "in all lands there will be good peace" (*Enūma Anu Enlil* 22), "there will be guilt" (MUL.APIN, tablet II, iii, Gaps B1–B7, Marduk), "there will be massacres in the land" and "important persons of the land will die" (*Enūma Anu Enlil* 63). See Chapter 8 for full treatment of all these texts.

The assimilation process was thorough, for in the *Brontoscopic Calendar* there are no anomalies or anachronisms that do not belong in central Italy, so, for example, no mention that "sesame and dates will not be plentiful" (*Iqqur īpuš* 89, Du'uzu). Neither is reference made to "irrigated land," donkeys, crown princes, or the pests of Mesopotamia. Societies were different, and Near Eastern politics was alien to Italy, so no parallels for "king will send messages of hostility to king" (*Enūma Anu Enlil* 22), or "a *muškênu*-class person will seize the throne" (*Šumma izbu*, tablet 20, omen 42v). "Decrease of barley and straw" (as *Enūma Anu Enlil* 16, 63, etc.) is never reprised in the Etruscan document – straw is not an important commodity in Etruria as it was in Mesopotamia – but in the *Brontoscopic Calendar*, wheat and barley may be linked, as "the wheat [variant: the barley crop] in the field will not be abundant" (*Iqqur īpuš* 89, Nisan).

OUTLOOK, POETIC OR FORMAL; CEREMONIAL PHRASING

Charles Guittard, remarking on the matter-of-fact, authoritative tone of the *Brontoscopic Calendar*, noted the narrow vocabulary of its apodoses, with four terms recycled for "predicts" and only four to express prosperity of disaster: "Le style est monotone, sans recherche ni variation: quatre termes reviennent systématiquement pour annoncer les présages: ἀπειλεί, δηλοῖ, ἔσται, σημαίνει;

quatre termes expriment la prospérité ou les désastres (εὐθηνία, ἀφθονία, φθορά, ὄλεθρος)."[32] Bezold noted parallels in modes of expression between celestial divination documents like the *Enūma Anu Enlil* and other categories of Near Eastern literature, such as haruspical texts,[33] where stock terms abound, especially for downfall, destruction, calamity, and so on, as in the brontoscopic literature.[34] Through its Byzantine Greek filter, the *Brontoscopic Calendar* also uses a small set of formulaic terms for suffering or well-being.

Thunder as the means of divine communication may be common to many cultures, but in formalized literature, the Etruscan document is reminiscent of the Near Eastern ideologies. The Assyrian term for thunder, *šasû*, has the original meaning of "to shout, make a loud noise, exclaim, proclaim, announce," hence thunder is the voice of Adad. As such, his voice is prophetic and can also be active, as in "may Adad thunder furiously over his city and turn his country into a hill of ruins."[35]

"THE SAME"

Bezold and Boll found that the expression for "the same" appeared frequently in Assyrian texts; it also was favored in later documents rescued by Lydus and embedded in *De ostentis*.[36] The Etruscan *Brontoscopic Calendar* uses, at six dates, the phrase "the same" or "the same again." For June (11–12), November (26–27), and March (5–6), the preceding prediction is a common one, about weather or war, but June has an additional omen of "glad festivities" and the predictions of December (6–7), April (10–11), and May (27–28) are quite different, with a sort of moralizing aspect that is not often encountered in the Near Eastern texts. This must be part of the *etrusca disciplina*, with concern for moral issues and the status of humans in the eyes of their gods: other types of *libri* would probably have been consulted to furnish the rites for expiation.

Sample Apodoses in the Etruscan *Brontoscopic Calendar* incorporating "the same":

December 6. If it thunders, the men shall be visited with visions of the faces of the gods, they shall experience a bad outcome.
December 7. …, it signifies the same for all.
April 10. If it thunders, of upright men there will be advances.
April 11. …, it signals the same things.

[32] Guittard 2003:462.
[33] Bezold 1911:47–49.
[34] Bezold 1911:54n81: "Sturz, Unfall, Schädigung, Schmälerung, Zerstörung, Vernichtung, Trauer, Betrübnis, Bestürzung."
[35] *CAD* Š II, p. 151 = CH xliii 78, here, Old Babylonian Cuthean legend.
[36] Bezold and Boll 1911:24.

May 27. If it thunders, there shall be prodigies, and a comet shall shine forth.
May 28. …, it shall be the same.

One additional prediction, for May 20, says, "If it thunders, for those in the East, prosperity, but for those in the West, not the same." Such apodoses are common in the Mesopotamian literature, but this is the only such occurrence in the *Brontoscopic Calendar.*

"The same" is also used in other Etruscan cultic literature, where we see it in the Zagreb *Liber linteus.* There, sounding oracular in tone, is the formulaic phrase *cntnam thesan*, "[during] the same morning," used at three dates, including animal sacrifices, as "during this same morning a sacrificial victim for Veive has to be slaughtered," *cntnam. θesan. fler. veiveś. θezeri* (11.14).[37]

CHOICE OF TERMS

As in the Etruscan *Brontoscopic Calendar*, very many entries of the Mesopotamian documents are repeated over the course of the year, and they tend to use a restricted palette of vocabulary. Thus, for example, "the harvest" is always (or not) "plentiful" – the prediction is not varied with synonyms such as "bountiful," "abundant," or the like. Perhaps this mundane repetition enhanced the credibility of the document: clearly no literary license had changed the tone or veracity of any pronouncements.

Some terminology is difficult to understand, or possibly intended to be obscure and exude moral overtones, as, for instance, in the statement above, at December 6, to be repeated for the following day, "men shall be visited with visions of the faces of gods." Bezold found a similar pronouncement in the *Enūma Anu Enlil*, tablet Sin 19.4, which he translated: "wenn bei seinen Aufgang des ŠUDUN-Sterns 'Augen' nach Westen gerichtet sind, das 'Antlitz' des Himmels zu sehen ist und gar kein Wind weht, dan wird Hungersnot eintreten,"[38] "if, when the *ŠUDUN*-star rises, its 'eyes' are aligned to the West, the 'countenance' of heaven can be seen, and almost no wind blows, then will come famine." The "countenance of Heaven," then, is a dangerous thing to see.[39]

POINTS OF DIVERGENCE FROM MESOPOTAMIAN EXAMPLES

Many common expressions of the Mesopotamian texts do *not* appear in the *Brontoscopic Calendar* but, in contrast, are clearly featured in later documents that Lydus published, such as treatises on astrology. It is evident that

[37] van der Meer 2007:150; also 7.12 and 11.18 (van der Meer 2007:117, 151).

[38] Bezold 1911:56n88.

[39] One thinks now of the menacing eye that seems to peer out of the black roiling clouds of the Underworld, behind the fanged, demonic charioteer of the 4th-c. Tomb of the Infernal Quadriga recently discovered at Sarteano; Minetti 2004:pls. 25, 27b, 29.

any original Mesopotamian texts from which the Etruscan document was derived had been extensively adapted for use in Etruria. This is paralleled in the Figulan text's "colophon," which says the calendar "was only for Rome." Possibly the Etruscan original had specified that it was only applicable for a particular *Etruscan* city (such as Tarquinia), simply to show that it was no longer the same as any other literature. One example of a topic frequently found in the old Mesopotamian texts and in the Late Classical divination literature is the prediction, "the people will sell their children/daughters for silver" – this never occurs in the *Brontoscopic Calendar*, but it does appear in other Classical divination texts cited by Lydus. In the Near East, it refers to economic collapse or famine, when a family might have to sell a child into slavery to buy food. Etruscan Iron Age society probably had no comparable tradition, but later (Hellenistic or Roman) scholars lifted such comments from the Mesopotamian texts to which they had access.

The *Brontoscopic Calendar* is a seamless adaptation, with much new Etruscan material integrated into it. The most obvious point of departure is the calendar itself: the basic annual format resembles the Mesopotamian Standard Calendar, but the system of all protasis events being the same – the perception of thunder alone – with a wide variety of different outcomes scattered throughout the year is subtly different from the template of the well-known Babylonian or Akkadian texts. The only decisive factor in any given omen is its date – yet the Mesopotamian hemerological and menological texts are more slanted toward advising the enquirer on which days are good or bad for the activities he or she wishes to perform.

Basic categories for predictions encompass food supply, agricultural and agronomic prosperity, health and disease, war and peace, rulers and citizens, and civil unrest. Yet within these categories, the Etruscan apodoses offer greater variety in the topics relating to women, social classes including "slaves" (see Chapter 7), and even in the species of animal pests and diseases that could attack the community. The Mesopotamian omens do not offer multiple predictions for common people, or for slaves, nor do they note as many specific diseases, as does the Etruscan *Brontoscopic Calendar*. In diseases (spotted diseases, wasting, plague – lethal and "not life-threatening"), and in animal species cited, like blind-mice and voles, and sometimes "the reptiles," the Etruscan document covers only conditions applicable – and important to – Etruria. No Assyrian king would have tolerated omen writing that seemed to favor the lower classes, just as Roman authors would not have inserted comments on the fortunes of women and slaves.

A number of diseases of interest to Mesopotamian divination, such as epilepsy and leprosy, are absent from the Etruscan *Brontoscopic Calendar*. In Mesopotamian predictions, *kurusissu* rodents eat the flax crops, and weather menaces date palms and sesame plantations: none had Italian correlates in the Etruscan brontoscopy. (Lydus did translate "sesame" in several of the

Hellenistic-Roman texts he recorded, so the omission of the topic here must derive from the earlier and more authentic original used by Figulus.)

COMPARISON OF ETRUSCAN *BRONTOSCOPIC* OMENS WITH MESOPOTAMIAN TEXTS

APODOSIS CATEGORY: WANT, SICKNESS, FAMINE

Bezold and Boll cite numerous instances in the Nineveh tablets and also Lydus' sources, including the *Brontoscopic Calendar*, which cite hunger, famine, lack of necessities, onslaught of sickness, and even epidemics.[40] The same Greek words are used in the Hellenistic-Roman astrological texts preserved by Lydus, and the picturesque phrases sound very similar to those of the *Brontoscopic Calendar*. Thus, humans will suffer a lack of grain/food/daily provisions (σίτου ἀπορία, ἔνδεια φθορά, ἐλλεῖψις τῶν ἐπιτειδείων). Φθορά occurs at Lydus *De ostentis* 49.20; 53.5; 53.25, and so on. "Bounty" (of grain), namely, σίτου ἀφθονία, also was widely used (Lydus *De ostentis* 50.21, 50.23, 49.12, 48.23, etc.)

Other predictions are more exact in suggesting causes of the crop failure, such as worms that ruin the grain supply, probably inspired by such phenomena as the Akkadian *akilum* pest in the *Enūma Anu Enlil*,[41] as at June 6: "some sort of wild pest that has sunk deep into the crops." The main foodstuffs cited in Mesopotamian omen literature are oil and dates, sesame (seeds or oil), wine, and grain, especially wheat and barley.[42]

Many diseases specified in the Mesopotamian texts do not appear in the *Brontoscopic Calendar*. These include "epilepsy (*bennu*) will seize him," "in this year there will be *asakku*-disease" (MUL.APIN, tablet II, iii, Gaps B1–B7, Marduk), "there will be *rapādu*-disease" (*Enūma Anu Enlil* 63), or "the Lamaštu demon will rage" (*Enūma Anu Enlil* 51). But a number of simpler predictions of epidemic, plague, disease, or destruction by disease were carried over to the Etruscan document (see Chapter 6 for discussion).

APODOSIS CATEGORY: THE DOWNFALL OF ONE IN POWER

All divination literature seems to include some predictions about the downfall of rulers or of citizens in powerful positions. These situations include the defection of a king's subjects or vassals, an uproar in the land or an outright revolt, the king falling into the hands of his enemies, the crown prince usurping/being threatened/winning/losing, or an outright usurper taking power, and perhaps then experiencing downfall. Bezold and Boll noted a sample of

[40] Boll and Bezold 1911:18.

[41] Bezold and Boll 1911:19.

[42] Bezold and Boll 1911:2. In the tablet texts, "barley" is such a staple that modern translators often render it simply as "crop" = *CAD* Š, vol. 2, 260–61, "crop" or "barley and flax."

the cases in which the Classical literature recorded by Lydus has preserved this.[43] The *Brontoscopic Calendar*, however, omits all potentially anachronistic or extraneous characters or situations: there is no mention of a palace, crown prince, usurper, or such inappropriate to a society of Mediterranean farmers and warriors with basic central organization (Chapter 7).

APODOSIS CATEGORY: CHILDBIRTH, REPRODUCTION

Of course, such family affairs as sickness, marriage, and childbirth were of interest to anyone in the ancient world, but in other literatures, we do not meet with the same expressions that we see in the Mesopotamian texts (e.g., *Enūma Anu Enlil*, Adad 3.24), and also in the *Brontoscopic Calendar* (at just two locations: "women in labor will have an easy delivery, but there will be abortion of cattle," June 2; "good deliveries for women," February 11).[44] Difficult births and spontaneous abortions are also covered with some regularity in both the Mesopotamian literature and the later texts embedded in *De ostentis*. There are just two, slightly different, entries in the *Brontoscopic Calendar* (June 2 and February 14, "loss of progeny").

APODOSIS CATEGORY: ANIMALS, GOOD AND BAD

Many citations in Mesopotamian texts are reprised in the Greek of the later documents that Lydus rescued and are also found in the *Brontoscopic Calendar*. The character of the apodoses is usually negative, and reasonable growth or prosperity is otherwise taken for granted. Many citations predict poor growth of livestock. The death or wasting of livestock by disease is picturesquely rendered as "a falling down" (πτῶσις) or "ruin" of the flocks (Assyrian *miqtim būli(m) (sīri)*), as at July 1 in the *Brontoscopic Calendar*. Bezold and Boll offer numerous examples that pair the Akkadian of the tablets with Greek expressions from the various documents in *De ostentis*.[45] In them and in the *Enūma Anu Enlil*, there is also the demise of "the beasts of the field," "the offspring of the cattle and sheep," and the cattle and horses – but nowhere are horses cited in the *Brontoscopic Calendar*. The poetic descriptions of birds as "the winged ones" and animals as "the four-footed" do recur in Lydus' other translated divinatory documents but are less frequent in the Mesopotamian sources.

In many *omina*, the fortunes of fish and birds (no doubt mentioned only for their value as food) are linked.[46] Of course, citations of both sea and river fish are eminently appropriate for Etruria. From a modern European or American vantage point, we may find mention of the sea and saltwater animals out of

[43] Bezold and Boll 1911:14–16, citing later texts in Lydus *De ostentis* 49–51.

[44] See also Bezold and Boll 1911:17–18.

[45] Bezold and Boll 1911:22–23.

[46] Bezold and Boll 1911:21–22.

place in Mesopotamian documents, but of course, many regions in antiquity were linked to the seacoast and foreign commerce. Further, any sort of Levantine mediation in the process of transmitting the divination calendar would have occurred in cultures dependent upon seafaring, and thus they would have been ready to accept such predictions. Again, somewhat poetic expressions describe the downfall, loss, and so on of the sea creatures, or a "lack of seafood" ("fish of the sea," ἐνάλιοι ἰχθύες, January 28).[47]

Wild animals occasionally attack men, but the picture of "lions and wolves running in the city," as proclaimed in the Mesopotamian texts (and some later brontoscopies but not the Figulus text), seems allegorical, like a reference to revolt. Wild boars, pigs, and dogs are also cited in Mesopotamian literature,[48] and they were common in Etruria, often depicted in art of all types (see Fig. 19), yet they do not appear in the *Brontoscopic* text – presumably the aristocratic boar hunt, like the horse, increasingly emphasized through the Archaic period, was not of concern to those who composed the *Brontoscopic Calendar*. Reptiles, τὰ ἑρπετά, as cited in the *Brontoscopic Calendar*, are also named in several Mesopotamian *omina*, where they should always be interpreted as snakes, even poisonous snakes. In some of Lydus' other, later, Greek texts, they are qualified as "wild animals and snakes," θερίων καὶ ἑρπετῶν.[49] (For environment, agriculture, and fauna, see Chapter 5.)

APODOSIS CATEGORY: THE REPTILES/SERPENTS/SNAKES

"The creeping things," usually translated "reptiles," are a curious feature of the Etruscan *Brontoscopic Calendar*. Section 58 of the Mesopotamian *Iqqur īpuš* gives "snakes" as the topic of predictions; for instance, "if a serpent sees a man/falls from a roof, in front of/behind a man, if a man sees a serpent." This is not identical to the Etruscan *omina* in which predictions are given concerning "the reptiles" or "poisonous snakes," but certainly the serpents present in Mesopotamian literature had some influence on the eventual Etruscan text. Labat has indicated that some at least of the snake-related protases were shared with the *Šumma ālu*.[50] He also explained that in Mesopotamian ideology, this would be a magical situation, for the use of the faculty of sight was an aggressive act, and "seeing" someone could be tantamount to possession, thus it was significant which of the parties, the snake or the man, saw the other first. The Assyrian term for snake is *sîru* (Sumerian ideogram MUŠ), and a *mušhullu* (ideogram MUŠ.HUL) meant "dragon." Sample predictions from

[47] See Bezold and Boll 1911:21.

[48] See Bezold and Boll 1911:19, and note 2.

[49] Bezold and Boll 1911:22.

[50] Labat 1965:124–27, tablet sources at 43.

Iqqur īpuš, section 58, are omens for Nisan and Ajar, and Shebat: "If at the beginning of the year, in Nisan, on the first day, or in the month of Ajar, the first day, any time during the day or night, a serpent sees a man: this man will die in the course of the year."[51]

APODOSIS CATEGORY: WIND, WEATHER, AND DIRECTIONS

The Mesopotamian divination texts predict weather, with a variety that was not approached by the Etruscan *Brontoscopic Calendar*, but several points of comparison are evident. The main categories in Mesopotamia are clouds/mist/overcast/darkness, rain and rainstorms, wind and windstorms, thunder and lightning, and earthquakes and mudslides. The thunder-omens linked to lunar eclipse phenomena in the *Enūma Anu Enlil* in some cases could been copied from the *Iqqur īpuš*.[52] The *Brontoscopic Calendar* includes a number of predictions of cold weather and winter travails that do not have close correlates in Mesopotamia.

Winds are powerful forces in several omens of the *Brontoscopic Calendar*, but their direction is not specified (in contrast to many Mesopotamian omens).[53] There are few traces of the designation of compass points or the four corners of the known world. The omen at May 29 predicts a "northern war," which could refer to many Etruscan or Italic localities. At May 20, "If it thunders, for those in the east, prosperity, but for those in the west, not the same" is reminiscent of the binary system of many Mesopotamian omen qualifiers. There is only one other reference to the west, at November 7, which states, "for those who are in the west, both humans and dumb beasts, diseases." But the east is clearly the favored direction, with four more omens, two citing a "king of the East," July 16, and "the king of the East shall be in danger," Jan. 9. At November 11, "men shall give blessings to the god, for the wind shall blow out of the east," and May 14, "there will be eastern war and great want."

Ampolo noted a Near Eastern parallel for the king of the East, namely, the king of Amurru,[54] but in most Etruscan cities, any citizen could probably have reeled off multiple places situated in the East. Not many polities could have been described as *west* of maritime Etruria, unless one counted Sardinia, Elba, or Sicily, but perhaps that meant that no one would suffer the lack of prosperity implied at May 20.

[51] See Labat 1965:124–27.

[52] See *Enūma Anu Enlil* 16 and *Iqqur īpuš* 88–91, also *Enūma Anu Enlil* 22, pt. 2, omen 2; Rochberg-Halton 1988:61.

[53] See Rochberg-Halton 1988:57–60; Bezold and Boll 1911:20.

[54] Ampolo 1990–91:188–89.

OTHER UNUSUAL EXPRESSIONS

One curious reference in the *Brontoscopic Calendar* occurs at November 24: "If it thunders, the watch post shall fulfill for the state good service against enemy tricks." This is difficult to place in a meaningful Etruscan political or social context without knowing much more. However, there might be a Mesopotamian analogy for it. As noted by Chadwick, the Assyrian kings maintained an extensive network of observation posts for celestial observation and military intelligence, whose professional staff sent hundreds of letters to the palace.[55]

CIRCUMSTANTIAL EVIDENCE: THE MILIEU IN WHICH THE *BRONTOSCOPIC* TEXT WAS CREATED

Recent studies of some late eighth- to early seventh-century Akkadian tablet texts may offer a parallel for the creation of the Etruscan *Brontoscopic* text. Ann Guinan (2002) has indicated an intriguing set of convergences between the theoretical omen compendia, especially the *Šumma ālu*, and a series of tablet texts of historical import, the *Prodigies of the Fall of Akkad* (CT 29 48–49), from the library of the great Assyrian scribe Nabû-zuqup-kēnu of Kalḫu (Nimrud) that had been removed to Nineveh. They refer to the destruction of Babylon ("Akkad"), probably during the campaigns of Tukulti-Ninurta (1243–1207 BC). The *Prodigies* may be associated with tablet 88 of the *Šumma ālu* series; they indicate that unusual or unnatural phenomena preceded the destruction of the city. The events presaging catastrophe are, in Guinan's words, "spooky images of horror and desolation: gates creak, moaning sounds come from the earth, curtains of a temple assume shapes, blood oozes from the ground."[56] Obviously, the thunder booms of the *Brontoscopic Calendar* are not the same as the protases of this series, but the precedents are clear for linking (un)natural phenomena to political and social change, up to and including the downfall of cities and kingdoms. The *Prodigies* show that the scholars of Neo-Assyrian, Sargonid Mesopotamia were accustomed to interpret historical situations in the light of the recorded, "theoretical" omen compendia.

Nabû-zuqup-kēnu (ca. 750–680 BC) is an interesting figure whose career coincidentally parallels the development of the Orientalizing phase of Etruscan material culture, during the reigns of Sargon II (721–705) and Sennacherib (704–681).[57] Colophons on his tablet sets enable scholars to trace his academic interests, from the *Enūma Anu Enlil* to the *Šumma ālu*, and *Prodigies of the*

[55] Chadwick 1989:188.

[56] Guinan 2002:12.

[57] See Robson 2008:145–50, and Guinan 2002:12–17.

Fall of Akkad, ending with numerology. Although Nabû-zuqup-kēnu probably never left Nimrud, his career furnishes us with a model of the scholarly milieu of Sargonid Mesopotamia at a crucial time for the spread of Oriental culture through the Mediterranean. Just one such person, armed with a library of tablet texts, could have inspired some inquiring minds from one of the Etruscan proto-cities, and a great textual tradition could have begun. A short list of the preferred texts/extracts would be: *Šumma ālu*, *Šumma izbu*, *Iqqur īpuš*, *Enūma Anu Enlil*, and perhaps the MUL.APIN (see Chapter 8). The tradition of haruspicy, probably already practiced in the pastoral communities of prehistoric Italy, could at this time have received an injection of Mesopotamian textual material.

CONTRASTS WITH LATER *BRONTOSCOPIA*

The *brontoscopia* generated by Ptolemaic and Roman or Byzantine scholars subsequent to the Figulan *Brontoscopic Calendar* all betray their post–Iron Age origins in references to the zodiac and by expressing the territorial concerns of the Hellenistic and Roman kingdoms. Even when they furnish some parallels to the Etruscan *Brontoscopic Calendar*, their internal heterodoxy distances them from it. One function of the *Enūma Anu Enlil* was to offer meteorological predictions, which became major concerns in Hellenistic and Roman astrological literature (e.g., Ptolemy's *Tetrabiblos* 2.11–13). Columella (*De re rustica* 11.1.31) noted this type of documentation, but he disagreed with it; Pliny (*Nat. Hist.* 18.57.207ff) actually attributed the field to the Chaldeans, the itinerant diviners-for-hire who had tainted the discipline, in many Romans' opinion. The disdain of Roman scholars for Chaldeans implies that, in the intellectual milieu of first-century AD Rome, a divide already was perceived between for-profit diviners and traditional priests of Etruscan religion. The *Brontoscopic Calendar*'s freedom from astrological taint[58] is one more hint of its greater antiquity.

Even the next oldest text, the Qumran papyrus, relies on the zodiac, showing that it is not a Jewish religious document but an instrument of scholarly divination, created within the Hellenistic tradition of philosophical writings. It cannot be said to derive *directly* from the Figulus document, nor can any of the others written in Latin or Greek – yet one wonders what stimulated the abrupt but copious start of such literature within one or two generations of Nigidius Figulus. Many later *brontologia* mirror some predictions found in the Figulus text, such as, for instance, Athens Codex 10, attributed to Heraclius (see Chapter 9), but all have some telltale (like "destruction of Arabs") betraying derivation from Hellenistic Near Eastern divination literature.

[58] Thesleff (1965:47) calls this "'Chaldean' horoscope astrology."

Clearly, the Figulan *Brontoscopic Calendar* differs intrinsically from later *brontoscopia* – and precedes all of them. Although many of its apodoses parallel those found in Mesopotamian tablet texts, the use of thunder as the sole trigger sets it apart. Mesopotamian literature has few simple links to thunder: most of the *Enūma Anu Enlil* "Adad" series encompass weather conditions that only occasionally include thunder. Only the Etruscan document is keyed purely to the occurrence or nonoccurrence of a thunder noise alone. So the organization of the document is at its core different in perception from Mesopotamian omen literature, and the apodoses themselves are never outré but are always plausible for the climate and society of Iron Age Italy. In fact, during the Iron Age period of heightened cosmic rays, excessive rain, and cyclonic storms, Etruria was more likely to embrace such information.

If the first-century brontoscopies attributed to Fonteius Capito (and perhaps the Augustan Claudius Tuscus) differ so radically in tone and zodiacal overlay from the Etruscan *Brontoscopic Calendar*, one obvious reason is that, even if they shared some remote Mesopotamian sources, they were formulated much later, in an atmosphere of Hellenistic philosophical enquiry and rationalization.

POSSIBLE BYZANTINISMS

The best argument for Byzantine redaction or tampering with the *Brontoscopic Calendar* text lies in a very few terms that may be recognized as Byzantine, post-Classical terminology. The most obvious is the term "the queenly city" at June 29, also used by Lydus at *De mensibus* 3.14, and elsewhere. The epithet could always be applied to either Rome or Constantinople.[59] Because Figulus specified that the calendar was applicable only to Rome, the original city seems the likeliest meaning – and in reference to a putative Etruscan original, this would have designated whatever city the text was composed for, whether Tarquinia or even Ruma.

Other Byzantinisms may be explained by flaws of translation: the term δῆμος would mean simply "the people, the citizen body, the state" – if it were pronounced in Periklean Athens – and eventually "the commons." But by the Byzantine period, it had acquired a more specific connotation of "factions" – the people involved in the affairs of the city, and those likely to engage in political action or even urban warfare. For now, we can only assume general meanings for most terms.

ETRUSCAN AFFINITIES: *PROPHECY OF VEGOIA* (APPENDIX A, TEXT 2)

The short text preserved under this title is highly relevant to the reading of the *Brontoscopic Calendar*: it survives in Late Republican Latin, because it

[59] See Ampolo 1990–91:192; *Th.L.Gr.*, s.v. *basilis* and Sophocles *Lexikon*, s.v. *basilis*.

was copied into the documentation of the professional guild of surveyors (*agrimensores*). Like the other long Etruscan texts, the Capua and Zagreb ceremonial calendars, it establishes its own authority with the incipit "this [was told] from Vegoia to Arruns Veltymnus." Cognizant of Roman reforms of the second to first centuries BC, many scholars believe it was a Roman forgery – but there were still men in the Roman scene who knew the *etrusca disciplina*, and at least a few who, like Figulus, could read Etruscan, so it seems unlikely that a brand-new document could have passed muster at that time. It seems more likely that an old, presumably Archaic, text was brought out and quoted or perhaps edited to furnish Vegoia's prophecy, all that survives of a set of "books," the *libri vegoici*, stored in the archives of the Palatine Temple of Apollo.

BOUNDARIES, DISEASES, AND SOCIAL CLASSES

The cosmogony of the Vegoia prophecy harbors elements similar to Near Eastern texts; the second part, the woes predicted for boundary cheaters (diseases, wounds, weakness, storms, whirlwinds, rain, hail, heat, mildew, civil unrest) are all evident in the Etruscan *Brontoscopic Calendar*. The Vegoia document, even if it was forged around 90 BC, predates Figulus' translating the brontoscopic text and establishes the existence, *prior* to the events of Caesar and Cicero, of such texts in Rome and central Italy. The striking warning of "the most awful diseases and wounds" prompts comparison with the Etruscan omen for October 20: "If it thunders, it warns there will be a festering wound, and for the many, extreme suffering out of the discord."

Some authors have seen in the Vegoia prophecy an Etruscan "nucleus" with references to a society like that of third- to second-century Etruria, characterized by conflict between *domini* and slaves.[60] As with the *Brontoscopic Calendar*, however, we are hampered by the lack of an Etruscan original. The significance of boundaries and limits, both geographic and chronological, is evident in the Perugia *cippus*; many Etruscan boundary stones exist: they were simply marked *tular* ("boundary") but represent formal religious rituals.[61] Even the carefully delineated cells of the Piacenza liver (Fig. 12) reflect this territorial symbolism.[62] The Vegoia prophecy, like Figulus' *Brontoscopic Calendar*, is an example of Roman use of extant Etruscan scriptures to support later political activity: the survival of both documents is probably due

[60] See Harris 1971:31ff.

[61] A marker buried at Marzabotto has surveyor's crosshairs. For basic references on *tular* inscriptions and land contracts, see Manthe 1979; *Scrivere etrusco* 1985; Agostiniani and Nicosia 2000. For the Perugia *cippus*, see *ET* Pe 8.4, Facchetti 2000:9–58.

[62] See Pfiffig 1975:112–15, as emphasized by Ingrid Edlund-Berry 2006:118–19. The basic study of the bronze liver is van der Meer 1987; see also de Grummond 2006:44–51. For Latin literary background, see Patterson 2000.

to such usefulness, when other categories of literature (e.g., *libri Acherontici*) have perished outright.

TRACES OF ETRUSCAN FORMULAE IN EVENTS OF THE FIRST CENTURY? *DE HARUSPICUM RESPONSO*

There may be hints of pre-Figulan Etruscan scriptures inadvertently preserved by Cicero in his speech *De haruspicum responso*, where the brief comments of the haruspices themselves, although rendered in Latin, appear in an oracular style alien to Cicero the author. (See Appendix A, Text 3; and Chapter 2.) The omen of a loud noise in the heavens over Latium, treated as a revelation, could be a thunder event. Many categories of concern in the *Brontoscopic Calendar* appear either in the haruspices' words or Cicero's interpretation: crops and provisions, slaves, commoners or lower classes (*deteriores*), the "rule of one man" and secret designs, plots and assassinations. Cicero quoted the actual response, noting, *nihil addo de meo*, "I add nothing of my own." On the topic of slaves, a cynical remark by Cicero seems to refer to the very genre of the *Brontoscopic Calendar*:

> Si examen apium ludis in scaenam caveamve venisset, haruspices acciendos ex Etruria putaremus: videmus universe repente examina tanta servorum immissa in populum Romanum saeptum atque inclusum et non commovemur? Atque in apium fortasse examine nos ex Etruscorum scriptis haruspices, ut a servitio caveremus, monerent.
>
> Had a swarm of bees come upon the stage or into the auditorium at the games, we should think it necessary to summon soothsayers from Etruria; and do we feel no alarm when not a man of us but sees such numerous swarms packed within the walls of a building? Indeed, were a swarm of bees actually in question, the soothsayers might perchance warn us, after references to their Etrurian books, to beware of our slave population.[63] (*De har. resp.* 12.25–26.1)

Compare the *Brontoscopic* entries at August 19, January 7, 15, and 25; Cicero in his speech refers to the revolts of Athenio and Spartacus, still chilling in Roman memory. It would seem that he had in mind the predictions of the *Brontoscopic Calendar* in order to make sarcastic reference to its frequent predictions of servile revolt and civil unrest.

Other common apodoses of the *Brontoscopic Calendar* fit Cicero's next interpretation (*De har. resp.* 14.31): movet me etiam in hac caritate annonae, sterilitate agrorum, inopia frugum religio Telluris et eo magis, quod eodem ostento Telluri postulio deberi dicitur, "my own devotion to the goddess Earth

[63] Translation of Watts 1923:348–49.

moves me, in this time of the high cost of grain, failure of harvests and scarcity of food, and even more because it is said that this omen indicates arrears in the offerings to Earth."[64] The high price, namely, a poor harvest (of wheat), crop failures, and scarcity of daily provisions are recurrent elements in most omen literature.

One of the sharpest points of reference to the Etruscan brontoscopic (or related) literature comes in Cicero's analysis of the social situation. His description of the "baser, rejected" elements cited by the haruspices sounds like the landless urban poor, familiar in first-century BC Rome, but there were already potential problems in the Archaic Etruscan cities, where urban laborers separated from the land and the patronage of nobility, and they would have been subject to hard times. At least thirty-seven entries, tallied conservatively, warn of discord, civil unrest, civil war, or slave revolt, or in some way cite the welfare of "men of lowly degree," "the common people," and so on (see Chapter 7). Cicero was dredging up ancient references with which to accuse his political enemies, particularly Clodius Pulcher: his assumption of the audience's familiarity with "Etruscan books" provides further circumstantial support for the use of Etruscan scriptures that were acknowledged to be of venerable age by the first century BC. Surely the authority for the haruspices' reply to the senatorial summons rested in their thorough research into the religious writings.

The parallels are too many to be coincidental between the *Brontoscopic Calendar*'s predictions and the brief passages Cicero has vouchsafed us from the haruspices' responses. We know of no other document (extant or known by title) that could have been consulted by the haruspices on the subject of celestial noise. Was Figulus perhaps prompted by the thunder events of 56 BC to make that text more accessible to his fellow citizens?

FEATURES OF THE TEXT: VOCABULARY, PHRASES, SYNTAX?

Few appropriate texts exist for comparison with the *Brontoscopic Calendar*, for it is no longer in its original Etruscan, nor in the Latin of Figulus, and so cannot be compared closely in vocabulary, syntax, and such with the Capua tile, Zagreb *Liber linteus*, or *Prophecy of Vegoia*.

Two terms in the Zagreb *Liber linteus* are paralleled in the language of the *Brontoscopic Calendar*: the *Liber linteus* alludes to "living beings" (*sveleri*, cf. the verb well known from tombstones, *svalce*, "he lived")[65] and to "crops" (*meleri*, from *mele*, "fruit"),[66] which was likely the original Etruscan term,

[64] Cf. Watts 1923:356–57.

[65] *Sveleri* appears at *Liber linteus* 2.8 and 4.4; see van der Meer 2007:67, 87.

[66] *Meleri* is cited at *Liber linteus* 4.4, and 4.17; see van der Meer 2007:84, 87, 90.

now rendered by the calendar's Greek καρπῶν, except in cases where "hard fruits" (ξύλινα κάρπα) or "tree fruits," namely nuts, are specified. The terms appear together in *Liber linteus* 4.4 and 4.17 in a longer formula that certainly holds cosmic/religious overtones:

> cisum. pute. t[ul. θans] hatec. repinec. meleri. sveleric. (4.3–4)
> cisum. pute. tul. θans. hatec. repinec/ meleri. sveleric. (4.16–17)

This probably said something like "a triple liquid(?) in/with a vase(?) lift up(?), O priest, for prosperity and well-being, for the crops and for the living beings."[67]

Some scholars have sifted through the Roman festival calendar for points of interest to putative Roman forgers; they find such holidays as the *Parentalia*, celebrated for about a week beginning on February 13 (at the sixth hour) and memorializing the family's dead.[68] It ended with the *Feralia* on February 21, in which sheep were sacrificed among other offerings for the dead. On February 12, the *Brontoscopic Calendar* predicts, "If it thunders, it threatens frequent death and unseasonable winds," hardly veneration for beloved family members. Many other significant dates in the Roman calendar have no possible corollaries in the *Brontoscopic* text. Although a wolf omen appears at June 9 (flocks overrun by wolves), there is no note of wolves running in the streets on or near February 15, the *Lupercalia*, and similarly the other major Roman festivals are not signaled even metaphorically in the *Brontoscopic Calendar*.

Some other Roman festivals have elements of primitive agricultural and pastoral cults, such as the *Robigalia* (April 25)[69] or the *Liberalia/Agonalia* (March 17).[70] The *Brontoscopic Calendar* indeed mentions grain crops destroyed by damp or wet weather (exclusively in June [4, 8] and July [7, 11, 18]), but on or near April 25 no omens touch upon either rust or grain. Likewise, the *Vinalia* of April 23, although Ovid links it to the death of the Etruscan king Mezentius and to first-fruits offerings, has no correlate in the topics of the brontoscopic omens. Wine is mentioned only twice in 360 days (scarcity at June 21 and October 7) and nowhere near the familiar Roman date.

Ilaria Domenici is also of the opinion that the *Brontoscopic* omens describing agricultural conditions, famine or plenty, and so on are too commonplace and repetitive to have been tampered with by Nigidius Figulus, and they thus constitute a faithful translation from the original Etruscan text.[71] She sees a difference, however, in the terminology of the predictions of social and political

[67] See van der Meer 2007:84, 87.
[68] See Scullard 1981:74–76; Fowler 1989/1969:306–10.
[69] Scullard 1981:108–10.
[70] Scullard 1981:91–92.
[71] Domenici and Maderna 2007:29–30.

import, particularly in the layered effect of several different terms for rulers and officials. The person in charge is called a "ruler" (δυνάστης at June 13, Sept. 26, Oct. 19, Jan. 8); at September 8, March 12, and April 19 he is "a powerful man in politics" (δυνατὸς τοῦ πολιτεύματος); other players are called "rulers" (κρατῶν, June 27, Oct. 24, Nov. 10, Jan. 26, March 7) or ἄρχων (at July 5 and 17). At October 19 there is a juxtaposition of both ruler and king, when thunder "warns the downfall of a ruler or the overthrow of a king, but it warns both discord among the common people and abundance." In Mesopotamia, separate terms would have acknowledged the tiered systems of multiple kings, vassals, and "great king," but for Etruria it could refer to two rival cities or territories. An outright king, βασιλεύς, is cited at several dates (July 16, Oct. 19, Nov. 16, Dec. 27, and January 9, 16, 19, 21), but a τύραννος (Sept. 25, Oct. 1) might have represented the "usurper" of the Near Eastern texts, rather than the popular demagogue that "tyrant" meant in Greece. There are also "powerful men" (δυνατοί) so frequently mentioned that they seem like the dominant political force (June 19, 24, Sept. 4, 25, 27, Feb. 16, March 30, April 9, 24), such as a council or oligarchy of aristocrats. In the stratified society of the calendar, "nobles" (εὐγεννείς) are separately designated (Aug. 24, Sept. 11, Oct. 17). At March 12, a general (στρατηγός) is the binary protagonist opposite the "powerful man in politics," but at November 28, a council (συγκλήτου) is implied when "many of the councilmen of the wealthier rank shall be ruined utterly by cowardice." There are also factions composed of commons, slaves, and so on, which Domenici and many others have read as first-century intrusions (or Figulan whole cloth) – but a case can be made for a stratified society of this complexity as early as the eighth century BC (see Chapter 7). Domenici reads in the "layering" of different terms – always assuming that Lydus' choice of Greek terms rendered the diversity of Figulus' Latin vocabulary – evidence of the multiple redactions and outright alterations that the text underwent over a long period of formation.[72]

Rare hints of the language of oral rituals and Archaic religion occasionally surface in historical accounts of early Rome. For instance, the phrases recorded by Julius Obsequens (*Prodigy* 20) echoing Livy's history (7.6.1–6) for the year 359 BC, sound reminiscent of the *Brontoscopic* predictions of disease. Describing the affliction caused by the opening of the Lacus Curtius sinkhole, Livy says: *vaporibus pestilentissimis multorum hominum corpora*, "with the most virulent vapors affecting the bodies of many men." Naturally, disease would infect *the bodies* of men, yet this resembles the language of some *Brontoscopic* apodoses. Compare the calendar's June 1, when "dangerous diseases will be inflicted upon bodies," or July 22, "but for the bodies, diseases around the head," and August 13, "plague upon the bodies of both humans and dumb animals." The purpose of the Roman account is to note

[72] Domenici and Maderna 2007:29–30.

the means of expiation (in this case the sacrifice of Marcus Curtius, which brought immediate cessation of plague). It is not a formulaic text like the *Brontoscopic Calendar*, but traces of ceremonial statements linger.

OTHER RELIGIOUS TEXTS IN ITALY: *IGUVINE TABLES*

Specialized, poetic terminology appears in the *Brontoscopic Calendar*, as in other religious texts of Italian origin. We do not meet the philosophical term "mortals" (βροτοί), as in Homer and Hesiod: instead, men are δῆμος, or κοινή/κοινοί when understood as "the community"; occasionally "the people" is translated ἄνθρωποι (Nov. 14 and 30, where they "gently undo" snakes). A binary system shows humans as "gifted with speech" (λογικοί) as opposed to "dumb animals," who are ἀλόγοι (Aug. 4). Animals get poetic appellation: livestock are "the four-footed" (τετραπόδοι, Aug. 8), and birds are τὰ πτηνά, "the winged ones" (June 15). The last term also appears in the *Iguvine Tables*, a set of requirements for sacrifices and other rituals, inscribed in Umbrian on seven large bronze plaques discovered in 1444 in Gubbio. In part they record laws for the priestly college of the *Fratres Attiedii*, and have at times been compared to a breviary – they are not composed in polished, "literary" language or style. Although written in the Umbrian language, some tablets use an Etruscan, retrograde alphabet and others a Latin hand, running left to right. Monumental copies of much older documents, they were made at various times from the late third through the early first century BC.[73]

The formulae of the *Iguvine Tables* tend to drop some terms and verbs that worshippers would have understood as implied in rituals, and this type of brevity is also seen in the *Brontoscopic Calendar*. Most of the Umbrian texts incorporate directions for sacrifices of oxen, heifers, sheep, piglets, and dogs; several order observation of birds before the ritual can commence. Tablet VIa, in Latin alphabet, calls on [Jupiter] Grabovius to "purify the warriors, the *curiae*, the heads (of family), the beasts, the fields, the fruits."[74] The sacrifices were begun by observation of birds in relation to the city boundaries: officiants were to speak out if they saw a hoopoe, crow, woodpecker, or magpie (all wild songbirds) during the augural ritual.[75] Although the tone, language, and character of these texts are quite different from the *Brontoscopic Calendar*, many of the basic aspects of life and worship cited in the *Iguvine Tables* mirror elements of the brontoscopic predictions and show many affinities with the more nearly contemporary ritual calendar of the Zagreb *Liber linteus*.[76]

[73] See Ancillotti and Cerri 1996A and 1996B for translations in Italian and English; also Wilkins 1994; Prosdocimi 1984; Pfiffig 1964; Poultney 1959; Devoto 1948.

[74] Ancillotti and Cerri 1996B:73; Beard et al. 1998:vol. 2, 8–10.

[75] Ancillotti and Cerri 1996B:72; cf. Ancillotti and Cerri 1996A:99–100.

[76] For environment and daily life in Umbria, see Ancillotti and Cerri 1996A:93–128.

VIEWPOINT, ACTUAL USAGE OF THE ETRUSCAN *BRONTOSCOPIC CALENDAR*

Whereas the flight of birds, appearance of clouds, or configuration of a victim's entrails can be observed in public, as scheduled rituals, thunder cannot be summoned (save by a Numa or Porsenna). A storm blows up without any human interference, and often in Italy without much warning, as the violent and beautiful summer electrical storms seen today over the Bay of Naples.

The affair of the haruspices' responses (Appendix A, Text 3) may bring us as close to the truth as possible: experts present themselves as prepared to offer an interpretation, and they base their responses on special literature, which they alone have been trained to interpret. (Compare scholars' responses to the Assyrian king, Appendix C.) This is different in type from the *Brontoscopic Calendar*, though – the heavenly noise omen of 56 BC was really a communication to warn mortals to correct and expiate their misbehavior, the message strictly a moral one. In contrast, the brontoscopic text is simpler: if thunder happens, the following physical conditions will or can result; few entries imply any sort of limitation (e.g., August 30: it threatens diseases in the city over which it [the thunder] is cast down). In Mesopotamia, too, dire omens could be countered by rituals of atonement.

Very curiously, there is no indication whatever in the Figulus document of directionality, of the exact source of the thunder having any bearing on the predictions, or even being noted. This fits Seneca's and Cicero's criticisms of Etruscan divination, that it was not cataloged according to variables for ease of interpretation. In contrast, many later *brontoscopia* are qualified with conditions like day versus night, or geographical region. Seneca, Lucretius, and others all seem concerned with the source of thunder and lightning, and with its mechanism, whether celestial or environmental.

Other moralizing omens appear for August and April:

> August 23. If in any way it should thunder, it signifies that the lightning bolt shall fall, and warns of slaughter.
> April 27. If it thunders, omens from the sky incredibly shall be revealed.
> 28. …, by shields the people shall be saved.
> 29. …, a zephyrus will prevail.
> 30. …, a shower of good things.

The lightning bolt must recall the doctrine preserved piecemeal by Pliny and Seneca: a warning bolt is sent to change mortals' behavior before it is too late. The April predictions – an omen foretelling the appearance of another omen – are unusual, and the "shields" is obscure: the original Etruscan term must have been difficult to interpret. April 28 was the period of the *Floralia* festival at Rome and included the moveable "Latin" festival, a major holiday, but there is no detail to anchor it to anything Roman. The Arval Brethren were involved in these festivals, but the Salii, whose cult had the famous sacred shields, the Ancilia, are linked to quite different dates, such as March 1 when the original Shield of Mars fell from the sky or was discovered in Numa's house.[77] Evidence is increasing to show the Etruscan origin or early shared worship within such cults as the Salii, but the *Brontoscopic Calendar* is not bound to a specific cult. And again, there is no link in date between the historical festival calendar of Rome and these cults.

If the viewpoint of the brontoscopy is not that of Roman cults, neither is it an ideological match for Mesopotamian thought and belief. The Akkadian and Babylonian documents always look down from the palace, their focus always on the kingdom. In the Etruscan *Brontoscopic Calendar*, the gods are not always favoring a ruler, even a good one, and there is a more egalitarian, or even proletarian, outlook, in which city factions, common people, and even slaves are perceived with some degree of sympathy. Likewise, its gods'-eye view encompasses the reptiles and winged ones along with the slaves and the women. As the analysis of the world depicted in the calendar will show, borrowing from Near Eastern texts was a deliberate strategy, but the finished product was unique to Etruscan culture and belief.

MEMORY, RECORDS, AND THE EMERGING STATE

When and where was the Etruscan *Brontoscopic Calendar* created and recorded? The close correlation between so many of its apodoses and Mesopotamian omen literature suggests to me that the original "document" was probably committed to writing close to the time of its creation – or else we should see greater divergence from the Babylonian/Akkadian models. The seamless insertion of many predictions specifically keyed to central Italy and to life in the Etruscan cities, such as spotted diseases, the fortunes of women and slaves, and civil unrest, look like the work of a single editor or group, such as a family or college of priests. I suspect that the notion of committing such texts to writing goes back to the inception of literacy in Italy and thus to the Late Villanovan/Early Orientalizing period, perhaps as early as the end of the eighth century, or around the time of the gearing up of Etruscan city cults – and city administration – as evidenced by the discoveries at Tarquinia Pian di Civita, Veii, and elsewhere (ca. 680 BC). Tim Cornell has noted that the practice of inscribing public documents was common in Archaic Rome, and that many of the texts posted were still visible through the Middle Republic.[78] The great historians of the second and first centuries BC certainly consulted

[77] Cf. Fowler 1899/1969:38–46; Scullard 1981:85–86.

[78] Cornell 1991:26–32.

a number of Archaic documents.[79] Food shortages, famines, military campaigns, and prodigies are recorded in Roman annals, for which there must have been various repositories; Etruscan cities must have set the standard for this.

Mary Beard has emphasized the significance of writing and display of documents in Classical religion – for instance, the erection of written monuments by pilgrims, the marking of votive offerings, and the use of writing in some forms of divination, such as the *sortes* of Fortuna at Praeneste.[80] These "alphabetic oracles" were said to have been found in a rock, and the writing on them rendered the actual prediction, so the letters were an indispensable part of the cult. Further, there were numerous monumental "public, written representations of the Roman religious and civil year,"[81] that is, the painted and carved *Fasti* posted on walls, which allowed the year to be seen as a whole and consulted in perpetuity, promoting a cosmic view rather than that of a single mortal in the midst of a life. Eventually, elite scholars would elevate this data to the level of literature, in Ovid's *Fasti*, Suetonius' *De anno*, and Varro's *Antiquitates rerum divinarum*. Some forms of divination must have been practiced since prehistoric times, but the strong links to the written word for certain varieties, like brontoscopy, imply that the formal practice of this particular discipline was adopted about the same time that the alphabet was acquired and avidly taken over by Etruscan men and women.

More immediately, of course, recording something in writing crystallizes it in a single form and language, which will soon become old-fashioned, archaism enhancing its venerable aura, as in the case of Roman cult. The hymn sung by the Salii in their ritual procession through the city of Rome had become, by the first century BC, virtually unintelligible – clearly, new initiates could only have learned the words by memorizing them from a written text. Likewise the hymn of the Arval Brethren was posted in written form in their sacred grove: "The preservation of these ancient texts would have been impossible without written documents; and indeed the inscriptions from the Arval grove explicitly refer to written copies of the hymn used by the priests in their chanting," according to Beard.[82] She suggests that the incomprehensibility of such texts was what led to the writing of commentaries, a genre that seems to have flourished during the first century BC. As she notes, there are obvious "implications for religious power and control that follow from the obscurantism enshrined in this writing," because such "unintelligibility could be an important defense of priestly or elite power."[83]

[79] Cf. Ogilvie 1958.
[80] *De divinatione* 2.85–87; cf. Klingshirn 2006:143–44.
[81] Beard 1991:54.
[82] Beard 2004:131.
[83] Beard 2004:132.

The shoring up of a state's hold on its people is also effected by the posting of laws, and it seems to have been at the end of the Archaic period that the new republic of Rome saw fit to publish its *Twelve Tables*, circa 450 BC, as a concession to the plebeians to ensure they were being treated fairly under the law.[84] Coincidentally, the *Twelve Tables* already encompass most of the classes of the social system and the concerns that appear, to an enhanced degree, in the *Brontoscopic Calendar*: the status of women, social hierarchy, land use, slavery, and health (in the *Twelve Tables*, the paternal right to reject the malformed). Authors from Cicero (*De republica* 2.36ff) and Livy (3.9–57) to Tacitus (*Annales* 3.27) say the laws were posted on bronze tablets, but Sextus Pomponius (*Dig.* 1.2.2.4) calls them ivory. Multiple or parallel exemplars must have existed, whence the originals were able to be reconstructed after the Gauls sacked Rome in 390 BC.

There is no indication of the original substrate on which the first exemplar of the *Brontoscopic Calendar* was recorded: it could have been almost anything, from wax tablets (for which there are ample Assyrian precedents and also Etruscan examples) to linen books, or bronze plaques. The Capua tile shows the use of terracotta as a medium by the end of the Archaic period, and stone, in Etruria used for epitaphs since the eighth century, had special significance too.[85] Even linen books could survive for centuries, it seems,[86] and for a sacred document we should perhaps not be surprised that a small number of copies progressed in a carefully preserved descent down to the first century BC, when Nigidius Figulus found the text that he translated into Latin. Not only Etruscans but many Italic cultures embraced literacy for religious purposes; for instance, in addition to the Umbrian *Iguvine Tables*, there is the Oscan Agnone bronze tablet, inscribed with directions for rituals, circa 250 BC.[87]

A curious anecdote repeated by Pliny (*Nat. Hist.* 13.86) seems to indicate the preservation of books (*libri*) believed to be from the early seventh century BC.[88] In Rome in 181 BC, farmers working on the land of a scribe, L. Petillius, below the Janiculum, found two stone chests. One, according to its inscription, held the body of King Numa Pompilius, the other *scriptura*: seven books in Latin and seven written in Greek. Apparently the Senate preserved the Latin works and ordered the Greek texts burned by the state *victimarii* (sacrificial assistants).[89] Written texts were certainly being created in Italy by the seventh century, but what this discovery really was is unknown – at least it

[84] See Warmington 1967:xxvi–xxxiii; text of the *Twelve Tables*, 424–515.

[85] Roncalli 1990. And there are other possibilities, as noted by Mauro Cristofani (1995:99–101), Francesco Roncalli (1980 and 1985), and others (Piccaluga 1994).

[86] See van der Meer 2007:10.

[87] See Bonfante 1990:55.

[88] See Livy 40.29.3–14; reprised by Valerius Maximus 1.1.12.

[89] See Shackleton Bailey 2000:24–27.

shows that Romans of Pliny's era were prepared to believe in the antiquity of arcane documents.

Augustus was certainly a master of the manipulation of religious formality for the exploitation of the state, and historians of his day noted another example of a ruler fully cognizant of the power of written state documents (and control of access to same). As recounted by Dionysios of Halikarnassos (4.62.4) and Valerius Maximus (1.1.13), the Duumvir Marcus Atilius was bribed to provide "Petronius the Sabine" the text of a book that "held secrets of the state rituals," apparently one of the Sibylline Books (*librum secreta civilium sacrorum continentem*). King Tarquinius Superbus had the duumvir sewn alive into a skin and thrown into the sea to drown, a punishment for heinous crimes like parricide. This tale, whatever the veracity of its details, places books similar to the *etrusca disciplina* firmly in the sixth century BC; they, like the *Brontoscopic Calendar*, were already being copied and used a century earlier.

SUMMARY

The climate of Iron Age Italy was different from that we share with the Republican Romans, and it probably reinforced interest in divination by thunder. A number of predictions in the *Brontoscopic Calendar* find nearly exact parallels in the great works of Mesopotamian divination (see Chapter 8) – yet no anomalous or anachronistic references (like date palms or donkeys) appear, as they do in all other (i.e., later) brontoscopies (see Chapter 9). A few hints remain that anchor this text firmly to other Etruscan cultic documents, such as the Zagreb *Liber linteus* calendar, and to the tone of other Italian religious writings, such as the *Iguvine Tables*. Actual thunder-divination and consultation of the *etrusca disciplina* is described (though for different purposes) by Cicero, in the event of the responses of the haruspices in 56 BC. A sound in the heavens was interpreted by the haruspices based on consultation of the books of *etrusca disciplina* – quite possibly the *Brontoscopic Calendar* itself, which around this time was translated into Latin by Nigidius Figulus, in a move to make the document more accessible. The next chapters will investigate the specifics of environment and society that are embedded in the *Brontoscopic Calendar*'s predictions and suggest that they derive from the background of Iron Age Etruria.

CHAPTER 5

WEATHER, FAUNA, AGRICULTURE, PESTS: "REPTILES WILL TROUBLE HABITATIONS"

Not the least of the things which have contributed to their luxury is the fertility of the land; for since it bears every product of the soil and is altogether fertile, the Tyrrhenians lay up great stores of every kind of fruit. In general indeed, Tyrrhenia, being altogether fertile, lies in extended open fields and is traversed at intervals by areas which rise up like hills and yet are fit for tillage; and it enjoys moderate rainfall not only in the winter season but in the summer as well.

Diodorus Siculus 5.40.5, citing Poseidonios[1]

INTRODUCTION: LAND, FLOOD, FAUNA, AND FAMINE IN THE CLASSICAL MEDITERRANEAN

Book 5 of Diodorus' *Library of History* deals with the Italian archipelago, but his brief commentary on the Etruscans tells more about the Greek mind than about "Tyrrhenians." (See Appendix A, Text 4 for his remarks on the social values and luxury of the Etruscans.) Yet Diodorus' description *is* true: the soil of Etruria is particularly fertile, both for vines and the all-important cereal crops of antiquity (cf. Varro *Res Rusticae* 1.44). As Paolo Giulierini reminds us, it is no coincidence that, in the legend of Tages, the prophet was said to have appeared from a plowed furrow (*sulcus*), nor that, during the Hannibalic War, Etruria was required to provision Scipio's fleet with grain (Livy 22.3).[2] The Gauls, in 380 BC, lean from several years of cold, wet weather, and poor harvests, marveled at the sight of Italy below their Alpine vantage point: it sparkled with grain fields, vineyards, olives, and figs.[3] The forests that remained in first-millennium Tuscany provided rich game, hillside forage for

[1] Quoted from Oldfather/Diodorus 1939:207–9.

[2] Giulierini 2005:66.

[3] Livy 5.33, Dionysios of Halikarnassos *Ant. Rom.* 13.10, Pliny *Nat. Hist.* 12.5, Diodorus 5.40.3, Plutarch *Camillus* 15.1–6; Camporeale 2000:176–77.

flocks and herds, and the sea and rivers abounded in fish and game birds. All these riches had been exploited in the Italian peninsula since the Neolithic period. From the Iron Age through the Republican first century BC, these resources furnished a well-rounded and pleasant lifestyle for the elite of Etruria – and for its common people. All of the elements of importance to the ancient population are cited in the *Brontoscopic Calendar*, to varying degrees, and the incidence of certain telltale commodities is an additional support for dating it to the Iron Age.

INCIDENCE OF OMENS RELATING TO FOOD, AGRICULTURE, ANIMALS, PESTS

The most important aspect of life attested by the *Brontoscopic* omen topics is surely the food supply, both animal and vegetable. The early days of each month (especially days 2–6, 12–17) are more likely to include predictions on the food supply. The last days of a month usually have fewer food omens; but no day is without a food prediction in *some* month. The first month, June, has the highest incidence (15) of food-related omens, whether good or bad. May comes next (13), with September, November, and March having twelve omens each. The other months show from eleven (July, Oct., Dec.), ten (Aug., Jan., April) to nine (in February, when there would have been less agricultural activity). Often these predictions are accompanied by sociopolitical or health-related situations. As in real life, some food shortages are linked to social unrest, and wars seem to reap a legacy of famine or epidemic – also a consequence of medieval and modern warfare.

A sampling of food and agricultural omens for the month of June:

1. If in any way it should thunder, there will be an abundance of fruits, with the exception of barley; but dangerous diseases will be inflicted upon bodies.

3. …, there will be a scorching and drying wind, such that not only grains but even the soft fruits will be parched through and through and shrivel up.

4. …, the air will be cloudy and rainy, so that out of a moldy dampness the fruit crops will rot.

6. …, just as the crops are maturing, some sort of wild pest that has sunk deep into them will waste them.

7. …, diseases will infect [men], but not many shall die. And although the cereal crops shall be successful, the soft fruits shall dry up.

8. …, it indicates wet weather, and ruin of the grain.

14. …, the atmosphere shall be burning hot, but there will be abundant harvest and good flow, not the poorest, of the river fish. Bodies, nevertheless, shall be utterly weak.

17. …, there shall be days of burning heat and destruction by mice, blind-mice, and locusts. Still, it brings abundance and at the same time murders to the people.
21. …, it means there will be a dearth of wine, but an increase in the other crops, and an abundance of fish.

Thunder can proclaim future abundance, good harvests, or plenty about as often as crop failures or famine. Many food predictions include details of an environmental reason for problems, such as storms or insects. The word I have translated as "crops" (καρπῶν) literally means "[of] fruits" or "soft fruits," but it was probably applied generically to mean crops or harvests, different from cereal crops. Occasionally a distinction is made between these and ξύλινα καρπά – "hard-tree fruits," that is, nuts.

REFERENCES TO ANIMALS

Over seventy apodoses in the *Brontoscopic Calendar* cite animals in some aspect: the health of herds or flocks, dangers to them (wolves or storms), the availability of animal fodder or of human food from fish (both salt- and fresh-water species), the well-being or death of birds and reptiles, and attacks by insects or other pests. The incidences of animal-related apodoses may form a pattern. It appears that in March there is a greater likelihood of negative events concerning animals, perhaps reflecting a seasonal trend: by winter's end, many wild beasts – and humans – are short on food supplies and causing problems or invading other territories. (Snakes emerging from hibernation are cranky and noticeable, and small mammals and birds may consume the buds of fruit trees or freshly sown seed. It is also the time of birth for many species and thus more mouths to feed.) Humans will tangle with animals more often at this time of year.

February has eight citations of animal issues, and five months have at least seven citations; April and May (also July) have just four. If we sort by the days of a given month, a slight pattern emerges: mid-month, especially from the 8th through the 17th, tends to show more incidences (there are no mentions on the 29th–30th of any months, for instance). The 14th of six months, and the 12th of five months show the most frequent indications. Curiously, September has no animal- or pest-related omens, yet this is precisely the time when flies abound, and other pests appear to take advantage of the harvested crops. September in Tuscany, however, can be a rainy time, which may cut down on encounters with animals.

Sample animal-related predictions for the month of March:

9. If it thunders, it threatens ruin of man and creation of wild beasts.
10. …, destruction to the four-footed.

11. …, it signifies heavy rain and the creation of locusts.
12. …, a powerful man in politics or a general is endangered; on his behalf, battles will be waged, and the wild beasts shall fall upon man.
13. …, there will be plenty but the wild beasts shall be destroyed and the fish shall increase; and reptiles will trouble habitations but will not be harmful.
15. …, it signals hot spells and drought and a great throng of mice and fish.
22. …, destruction of birds, but a plenty of daily supplies.
28. …, there will be a plenty of marine fish.

The prediction of March 12, close to the Ides, might be seen as an insertion doctored for allegorical impact – but the haruspex Spurinna,[4] who warned (and was rebuffed by) Julius Caesar, would have been reacting to some precedent in the *etrusca disciplina* – and why not the *Brontoscopic Calendar*? The imagery of wild beasts running in the streets occurs in Mesopotamian literature as an icon of the destruction of a city falling victim to enemies or criminals. Wild beasts, locusts, and mice obviously threaten humans or their food supply. The fortunes of fish and birds are good indicators of the health of an ecosystem, and thus of human prospects; they might be cited merely as sources of human food – or perhaps in their own right, as elements of the universe. The "four-footed," like "the winged ones," retain a poetic, metaphysical, or slightly ambiguous tone in some predictions. Some cases take a godlike viewpoint of humans and animals sharing the same disease, famine, or ruin, predicted "for humans and dumb animals alike" (Aug. 4, 13; Nov. 5, 7; March 17). March 17 is the most dramatic: "If it thunders, something unexpected will befall the people; ruin upon ruin for men and four-footed beasts." (Was this quoted to the Caeretans after they stoned to death the captives of the Battle over Alalia?) (For zoonoses, see next chapter.)

THE ECONOMY AND ENVIRONMENT OF THE *BRONTOSCOPIC CALENDAR*

Details of food species and related events seem to reflect a period when the agriculture of Tyrrhenian Etruria was much more diverse than in the first century BC. We see references to both barley and wheat, when, even before the Hellenistic period, barley, considered fodder for livestock or slaves, was eschewed by all humans who could afford wheat for porridge or bread. Barley appears six times, wheat three, and both together three times; there are also references to grain supplied from abroad – a situation common by the first

[4] See Haack 2006:110–12.

century BC (and in Athens, as early as the fifth century), but not detectable archaeologically in Iron Age Etruria.[5]

Acorns are mentioned also – for the Roman authors, a commonplace allusion to an idyllic, prehistoric time, and archaeological finds confirm their consumption by humans by the Early Bronze Age. Fruits, soft fruits, and "hard fruits" or perhaps "dry fruits," nuts that could be preserved, are noted on several occasions, but grains and pulses are cited specifically in fifteen apodoses. Such diversity is well attested in the archaeological record of Italy from the Neolithic period on; yet by the times of Cato, Varro, and Pliny, Etruria was no longer self-sufficient: so many landowners had converted family farms to produce surplus "cash crops," such as wine for export, that traditional commodities had to be imported into Italy.

This situation is one more hint that the *Brontoscopic Calendar* was composed during an earlier age. It seems as if, when it was created, viticulture for commercial surplus was not essential to the economy of central Italy, for there are just two citations of wine – June 21: "…, it means there will be a dearth of wine, but an increase in the other crops, and an abundance of fish," and October 7: "…, pulses will be plentiful but wine less." The antiquity of wine making in central Italy has been pushed back to the Iron Age, and perhaps earlier, by finds of grape pips in sites such as Bolsena-Gran Carro (below), by vases identified as banquet kraters,[6] and by an epigraphic find, a Faliscan inscription – *vinum* – dated to the seventh century BC.[7]

Only one other reference cites pulses, May 5: "…, there will be an interchange of hardships in political affairs, and wheat more plentiful than barley. The pulses, however, will be ruined." These kinds of combinations may be found in Mesopotamian prototypes, but they fit perfectly into first-millennium Italy, where cereals and pulses were mainstays of the diet. There are no citations of oil or olives, possibly used for medicine rather than cuisine, since the Early Bronze Age, yet in Etruria by the Archaic period olives were prized commodities suitable for funeral and wedding gifts.[8]

The concerns of almost anyone living on the land in the ancient Mediterranean world were similar: Would there be enough food for families, could they keep from losing it to pest infestation, would the flocks stay safe

[5] Background on cereal cultivation is furnished by Pliny *Nat. Hist.* 18.48–116: barley (18.71–75, 78–80); millet (18.53–55, 100–101); emmer wheat/*alica* (18.109–16); wheat varieties (18.66–70, 81–95). For discussion of environment, rural life, and fauna in the Umbrian *Iguvine Tables*, see Ancillotti and Cerri 1996A:93–128.

[6] Delpino 2000 and 2007.

[7] Wallace 2008:128. The word recurs several times in the ceremonials of the Hellenistic-era Zagreb *Liber linteus*.

[8] See the Caeretan, 6th-c. Tomba delle Olive; Cianferoni 2005:86–91 and 118, no. 81, bronze bowl with olives; similar bowl carried in wedding(?) procession on Murlo 6th-c. frieze plaques (Haynes 2000:123, fig. 103).

and healthy, and would the people keep free of disease?[9] In good times, life in an Iron Age Tyrrhenian village could have been pleasant, but the specter of famine or loss would never have been far from mind. Mice, locusts, and other pests threatened the flocks and their fodder, and wolves were another danger for humans and flocks. Weather-borne "plagues" could affect both humans and domestic animals, and also wild birds. Archaeology has confirmed the picture painted by the ancient authors of relatively small, self-sufficient communities cultivating cereals and pulses for their staple diet and rearing small numbers of sheep, goats, pigs, and cattle for food and sacrifice. The diet was supplemented with hunting, fishing, and collecting of shellfish, turtles, and birds or eggs, along with fruits, nuts, and berries from wild plants and trees.[10]

A VIGNETTE OF LIFE IN IRON AGE ETRURIA: THE DIET OF THE RISERVA DEL FERRONE NECROPOLIS "POPULATION"

Studies of human bones from Etruscan sites, such as the seventh-century community at Riserva del Ferrone in the Tolfa mining district, confirm our assumptions about the ancient diet: relative levels of elements in the bones show a reliance on vegetable sources of protein, rarely supplemented with meat. Anemia has also been found in some skeletal populations, yet this can be the result of illness, frequent childbearing, or parasitic infestation rather than an iron-deficient diet.[11] At Riserva del Ferrone, individuals with signs interpreted as anemia enjoyed the same life expectancy as their neighbors, so, as at Pontecagnano to the south, it may have conferred protection against the effects of malaria.[12]

Etruscan populations of the Archaic through Hellenistic periods show the same profile of nutrition and health as established for Iron Age villages. The basic foods found in habitation sites and votive deposits were various wheats (including emmer, einkorn, or spelt), barley, vetch (used for animal fodder), peas, broad beans, millet, fruits, and nuts. The main daily dish was probably something like the *puls* described by Latin authors, a sort of porridge made from emmer by roasting, pounding, grinding, and then boiling the grain, with other ingredients added for flavoring. The supply of fresh, pure water is not cited directly in the brontoscopic omens, but obviously it was of great importance, and early settlements developed various means of maintaining

[9] Rome had to turn to Etruria to supply grain during at least four periods of famine that Dionysios of Halikarnassos and Livy recorded for the 5th c. alone; see Giulierini 2005:66.

[10] For a brief survey of Etruscan foods and cuisine, with special reference to literary sources, see Sassatelli 1999.

[11] Cf. Haynes 2000:8–9; see next chapter.

[12] See Rubini et al. 2000:508.

a supply. The calendar notes periods of dangerous drought (at 15 dates); wet weather, rain, or floods appear at over twenty entries – not surprising for the Hallstatt Minimum climatic phase that generated cyclonic storms also.

Riserva del Ferrone yielded evidence of fifty-four individuals buried in twenty family tombs of the second half of the seventh through early fifth centuries BC.[13] The skeletons ranged in age from infants of less than one year and children three to four years in age to adult men of fifty to sixty years, with several men and women in their teens, twenties, or thirties. Life expectancy was estimated at twenty-seven to thirty-four years for women and thirty-seven to forty-four for men, with only a few individuals attaining forty-five to fifty years (women) or more than fifty (men). The skeletons of twenty-eight adult men and women were analyzed for relative levels of calcium, strontium, and zinc, and, in contrast to some third-world areas today, there was no variation between men and women in diet, particularly in the ratio of plant to animal foods consumed. This was also the case for the Iron Age Tarquinia, where the bones of women buried in the Monterozzi necropolis show the same nutrition as their male counterparts, apparent confirmation of the better status of women in early Etruria.[14] The balanced diet at Ferrone was essentially the case throughout Italy.

SPECIFIC CROPS, ANIMALS, PESTS: BACKGROUND

Hiberno pulvere, verno luto
grandia farra, camille, metes.
From winter dust, from summer mud,
You shall reap, boy, splendid grain harvests.

This old Latin proverb, perhaps from a *Canticum Rusticum*, was preserved by Varro in *De lingua latina* (6.21), and it serves to illustrate the importance of grain and the farmer's dependence upon the climate for his fortune and life.[15] Republican Rome cherished such aphorisms that seem to derive from orally transmitted farmers' wisdom.

Weather is obvious in the brontoscopic omens: naturally we expect wind and rain with the thunderstorms that invoke the calendar's predictions. At least twenty-three apodoses cite rain or wet weather, but another fifteen warn of drought and/or extremely hot weather. From June through September, several omens warn of winter weather – implying a delay in the outcome of predictions of three months or more. Strong or damaging winds appear in

[13] See Brocato 2000; Rubini et al. 2000. Some Faliscan men and women of early 7th-c. Narce lived beyond age 65; see Becker et al. 2009:47–61.

[14] Rubini et al. 2000:505–6.

[15] See Speranza 1974:3–4.

approximately eighteen entries, six of them in February alone; November has no drought or rain omens, but two wind predictions. February 20 is graced with good weather, a rare prediction. (The second and sixth days of the month seem to lack weather-related omens.)

The most famous, abiding, and valuable category of foodstuffs/crops for the history of Italy, and the concerns of divination, was cereals, for which Etruria was famous. The sites described below all illustrate the primacy of grain, whether barley, wheat, or a mixture of species, all of which had been common to the peninsula for millennia. The provisioning of armies and navies and the daily needs of settlements were predicated upon grain. Long before Rome "assessed" the Etruscan cities to provision Scipio's fleet, the commander Decius had forced Tarquinii to provision his army with grain (Livy 9.41); throughout its history, Rome periodically relied upon grain from Etruscan Veii, Chiusi, and so on to bail it out of famine. The major grains were barley and wheat in several varieties, both hard and soft, for cooking porridges or baking bread. Most people got only cereals, in the form of *puls*, to the extent that Italians were nicknamed *pultiphagi* by outsiders (cf. Plautus *Mostellaria* 828 and *Poenulus* 54).[16]

Martial's comment on ripening wine in Chiusine jars that previously held *puls* (13.8.1) has been interpreted by some scholars as implying that the best porridge grain came from the fields of Chiusi. In the same vein, Juvenal's (11.109) comment on serving *farrata* (a spelt porridge?) in an Etruscan bowl (*Tusco catino*) has sometimes been construed to say that porridge was the national dish of Etruria. By the time of Festus (81 L) *farrago* was a mixture of feed grains for livestock, but it had originally been the basic sustenance of men – whence presumably its connotation of a spicy mixture, "satire." The farmer plowing his field, and the priest performing a ritual plowing for city foundation, were icons, as in the life of Camillus.[17] A fine, fourth-century bronze votive figurine from Arezzo of a man plowing probably commemorates both aspects.[18]

Excavations have furnished surprisingly rich documentation for the storage and consumption of fruits, both soft and hard, in settlements and tombs. Etruria gathered and tended figs, plums of several varieties, pears, pomegranates, Cornel fruits, and "hard fruits" – hazelnuts, walnuts, and chestnuts; legumes such as fava beans, peas, and lentils supplemented the protein of grain.[19]

[16] Other references to *puls* are Cato *De agri cultura* 85; Columella 2.9.19; Pliny *Nat. Hist.* 18.83–84, 18.149; Varro *De lingua atina* 5.105; Valerius Maximus 2.5.5.

[17] See Giulierini 2005:66–69, for more background.

[18] Barker and Rasmussen 1998:190, fig. 70; see 179–99 for the technology of Etruscan farming.

[19] See Giulierini 2005:69–70.

Two famous experts on farming were the Etruscan Sasernas, father and son, whose book on agriculture has survived in part through Latin translations. Short passages quoted by the Roman agricultural authors, Varro and Columella, and comments in Pliny's *Naturalis Historia*, must serve to indicate the type of information they published.[20] Varro and Columella knew the treatise as *De agri cultura* (Varro *Res Rusticae* 1.2.26, 2.9.6, Columella *De re rustica* 1.1.5) and cited it for the carrying capacity of land and the area of land that could be plowed by a yoke of oxen.[21] Some passages sound like charms to get rid of warts and such,[22] a mingling of popular religion with agriculture that was continued by Cato and the others. A hint of earlier Etruscan expertise, corroborated by the archaeological record, comes in a fragmentary comment on the use of a variety of pulses as "green manure" to restore the nutrients to a field used for cereals.[23]

Italy did not wait for the arrival of Phoenician traders or Greek colonists in the eighth century to begin the domestication and cultivation of vines for wine production. Wild grapes were used in the Neolithic Age (grape pips were found at about 10 sites), and by the Middle Bronze Age (mid-2nd millennium) at least thirty sites show systematic gathering and use of grapes, with some possible domestication.[24] At San Lorenzo a Greve, near Florence, the *Early* Bronze Age settlement held an underground room or pit with many grape pips: most were wild but some were from domesticated vines.[25] Early grapes might not have become wine, but by the eighth century ritual consumption of wine used special vases patterned on Greek geometric mixing bowls.[26] By the sixth century, Etruscan cities were shipping assorted vintages across the Mediterranean to France and beyond. The two mentions of wine in the *Brontoscopic Calendar* seem to predate this time of the great "wine culture" of the Classical Mediterranean. The *Iguvine Tables*, inscribed in the late third century (and later) refer to Archaic Umbrian ceremonial wine offering.[27]

FLOCKS AND HERDS, THE BIG THREE: SHEEP/GOATS, CATTLE, SWINE

The majority of citations of animals in the *Brontoscopic Calendar* refer to domestic flocks or herds. Cattle in Classical literary sources are considered

[20] Now gathered by Feliciano Speranza in volume 1 of *Scriptorum Romanorum De Re Rustica Reliquiae* (Speranza 1974:33–45).

[21] See Speranza 1974:38–39, nos. 3–4a.

[22] Cf. Speranza 1974:44–45, fgts. 14–17.

[23] Speranza 1974:40, fr. 6, apud Columella 2.13.1.

[24] Cianferoni 2007; Forni 2007.

[25] Aranguren et al. 2007.

[26] See Delpino 2007.

[27] See Ancillotti and Cerri 1996A:110.

the most valuable of livestock species, conferring status on those able to own them because they require the most land, fodder, and water. Most Iron Age villages probably had few cattle, perhaps just maintaining an ox team for plowing. All herds would have been in pasture most of the time, thus exposed to dangers and weather, but finding their own food: many would be slaughtered before winter. The importance of sheep (and goats) to the lives of Etruscans is evident and symbolized in the practice of haruspicy, originally confined to the livers of sacrificed sheep or lambs. The symbolic status attached to the matron's spinning and weaving textiles hinges on wool production (cf. Figs. 21, 22).[28] Domestic swine were kept for meat and are attested at most sites by bones with butchering marks and seen in artistic representations (see Fig. 19).

In Pompeii and environs, the trash pits of houses hold remains of meals, as do sacrificial pits in sanctuaries, and there are also pathetic finds of domestic animals trapped in houses and stalls at the time of the eruption. All the major groups of cattle, sheep, goats, and pigs were well represented, along with horses, donkeys, and mules.[29] King suggested that, although pork was the most common meat, the middle-class population at Pompeii ate older sheep, but aristocrats indulged a taste for lamb.[30] Roe deer (*capriolo*)[31] are attested, and gardens held remains of small fauna: wood mice, house mice, water and common vole, garden, common and edible dormouse, hedgehog, hare, rabbit, stoat or weasel, black rat, mole, and pine vole and pine marten (at Boscoreale).

The horse, however, is a telling omission. Although of limited use to ancient Mediterranean societies, the horse, of the greatest symbolic and social significance, is missing from the *Brontoscopic Calendar*'s predictions. Horses did not factor in Mesopotamian cities, but in Italy they came to reflect the prestige of rulers, beginning in the later eighth century. Domesticated equines were raised in Italy ever since the Eneolithic period, but for meat rather than traction.[32] The impetus for chariotry came from contact with Levantine travelers who introduced Cypro-Phoenician equipment and stock.

THE "WINGED ONES"

Birds are mentioned several times in the *Brontoscopic Calendar*, sometimes under the poetic term "the winged ones" – we do not know the Etruscan word, but the Greek term πτηνός has been used at least since the time of

[28] Turfa 2005:8, 19–20, 160–62; Gleba 2008:72–75, 171–87; cf. Barber 1994.

[29] King 2002:408–10, 415–16, 437–39, 443–45.

[30] King 2002:439, 445.

[31] Watson 2002:416–17.

[32] At Querciola, Remedello, Conelle di Arcevia, etc.; see De Grossi Mazzorin 1995A: 175; De Grossi Mazzorin et al. 1998; Azzaroli 1998.

Aeschylus (*Choephoroi* 591) and appears in other tragedies (Sophokles *Ajax* 168; Euripides *Ion* 504) and in Aristophanes (*Birds* 1707 – πτηνὸν ὀρνίθων γένος, "the winged tribe of birds"). Usually in the Etruscan document they are wildfowl, probably waterbirds, who would be hunted or collected. There probably were "barnyard" fowl, ducks, or geese, in Iron Age settlements by the eighth century, when their images become popular, perhaps implying the domestication of the mallard (*Anas platyrhynchos*) and greylag goose (*Anser anser*).[33] Strabo (5.2.9) knew that the lakes near Chiusi and Perugia still had waterfowl.[34]

The earliest domestic chicken wing was a funeral offering found in an Early Iron Age Latial hut urn from Monte Cucco.[35] Chicken eggs appear in sixth-century tombs at Vulci and Cerveteri, and chickens have been identified in the murals of Tarquinia.[36] Chicken eggs became common funerary symbols (cf. perhaps the couple on the Tragliatella *oinochoe* (Fig. 20)). In the Vesuvian sites, most bird bones came from meals (especially at Boscoreale) or sacrifices, but a wild turtledove nest with eggs and bones and an owl skull indicate the close proximity of wildlife in Roman towns.[37] Pompeian art is replete with pet and wild songbirds and game species, from a mynah imported from Asia to a host of finches and pigeons, chickens, and ducks.[38]

FISH OF ALL KINDS

Fish were important as food and omens, cited for twenty-one dates (June 2, 14, 15, 21, Nov. 15, 21, Dec. 2, 3, 10, Jan. 6, 12, 28, Feb. 8, 24, March 13, 15, 28, April 12, May 7, 12, 22.)

Sample omens mentioning fish in the *Brontoscopic Calendar*:

> June 14. If in any way it should thunder, the atmosphere shall be burning hot, but there will be abundant harvest and good flow, not the poorest, of the river fish. Bodies, nevertheless, shall be utterly weak.
>
> 15. …, the winged creatures shall be injured during the summer, and also the fish shall perish.
>
> 21. …, it means there will be a dearth of wine, but an increase in the other crops, and an abundance of fish.

[33] Cf. Watson 2002:364–66. Ducks and geese in the 4th-c. Tomba Golini of Orvieto and Tomba dei Rilievi at Cerveteri; see Giulierini 2005:72.

[34] For fine illustrations of the bird species present in ancient Etruria, see the Web document Mazzuoli 2007.

[35] Castel Gandolfo, first half of 9th c. BC; Bowmer and Molleson 1986:229, no. 163.

[36] Fifth-century Tomba del Triclinio; see De Grossi Mazzorin 1995A:176.

[37] Watson 2002:358.

[38] See Watson 2002. The Tarquinian Tomb of Hunting and Fishing depicts a bird feeding a fish; Steingräber 1986:pl. 49.

November 15. If it thunders, the fish especially plentiful, but it shall plague the water-bound beasts. The commonwealth rather better.
December 3. …, men will excessively consume their flocks because of a dearth of fish.
10. …, it threatens slaughter for men from diseases, but the fish shall be abundant.
March 28. …, there will be a plenty of marine fish.
May 22. …, it signals heavy rains and destruction of marine fish.

In the *Brontoscopic Calendar*, the majority (12 plus 2 for marine fish) of omens citing fish promise abundance; five predict dearth of fish, plus one a dearth specifically of marine fish. One predicts abundance of river fish. Thus a clear distinction is made between river/freshwater fish and sea fish, giving us the impression of an extensive fishing "industry," a wide territory included in the theoretical universe of this text. The role of fish in the ancient diet is often hard to appraise because of the vicissitudes of preservation. Species attested at Bronze Age sites include pike, tench, chub, and rudd (*Scardinius erythrophthalmus*).[39] At the northern Etruscan site of Forcello di Bagnolo di San Vito (Mantua), both fish (pike) and fowl were attested.[40] Mollusk and mussel shells have been identified in the tombs of Caere and Tarquinia; a plate with fish vertebrae is under study at Pyrgi.[41] Gianfrotta has considered the numerous species sought in the lakes and lagoons of Etruria and the Tyrrhenian Sea: rays and skates, and the species common to brackish waters, including female eels, eels, bass, gilthead bream, grey mullet, goby, prawns, mullet, and sole ("capitoni, anguille, spigole, orate, cefali, ghiozzi, gamberi, triglie, sogliole").[42] Studies of minerals in the bones of the Archaic-period Paleovenetic population of Este demonstrate that they were consuming *both* salt- and freshwater fish.[43] The Roman fishery and breeding industries developed on the sites of earlier Etruscan enterprises, and on lakes Bolsena, Vico, and Bracciano, saltwater species were stocked for market (Columella *De re rustica* 8.16). Other fish farms were built along the Tyrrhenian coast, and a large fishing fleet operated out of Pyrgi.[44] Lookout towers were built for catching tunny and swordfish at Monte Argentario and Populonia (Strabo 5.2.6 and 5.2.8).

Archaeological finds illustrate the gathering of freshwater mollusks (also turtles) in northern rivers, and harpoons, fish hooks, and other equipment are found in Late Bronze and Iron Age sites, such as Fiavé and Barche di

[39] Menotti 2005:214.
[40] See Menotti 2005:192; Scarpa 1986:184–92.
[41] Gianfrotta 1987:56.
[42] Gianfrotta 1987:57–58.
[43] Bartoli et al. 1999:224.
[44] Athenaeus 6.224c; see Gianfrotta 1987; Camporeale 2000:45.

Solferino.[45] Ninth-century burials at the coastal site of Sasso di Furbara, between the future cities of Caere and Tarquinii, include the marine sail-boats of fishermen, used as their coffins, evidence of the exploitation of saltwater fish, as noted in the calendar.[46] Pascucci (1998) identified jars developed by First Iron Age, in the region of Civitavecchia and Vulci, for the storage and shipping of dried fish. They attest a sophisticated early system of exchange and interchange among coastal and inland settlements in southern Etruria, paralleling salt trading. Stone fishing weights occur in coastal and lakeside sites (Cerveteri, Pyrgi, Santa Marinella, Talamone), as do tools for making and mending nets. A fisherman's profession was fraught with dangers: a painted plate from Acqua Acetosa Laurentina shows a boatman grappling with (trying to harpoon?) a huge and belligerent fish (ca. 680–650 BC).[47]

Funeral offerings of the ninth to eighth centuries in the Roman Forum necropolis illustrate the ceremonial consumption of fish. In Tomb CLXXVI, a bowl containing two skeletons of river fish, European barbel (*Barbus fluviatilis*), was deposited with other foods (small broadbean and wheat).[48] Other burials there and elsewhere in Latium (Cures Sabini, Ficana) show a variety of fish offerings, including gray mullet.[49]

The cities of Vesuvius furnish additional evidence, including factories and shops dealing in fermented fish sauce (*garum*) – and fishponds in many homes. It seems reasonable to extrapolate the Etruscan situation from this later evidence. Murals and mosaics, in addition to bones or scales, show that Romans were familiar with a wide array of species, from garfish, wrasse, red mullet, sea bass, flounder, mackerel, ray, sea bream, sardines, anchovy, and John Dory to conger, lamprey, and moray eels.[50] Seafood is also amply attested in the shellfish and other marine invertebrates identified at Pompeii.[51] The importance of fish to the economy of central Italy was reflected in Etruscan art (Tomb of Hunting and Fishing, Tarquinia, 530–520 BC); the coins of Populonia, Vetulonia, and Volterra depict sea creatures also.[52]

[45] Menotti 2005:179–80.

[46] See Brusadin Laplace and Patrizi Montoro 1977–78.

[47] Martelli et al. 1987:92, 263, no. 39; see Gianfrotta 1987 for full documentation.

[48] The barbel is a species of the carp family, rather bony to eat, and only living in fresh water; Davidson 1999:58.

[49] De Grossi Mazzorin 1989:128–29.

[50] See Reese 2002a.

[51] Reese 2002b.

[52] See Giulierini 2005:73–77, for background. For a selection of illustrations of edible sea life from Classical art, see Donati and Pasini 1997; the Iron Age Etruscan Pittore delle Palme has rendered fine bluefish among others; Donati and Pasini 1997:57.

WILD GAME

Bronze Age sites already show a plethora of hunted species used in cuisine: deer, roe deer, wild boar, hare, lynx, fox, marten, badger, otter, beaver, tortoise.[53] Complete ritual funerary meals are implied in food remains containing multiple ingredients at Orientalizing Bisenzio. Tomb 10 at the "Buccacce" site held the bones of a hare together with hazelnut shells; other necropolis sites, such as Chiusi (Poggio Renzo), had similar meat offerings within vases.[54] At Ficana and Cures Sabini, bone finds show the consumption of at least one beaver and otter respectively.[55] The hunt by the Archaic period had become an aristocratic symbol of those who controlled the land and had the time for bloodsports.[56] The *Brontoscopic Calendar* holds few allusions to wild game (e.g., January 22: "…, there will be plenty, but also there will be an abundance of mice and of deer").

Τὰ ἑρπετά – REPTILES

Sample references to "creeping things/reptiles" in the *Brontoscopic Calendar*:

> July 28. If in any way it should thunder, there will be a dearth of water and a plague of poisonous reptiles.
> August 11. …, there will be a good harvest, yet the downfall of reptiles and harm to men.
> November 14. If it thunders, poisonous snakes shall somehow be gently undone by the men.
> February 24. …, it threatens good health for men, but destruction for both fish and reptiles.
> March 13. …, there will be plenty but the wild beasts shall be destroyed and the fish shall increase; and reptiles will trouble habitations but will not be harmful.

Eight dates cite reptiles, some poisonous and/or threatening, others facing destruction (July 13, 28, Aug. 11, Feb. 14, 19, 24, March 13, May 10). In Italy, "reptiles" could be poisonous snakes[57] or beneficial, like the majority that hang out around settlements to consume the insect and rodent pests that human habitation attracts. Τὰ ἑρπετά, "the reptiles," even poisonous ones, need not have been snakes, however, because the Greek term encompasses all "creeping things" – we might say cold-blooded creatures in general.

[53] Menotti 2005:214.
[54] Galli 1912:458, no. 24, and note 1.
[55] De Grossi Mazzorin 1989:140.
[56] See Giulierini 2005:72–73; Camporeale 1984.
[57] Cf. Pliny *Nat. Hist.* 8.85–87; Bodson and Orr 2002:346.

Reptiles were usually considered good or natural elements of nature.[58] Turtles and tortoises, although eaten, would not have registered negatively, nor would frogs and toads. The poisonous reptiles of July and February (likely to be cranky if seen at all) may be vipers, who have a long history in Etruscan art and mythology.[59] Pompeii, Herculaneum, and Boscoreale have furnished remains of several native species that could all have been present in Etruria at any time in the first millennium BC: toad/frog, Herrmann's tortoise (some pets, some butchered), gecko and wall lizard (*Podarcis*), rat snake, and other native snake species.[60] Pet colubrid snakes (rat and water snakes) were appreciated for pest control (Pliny *Nat. Hist.* 29.67, 29.72). Orr notes "the universal appeal of snakes in practically every world religion," including Roman domestic cult where "the serpent was the 'genius' that not only protected the place from harm but also represented the procreative force of the humans who lived there."[61]

Honey is mentioned once (plentiful at August 6) and its production in Etruria has been documented in the tombs of Verucchio and elsewhere, and in finds of honeycombs in vessels in habitations[62] and tombs, such as the "princely" seventh-century Casale Marittimo Tomb A.[63] It is possible that it was used for mead, because traces of a cereal-based, honey-sweetened beverage have been identified in some Etruscan vessels.[64]

PESTS: INSECTS, MICE, CROP DISEASES

The effects of pests and diseases on the food supply cannot be underestimated: one season's bad harvest or the loss of stored cereals could mark the difference between survival and starvation for many communities.

Selection of *Brontoscopic* omens relating to animals, pests, and food supply:

> June 6. If in any way it should thunder, just as the crops are maturing, some sort of wild pest that has sunk deep into them will waste them.
> 17. …, there shall be days of burning heat and destruction by mice, blind-mice, and locusts. Still, it brings abundance and at the same time murders to the people.
> November 6. If it thunders, borers will ruin the grain.
> 16. …, the creation of locusts and field-voles, to the king, danger, and there will be an abundance of grain.

[58] See Bodson and Orr 2002:328–29, 339–42.

[59] See Hostetler 2004–7.

[60] Bodson and Orr 2002:332–35, 346.

[61] Orr 2002:350–51.

[62] Camporeale 2000:177.

[63] Esposito 2001:90.

[64] See McGovern 2007. For background on Roman apiculture, see White 1970:331; Larew 2002:318–19; Varro *Res Rustica* 3.16; Pliny *Nat. Hist.* 11.11–70.

December 17. …, it threatens that small locusts shall be born, yet there will still be plenty.
April 22. …, it will be the destruction of the flies.

The predations of large creatures such as wolves (June 9) are readily apparent; insects such as flies, locusts, and borers (also worms) still cause great damage to the food supply.[65] Similar pests are featured frequently in Mesopotamian literature, at, for example, *Enūma Anu Enlil*, tablet 51, line 6: "If the mongoose […]: the *kurusissu* rodent will eat the flax." (See Chapter 8.)

At the early seventh-century Verucchio necropolis, anomalies in tree rings have prevented the accurate dendrochronological dating of the wonderful furniture (see Fig. 21) deposited there, but this has reminded some scholars of insect pests, namely, seasonal borers such as cockchafers.[66] At Etruscan Forcello di Bagnolo San Vito, food stores of seeds were attacked by bruchid weevils, also found in carbonized beans stored in the Pompeian House of the Ship *Europa*, and various beetles were identified in hay stored at the Oplontis villa.[67] Columella (*De re rustica* 1.6.10–17) describes methods of dealing with weevil infestation, still a problem in the Roman period. Pliny (*Nat. Hist.* 18.149–62) described several threats to crops in his day, ranging through diseases, weeds, and animal pests: maggots, beetles, caterpillars, slugs breeding, weed infestation by vetch, starlings and sparrows eating millet, and so on. Of the species noted in the *Brontoscopic Calendar*, grasshoppers or locusts are illustrated in Pompeian art: Pliny (*Nat. Hist.* 11.101–7) says the swarms came from Africa, and Rome had to resort to the Sibylline Books for remedies (note the link to divination texts).[68]

The snail genus *Lymnaea*, found in Italy, can be a vector for the liver fluke parasite, devastating to flocks (see next chapter). Mice and blind-mice (field mice that proliferate at harvest time), the "mice of dry land," and field-voles (identified in Bronze Age Nola-Croce del Papa) pose obvious problems for food storage. This may be why domestic or commensal cats seem to have been introduced very early in Italy: in the late ninth century, the oldest known domestic cat in Europe perished in a fire that destroyed the large hut discovered at Fidenae.[69] Other Latin sites (Ficana, Cures Sabini, Acquafredda) had domesticated cats during the seventh-century Archaic period.[70]

65 See, for instance, Pliny's data on protecting stores of grain, discussed by White 1970:189, 196–97.

66 See *Cornell-2000*; however, according to Peter Kuniholm (2007), the Verucchio case may result from coppicing, a technique of harsh pruning still practiced in some places. I am indebted to him for discussing this with me.

67 Larew 2002:317.

68 Larew 2002:322–24.

69 De Santis et al. 1998:44–45, figs. 62–63; Turfa 2006F:note 10.

70 See De Grossi Mazzorin 1989:139–40.

The same species are still found in Tuscany, where restoration of natural habitats is bringing some ecologies back from the brink of extinction; Web sites maintained by provincial agencies, the Italian Ministero dell'Ambiente, and *SIRA Conoscenza Ambientale in Toscana* publish lists of species, offering a sample of what the Etruscan environment must once have been.

BACKGROUND ON AGRICULTURE, AGRONOMY IN PREHISTORIC ITALY

Archaeological evidence from all over Italy, from the Neolithic through the Roman Empire, can inform our picture of life in Etruria during the Late Iron Age–Archaic period, when the *Brontoscopic Calendar* was first used. During the Neolithic period and Bronze Age, a pattern of daily life and seasonal cultivation and celebration was set that continued through the first millennium to affect the life of country- and city-dwellers alike. Its effects are still apparent in Tuscany and Europe.

The environment dictated a certain efficient way of life, with fertile volcanic soils creating hillside pastures and terraces for vineyards; well-watered valleys and broad fields still yielded rich harvests of grain into the Roman period. Life could be terribly disrupted and threatened by climate changes, drought, or flood, so its inhabitants took measures to offset troubles. In the Po region where flooding was common and could become devastating once every decade, Bronze Age builders contrived special water canals, wooden coffer-wall systems, and villages on platforms, the *palafitte* (2nd to early 1st millennium BC).[71]

The Neolithic Revolution had introduced an array of domesticated plants: barley, emmer, einkorn, soft and hard wheats, and peas, vetches, and lentils. Varieties diminished but hardy types continued in cultivation; anyone consulting the *Brontoscopic Calendar*, whether in the Iron Age or the Late Republican period, had a similar food supply. Waterlogged sites such as La Marmotta in Lake Bracciano illustrate the diversity of food supplies, with finds of spelt, barley, and several types of wheat, and also acorns shelled for food.[72]

In northern Italy, Neolithic villages show the full spectrum of wild and cultivated foods: hazelnuts (*Corylus avellana*), cornel berries (*Cornus mas*, a type of dogwood), apples (*Malus sylvestris*), wild plums, blackberries, raspberries, strawberries, elderberries (*Sambucus nigra*), blackthorn berries (*Prunus spinosa*), and wild grapes; poppies may also have been cultivated for their seeds.[73] Neolithic farmers developed means of cultivating fruit trees, such

[71] See Turfa, "The Etruscan *Brontoscopic Calender* and Modern Archaeological Discoveries," *Etruscan Studies* 10, 2004–7: 163–73.

[72] See Boscato et al. 2008; Costantini et al. 1995:328.

[73] See Menotti 2005:178–89; 213.

as pear (*Pyrus communis*) and fig (*Ficus carica*). Acorns have been attested at various sites.[74] Some of the second- to first-millennium *palafitte*, such as Fiavé, actually have shelled acorn meats preserved inside vessels used to prepare human food.[75] The introduction of iron plowshares led to intensified cereal cultivation – and increased population.[76]

VILLAGE LIFE

Within a small village of thatch-roofed huts at almost any time during the third through first millennium, small communities carried on their lives with some members tending gardens near home, others hunting or gathering the fruits of field and forest, and some following flocks that had to be driven to upland pastures in summer and down nearer the coastal plains in winter.[77] The usual ratio of species was a high number of sheep/goats with fewer cattle and pigs. In fact, the same basic species of farmed and hunted creatures were still attested at Pompeii in 79 AD. In the Etruscan period (especially the 4th to 1st c. BC) votive models attest to the prevalence of sheep, cattle, and pig.[78] As Horden and Purcell have indicated, the Mediterranean terrain varies greatly over rather small distances, making some crops and livestock profitable for one village but not feasible for the next.[79] The ages and sex of slaughtered animals demonstrate that milk and wool, secondary products, were more important than meat, which would have been a rarity, except for seasonal slaughter before winter.

Food offerings deposited in caves and rock shelters during the Bronze Age and Iron Age included grains, legumes, and animal foods, such as stews or cooked joints of sheep/goat, pig, and ox – in other words, domesticated species and the products of agricultural toil.[80] Many sacred sites were located near sources of living water, such as springs and streams (some sulphurous, some fresh), and this phenomenon continued in historical Etruria.[81] At first-millennium Satricum, beans and grains were burned and offered in a *mundus* pit under the ritual hut.[82] Jacopo De Grossi Mazzorin and his colleagues noted a distinct increase in pastoral activity at the turning point of the Final Bronze Age, when flocks increased substantially as a result of changes in social organization,[83] as livestock became the "mobile riches" of an emerging elite class in

[74] E.g., Archaic Bagnolo San Vito; see Menotti 2005:212–14.
[75] See references of Costantini et al. 1995:330.
[76] Camporeale 2000:176.
[77] For background on the practice of transhumance, see Holden and Purcell 2000:63–64, 550–51.
[78] See Söderlind 2004; also Santillo Frizell 2004 on sheep dips.
[79] Horden and Purcell 2000:59–65.
[80] Barley, wheat, beans, millet, seeds, and nuts/acorns; for the variety, see White 1970:198.
[81] Miari 1995:11, 20.
[82] Miari 1995:19.
[83] De Grossi Mazzorin 1989 and 1995A; De Grossi Mazzorin et al. 2004.

many ninth-century Tyrrhenian communities.[84] Domestic animals invariably outnumber hunted wild species by a huge margin: it appears that hunting was too precarious ever to be relied upon.[85]

MEALS

Bread, the staple of Rome's urban population in the first century BC, was not so common in earlier times. Late to Final Bronze Age and Early Iron Age sites have furnished pottery containing burnt food: a "polenta" of barley and wheat mixed with vegetables at Barche di Solferino and Polada, and flat breads or porridge elsewhere. Bog sites preserve loaves or dumplings, as at Ledro and Lazise.[86] No one relied on meat as more than an occasional treat or supplement to the everyday fare of cereals.[87] The tale of the Etruscan swineherd who taught his pigs to follow the music of his flute (Polybius 12.4) seems immortalized in the silver Plikaśna *situla* from Chiusi (Fig. 19).[88] The display of meats, both domestic and wild, in the funeral larder of the Tomba Golini (Orvieto, 4th c.) shows that meat consumption continued to have special connotations and privileged contexts, potentially a discriminatory, socially divisive activity (see Chapter 7).[89] The quantities of hazelnuts and acorns at Bronze Age sites call to mind the poetic association of acorns with the days of noble savages.[90] Deposits of large quantities of acidic, slightly bitter cornel fruits (dogwood fruits or cornelian cherries) may indicate their use in fermented beverages – some *palafitte* have thick strata of fill composed entirely of cornel pits.[91]

Some information on daily meals may be extrapolated from funerary meals found in Etruscan and earlier tombs, but it should be borne in mind that rituals and religious beliefs undoubtedly contributed to the selection and rejection of foods on these special occasions. At nearly contemporary Casale Marittimo in the region of Volterra, an important person was buried with imported incense, hazelnuts, apples, grapes, and a pomegranate, the oldest yet identified in Italy; the fruits, and also a honeycomb, were all in a bowl.[92]

[84] See De Grossi Mazzorin 2004: 229.

[85] De Grossi Mazzorin 1989:126–28, 140.

[86] Bread loaf of the wheat *Triticum monococcum*; see Menotti 2005:180–81.

[87] See Cato, and de Martino 1993. On cultivated vegetables, see Pliny *Nat. Hist.* 18.126–48. See White 1970:199–219, for Roman data on fodder and pasturage; fodder grains: Pliny *Nat. Hist.* 18.142–48.

[88] Haynes 2000:109.

[89] Barbieri 1987.

[90] Menotti 2005:213. Roman literary references to the acorn, *glans*, are Cato *De agri cultura* 54.1; Vergil *Georgics* 1.305; Horace *Satires* 2.4.40; Columella 4.33.5. Holm-oak (*Quercus ilex*) with tastiest acrons, native to Italy; Davidson 1999:3.

[91] Menotti 2005:213; Aranguren 2007:94, with further references. In modern Turkey they are still prized for jams; see Davidson 1999:253.

[92] Esposito 2001:87–98.

(Locally made wooden objects attest other food species in the vicinity of the town: cornel, pear, oak, sorb.)[93]

VERUCCHIO

At Adriatic Verucchio Podere Lippi Tomb 89/1972, the burial, at twenty-five to thirty years of age, of the warrior-priest,[94] identified by his throne, footstool, and fine garments, held the bones of a three- to four-month-old lamb burnt on the funeral pyre.[95] Bones from at least four older sheep represent cuts of meat also cremated; several wild animals were apparently thrown whole into the flames, including duck (migratory widgeon?) and one large bass or sea bass.[96] Anything carried uphill to this necropolis must reflect a considered, ideological choice. All the *Brontoscopic Calendar*'s categories of animal species were represented: domestic flocks, the "winged ones," and fish (in fact, a species considered both salt- and freshwater). Although plant remains were less well preserved (with the exception of pollen), the families of Verucchio also commemorated their cultivated crops in the offerings they brought to funerals. In Tomb 89, a bronze jar held the remains of a drink made from cereals; the pollen of barley, wheat, and hops found in the sediment implies beer.[97] The profile of tree pollen in the Verucchio vase indicates local brewing, and the proximity of town and forest, circa 680 BC; other pollen attests walnut, pine nut, plum, white mulberry, and grape.

In the "princely" Podere Lippi Tomb 85 (also ca. 680 BC), a meal was set out in bowls on three small, ornate wooden tables: bread and sweets made with honey, a hare, and a fish; fruit and nuts (hazelnuts, grapes, and other seed-bearing fruits), and (probably) drinks were placed, in four cups, on the third table.[98] These are all the categories of food represented in the *Brontoscopic Calendar*: the products of hunting and fishing, tree fruits, both soft and hard, and the cultivated vines and/or grains brewed into beer, mead, or wine.

The picture of foods, cultivation, and gathering at Verucchio is similar to that seen in southern Etruria and beyond, with barley and other cereals, especially wheats (not bread wheat), predominating. Pulses are represented by small broadbean (*Vicia faba*), lentil (*Lens esculenta*), and pea (*Pisum sativum*); linseed (*Linum usitatissimum*) appears first in the fifth or fourth century

[93] Esposito 2001:95–98.

[94] "Guerriero e sacerdote"; see von Eles 2002.

[95] Bertani 1995:44.

[96] Pliny's *lupus*, a bottom-dwelling fish often found in lagoons; see Farello 2002.

[97] Also some millet and rye; Marchesini and Marvelli 2002:305. A vessel in the so-called Midas Tomb at Gordion, now believed to predate the historical Midas, held a mixture of barley beer, grape wine, and honey mead, and McGovern (2007), who analyzed it, suggests similar beverages for Iron Age Italy. I thank Pat McGovern for sharing many publications with me.

[98] Bertani 1995:55; see Menotti 2005:193.

BC. Nuts are an important staple, with hazelnut (*Corylus avellana*), walnut (*Juglans regia*), and chestnut (*Castanea sativa*) at numerous sites, especially from the end of the seventh century on, and plum (*Prunus*) and cornel berries (*Cornus mas*) are equally well exploited in the Archaic and Classical periods. Domesticated vines are attested at most northern sites by the fifth century BC (and earlier in central and southern Etruria). The array of plant foods illustrates the alternation of crops, especially of legumes, which replenish the soil, with cereals, which will deplete it.[99] The same set of legumes and grains also appears, dried and stored, in habitations.

NORTHERN FISH AND FOWL

The domestic chicken (*Gallus gallus domesticus*) was raised in Felsina-Bologna at least as early as the last quarter of the eighth century BC. This falls very close to the earliest attested chickens in Greek contexts,[100] but in Latium, some burials of the early ninth century seem to have had chickens, making the evidence earlier in Italy than in Greece. As in the south, birds' eggs were a major funerary offering, but not all were from domestic chickens. Fish and shellfish were offered in tombs at Bologna, Verucchio, and (5th-c.) Spina; species include those of freshwater (loach, *luccio*), brackish water (eel), and saltwater (ray, *razza*). The freshwater species appear in early domestic contexts, at Castenaso, Arginone, and Forcello di Bagnolo San Vito. Interior sites show trade in salt- and freshwater shellfish.[101] Even today, there is a healthy skepticism about consuming seafood so far from the coast; so the *Brontoscopic Calendar*'s citations of saltwater fish (March 28, May 22) probably speak to a Tyrrhenian, coastal region for its creation.

NOLA-CROCE DEL PAPA: AN EARLY BRONZE AGE POMPEII

A precious time capsule of Early Bronze Age village life was preserved by the Avellino eruption of Vesuvius that occurred 3780 years BP (ca. 1780 BC). The lifeways of Bronze Age communities throughout central Italy were remarkably similar to those of proto-urban Etruria, so a brief summary will illustrate the character of life in early Etruscan communities. At Nola-Croce del Papa, a small village of oval huts with deep eaves and thatched roofs, was destroyed by a pyroclastic flow and then inundated with mud, which preserved the forms of huts, baskets of food, and trapped animals. A watchdog huddled in fear under the eaves of a hut, where nine goats (all four months pregnant)

[99] Bertani 1995:44–45; Pliny (*Nat. Hist.* 18.117–25) discusses legumes.
[100] See Bertani 1995:51–52, with references.
[101] Shells only appear in tombs at Spina, however; Bertani 1995:53.

were tethered to a fence and within a small pen in the midst of the houses.[102] Thousands of footprints show a mass exodus of men, women, children, and animals (sheep, goats, cattle, pigs), who apparently survived the cataclysm (at nearby San Paolo Belsito, a man and woman did not escape; see next chapter).[103] In the aftermath of the eruption, much of Campania experienced famine, disease, and social collapse.[104]

Circumstantial information on human health abounds in this context. Although the villagers kept livestock, their mainstay was always the cultivation of cereals, as confirmed by nutritional studies of skeletons.[105] In the midst of the huts was a threshing floor, an emblem of the centrality, for this society, of grains, agriculture, and communal activities; wells and waterholes were in the same area.[106] Piles of trash strewn between the huts included remains of meals, especially the butchered bones of beef, pork, and sheep/goat (of relevance for health issues, a large cut of beef was stored outside one hut in a container of organic material).[107]

Huts held storage lofts and containers full of food, including almonds, grain, and ground flour. Cuts of dried, butchered meat (beef, pork, mutton) were hung from roof poles, and as many as 100 containers full of different foods were stored in rear, apsidal rooms. Casts of grains and other foods, such as mushrooms, were preserved within baskets and bags: barley and wheat (*monococcum* and *dicoccum*), fruit, nuts (almonds), olives, and acorns, buffers against times of want, but also inviting to pests that could spoil the food supply. Impressions of oak leaves from the nearby forest were preserved in the ash; carbonized wood from the huts includes fig, beech, and black hornbeam, all of which, according to Claude Albore Livadie, "give the impression of a human landscape with fruit trees, pasture, and cultivated fields not far from a mixed beech woodland."[108] Such environments, like that assumed in the *Brontoscopic Calendar*, were the reality for most pre- and protohistoric communities throughout central Italy. Storms that blew down trees (Sept. 24, Oct. 3) would be of great consequence to people who depended upon wild fruits and nuts for dietary staples.

Some of the footprints of fleeing livestock show an interesting biological feature that would not otherwise have been known: a genetic condition that caused paw-like growth of the hooves. This condition in itself might have had little

[102] Albore Livadie 2002 and Mastrolorenzo et al. 2006:4367–68. Numerous finds of cows, goats, equines, sheep, and pigs sheltered in houses and villas at Pompeii and Herculaneum at the time of the Plinian eruption of 79 AD attest to the continuation of such practices over millennia (cf. King 2002:409, 416, 422–25, 438, 444–45).

[103] Fedele and Petrone 2000.

[104] Mastrolorenzo et al. 2006:4368.

[105] Mastrolorenzo et al. 2006:4368.

[106] Albore Livadie 2002.

[107] Albore Livadie and Vecchio 2005:586.

[108] Albore Livadie 2002:942.

impact, but it does attest to inbreeding and possibly other health problems of livestock. A toad and lizard were found where they had died in the roof thatch of one of the huts, showing the commensality of such "reptiles" with man; numerous amphibians were living in the wells and watering holes, and land snails and other potential pests, such as the common vole and wild rat, flourished.[109]

POGGIOMARINO (CAMPANIA): MIDDLE BRONZE AGE THROUGH SEVENTH CENTURY BC

Another site near the Sarno River confirms the interpretation of Early Bronze Age data. Poggiomarino in its early stages shows affinities with the famous "Terramare" culture of northern Italy. The town continued until centuries of depletion of the soil by agriculture, compounded by the environmental crisis of the Hallstatt Minimum, resulted in the silting up of the village's fine system of canals. It was abandoned during the first decades of the sixth century BC.[110] The building and maintenance of canals, walkways, and platforms, with terraces of earth and wood, demonstrates a well-organized society. Well-preserved, modern-looking wooden mangers were used for animal feeding.[111] Surely, the farmers would have called their contents "fodder," like the category noted in the *Brontoscopic Calendar* (Aug. 12, Nov. 17, Jan. 12). Fauna is attested in butchered and cooked bones of cattle, sheep/goat, and pig, with some deer, wild boar, wild goat, bear, wild cat, fox, marten, and otter. A very few equid bones, several dogs, turtles, birds, micromammals (mice? – research is ongoing), and snails complete the picture, with fishing attested by equipment such as hooks and net weights. Wild hazelnuts and acorns were gathered along with oak galls, probably used in cooking.[112] Beams and objects of oak and white fir, including two canoes, illustrate the proximity of the forest to this wetland oasis. Although Campanian rather than Etruscan, Poggiomarino is an intriguing example of the complexity and prosperity of Tyrrhenian Italian settlements during the Late Bronze Age and First Iron Age.

SORGENTI DELLA NOVA: PROTOVILLANOVAN (1100–900 BC)

At the well-watered site of Sorgenti della Nova, in the territory of the future Vulci, a very large and well-organized community of about 15 hectares grew

[109] Albore Livadie and Vecchio 2005:582, 586. Several species of terrestrial snail were found in levels of 79 AD at Pompeii and Oplontis, including some edible varieties; Reese 2002B:312–13.

[110] Albore Livadie and Cicirelli 2003:124.

[111] Albore Livadie and Cicirelli 2003:122, fig. 16.

[112] See Albore Livadie and Cicirelli 2003:120–22. At Pompeii and Herculaneum these galls, made by Cynipid wasps, were stored in bags and *dolia* in shops for use in medicine or tanning and dyeing; Larew 2002:319–20.

as a cluster of about 300 large, comfortable, closely situated huts, home to perhaps 1,500 people during the eleventh to tenth centuries BC.[113] Communal structures incorporating artificial caves held large ovens, storage rooms, and animal pens. It appears that public works, in the form of some roadways and small bridges or water culverts were constructed and maintained by the villagers. Supplementing this material with finds at other sites of the Protovillanovan period and slightly earlier, we get a picture of an environment rich in trees (maple, oak, elm, poplar, willow, beech, oak, Cornel), with a variety of wild foods plus grains and legumes cultivated near the settlements (*Triticum dicoccum* and other wheats, barley, millet, fava bean).[114]

In addition to common species, villagers caught tortoise, badger, wolf, bat, and *glis glis*, the edible dormouse.[115] Most faunal remains were found in artificial caves with fenced entrances, either for sacrificial use or pig breeding: over 75 percent are swine, with a high incidence of piglets and fetuses.[116] Sorgenti della Nova is defined as a proto-urban settlement, with fine evidence of a sophisticated infrastructure and central organization seen in communal public works.[117]

PROTOVILLANOVAN DIET

For the beginning of the first millennium (Protovillanovan period) an unusual find proves unequivocally that the population of central Italy ate barley and other grains as the basis of their diet. A burial in a log coffin at the site of Celano in the Fucino Basin (Abruzzo) preserved the remains of a woman who had died at sixty to seventy years of age. Her coffin was lined with straw dried from cereal plants, and within the abdominal cavity were remains of feces showing that the woman had consumed a meal of grains, apparently wheat and barley (orzo, grano).[118]

PROFILES OF IRON AGE SITES

At the northern Villanovan site of Castenaso (9th–8th c. BC), a picture emerges of a rather messy but active and well-fed village: numerous bones of wild and domesticated animals are from kitchen middens, but others seem never to have been eaten, and at times there must have been rotting carcasses, some as large as an ox, lying around the settlement.[119] (Dogs and rats gnawed on the bones.) Cattle, pigs (many butchered as piglets), sheep/goat, dog, and a small

[113] Negroni Catacchio 1995 and 1982:18–19.
[114] See Negroni Catacchio 1982:31, fig. 9.
[115] Negroni Catacchio 1982:32, fig. 10.
[116] See De Grossi Mazzorin 1995A:169.
[117] Negroni Catacchio and Domanico 2001.
[118] Capasso 1986.
[119] See Farello 1994.

number of horse, deer, boar, goose, and duck bones were preserved. Toads, water turtles, and land snails were not eaten but living in the vicinity, attesting to a rather wet climate (the Hallstatt Minimum).

An Iron Age settlement at Roverbella shows an essentially vegetarian diet of pulses, cereals (barley, durum wheat, emmer, and einkorn), small broad-beans (*Vicia faba minor*), vetch (*Vicia ervilia*), small-seeded cereals, such as millet and Italian millet, and poppy seeds. Spelt was attested, but it was more common in Europe than Italy. The excavators interpret these as the remains of the previous year's harvest, nearly depleted when the settlement was abandoned; the presence of both pulses and cereals may show that the farmers were practicing crop rotation, as indicated in the agricultural treatise of the Sasernas. The farmers of Roverbella harvested legumes in early summer and sowed millet to harvest at the end of summer.[120] It appears that this settlement only stored a year's worth of food at a time, but today subsistence farmers prefer to keep a three-year buffer to reserve seed-grain against failed harvests.[121] If Etruscan Iron Age villages did not lay up stores against bad years, they would have felt even more keenly the fears engendered by brontoscopic omens of famine, storms, floods, and drought.

The now submerged ninth-century settlement of Gran Carro on Lake Bolsena offers rare evidence for central Etruria of organic materials and foodstuffs.[122] The site was abandoned at the beginning of the eighth century BC – just when the meteorological effects of the Hallstatt Minimum were being felt, and the lake level rose too high.[123] Some carbonized *Triticum dicoccum* (*farro*, spelt) illustrates the presence of basic cereals. Other foods include seasonal fruits and cultivated crops: small broadbean (*Vicia faba minor*), cornel fruit (*Cornus mas*), hazelnut/filbert (*Corylus avellana*, nocciolo), wild plum (*Prunus spinosa*, prugnolo/pruno selvatico) and *Prunus insititia* – pruno selvatico or damascene plum. Oak is represented by wood fragments and broken acorn shells prepared for human food.[124] Vases held remains of plums, carbonized spelt, chestnut meats, and cornel berries (more than 95% of fruits recovered), which furnished a good diet of complementary proteins and carbohydrates, vitamins, and minerals.[125] Well-preserved grape pips of *vitis vinifera* are conclusive evidence of the domestication of grapes.[126] The pattern revealed is of mixed deciduous forests adjacent to open grain fields, with inhabitants relying heavily upon seasonal wild plant foods.[127] About 68

[120] See Menotti 2005:214–15.
[121] Cf. data gathered from modern Greek farmers by Hamish Forbes 1976.
[122] See Costantini et al. 1987 and 1995.
[123] See Tamburini 1995B:362.
[124] Costantini et al. 1995:330.
[125] See M. Hopf in Costantini et al. 1995:336–38.
[126] Costantini et al. 1995:330–31.
[127] Costantini et al. 1987:65.

percent of faunal remains are domestic species (including two dogs and one horse), 32 percent wild: one bear, one wolf, and some deer, which is more wild meat than neighboring communities enjoyed.[128] Excavators suspected that certain storage jars were made especially for storing dried fish.[129]

San Giovenale, one of the first Etruscan sites studied for the identification of animals and plants, has set a standard for excavation reporting. The deposits range from the Protovillanovan to mid-Republican eras, with the usual panorama of bones: cattle, sheep/goat, and pig, wild goat, hare, and tortoise. Barley, spelt, and fava beans are identified; grindstones were very coarse, implying that grains were processed for porridge rather than flour/breads.[130] Similar finds are noted at the seventh- to sixth-century site of Acquarossa near Viterbo: barley, spelt, oats (*Avena*), and some legumes (pea, chickpea, or vetch), with the usual assortment of cattle, sheep, pig of all ages, and small numbers of bones from deer, horse, and donkey.[131]

The eighth- to seventh-century Latin settlement at Satricum, in addition to some ritual deposits of Villanovan and Archaic date, has actually furnished pits filled with cooking debris and trash left after meals.[132] Marks of gnawing on animal bones hint at the level of sanitation and presence of pests and potential disease vectors in the village. Vegetable remains were less well preserved, but an eighth-century impasto cup still held grains of bread wheat (*triticum durum*), barley (*hordeum*), and horse bean (*vicia faba*); more wheat was found in seventh-century deposits. The trash pit of a sixth-century courtyard house still held traces of bitter vetch, sorrel, dandelion, buttercups, horse beans, wild grasses, and food plants with the accompanying weeds.

EVIDENCE FOR ARCHAIC ETRURIA

Forcello di Bagnolo San Vito (Mantua) was founded by Etruscans at the end of the sixth century BC, constructed with a grid plan and raised flood rampart. The large, three-room "house of the loom weights" included a fireplace with oven and a larder, with pots containing seeds, a heap of shelled small broadbeans, and many small jars with handfuls of pulses: lentil, pea, vetchling, vetch (thus, bitter vetch, normally grown as animal fodder, was used here as an ingredient in human food). Cereals included emmer, soft- and hard-grained wheat, einkorn, and spelt, all carefully shelled and cleaned. Many seeds show the borings of a weevil, *Bruchus rufimanus*, a rare indication of the troubles that threatened the early farmer.[133]

[128] Costantini et al. 1987:66–69; Tamburini 1995B:359.
[129] Tamburini 1995B:360.
[130] See Pohl 1987.
[131] See Scheffer 1987; Sorrentino 1981A and 1981B; Helbaek 1981.
[132] See Bouma et al. 1995:187–88.
[133] See Menotti 2005:215–16.

The fifth-century necropolis at Bozzolo, near Mantua, furnishes valuable evidence of the trials of Etruscan farm life. The nineteen skeletons of men, women, and children (aged 4 to 8) buried there show the effects of hard work and a coarse diet: severe wear and loss of teeth, arthritis, osteochondritis (from repeated microtrauma of the upper limbs, i.e., hard labor), and traumatic injury – one person suffered a cranial fracture and underwent trepanation to relieve the pressure but did not survive. Some of their problems are attributed to a diet rich in cereals but poor in iron. Clearly, the people of Bozzolo worked hard, and had a life expectancy of fifty to sixty years, with a mean age of twenty-six (five are said to have suffered rheumatoid arthritis, including one 8-year-old child.)[134]

EVIDENCE FROM POMPEII AND HERCULANEUM (79 AD)

The finds of "necessities of life" in the settlements surrounding Vesuvius are too important to overlook. Many of the same crops and techniques were still employed in nineteenth-century Italy and were demonstrably prevalent during the period of Etruscan autonomy. Fodder plants (important for fallowing fields and crop rotation), cited several times in the *Brontoscopic Calendar*, have actually been found, carbonized, in many mangers, stalls, and barns.[135] The manger in the stable at Pompeii House IX,12,6 held hay/fodder cut in a field of mixed oats, barley, turnips, and legumes; many bins also held cuttings from rather woody plants. The hay at the Oplontis villa held traces of over 100 species of plants, including grape, olive, flax, vetch, and many weeds or wildflowers.[136]

Other foods attested at Pompeii and in the *Brontoscopic Calendar* and the Etruscan settlements include chestnuts (cultivated, as in Columella *De re rustica* 4.33.4 and 5.10.14) at Oplontis, Herculaneum, and Pompeii where, in the Tavern of Verecundus, they were eaten as food.[137] Small broadbeans (*Vicia faba*) were found in many cooked dishes and storage containers in houses, *tavernae*, and even the *lupanar*; leaves and stems were used as fodder. Lentils were abundant,[138] and they remained an important food crop for Rome.[139] Cereals were well represented: stored and cooked wheat (and bread), barley, and millet.[140]

[134] See Menotti 2005:216–17.
[135] Borgongino 2006:55–56.
[136] See Ricciardi and Aprile 1988.
[137] Borgongino 2006:64–65.
[138] Borgongino 2006:103–6.
[139] See Jashemski 2002:120–21, for ancient sources.
[140] See Borgongino 2006:98–102, 109–10, 122–26, 140–46.

Tree fruits, both soft and hard (nuts), were also readily consumed and carefully stockpiled in Pompeii and Herculaneum, from pine nuts, almonds, walnuts, and hazelnuts[141] to plums/prunes stored in glass and clay vases, and pomegranates, stored in layers of straw at the villa of Crassius Tertius at Oplontis.[142] Apples were certainly used, but they had not yet been harvested at the time of the eruption.[143] At the Pompeian Temple of Isis, fruit was a burnt sacrifice: walnuts, hazelnuts, pine nuts, chestnuts, figs, grapes, and dates, only the last being a foreign import.[144] Insect pests, snails, mice, toads, lizards, and such are all attested in the Vesuvian destruction debris, and they again illustrate the species referenced in the *Brontoscopic Calendar*.[145] Residue inside a *dolium* at a farm outside Pompeii incorporated many species considered noxious, and Annamaria Ciarallo (2002) suggested it was the antidote compound *mithridatium*. It included owl, toad, chicken, mouse, skink, and three frog species, along with poisonous and other plants, such as *cannabis sativa*, cypress, fig, fir, grape, malva, ranunculus, and verbena.

CONCLUSION

In the rich, volcanic soil and terrain that fed abundant flocks, herds, and wild things, many factors conspired to foster the anxiety of a community, and those anxieties only heightened as cities and urban populations grew. Secure storage of the food supply and preserving it from pests and decay was always a problem. Reliance on grains in agriculture meant that a year of failed harvests could kill, just as a good year could support a large family with surplus left to barter. Living so close to the domesticated livestock, not to mention wild animals, meant increased likelihood of infectious diseases, which, if they came into an unexposed population, could have wreaked havoc. Evidence from Nola-Croce di Papa shows that exposure to pathogens was enhanced by village or city life, because trash was lying around the houses, and water supplies could be easily contaminated, spreading disease and parasites. Many communities were also at risk for heavy metal poisoning from metallurgy.[146] Life could be very good in Etruria (as it is today), with abundance of the "necessities of life" predicted in the *Brontoscopic Calendar* (June 16, July 12, Aug. 2, 17, Oct. 5, 28, Dec. 20, March 16). But no one was truly safe, and thus the desire for forewarning of troubles is quite understandable.

[141] Borgongino 2006:129–32, 106–7, 114–18, 111–14.
[142] Borgongino 2006:135–37, 107–9.
[143] See Jashemski 2002:124–25; one apple was identified in the garden of Poppaea's villa.
[144] Borgongino 2006:152.
[145] See Larew 2002; Reese 2002b; Bodson and Orr 2002.
[146] See Harrison et al. 2009 and references therein.

CHAPTER 6

HEALTH AND DISEASE: "PLAGUE, BUT NOT EXCEPTIONALLY LIFE-THREATENING"

HEALTH AND DISEASE OMENS IN THE *BRONTOSCOPIC CALENDAR* AND ITS PREDECESSORS

More than 100 predictions in the *Brontoscopic Calendar* relate to human and animal health, disease, or well-being; they are indicative of the attitudes of Etruscan society toward infectious disease and the conditions that could foster it.[1] The entries of "epidemiological" interest may be sorted by topic: good health, weather triggers, malnutrition, enteric and respiratory diseases, skin lesions, plague-like severe illness, zoonoses, and socially factored diseases. A certain number of health-related omens have parallels in Mesopotamian literature, but many do not – they reflect conditions and dangers of peninsular Italy rather than Mesopotamia and probably derive from oral traditions in Bronze Age and Iron Age Italy. The importance of such information and the urgency of remembering it could well have fostered orally transmitted rituals or myths, as postulated by Elizabeth and Paul Barber (2004), and by Adrienne Mayor (2000 and 2003) for a number of Classical and other myths. Such an oral tradition in play in Etruria during the early eighth century BC could easily have been melded with information learned from Near Eastern or Levantine omen texts.

The omens relating to health, disease, or occasionally "destruction" of humans and/or animals fall on all days and months, but certain days seem punctuated with more woes than others. Five or six months all have problems of health or life on the following days: 5, 7, 8, 12, 13, 14, 22. The clustering around mid-month might be because this was a frequent time for lunar eclipses, as seen in the Mesopotamian omen series. January, February, and December show the greatest number of health-and-welfare-related predictions, whether

[1] This chapter supersedes Turfa (in press), my initial survey of the disease in the *Brontoscopic Calendar*, presented at the conference on Ancient Medicine sponsored by the University of Stockholm, Medelhavsmuseet and Karolinska Institutet, Stockholm, November 2001.

for good or bad, with a tally of at least fifteen, twelve, and ten apodoses, respectively. May, June, and September had the least (4, 3, and 4 items). If we assume that the results of a prediction fell within the same month as the omen trigger (thunder), this fits a pattern recognized for modern diseases, with good weather provoking fewer troubles, and cold or stormy weather (meaning poor food supply, cramped indoor conditions, etc.) promoting health crises in the depth of winter.

The health of women is cited specifically in the *Brontoscopic Calendar*, and there are predecessors within the Mesopotamian texts.[2] At June 2, thunder means that "women in labor will have an easy delivery, but there will be abortion of cattle, yet there will be an abundance of fish," and at February 11, "good deliveries for women" are predicted – both are good omens, and they could be related to the air pressure drop during thunderstorms. (Abortion of cattle may indicate a separate disease, as discussed later in this chapter.) The fact that childbirth was considered important, and the loss of a newborn a special tragedy, is occasionally seen in the funerary rites that followed, as seen, for instance, in a third-century burial of a woman and baby who died at birth at Etruscan Norchia: the mother was given a variety of special offerings, including ribs of lamb and *two* pairs of fashionable *tyrrhenika* sandals.[3] More telling for the period of formation of the *Brontoscopic Calendar* is the Late Villanovan–Early Orientalizing tomb of a Faliscan lady (probably buried at Narce) – her urn is in a rare stone receptacle along with a quantity of valuables, some miniature.[4] These were probably gifts/markers of her neonate infant, whose tiny bones were mingled with hers. Other elite children at Narce were buried with miniature sets of jewelry and banquet vases, suggesting their importance to the Villanovan community.[5]

ANCIENT PERCEPTIONS OF DISEASE AND HEALTH

The *Brontoscopic Calendar*, being essentially divinatory in nature, portrays disease, like weather, famine, or prosperity, as sent by the gods. The faithful undoubtedly interpreted disease as a punishment or warning from the gods, as explicitly stated in the *Prophecy of Vegoia* (Appendix A, Text 2). It may seem curious that the calendar refers to "plague" (λοιμός at Jan. 29, Feb. 15, April 15, and May 9, or ὄλεθρος at Aug. 13, 20, Dec. 8, 24), when the other omens, apart from one citation of "consumption" (Feb. 2), merely refer to "disease" (νόσος and its derivatives account for 37 entries). Λοιμός is the characteristic term for

[2] Akkadian predictions are usually negative: lunar eclipse and Jupiter omens foretell miscarriage or death in the *Enūma Anu Enlil* 22 and 63.

[3] Barbieri and Becker 1996–97.

[4] De Lucia Brolli 2004.

[5] See Turfa 2005:20–21, 65–66, 132–35, Narce Tomb 102F.

"plague" or epidemic, usually translated into Latin as *pestis*. The other term, ὄλεθρος, is used in Greek literature for "destruction, ruin, death."

We are conditioned to assume "plague" was a pandemic like the Black Death of medieval Europe, but, in the absence of Etruscan literature, it may be appropriate to extrapolate from the Hippokratic early texts. W. H. S. Jones (1923/1995:lv–lvi) noted that ancient doctors had to describe diseases by their distinctive array of symptoms (such as "tertian" or "quartan" fever, for instance, that manifested every "third" or "fourth" day).[6] Epidemics did occur sporadically and often probably were triggered by the stress, dietary deficiencies, and the mingling with foreign groups that invariably happen during wars.

Most historians cite a small number of well-known events: the plague of Homer's *Iliad*, the Homeric and Hesiodic disease season of the Dog Star, the plague that befell the Philistines when they captured the Ark of the Covenant, and possibly plague in the form of an angel that struck the camp of Sennacherib at Jerusalem (2 Kings 19:32–36). The Plague of Athens at the end of the fifth century linked pestilence with war, as did the diseases of 413 and 213 BC, when the Athenian and then the Carthaginian siege expeditions were turned away from Syracuse by sickness. Early Rome also knew pestilence, with several incidents in the course of the fifth century BC. Livy (1.31) cited a plague in Rome during the reign of Tullus Hostilius (672–640 BC) that had been presaged by a rain of stones and a heavenly voice on the Alban mount. Men of military age were afflicted, and the king himself was so ill that he consulted the commentaries written by Numa Pompilius and tried to perform their secret rituals. His failure resulted in his being struck down in his house by a thunderbolt and killed.

Greek Iron Age literary evidence (Homer and Hesiod) is contemporary with the formation of the Etruscan cities and cults. For instance, Apollo sends arrows of pestilence through the Greek camp at Troy in response to Chryses' prayer (*Iliad* 1.8–26, 33–108). Hesiod also sometimes portrays disease as meted out by the gods as punishment, rather than as a force of nature. In *Works and Days* (240–44), those who practice violence are punished by the gods: "On account of a sinful man, Zeus afflicts a whole city with both famine and plague, so that men perish and the women do not bear children."

Through the Early Republic, pestilence often struck Rome and accompanied conflict: in 463 BC, disease caused the Volscians and Aequi to abandon their attack. A plague in Campania followed the eruption of Vesuvius in 79 AD, and from the second century AD on, the Roman Empire knew periodic plagues, with the so-called Antonine and Aurelian plagues, and then the plague of Justinian, usually identified as *Yersinia pestis*, bubonic plague. The dense, large populations of Mesopotamian and Levantine cities may have harbored some endemic infections, but I will argue that those diseases cannot be

[6] Jones 1923/1995:lv–lvi.

at issue among the smaller populations and more temperate climate of Iron Age central Italy. Malaria might account for some recurring disease, according to the recent studies of Sallares (2002) (also see the section on malaria/nonlethal disease later in this chapter).

Dionysios of Halikarnassos (*Ant. Rom.* 1.23–24), citing Myrsilos of Lesbos, tells of the Pelasgians, who became the Tyrrhenians and then were driven out of Etruria, "two generations before the Trojan War" (1.26.1). Having conquered a prosperous region, they were stricken with a series of catastrophes: crippling drought, loss of crops, of drinking water and of livestock, abortions in cattle and women, adults afflicted with unusual diseases and unexpected deaths. An oracle explained: they had neglected to honor an old vow made for relief from famine – they offered animals and crops to the gods, but neglected to dedicate their children as first fruits! The history (and identification) of the Pelasgians is a separate issue,[7] but the tale serves to place at least as early as the Late Bronze Age the cultic association of physical afflictions and disease with moral transgressions, and the need for expiation.

PRODIGIES AND ACCOUNTS OF UNUSUAL PHENOMENA

Other texts, from Roman archives, attest to concern about the infliction of sufferings by angry gods. The prodigy accounts, culled by the fourth-century AD author Julius Obsequens mainly from now lost portions of Livy, illustrate the religious significance of diseases and bodily afflictions, along with the appearance of deformed or bizarre creatures and phenomena, such as rains of stones or blood and apparitions in the skies. (Today Obsequens is lauded by UFO societies; and the Roman-era births of hermaphrodites and talking livestock would have fit nicely in the framework of the Mesopotamian *Šumma izbu*, but they do not appear in the *Brontoscopic Calendar.*)[8]

Note the syntax of the description by Obsequens (Prodigy 20; and Livy 7.6.1–6) of the affliction caused by the Lacus Curtius sinkhole (4th c. BC): *vaporibus pestilentissimis multorum hominum corpora*, "afflicting the bodies of many people with the most virulent vapors." What else would disease infect but the "bodies of men," yet this resembles the language of some *Brontoscopic* apodoses, a poetic statement presumably meant to be intoned aloud to an audience. Compare June 1, when "dangerous diseases will be inflicted upon bodies," or July 22, "but for the bodies, diseases around the head," and August 13, "plague upon the bodies of both humans and dumb animals."

[7] See Briquel 1984.

[8] Jacobs 2010 (received too late to include here) suggests some affinities between Cicero's *De divinatione* and the *Šumma izbu*: much more could be sought in Late Republican intellectual culture.

TABLE 6.1. *Sampling of prodigia recorded by Julius Obsequens and others*

Source	Approximate year (all BC)	Prodigy
Obsequens 1	738 Romulus	Rain of blood in Rome Plague *in urbem*, causing rapid death *Sterilitas* of fields, loss of crops Causing great famine (Romulus disappears in storm cloud)
Obsequens 2	705 Numa year 8	*Morbus pestifer* in Rome and throughout Italy (shield falls from heaven, Salii given custody)
Obsequens 3 Livy 1.31	643 Tullus Hostilius	Rain of stones on Alban mount Voice heard in grove warns Albans to sacrifice Serious *pestilentia* affects king also Hostilius killed as lightning burns down his house
Servius *Ad Aen.* 2.140 Libri fatales	Later 6th c. Tarquinius Superbus	Miscarriages or pestilence expiated with the Ludi Taurei
Plutarch *Publicola* 21.1	504	Miscarriages expiated by games for Pluto and Apollo Authority of Duoviri
Dion. Hal. 6.17.2–3 Pliny *NH* 35.154	496	Famine causes Duoviri to institute the *Cerealia*
Dion. Hal. 9.40.1–4	472	*Incestum* of Vestals Pestilence, signaled by *manteis/pontifex*
Obsequens 13	464	Sky burning, a year of pestilence for humans and flocks
Obsequens 14 Livy 3.5.14	462	Sky on fire; unhealthy season in city and country Livestock kept in city due to war, "cattle suffered as much as men" 3-day *feriae* for prayer
Livy 5.31.5 Dion. Hal. 13 fgt. 4	392	Famine and pestilence
Livy 5.35.5		Pestilence, sacred books are consulted
Livy 7.2.1–7 Augustus *CD* 2.8 Orosius 3.4.1–6	364	Pestilence Pontifex decrees *lectisternium*, *Ludi scaenici* held
Obsequens 20	359	Sinkhole opens in Forum Pestilential vapors released by sinkhole – plague only abates in response to Marcus Curtius' *devotio* ritual
Livy 7.27.1	348	Pestilence, expiated by 4th *lectisternium*

TABLE 6.1 *(continued)*

Source	Approximate year (all BC)	Prodigy
Obsequens 23 Livy 10.31.8	295	First lightning prodigy: many in army of Appius Claudius are struck by lightning Rain of earth *Pestis* in Rome and throughout Italy Decemviri officiate
Livy 10.47.6 Valerius Max. 1.8.2 Ovid *Meta.* 15.622ff	293	Pestilence leads to importation of Asklepios cult (see MacBain 1982:86)
Augustus *CD* 3.17 Orosius 4.5	266	Pestilence provokes restoration of Roman shrines
Obsequens 27	265	Many *prodigia*, including rains of blood, earth, milk *Pestilentia crudelissima* in city [Gurges' sac/relief of Volsinii in revolt of *libertini*]
Livy per 49 Censorinus *De die natali* 17.8, etc. Pliny *NH* 18.286	249–238	Lightning omens, war, disease Dearth causes establishment of *Floralia* *Instauratio* of *Ludi saeculares* (see MacBain 1982:87)
Livy 27.23.	208	Many omens throughout Italy, including: Vulture in temple of Jupiter at Caere Lake Bolsena stained with blood *Pestilentia gravis incidit in urbem agrosque quae tamen magis in longos morbos quam in perniciabiles evasit* Romans offer prayers, games in expiation

Source: See Schlesinger 1959.

OTHER SOURCES ON THE HEALTH OF ETRURIA

Fortunately, our picture of the Etruscans' approach to health and disease may be augmented by a unique category of evidence, the later votive deposits (ca. 300 to 1st c. BC), in which large numbers of terracotta models of body parts were offered. The level of physiological knowledge expressed in the anatomical sculptures is usually minimal: hands, feet, eyes, and noses are self-evident, but even large polyvisceral plaques are only schematic, rationalized versions of actual organs, subject to the same aesthetic rules as Classical art. Most organs could have been modeled from butchers' displays.[9]

The only models of internal organs that resemble human originals are testicles and those that depict a pregnant, laboring uterus, sometimes with

[9] Tabanelli 1962, 1963:3–69; cf. Turfa 1994.

additional features (urinary tract, fibroids, bicorporate uterus) added to show the artist's or patron's understanding of human health. In fact, there are numerous indications in Etruscan religion of a view of human welfare and traditions of medical knowledge that are alien to those of Greece (Hippokratic or otherwise). The uterus models, thank offerings for cures of obstetrical-gynecological problems, show that postmortem C-section was known in Etruria by circa 300 BC: this is not attested outside Italy. The Roman *Lex regia de mortuo inferendo*, attributed to Numa Pompilius (715–672 BC), and still on record in the era of Justinian, required that an attempt be made to remove a fetus if the mother died. It is very likely that an Etruscan law or precept was the origin of this tradition.[10] Certainly the acknowledgment of Numa's law attests to the use of written documents early in the seventh century in Rome. The large quantities of such models in more than fifty Etruscan and Latin sanctuaries are further indication of the social status of women in Late Republican society. Large numbers of swaddled infant images were also offered, indicating that the gods were indeed thought to control conception and childbirth in Classical Etruria. Two model uteri deposited at Vulci name Vei (Ceres), not Tinia (thunderer, erstwhile gestator of Menrva and Fufluns) or Thalna, the birth goddess.[11]

We may come closer to what Etruscans actually thought about their health, and its dependence upon the cosmic forces of nature, in the rare passages of religious texts, especially the *Prophesy of Vegoia* (Appendix A, Text 2).[12] "The most severe diseases and wounds" (*pessimis morbis et vulneribus*) are wielded as punishment by the gods who defend the cosmic – and thus also terrestrial – boundaries. Those who displace the boundary markers will suffer sickness and idiopathic wounds, and their limbs/bodies will become weak. The identities of "the gods" who dispense the punishment are not revealed.

Vegoia's warning linking *disease* to the transgression of boundaries might express a perceived territorial component to some forms of human illness: early Etruscans could have seen that people who traveled to distant regions sometimes contracted more severe or different diseases, or that visitors from afar brought new diseases with them. Even crop or pasture rotation, or a move to a different, better-looking, better-watered pasture could have exposed people and flocks to new afflictions, such as parasites. The *etrusca disciplina* was probably being codified in just the period, the eighth to seventh centuries BC, when foreign traders established regular routes and Levantine and Greek colonists founded permanent settlements in Campania. And some colonists married Etruscan and Italic men and

[10] See Turfa 1994:227–32, also 1986, 2006A, and 2006B:75–81.

[11] Colonna 1988:33, fig. 11; Turfa 2004:363, no. 302.

[12] Valvo 1988; Heurgon 1959 and 1964:228–30; Schultz 2006:222–23.

women, to judge from the evidence of the Pithekoussan graves (discussed later in this chapter), which also show very high infant mortality in the first generation born to the settlers on Italian soil.

METHODOLOGICAL CONSIDERATIONS

Because we do not know the original Etruscan, or the Latin of Figulus' translation, identification of the diseases cited in the text must be somewhat speculative. The language of the original, never intended as a scientific text, could have been 700 years old when Figulus rendered his Latin version. Few symptoms are specific enough for diagnosis; occasionally the omen contains additional clauses possibly relating to predisposing conditions of weather, wind, famine, drought, or war. Another issue for the interpretation of omens as common natural phenomena, linked to weather and seasonal conditions, is the determination of the time elapsed between the hearing of thunder and the arrival of its consequences. This is never specified in the text and so cannot be correlated with the incubation periods of diseases. The span of three months or less, the length of one season, figures in many prophetic *schemae* in Mesopotamia and the Classical Mediterranean. Some events, like "the summer" or "the harvest," would develop over a longer period of time, but the rapid barometric drop of an actual thunderstorm might be enough to trigger "good deliveries for women" (June 2, February 11) – few other cases are so clear-cut. (In Roman prodigies, the time lag could vary so greatly that causality would hardly seem at issue: some *prodigia* associated with Hannibal supposedly played out more than a year after the prediction was given.)

DISEASE AND POPULATION DENSITY

Some diseases may be eliminated from consideration, on the grounds of density dependence or known post-Archaic origin. Diseases that only appeared *after* the Classical period include yellow fever, bubonic plague, cholera, leprosy, and rabies.[13] Those of tropical climates such as schistosomiasis or filariasis are not likely candidates either.[14]

Smallpox might be included, but see "Other Poxy Diseases," later in this chapter, for this hotly contested issue. I have found no grounds to consider treponemas in this analysis. Post-Archaic skeletal material from the

[13] Mesopotamian texts furnish one possible prediction of rabies: *Enūma Anu Enlil*, Lunar Eclipse tablet 22, part II, section IX, paragraphs 1–2: "[If an eclipse occurs in Kislimu from the 1st to the 30th day]: Dogs will rage"; see Rochberg-Halton 1988:267.

[14] For brief surveys, see Ortner and Theobald 1993, Stannard 1993, Aufderheide and Rodríguez-Martín 1998.

countryside of the Greek colony of Metaponto has been interpreted by some authors as evidence of endemic syphilis (congenital treponematosis), but this remains controversial.[15] Instead, taphonomic conditions probably account for lesions said to be present in an incredibly large percentage of the population, even young children. The porotic hyperostosis and anemia evident in some skeletons may have had quite different causes.[16] A related study found no evidence of the syphilis antigen in soils or bones.[17]

The population of an Iron Age village or even an Archaic city in Etruria can never have risen to the density needed to sustain most well-known epidemic infections, because populations invariably fall well below the necessary threshold, usually by an order of magnitude.[18] The exceptionally large Protovillanovan settlement at Sorgenti della Nova, with an estimated 300 huts, probably housed around only 1,500 people during its main period of habitation.[19] Although dense habitation and shared water sources, ovens, and storage facilities could make for some problems of sanitation and public health, a population of 1,500 cannot support a high enough birthrate to maintain the familiar epidemic or endemic diseases.[20]

Significant population increases occurred during the period of urbanization of the seventh to sixth centuries BC, but even that would not have been sufficient to foster a reservoir for permanent endemic disease. Cornell has evaluated the rough figures based on historical sources, defensive circuits, and the carrying capacity of adjacent farmland, which show that, at Volsinii, Caere, Tarquinii, Vulci, Veii, or *Roma quadrata*, populations peaked in the late sixth century only at 25,000 to perhaps 40,000.[21] Rasmussen and Perkins have worked up figures for the Albegna valley region and extrapolated an estimate for the population of Tyrrhenian Etruria in the fifth century BC, arriving at approximately 282,480.[22] Steingräber (2001) suggests populations as low as 5,000 (for Chiusi). It seems likely that the major cities had populations of 20,000 or more by the fifth century, and that the intercommunication of the maritime Etruscan cities could have produced enough people to support some of the epidemic diseases.[23] Even at the upper limits of scholars' estimates, these cities, though they periodically would reel from the arrival of some new

[15] Henneberg and Henneberg 1998:527–37.

[16] Compare Aufderheide and Rodríguez-Martín 1998:154–71.

[17] Jeske-Janicka and Janicki 1998.

[18] For dynamics of transmission of infectious diseases, see Anderson and May 1992; Evans 1998.

[19] See Negroni Catacchio 1982:18–19.

[20] For background on ancient health as compromised by the stresses of civilization/urbanized life, see Cohen 1989:105–29.

[21] Cornell 1995:204–8; see also Heurgon 1964:145–48. Heurgon based his estimate of 25,000 for the population of Caere on the Banditaccia tomb census of the late 6th c.

[22] Rasmussen 2005:86–88; Perkins 1999.

[23] Cf. Becker 2007.

infectious disease, could still not have sustained such diseases as plague. For comparison, the estimated population of *average* Greek cities [not Athens] would probably have been 2,500 to 4,500 during the Classical period, and the earlier population of the Phoenician colony at Toscanos in Spain (8th–7th c. BC) was estimated by Niemeyer at 1,000 to 1,500 within its walls.[24]

THE USE OF OSTEOLOGICAL AND RELATED STUDIES

Studies of bones from Etruscan and Italic necropoleis are beginning to develop a picture of the general health and welfare of Archaic Italians, but the number of individuals studied is not yet sufficient to support generalized conclusions. Capasso and Di Tota (1997) attempted a study of bones from various sites and periods now stored in the Florence Museo Archeologico, finding only a tiny number of pathological conditions, most of which were dental or related infections. The one severe lesion they were able to identify was probably a case of Pott's disease from the region of Perugia (see the section on tuberculosis later in this chapter).[25]

Mancinelli et al. (1997) studied 2,775 skeletons from central Italian necropoleis of the ninth century BC through third century AD, furnishing preliminary mortality tables based on a variety of populations. They caution that data are affected by the low number of children, suggesting that most people who reached twenty years of age could expect to live another seventeen to twenty years, in all periods. The situation for Tarquinia remained essentially static from the eighth through the fifth century BC. Analysis of enamel hypoplasias confirms a higher mortality rate for individuals who experienced more childhood stress from disease and/or poor nutrition. They found some indication that health and longevity were better prior to about 700 BC and then declined slightly as urbanization spread in southern Etruria; the urban trend reached the Adriatic slightly later, and only then did mortality take a downturn there.

BACKGROUND: ARCHAEOLOGICAL EVIDENCE FOR THE HEALTH AND WELFARE OF EARLY ETRUSCAN AND ITALIC POPULATIONS

In the absence of Etruscan literature, we must rely on studies of burials, rare instances of artistic depiction or intimation of human infirmity, or indication of relative hygiene and nutritional background for early populations. Few skeletons yield evidence of cause of death, but it seems likely that the burial of multiple

[24] On Greece, see Raaflaub 1991:567, 570–72; on Phoenicians, see Niemeyer 1990:484; 1986: 117.

[25] For a review of past scholarship on paleopathological evaluation of Etruria, see Mallegni and Vitiello 1997. For a critical evaluation of skeletal studies as related to nutrition and health, cf. Cohen 1989:118–22.

individuals simultaneously or within short intervals, as at Osteria dell'Osa, may be the result of infectious disease moving through the population.[26] A few of the most important sources for this information are briefly presented below.

SECOND MILLENNIUM: SAN PAOLO BELSITO (CA. 1780 BC)

One of the earliest relevant finds was preserved when two individuals were killed by a plinian eruption of Vesuvius early in the Bronze Age at San Paolo Belsito. The skeletons of a fortyish man and a young woman (probably 21) were discovered where they collapsed as they fled their village. The bodies and a nearby village (Nola-Croce del Papa) were covered by lapillae. The skeletons showed the effects of rugged life and hard work.[27] The woman, about 1.5 m tall, showed childhood want or illness in a light level of hypoplasia of her tooth enamel; anemia was a chronic condition that had already caused porotic hyperostosis and *cribra orbitalia*.[28] The excavators estimated that she had given birth three or more times, according to depressions on the ischiatic incision of her pelvis.

The man was relatively tall (approx. 1.72 m) and well muscled, but he showed a slight degree of anemia/*cribra orbitalia*. His tibias were slightly bowed from a mild case of rickets, and he had stress-induced arthrosis in both feet and knees, and periostitis on the diaphyses of his tibiae and femurs – the result of hard and constant work. The authors suggested that at this time the Italic inhabitants of Campania were making the transition to a heavy reliance on cereals as the staple of their diet, so they were consuming less meat, experiencing poor iron absorption, and thus developed anemia. Such rare finds of people killed by catastrophe in the prime of life rather than those who died of natural causes and were buried in funerary rituals, can give a more balanced, albeit depressing, picture of a population's health and well-being.

A MIDDLE BRONZE AGE CAMPANIAN POPULATION: SANT'ABBONDIO (POMPEII)

A population of sixty-two individuals, from a mid-second millennium necropolis excavated near Pompeii, at Sant'Abbondio, showed relatively little trauma,[29] perhaps evidence that the weapons display behaviors of this period

[26] Shared burial, possibly evidence of infectious disease felling assorted members of a group, is also found in Verucchio Tomb 89, and an early 7th-c. tomb at Vulci; see Becker et al. 2009:44–46.

[27] Fedele and Petrone 2000:55ff.

[28] *Cribra oribitalia* are alterations of the skull adjusting to anemia. For background on these conditions, see Fornaciari 1997:467–68. Consensus does not exist on the value of methods of estimating fecundity by study of the pelvic bones.

[29] Tafuri 2005:18–19.

did indeed deter actual violence.[30] At least twenty-five individuals had enamel hypoplasias formed in their teeth at three to four years of age, in response to the stress of weaning, which invariably brought nutritional problems and heightened susceptibility to infections. The hypoplasias were evenly distributed between males and females, likely evidence that boys and girls were treated equally as regards their early diet/upbringing. Twenty-five people had dental caries, interpreted as good indicators of diet: the bacterial breakdown of carbohydrates (from cereals) causes tooth decay.[31] Over half, both male and female, had multiple caries, especially young adults. More people here had cavities than anemia, so perhaps they carried thalassemia that protected sufferers from malaria.

Tafuri's biochemical studies reveal past diet and nutrition, with implications for the society of this group.[32] The teeth of several men showed higher levels of lead; they probably drank from a different water source, spending significant amounts of time away from their homes in the process of transhumance.[33] Some women buried in the western part of the necropolis showed different levels of minerals, implying that they had spent their formative years elsewhere. They were buried by the same rituals, and their nutrition and health were comparable to the other women, so their biochemical differences may reflect the practice of exogamy, women moving to a new community when they married.[34] They might also have maintained a slightly different diet from their female in-laws, too: perhaps an additional source of social unease.

Those interred in the eastern necropolis showed more signs of physiological stress in enamel hypoplasias and *cribra orbitalia*, and Tafuri wondered if these were the family burial plots of a poorer segment of the village's population.[35] If so, it marks an early beginning to the differentiation of classes in Italian society, because the second phase of the Iron Age or Villanovan period is characteristically viewed as the stepping off point for class hierarchy in *Etruscan* society.

Facchini and Brasili (1997) identified past skeletal studies from the necropoleis of Bologna that showed a high degree of homogeneity of skeletal characteristics across time and ethnic boundaries among the populations of the Po region, Adriatic, Tyrrhenian Etruria, and Italic central-southern Italy during the first millennium BC, especially the ninth to mid-seventh centuries. Given this situation, and the relative uniformity of climate and landscape across

30 In another part of the Mediterranean, Bronze Age Greek Lerna, the battered male and female population might have disagreed; but see Angel 1971.

31 Tafuri 2005:16–20.

32 Tafuri 2005:55–70.

33 Tafuri 2005:67–70.

34 Tafuri 2005:65–66.

35 Tafuri 2005:21.

central Italy, it seems reasonable to extrapolate some general impression of health and welfare conditions for the Iron Age recipients of the *Brontoscopic Calendar* from somewhat scattered evidence.[36]

IRON AGE LATIUM: OSTERIA DELL'OSA (GABII) IN NINTH TO EIGHTH CENTURY BC

Osteria dell'Osa, predecessor of Latin Gabii, was ideally situated near ridge roads and rivers linking *Latium Vetus* with Etruria (Veii) and Campania.[37] Interdisciplinary studies of the Iron Age necropolis have furnished a wealth of information about kinship groups, social stratification, industrialization, and the urbanization of central Italy.[38] Several tombs had been disturbed by the families for a second burial after only a short interval.[39] In *Fossa* grave 482 (830–770 BC) a man's body buried on the floor of the inhumation trench (grave 483) had later been propped up by the shoulders and left resting against a *dolium* containing the cremated bones of a woman. Might the short intervals between burials of family members of a different sex and age be the result of infectious disease passing through the settlement?

These ninth- to eighth-century Latins were studied by Marshall Becker and Loretana Salvadei (1992). The man in burial 483 was about thirty years old, and his skeleton shows no sign of pathological conditions. The cremation urn (no. 482) held the remains of an elderly person, possibly a woman, over sixty years of age; no evidence was found of her cause of death.[40] Dental caries were not common and only developed after thirty to forty years of age, presumably because of a diet less reliant on carbohydrates/sugars; wear on the teeth indicates biting tough foods. Enamel hypoplasias were very common signs of multiple phases of stress from disease and/or poor nutrition during childhood.[41] Age at death indicates the community's good general health. Roughly 20 percent died between twenty and thirty years, 20 percent died between thirty and fifty, and 25 percent died aged fifty or older – but only a small percentage of children died in any given year.[42] Old people experienced age deterioration of the bones, such as arthritis, but, as elsewhere, there is little in the sample to attest to violence or plagues as cause of death, except perhaps

[36] Small populations of the 9th–7th c. at Ripaie and Guerruccia, near Volterra, had little skeletal pathology and few cases of iron deficiency anemia; see Romagnoli 1997:455.

[37] About ninety burials at Castiglione belong to a simpler village in the vicinity. Although two graves have been dated by radiocarbon to the early 9th c., no information on health of the skeletons has been published as yet. See Nijboer et al. 1999/2000:171–74.

[38] See Bietti Sestieri 1991 and 1992, and 2010.

[39] See Bietti Sestieri 1992:121–25.

[40] Bietti Sestieri 1991:99.

[41] Becker and Salvadei 1992:64.

[42] Becker and Salvadei 1992:108.

the close temporal spacing of some burials. The burial rituals do furnish rich evidence of social stratification.

THE EIGHTH-CENTURY COLONIAL SETTLEMENT OF PITHEKOUSSAI

The industrial and commercial settlement of Pithekoussai, established by 775 BC, was a joint venture of Greeks from several cities and Levantine people, probably (north) Syrian and Phoenician ethnics.[43] Burials of the eighth century also show rituals and artifacts that indicate close relations with Etruscans and Italic peoples from Campania and as far away as Daunia or Apulia.[44] Approximately 1,500 graves of the eighth to seventh centuries excavated in the San Montano necropolis represent less than 10 percent of the total Late Geometric population of the colony, which thus would have been a very respectable 10,000–20,000 souls. Conspicuous tombs of infants interred in transport amphorae demonstrate a high infant mortality during the eighth century.[45] Most family plots encompass *several* burials of young children.

Studies of the cremated bones of over 100 individuals showed a ratio of men to women of approximately 2:3.[46] Some of these women, according to grave goods, and harmless genetic dental anomalies like bifurcate tooth roots, were almost certainly the Italian (Italic or Etruscan) wives of Levantine or Greek colonists. Adults show reasonably good health: many were elderly (65–70 years), and few had died as teenagers or in their twenties. One woman showed signs of anemia, which Becker attributed to possible malaria or thalassemia.[47] The high infant mortality was probably the result of malaria, to which these new citizens had not yet developed resistance. Hoeppli (1971) pointed out a correlation between the movement of people in early navigation and the spread of parasitic infections, and this must also have affected the western colonies of Phoenicians and Greeks.

LATE IRON AGE: EIGHTH- TO SEVENTH-CENTURY NARCE AND VULCI

A survey of the health of a group of burials from Iron Age Vulci and Narce (ca. 700–600 BC) now in the University of Pennsylvania Museum supports the picture deduced from the other regions.[48] Two women (who died at 16±2 and 21±5 years of age) were buried at Vulci at the beginning of the seventh

43 Ridgway 1992:45–77, 109–20; Becker and Donadio 1992.
44 See Coldstream 1993.
45 131 were excavated: Ridgway 1992:51.
46 Becker 1995; Becker and Donadio 1992.
47 Tomb 150; Becker 1995:280.
48 See Becker et al. 2009:41–61, also chap. 10.

century, one apparently with her husband.[49] No cause of death was evident in their bones; we have assumed that they died of infectious disease, but there could have been many other causes. Of five skeletons of the late Iron Age (ca. 680 BC) found at Narce, four were adult women (30 to 65+ years of age) and one a man (45 years). Again, apart from some arthritis in older folk, and some bad or missing teeth, there is little pathology. Dental problems may have been increasing with improvements in agriculture and a higher-calorie, tastier carbohydrate-based diet for many dwellers in the proto-cities of eighth- and seventh-century Etruria. Through the seventh and sixth centuries, most women of whatever age seem to have been tiny in stature, yet robust in the bones of arms and legs, as a result of constant activity, walking and working (weaving, grinding grain, etc.), and enjoying good diets.

SIXTH-CENTURY CHIUSI

A family burial in the region of Chiusi circa 600 BC held four urns with elderly and middle-aged women (35±10, 50, and two over 65 years), one set mingled with a previous cremation of a child ten to twelve years old.[50] The death of a child of ten to twelve years seems like an issue of infectious disease, but it is too soon to speak definitively to statistics of mortality and morbidity in Iron Age Etruria.

VERUCCHIO: EARLY SEVENTH CENTURY

The northern Adriatic site of Verucchio is now famous for the preservation of its male and female "princely" burials, circa 680 BC. At least 168 individuals have been identified from cremated bones in an estimated 330+ tombs. Because of the cremation process, only such problems as arthritic conditions are identifiable, but the age census is telling.[51] These are affluent and, we may assume, well-fed families, but the infant mortality rate seems high: forty skeletons (approximately 24% of the cemetery population) are infants or small children, ranging in age from birth to seven years; another eight are seven to fourteen years of age. Although there is no conclusive evidence preserved in these burials, it would appear that infectious disease was taking its toll of the children.

TARQUINIA PIAN DI CIVITA SANCTUARY SITE

A different example is the boy buried during the ninth century BC in the sacred area of Pian di Civita at Tarquinia. His burial, and the others found in this area, must be considered unusual, and of cultic significance, for they are not

[49] His bones were not preserved; see Turfa 2005:106–7, nos. 39–40.

[50] Becker et al. 2009:77–79, no. 17.

[51] Onisto 2002.

in a necropolis and, unlike family burials, were not cremated. The boy is the earliest, and he had survived three episodes of malnutrition and/or disease (enamel hypoplasias) but was well nourished by the time of his sudden death from a brain aneurism.[52] Fragments of infant burials (eighth to seventh centuries) show congenital deformities that may have resulted from inbreeding, disease, or poor maternal nutrition. The unusual ninth-century boy's teeth and bones seem to show periods of want or illness for someone apparently deemed special by his community, whether or not we decide that his congenital aneurism or epilepsy conveyed to him the demeanor of a prophet.[53]

TARQUINIA AND MONTI DELLA TOLFA: MONTEROZZI AND FERRONE NECROPOLEIS

For the Villanovan-era Tarquinian necropoleis, recent forensic studies have identified a relatively elevated number of burials of infants and children, raising issues of general nutrition and infectious disease in urban populations.[54] Relatively good nutrition in the *Archaic* urban population of Tarquinia is also implied in the findings of Vargiu and Becker (2005) for eleven cremation burials. One middle-aged individual had an arthritic hip and an old foot fracture, but four of the deceased were children.

Bartoli and colleagues surveyed the strontium levels and zinc-to-calcium ratios of skeletons of the seventh to sixth centuries buried at Ferrone (Monti della Tolfa region) and of the sixth to second centuries at Tarquinia Monterozzi.[55] They found close correlation between the two locations and across the centuries, with strong reliance on vegetable foods rather than red meat or milk products. Baggieri (2005) found relatively good (skeletal) health, except for some arthritis and problems because of bad teeth among fifty-six people buried at Tarquinia Monterozzi. His tentative diagnosis of a case of thalassemia in one of three burials at San Giovenale would be a first, though not an unexpected discovery, linking the region to malarial conditions.[56]

LATER IRON AGE: CASTENASO AND BUDRIO (BOLOGNA REGION)

An eighth- to seventh-century village at Castenaso outside Bologna was abandoned in the sixth century BC – it offers much evidence of industry, and

[52] Fornaciari and Mallegni 1997; cf. Mays 1998:156–61, with bibliography. A later young boy is said to have been killed by a blade or pin thrust through his cervical spine; he was beheaded: this find has yet to be fully published (Bonghi Jovino 2010:166).

[53] Further on this site, see Bonghi Jovino 2010; Leighton 2004:40–41, 70–78; Bonghi Jovino et al. 2001; Chiesa 2001; Bonghi Jovino and Chiaramonte Treré 1997; Chiaramonte Treré 1995.

[54] See Trucco et al. 2005.

[55] Bartoli et al. 1997; cf. Brocato 2000.

[56] Baggieri 2005:707.

features pits for food storage, water supply, drainage, and disposal of trash.[57] Still, the health of the population was not optimal. A woman of about forty years buried in the odd "structure η" during the last phase of habitation of the site has been interpreted as rare evidence of a lower class in the village.[58] Her body was marked by a hard life, with multiple caries and lost teeth. Her arms and legs (humerus, tibia) showed growths and possibly infection from prolonged hard work, often seen in agricultural populations. The preauricular sulcus of her pelvis was said to show evidence of multiple pregnancies and births. At Budrio, we see Iron Age and Archaic Etruscans leading reasonably long lives, but never free of daily effort and its pains. A burial near the enclave of Villanovan IV B2 huts (ca. 600–550 BC)[59] held an older man (50–60 years) of reasonable stature (159 cm) whose skeleton showed the effects of hard work in fused thoracic vertebrae and a cervical spine with osteoarthritis.[60]

LATE IRON AGE–ARCHAIC: PONTECAGNANO, AN ETRUSCAN OUTLIER

Fornaciari (1997) and others have studied the burials of the Campanian site of Pontecagnano, inhabited from the Eneolithic period through the third century BC. The phase of Etruscan occupation, seventh and sixth centuries BC, has yielded 275 burials: more than half were children under fifteen years old, an unusually high level of childhood mortality. More than 8 percent may be diagnosed with porotic hyperostosis, in contrast to only about 3 percent of the Oscan (Italic) citizens who settled there in the fifth to fourth centuries BC. In the seventh century, one child of about twelve had a severe spinal deformity, for which Fornaciari suggested thalassemia.[61] Mineral ratios in the skeletons show reliance on vegetable foods; anemic individuals received the same diet as everyone else, so their condition might be interpreted as congenital.

A number of people showed the enamel hypoplasias of childhood stress. (In third-world populations, such childhood infections are gastrointestinal or respiratory, related to parasites or weaning.) Anemic adults, like their cohorts, were reasonably tall at around 165 cm for men and 153 cm for women. Lifespan figures, slightly better for men than for women, were not adversely affected in those with *cribra orbitalia* or porotic hyperostosis – again, this possibly illustrates the prophylactic edge that congenital anemia might have given against attack by malaria, and it implies that malaria had been endemic for a long time.[62]

[57] Forte 1994; comparable to Casteldebole.
[58] See Mainardi and Pacciani 1994.
[59] Losi 1994.
[60] Giusberti 1994.
[61] Fornaciari 1997:470.
[62] Fornaciari 1997:472.

SPECIFIC CONDITIONS, DISEASES IN THE *BRONTOSCOPIC CALENDAR*: "DISEASE"

Of the 360 entries in the *Brontoscopic Calendar*, approximately 125 are concerned with health, disease, or the appearance or destruction of pests and dangerous animals. Just eleven citations predict *good* health or well-being, such as "an end to disease" (June 23), "good health for men for a full year" (August 9), or "welfare of women" (November 9). About eighteen entries simply citing "disease" are too vague to permit diagnosis. The archaeological evidence for infectious disease in Classical or Iron Age Italy is meager thus far, but DNA research can now detect some pathogens in bones and teeth.[63]

Mesopotamian texts are rich in diseases: "destruction" often is translated plausibly as "plague." For example, *Enūma Anu Enlil*, tablet 65: "If Jupiter comes close to the Eagle: there will be epidemic among cattle, there will be an epidemic in the land, it will not rain for one month."[64] Likewise, the *Šumma izbu* (tablet 5, omen 90) warns: "If a ewe gives birth to a wolf – there will be plague in the land; madness; destruction of the land; disease of the herds."[65] The Mesopotamian texts also cite generic "disease" or "pestilence" (e.g., *bibbu*), which can kill cattle, sheep, or humans.[66] Naturally, the disease panel of Iron Age Italy differed from that of the Near East. Sometimes the Babylonian/Akkadian texts use special qualifying terms for it, such as *rapādu*-disease;[67] at other times, the demon Lamaštu is said to rage (for three years),[68] or Gula the goddess of health sends afflictions.[69] Most diseases are not cited in the *Brontoscopic Calendar*. For instance, the great cuneiform text series say: "epilepsy (*bennu*) will seize him,"[70] and "in this year there will be *ašakku*-disease,"[71] "a grievous malady, an evil disease" caused by a demon, but according to the *Chicago Assyrian Dictionary*, the term was not recorded in medical texts or lists, and should be understood as poetic.[72]

63 See Greenblatt and Spigelman 2003 for current research, especially Herrmann and Hummel 2003.

64 Group B, Tablet 80-7-19,90obv.6' and 7' = Reiner 2005:69.

65 Leichty 1970:81.

66 See *CAD*, vol. 2 "B" [1965] 217–19.

67 *Enūma Anu Enlil*, tablet 65: The *Chicago Assyrian Dictionary* (*CAD*, vol. 14 "R" [1999], 147) has accepted the suggestion of Sjöberg (1973:141n166) that it refers to Listeriosis or "staggers" in which infected sheep circle with their heads tilted to one side. *Rapādu* disease appears in Near Eastern medical contexts and other omens seem to show it afflicting or killing humans also.

68 *Enūma Anu Enlil* 51. See Chapter 8.

69 C.f. Black and Green 1992:70, 101, 115–16.

70 See Stol 1993; *CAD*, vol. 2 "B" [1965], 205–6.

71 MUL.APIN, tablet II, iii, Gaps B1–B7, Marduk; see Stol 1993.

72 *CAD*, vol. 1, part II "A" [1968], 325–27.

"HEALTHFUL LEANNESS"

Good health is often related to prosperity, but one surprising comment occurs at December 29, "If it thunders, it signifies the most healthful leanness for bodies." This calls to mind the attenuated bronze figurines of the fourth to third centuries BC that inspired Modigliani and became collector's items in the early twentieth century.[73] Might there be more than sculptural style implied in bronze votives such as the famous Ombra della Sera of Volterra and its cousins from the votive deposit of the Sanctuary of Diana at Nemi (see Fig. 3)?[74] They represent goddesses, men, women, haruspices, and clothed or nude youths, some making libations.[75] Some earlier bronzes in the Arezzo Fonte Veneziana deposit[76] already show the distinctive elongation, taking this perception back at least to the Late Archaic period (early 5th c. BC). Similar style characterizes Late Archaic votives at other sanctuaries with "living waters," such as Bologna Villa Cassarini, Marzabotto, and Fiesole.[77] Even the minuscule black silhouettes of humans cavorting in the dead bushes of the Underworld, depicted in the Tomba dell'Orco (Tarquinia, mid-4th c. BC) might represent some aspect of this concept: the figures, whatever they are, seem happy and energetic in their pared down form,[78] quite the opposite of the Roman trope of the *obesus etruscus* (Catullus 39.11; Vergil *Georgics* 2.193).

PREDICTIONS OF GOOD HEALTH

Several other correlations, some qualified with specific social conditions such as concord, also allude to good health or well-being, even in situations of want:

> June 23. If in any way it should thunder, it announces good cheer, a putting aside of ills, and an end to disease.
> August 9. If in any way it should thunder, it proclaims good health for men for a full year.
> March 16. If it thunders, healthful the year, yet lacking in necessities.
> April 5. If it thunders, it signals a hot, early summer but a healthful year.

[73] See Terrosi Zanco 1961.

[74] Also illustrated by Haynes 1985:122, 243, 322, no. 199; Cristofani 1985:172–73, 273–74, no. 66, 178–80, 275, nos. 73–75, 184, 276, no. 79.

[75] Cf. the Volterran water boy in Cateni 1991:pl. 85.

[76] E.g., Cristofani 1985:96, 253, no. 3.22; Zamarchi Grassi 1985:177–78, no. 13, nude youth holds bird and fruit.

[77] See *Santuari d'Etruria* 1985:92–94, 113–15; DeMarco 1985.

[78] See Steingräber 1985:330–31, color pl. 130.

WEATHER, WIND

On seventeen dates, the calendar prophesies suffering brought on by weather conditions, ranging from "unhealthy drought" (July 20), to "a disease-bearing wind will blow" (Feb. 28), or "the air shall carry plague" (Feb. 15), and "heavy rains bearing disease" (April 8). Occasionally, the weather is kinder, as at October 4, when "the mildness of the air will be healthy." A simpler example is August 7, "If in any way it should thunder, it signals harsh winds and diseases at the same time." It appears that the two conditions were deliberately linked. Likewise, in Mesopotamian omens, storm and wind can bring disease, as they often bring the destruction of crops. In the *Šumma ālu*, tablet 79, "If it rains frogs in a city, there will be plague in that city." In the *Enūma Anu Enlil*, tablet 15, "If an eclipse begins in the south and the east wind blows: Destruction of Elam and Guti; ... the king of Akkad will die; Adad will trample; there will be pestilence."[79]

The Hippokratic Greek work of clinical observation, *Airs Waters Places*, makes exceedingly precise correlations between types of weather and winds and the general health and character of the natives of different regions.[80] For example: "A city that lies exposed to the hot winds" fosters barrenness and abortions in women, convulsions and asthma in children, and in men "dysentery, diarrhea, ague, chronic fevers in winter, many attacks of eczema ... inflammations of the eyes ... but besides, they are liable to any epidemic disease that prevails through the change of seasons."[81] The Etruscan prediction, in contrast, and like its Mesopotamian predecessors, is not at all theoretical or philosophical.

Biometeorology is a well-developed discipline in modern climatology and medicine.[82] Certain weather conditions are known to trigger human problems such as asthma. In fact, the connection with *electrical* storms has been visited in recent scholarship. Wind-driven allergens, ozone produced by lightning strikes, and changes in temperature or barometric pressure have all been found to exert an effect on sufferers. In a recent study undertaken in the UK, in a twenty-four-hour period encompassing a major electrical storm, a distinct correlation was established between the severe storm and certain types of asthma.[83] Similar problems must have affected ancient asthmatics, who lived closer to soot from wood fires and to allergens or toxins from industrial and agricultural processes.[84]

[79] Tablet 15, col. iii, sections 6′ and 7′ = Rochberg-Halton 1988:74–75.

[80] Much of the Hippokratic *Corpus* was composed after the lifetime of its namesake, but this work is a genuine 5th-c. treatise; see Jones 1923/1995:xxiv–xxv.

[81] *Airs Waters Places* 1.3; Jones 1923/1995:75–77.

[82] Cf. Wood 2002; Tromp 1980.

[83] Davidson et al. 1996; Celenza et al. 1996.

[84] Cf. Harrison and Bartels 2006:21, 24.

Europe and Italy experienced the Hallstatt Minimum, a cool, wet period that peaked around 800 BC (see Chapter 3); concomitant changes in humidity and temperature must have affected many diseases and their vectors. In the Mediterranean and Near East, warm weather and harvests bring a seasonal onslaught of flies that can spread epidemics of enteric disease. Excessive rains or floods can also foster the transfer of a variety of waterborne infections, but not density-dependent cholera and typhoid.[85] Swampy or wetland conditions, even just a rainy summer, could stimulate infections of *Fasciola hepatica*, the sheep liver fluke (see the section on parasitic infections later in this chapter).

MALNUTRITION OR POOR NUTRITION

At five entries, the calendar portrays disease as following famine, want, or bad harvests:

> January 24. If it thunders, it signifies disease following want.
> August 2. If in any way it should thunder, it threatens both diseases and at the same time a dearth of the necessities of life.

Two dates might be taken together:

> September 13. If in any way it should thunder, it threatens grave famine.
> September 14. If in any way it should thunder, it threatens diseases.
> September 18. If it thunders, it signals both famine and wars.

Well over 100 predictions deal with crops, good or bad harvests, or crop failures caused by rain, drought, wind, or pests. As in many regions today, the success of the harvest must have been people's foremost concern. As Garnsey has indicated, for agrarian populations, there is always another level of hunger distinct from outright starvation, the condition of substandard nutrition.[86] A poorly nourished population can survive periods of dearth, unless something else, like disease or war, strikes them simultaneously. The lack of optimal nutrition affects the immune system, and deaths will result from diseases that might otherwise have been trivial.

Malaria and tuberculosis cause greater mortality in populations compromised in nutrition or immune response, especially small or weaning children and pregnant women. Winter vitamin deficiencies may have been common in the ancient population (as in medieval cities) and would have

[85] For modern conditions, see Ocean Studies Board 1999:43–58, and Hunter 1997:passim.
[86] Garnsey 1999:34–61 et passim; on modern conditions, see Dirks 1993.

led then, as now, to respiratory infections. Dental enamel hypoplasias can indicate malnutrition as a chronic situation in the Classical Mediterranean.[87] Most sufferers would have been unaware of the problem: they consumed adequate amounts of food/calories – but their cereal-based diet prevented the absorption of sufficient iron or other minerals and vitamins. In a number of the Iron Age necropoleis, skeletons such as the ninth-century boy at Tarquinia-Civita show this situation, and also Harris lines on long bones show that ancient Mediterranean populations experienced periodic physiological stress.[88]

DEFICIENCY DISEASES: ANEMIA

Many skeletons of the historical periods in Italy show conditions, such as *cribra orbitalia*, that in modern patients betray serious iron deficiency conditions. (Because red blood cells are produced in bone marrow, the skeleton will show the effects of this extra stress, and porotic hyperostosis, changes in the growth of the bone, is visible. The most recognizable symptom appears as a pitted surface on the interior of the orbits of the skull (*cribra orbitalia*).[89] Another anemic condition prevalent in ancient Italy was thalassemia, a hereditary disorder which, like sickle-cell anemia, is believed to have evolved as a defense response to endemic malaria. The low level of hemoglobin and short lifespan of thalassemic red blood cells is not conducive to infection by malaria.[90] Some of the individuals noted above also show the effects of iron and/or other deficiencies resulting from less than adequate nutrition or insufficient ability of the body to absorb vital nutrients. Tafuri noted that iron deficiency appears in the Neolithic period, as if the change in diet to plant foods was causing problems for some people.[91] Three-quarters of the cases of *cribra orbitalia* were found in women of childbearing age, and Tafuri, following Borgognini Tarli, suggests that their anemia was a physiological problem, caused by the stress of repeated pregnancies and lactation.[92] Other studies of deficiency in ancient populations point to infectious disease[93] and (outside Etruria) to poor hygiene or insufficient management of waste in Bronze Age settlements.[94]

[87] See Cohen 1989:chap. 7, especially 112–20, and 119–21, on episodic stress and bone studies; also Aufderheide and Rodríguez-Martín 1998:422–23.

[88] Fornaciari and Mallegni 1997:100–102; see Mays 1998:156–61, for indicators of childhood stress episodes.

[89] See Roberts and Manchester 1995:165–71, fig. 8.3.

[90] See Roberts and Manchester 1995:169–71.

[91] Tafuri 2005:15.

[92] Tafuri 2005:20; Borgognini Tarli 1992.

[93] Weinberg 1992.

[94] See Stuart-Macadam 1998 and 1989.

ENTERIC DISEASES

Enteric diseases and gastrointestinal parasites must have affected most ancient populations; for instance, whipworms (*Trichinuris*) have been found in the body of the Alpine "Iceman" of Chalcolithic date, around 3300 BC;[95] roundworm (*Ascaris*) and tapeworm (*Taenia*) are even older.[96] *Fasciola hepatica*, the sheep liver fluke, has been identified in Neolithic and Mesolithic samples from France and Germany.[97] Diarrhea, a major cause of childhood death today, is cited only once, at December 12, when, "If it thunders, it threatens diseases from diarrhea (γαστρορροίας νόσους)." (The actual cause of death is usually dehydration and electrolyte depletion.) Although not the terminology found in the Hippokratic early corpus (διάρροια),[98] we cannot be certain of the exact vocabulary of the Etruscan original. These complaints might have been taken for granted in children, but they would presumably have attracted comment if symptoms were severe, as in dysentery, or if the disease affected much of the adult population. Modern groups are infected with *Giardia*, *Salmonella*, *Shigella*, or *Campylobacter*, usually in warm seasons.[99] It may be that food types and cooking methods (smoking, salting, drying instead of pickling) in Iron Age Italy were less likely to foster the pathogens that cause "food poisoning." Nataro et al. (2003) emphasize that the evolution of *Salmonella*, *Shigella*, and *E. coli* occurred so long ago that we should assume that these bacteria were present in any ancient Mediterranean or Near Eastern humans, but other diseases, such as cholera, almost certainly were not.

The size and density of the vast middens of prehistoric Italian settlements, such as the famous Terramare of the Po region, must have provided breeding grounds for numerous bacterial pathogens.[100] It is possible that the early drainage and farming innovations of Villanovan and Etruscan culture, seen in the *cuniculi* of Veii and the *cloaca maxima* of Etruscan Rome, actually stemmed some sewage-related problems. Drainage works at Protovillanovan Sorgenti della Nova point to early resolution of some problems of sanitation and drainage, and presumably an improvement in health.

[95] Dickson et al. 2000; Aspöck et al. 1996, believed to be the oldest finding of whipworm infestation in a human.

[96] For the array of parasites in ancient humans, see Gooch 1983:209, table of helminths, updated by Aufderheide and Rodríguez-Martín 1998:222–46; Patterson 1993B:603–4 and 703; T-W-Fiennes 1978:129–30 (hydatid disease).

[97] Dittmar and Teegen 2003; Aspöck et al. 1999; Bouchet 1997; Turfa and Gettys 2009.

[98] Or for the severe disease, known to us by the same name, δυσεντερία, see Jones 1923/1995:lviii–lix, citing *Peri pathōn* 23 and 25; cf. Sallares 1991:231n322.

[99] DuPont 1993.

[100] For the possibilities, see Cocchi Genick 1998 and 1996; cf. T-W-Fiennes 1978:13–14.

PARASITIC INFECTIONS

At November 13, "If it thunders, a wealthy yet sickly period threatens, tormenting bodies with internal worms." This is the only unequivocal reference to internal parasites and seems to express the folk belief implicating rich diet in the propagation of worms. The entry of the previous day may be related, with the prediction that "it indicates insomnia for some time for men." Insomnia is a classic symptom of serious beef tapeworm infestation (*Taenia saginata*).[101] These three entries, in mid-November and December (Dec. 12, diarrhea), could represent gastrointestinal flare-ups when the population would have been reliant on stored foodstuffs, which might be contaminated – "a wealthy period."

Parasites troubled human populations in Europe and Italy from at least the Neolithic period; Patterson (1993B) published evidence of roundworms and hydatid disease in prehistoric populations, and Bouchet (1997) described evidence of intestinal parasites in Paleolithic and Neolithic France.[102] The prominence of haruspicy, divination by examination of sheep livers, may be explained in part by its relative accuracy in predicting some outbreaks of human infection and diminished food supplies for Etruscan and Italic settlements.[103] The sheep liver fluke, *Fasciola hepatica*, has an elaborate life cycle that requires wet conditions in pastures and the presence of snails and sheep. Sheep sacrificed while still whole in appearance may show distinctive lesions in the parts of the liver, which on the Piacenza bronze model are dedicated to the gods of the earth and Underworld. Humans who eat the sheep, or eat greens from the same fields, will sicken several weeks later than the quadrupeds, and haruspices who predicted this would presumably be respected for their skill.[104]

"COUGH"

Few predictions in the *Brontoscopic Calendar* describe disease symptoms with enough precision to permit diagnosis; coughing sickness (βηχώδη νόσον) and skin lesions are the most obvious. Four entries include coughing or a "dry cough" (βῆχα ξηρόν). At July 8, "ruin for the herds and a dry cough shall infect" seems to describe a zoonosis affecting both humans and cattle, perhaps

[101] See Aufderheide and Rodríguez-Martín 1998:240–41; Patterson 1993A; T-W-Fiennes 1978:12.

[102] See Gooch 1983:209, for socioeconomic effects of endoparasites.

[103] Turfa and Gettys 2009.

[104] For background on liver fluke, see Behm and Sangster 1999; Dalton 1999; Dittmar and Teegen 2003 present evidence of third-millennium infestation in Germany. Pliny (*Nat. Hist.* 30.44–46) records consumption of snails for medicinal purposes.

anthrax or brucellosis, even though today cough is not their most common symptom. In January (Jan. 6, 20), cough is twice linked with an abundance (of imported goods, or fish). The mention of an ailment in the calendar must imply that a noticeable, widespread, or serious condition was expected. It is debatable whether the rhinoviruses, causing today's colds, could have been supported in the small settlements of early Italy. Candidates for the "coughing disease" of fall and winter might be bronchitis or pneumonia. The association with "imported goods" at January 20 might suggest some foreign source, perhaps an endemic disease coming from a dense urban population, such as in the Levant.

TUBERCULOSIS?

On February 2: "If it thunders, wheat in less supply, but barley better, and an increase in livestock, but there will be a wasting away of humans." Although four omens refer to cough, only this one uses the term φθίσις, "consumption," the designation for tuberculosis, one of the most frequently cited diseases in the Hippokratic *Corpus*.[105] The comment at October 21, that "there will be coughing sicknesses and oppression of the heart," certainly sounds like the psychological effects of tuberculosis. The disease is documented in Italy since the Neolithic period (ca. 4000 BC): skeletons excavated near the Arene Candide caves exhibit Pott's disease, a tubercular infection (and collapse) of the spine.[106] (There is little physical evidence from the Near East apart from the Early Bronze Age Levantine site of Bab-edh-Dhra.)

Capasso and Di Tota have indicated a lesion of the sacrum and lumbar spine from Pott's disease in a woman buried at Pozzuolo (Perugia) and considered to be Etruscan.[107] (They further interpreted as tuberculosis bony lesions of the spine and ribs in a middle-aged man and woman among the victims found on the shore at Herculaneum after the eruption of 79 AD.[108] They emphasized that, even without consuming cow's milk, most Romans would have been exposed to the bovine sources of this disease.) Aufderheide and Rodríguez-Martín have suggested, on analogy to the modern incidence of this syndrome in only 1 to 3 percent of sufferers of tuberculosis, that in each ancient population as many as 100 to 300 people were infected with the more common *respiratory* symptoms of tuberculosis without skeletal damage.[109]

105 See Jones 1923/1995:lvi–lvii.

106 Formicola et al. 1987.

107 Capasso and Di Tota 1997:553–54.

108 Capasso and Di Tota 1999.

109 Aufderheide and Rodríguez-Martín 1998:118–40. See the lengthy discussion by Grmek 1989/1991:177–209; also Roberts and Manchester 1995:135–41, and 149 on the related medieval phenomenon of leprosy. For background, see Johnston 1993; Comstock and O'Brien 1998. For compromised productivity, see Mckinney et al. 1994.

The weather of January and October, including thunderstorms and drops in barometric pressure, might evoke distressing symptoms in many sufferers.[110] Diphtheria, transmitted in cold months and from milk, is not attested until the sixth century AD.[111]

SKIN LESIONS

Examples of skin problems:

> July 27. If in any way it should thunder, it threatens subcutaneous eruptions to bodies.
> October 22. If it thunders, it threatens for the people, bad conditions and spotted diseases.
> December 5. …, it threatens mangy diseases.
> February 18. …, it threatens a heavy wind and eruption of pustules on bodies.
> February 26. …, it threatens hot weather, and a lack of water, and scabs on bodies.

Five entries in the *Brontoscopic Calendar* speak of skin lesions or infections – "spotted diseases" (νόσους ποικίλας), pustules or scabs on bodies (ἀνάστημα τοῖς σώμασιν, ψώρωσιν τοῖς σώμασιν), subcutaneous eruptions (ἐκβρασμὸν τοῖς σώμασιν), and "mangy diseases" (νόσους ψωρώδεις), presumably of humans.

Vulcanism, a potential "thunder" noise, is a source of irritant particles that cause skin infections and breathing troubles, especially if a heavy wind comes with "the eruption of pustules on bodies" (Feb. 18). Major eruptions occurred in the Italian archipelago at least thirteen times between 1080 and 450 BC, six of them between 800 and 600 BC.[112] Most modern deaths from volcanoes result from starvation following destruction of crops, or from burns, asphyxiation, and such caused by pyroclastic flows, lahars, and tsunamis, but less drastic effects are well known.[113] In modern Hekla, Iceland, sheep died in large numbers after grazing on plants that had been covered with fresh volcanic dust.[114] Many past eruptions, like Tambora, covered and killed growing plants, resulting in famine. Other skin irritations surely circulated in Iron Age Italy; Imperial Roman evidence tends to indicate compromised personal hygiene in urban settings. For instance, a louse egg was found in the hair of an affluent woman who died at Herculaneum in 79 AD.[115]

110 For instance, Smit et al. 1999.
111 Cf. Carmichael 1993A, Aufderheide and Rodríguez-Martín 1998:266; T-W-Fiennes 1978:27–28.
112 Simkin and Siebert 1994:165, 183–84.
113 See Baxter 1990. See Blong 1984:70–131, 311–43 for effects on agriculture, animals.
114 Vitaliano 1976:256–57.
115 Capasso and Di Tota 1998; Capasso 2001:466–69, 472, color photo p. 68, fig. 77: Erc 52.

CUTANEOUS ANTHRAX?

Cutaneous anthrax seems another strong possibility for the "spotted diseases" of October 22, in a population that, ever since the Neolithic period, was heavily engaged in sheep breeding, manufacture of woolen textiles, and use of sheep for sacrifice, with officials handling their internal organs for divination. Although sometimes known as the "woolsorters' disease" that causes alarming, black, spreading lesions on the skin, anthrax in its pulmonary and gastrointestinal forms is often fatal to humans and animals.[116] When cutaneous lesions are not treated, up to 20 percent of those infected can actually die from complications like septicemia.[117] Modern studies find that lesions appear within twelve days of exposure to the spores. Other spores inhaled or eaten in infected animals can incubate in two to five days, and mortality rates can reach 75 percent.[118] This sounds similar to the "worst diseases and wounds" (*pessimis morbis et vulneribus*) predicted by Vegoia's prophecy. Spores can survive in the soil for a few years, so recurrence could have been a seasonal or perennial problem in agricultural regions of Etruria.

Some curious references in Near Eastern texts might also relate to anthrax. In Egyptian papyri, cutaneous diseases are said to be sent by "the god above." Some were equated with "the disease of Asia," "the disease of Canaan," or "the disease *tmyt*," and Semitic roots have been identified in the Egyptian medical terms. The disease of Canaan is described as black "like charcoal."[119]

OTHER "POXY DISEASES"

Smallpox (*Variola*), although occasionally diagnosed in some ancient individuals,[120] cannot be demonstrated in the Mediterranean region until late antiquity. It must be sustained by a steady birthrate in a dense population, exposing new subjects before the original victims' lesions disappear.[121] The rashes are produced by typhus, rickettsial pox, or *Tinea*, and the fungal infection of "ringworm" cannot be attested for ancient Etruria. Measles, only sustainable

[116] Acha and Szyfres 2001:21–28.

[117] For a survey of the disease, see Wilkinson 1993A; Brachman and Kaufmann 1998.

[118] The literature on anthrax has increased vastly since 9/11, but it has not been directed toward paleopathological studies. An epidemic in Zimbabwe between 1979 and 1985 affected more than 10,000 people, but modern conditions have changed the situation somewhat; see The Working Group on Civilian Biodefense, "Anthrax as a Biological Weapon: Medical and Public Health Management," *Journal of the American Medical Association* 281(18) (1999):1735–45.

[119] See Bardinet 1988:19–24; Stol 1993:129. Stol suggested this was true leprosy (Hansen's disease), but in the *Enūma Anu Enlil*, Ishtar 84:2, it is realistically rendered *epqu* ("scales"). Reiner and Pingree 1981:48, text VI. For leprosy, see also Bardinet 1988 – but there seems to be no evidence in classical Etruria.

[120] Particularly the mummy of Ramses V; see Jeziorska 2008:91.

[121] Aufderheide and Rodríguez-Martín 1998:201–7. Cf. Crosby 1993A, who suggests that the Antonine or subsequent plague(s) of Rome might have been smallpox.

by a population of 200,000–250,000, just might have been possible in early Italy as a single epidemic event. A recent study of measles has suggested that the communicating population of the many closely linked Bronze Age cities of the Tigris-Euphrates region probably sufficed to keep the disease active.[122]

Some authors attribute the plague of Athens (Thucydides, bk. 2) to measles; others associate it with the disease that ended the Carthaginian siege of Syracuse in 396 BC, but the arguments are not sufficiently compelling to warrant extrapolating to Iron Age Etruria. In fact, three different epidemics appear in the history of Syracuse (in 415–413, 396, and 213–212 BC), but only the event of 396 was described with a seemingly "clinical" attention to detail.[123] The "childhood diseases" of twentieth-century Europe and America all began in previous eras as scourges of the adult population, causing high mortality before the population adapted to them.[124]

Columella (*De re rustica* 8.5.20–22) described a severe disease of chickens, pip, (Latin *pituita*) that causes pus in the eyes, cheeks, and so on during cold weather and again in warm weather if the birds drink contaminated water or eat contaminated windfall fruits. Today it can affect some human handlers with dermatological symptoms. Chicken pox was entertained by Grmek as a possible infection in the Classical world, because its correlate, *herpes zoster* ("shingles"), was actually described in a text of the first century AD, but there is no further evidence for this affliction.[125] Because a high proportion of ancient people almost certainly died of acute infectious diseases, DNA studies may one day resolve this question of diseases of the soft tissues.[126]

"PLAGUE"?

Sample references to epidemics in the *Brontoscopic Calendar*:

> August 13. If in any way it should thunder, there will be plague (*ὄλεθρος*) upon the bodies of both humans and dumb animals.
> October 29. If it thunders, a year of serious disease (επίνοσος).
> December 8. …, virulent disease (ὄλεθρος); out of it, though, will be an abundance of crops but a plague on the flocks.
> December 10. …, it threatens slaughter for men from diseases (νοσημάτων), but the fish shall be abundant.

[122] Cliff et al. 1993:46–51, with reference to McNeill 1976. Evolutionary links have been indicated between Rinderpest, canine distemper, and human measles; see Cliff et al. 1993:28–30, 47; also Aufderheide and Rodríguez-Martín 1998:207–9.

[123] For a literary review of these events, see Villard 1994.

[124] Cf. Ewald 1994 with earlier bibliography; Black 1975.

[125] Grmek 1991:335–36.

[126] See Becker 1999; Jones 2001:217–33; Greenblatt and Spiegelman 2003.

December 24. …, it threatens civil wars for the city, and a plague (ὄλεθρος) on the beasts of the woods.
January 29. …, the condition of the air oppressive, and disease-bearing (λοιμικόν) for all.
February 15. …, the air shall carry plague (λοιμικός), creation of both wild beasts and mice.
April 15. …, it signals a plague (λοιμός).
May 9. …, it signals plague, but not exceptionally life-threatening (λοιμός).

In the surviving Greek version of the *Brontoscopic Calendar*, several predictions mention "plague," usually indicated by the term λοιμός, as in the Classical Greek authors. I have also translated as "plague" the term ὄλεθρος, meaning literally, "destruction/ruin/death." Destruction in such a context can only mean death by disease, yet this cannot be *Yersinia pestis*, the density-dependent bubonic plague, known in Justinian's Constantinople, and thus horribly relevant to Lydus.[127] (The so-called Antonine or Aurelian plagues of Imperial Rome are controversial but could have been *Yersinia pestis.*) The high humidity and low altitude of Etruscan coastal cities could have fostered a great variety of infectious diseases, including tuberculosis, polio, malaria, asthma, brucellosis, and typhoid, all common to the postantique Mediterranean and southern Europe. The *Brontoscopic Calendar* here refers to dramatic symptoms and presumably many casualties by way of secondary infections, even if the original disease was not in itself lethal.

It is tempting at August 13 to recall the famous plague of the *Iliad* (1.8–26, 33–108) in which the "bad disease" sent by Apollo first attacks the mules and swift dogs, and then causes men to fall ill. It sounds like a zoonosis. (Even if it was a poetic fiction, the Homeric audience must have felt that such an epidemic was possible.) Roman poets also cite afflictions, and very often the epidemic (λοιμός, *pestis*, or *pestilentia*) is said to affect animals and humans.[128] Epidemics are noted by Columella, Lucretius, Ovid, Seneca, Manilius, and Silius Italicus, all Late Republican or Imperial in date. Much closer in time and space is the tale of the aftermath of the battle over Alalia (ca. 535 BC; Herodotus 1.167) when the Caeretan fleet brought back Phokaian prisoners of war, who were stoned to death on the beach by the Caeretans. As the bodies decomposed there, anyone, human or animal, who passed nearby became crippled and paralyzed. The Caeretans sent to Delphi, and the oracle told them to institute sacrifices, funeral games, and rites for the unjustly killed Phokaians, after which the plague ceased. This sounds like a zoonosis with

[127] See Aufderheide and Rodríguez-Martín 1998:195–98; Park 1993:612–16; Carmichael 1993B:628–31; Poland and Dennis 1998. Identification of the 14th-c. and other plagues is still debated; see Lerner 2008, for instance.

[128] See Aygon 2003:276.

neurological symptoms associated with the dead bodies, but there is no clear candidate for a pathogenic cause.

BRUCELLOSIS

The terms used in Lydus' Greek translation (ὄλεθρος, λοιμός, λοιμικός ἀέρ), like Latin *pestis* or *pestilentia*, in the medical authors referred to any severe disease affecting a large number of people, with a relatively high mortality rate.[129] Brucellosis, Malta, or undulate fever could have spread through unpasteurized milk, as it did in the time of Hippokrates, and especially during the nineteenth century, when it caused numerous deaths.[130] Those who survived an attack (like Napoleon) experienced lifelong ill health with recurrent ("undulating") debilitating fevers. Other symptoms include night sweats, fatigue and weight loss, respiratory or gastrointestinal problems, eye infections, and later complications like arthritis and osteomyelitis.[131] *Brucella* bacteria, especially well known in Mediterranean Europe, Greece, and Turkey, are found worldwide in domestic sheep, goats, swine, and cattle. Luigi Capasso (1999) has identified evidence among the victims at Herculaneum of the results of brucellosis infection. As with other diseases of soft tissues, evidence is rare in skeletons.

Today, at least a half a million cases are diagnosed worldwide annually, especially in areas that rely heavily on breeding sheep and goats; incidence increases with large-scale importation of livestock.[132] It was a serious disease in the Balkans in the aftermath of the war in Kosovo.[133] In addition to consuming unpasteurized milk, humans contract the disease through contact with the eyes or broken skin. Its prime symptom in flocks and herds is abortion, and workers who handle fetuses or fluids or work in slaughterhouses are especially susceptible. The incubation period for *Brucella melitensis* is usually from one to three weeks, so ancient observers could have intuited the connection between the sick or aborting animal and the sickening humans.

A curious aside on the possibility of brucellosis causing abortions may lie in the reference in Roman history to the origins of the *Lupercalia*: an epidemic of spontaneous abortions among the Sabine women in Rome prompted Romulus to establish an expiation ritual, having them beaten with goatskin thongs. In 276 BC, an epidemic of miscarriages among women and abortions

[129] Patrick 1967:238–46.

[130] See Aufderheide and Rodríguez-Martín 1998:192–93; Etxeberria 1994; Young 1983; Wilkinson 1993B; Young and Hall 1998; T-W-Fiennes 1978:95–96.

[131] Kraus et al. 2003:185–87.

[132] Acha and Szyfres 2001:40–67.

[133] Hadjichristodoulou et al. 2000.

among livestock added another layer of meaning to the flagellation ritual performed by the Luperci.[134] Although the use of goatskin thongs seems more like what would *spread* the disease, the link to this species is telling, and another indication of the zoonotic situations that probably affected all central Italian villages.

POLIO OR TYPHOID AS POSSIBLE "PLAGUE"

Poliomyelitis has been present in Europe at least since the Neolithic period, and it is attested in Italy since at least the sixth century BC.[135] Rather than manifest as a plague-like epidemic in antiquity, it would have behaved more as it does in the underdeveloped world today, coming as an immunizing gastrointestinal infection in small children, rather than a paralyzing killer of teenagers.[136]

Typhoid, perhaps the disease that the Emperor Augustus would famously survive, has been considered endemic in the Mediterranean by Grmek.[137] A seasonal killer in late antique Rome, it was fatal to 10 to 15 percent of the population, striking young adults especially hard.[138] It is possible that the Hippokratic authors (*Epidemics* I and IV) described at least six different cases of typhoid or paratyphoid; the infection can also cause spotted skin lesions in addition to its dramatic, plague-like symptoms.

Another possible cause of "plague" could be typhus, which also produces a rash and has also been linked by some scholars to the plague of Athens. Transmitted by lice, it thrives in cold climates and peaks in winter, with a 5 to 25 percent mortality rate, but it usually (in modern studies) spares those under fifteen years old.[139]

MALARIA

Sample references to nonlethal disease in the *Brontoscopic Calendar*:

June 7. If in any way it should thunder, diseases will infect [men], but not many shall die.
December 23. If it thunders, it signifies a disease for men but a harmless one.
January 17. If it thunders, it threatens nondangerous diseases.
January 27. If it thunders, it signifies nonthreatening diseases.

[134] Ovid *Fasti* 2.425–53; see Wiseman 1995:84, 127.
[135] Aufderheide and Rodríguez-Martín 1998:212; Wyatt 1993.
[136] Cf. Melnick 1997.
[137] Grmek 1989/1991:89, 346–50.
[138] Cf. LeBaron and Taylor 1993; Levine 1998.
[139] See Aufderheide and Rodríguez-Martín 1998:190–91; Harden 1993; Carmichael 1993C; for ancient literary sources on the plague of Athens, see Longrigg 2000.

Disease (νόσος or νοσήματα) that is "not dangerous" or nonthreatening (ἀκίνδυνος) and "will not be serious," or "not many will die," sounds like modern childhood diseases, but they were not present in the relevant period. If such virulent diseases had appeared, they would have killed many adults who had no immunity. The predictions might instead refer to one major Mediterranean affliction, malaria.

The name "malaria" – "bad air" – is a reminder that the disease has been associated with weather conditions in many cultures.[140] Recent studies suggest that Tutankhamun and other Egyptians, from an Early Dynastic child to 18th Dynasty royalty, suffered from the severe malaria *Plasmodium falciparum*, which kills 15 percent of those infected, especially the young, but heterozygous carriers of sickle-cell or thalassemia genes are partially resistant to it.[141] The likely occurrence of malaria in the Bay of Naples could explain the higher than average infant and child mortality rate in eighth-century colonial Pithekoussai. The second, hybrid, generation of children born in the colony may have inherited some resistance.

Malaria can cause periodic suffering and weakness but is seldom immediately fatal. (It could have been the stimulus for the June 14 omen, "bodies shall be utterly weak.") If thunder means a period of rainstorms, then puddles could support the growth of the mosquito vectors of malaria. Fresh malarial infection is unlikely during the winter, because eggs require temperatures between 20 and 30 degrees C, with humidity above 60 percent.[142] Those already infected would suffer complications from any other conditions they happened to harbor. Scheidel (1996) presents a study of the dates of death recorded in Christian epitaphs from the Roman catacombs that shows a "conspicuous peak of mortality" from July through October, no doubt the result of infections that attacked people already weakened by malaria.[143] Oaks et al. (1991) noted that rainfall can stimulate an "explosion" of *Anopheles* mosquitoes, and road building and agricultural activities (which escalated in Iron Age-Archaic Etruria) could have increased their spread exponentially, even if the famous *cuniculi* and other waterworks did actually drain some swampy and potentially malarial areas.

The debate on malaria as a background disease in the ancient Mediterranean has been considerably blunted by the intensive study presented by Sallares (2002). A past study suggested that malaria was *not* epidemic in Europe until about 500 BC, because populations were too low to support the establishment of endemic *foci*, and the *Anopheles* vectors would have been too weak. T-W-Fiennes noted that the infection is eliminated in eighteen months in

[140] See Aufderheide and Rodríguez-Martín 1998:228–38, esp. 234.

[141] Hawass et al. 2010. See Becker and Donadio 1992; T-W-Fiennes 1978:109.

[142] On etiology and symptoms of malaria, see Dunn 1993.

[143] Lowest rates occurred in June and February: Scheidel 1996:139–63, esp. 148–53, supplementing Sallares 1991:237–38; see now Sallares 2002.

those who recover.[144] Today, *Anopheles* mosquitoes are found in Tuscany and the Pontine marshes.[145] Under such circumstances, the disease should have been erratically transmitted and mild.

The issue is not clear-cut, however; Hippokrates had recognized the symptoms of what sounds like malaria by 400 BC.[146] Sallares makes a strong case for malaria behind the varieties of fever known in Imperial and Late Republican Rome: "There is no doubt that *P. falciparum* malaria was present and well known in all its varied manifestations in Rome at the beginning of the Imperial period."[147] He maintains that from at least the fifth century BC, when relevant literary accounts begin, malaria in the Etruscan Maremma and farther south was accepted as so common that it was never noted as a new or threatening illness.[148] The concentration of thalassemia in Italy and Sardinia suggests a very early advent for malaria to have provoked such a strong genetic response. Shaw (1996), in a thorough comparison of Imperial Rome with medieval and modern Italian data, strongly suggested the "great killer" tuberculosis and its attendant infections as the reason for the mortality figures in Rome. The diseases that Etruscans would have feared in response to thunder-omens could have been much simpler and truly transient problems, but the issue cannot be resolved without well-preserved forensic evidence.

THE AFFLICTION OF PLENTY?

Reference to a good harvest:

> August 11. If in any way it should thunder, there will be a good harvest, yet the downfall of reptiles and harm to men.

Several predictions associate prosperity or a good harvest with disease (June 10, Aug. 11, Dec. 13, Feb. 5, March 14). Modern studies in Libya found that bacterial infections rise dramatically during the fall date harvest, when flies multiply and spread them.[149] The *Brontoscopic* entry at June 7, noting good cereal crops and diseases that kill few people, even offers the possibility of something like favism, which might provoke serious reactions in young children, but it would not kill many. The *benefits* of this congenital affliction, as with thalassemia, would only accrue in conjunction with the threat of malaria, though.[150]

144 T-W-Fiennes 1978:108–12.

145 Bruce-Chwatt and de Zulueta 1980:14 et passim; on the arguments against early malaria in Italy, see Sallares 2002:17, and note 24.

146 *Airs Waters Places* 24 = Jones 1923/1995:lvii–lviii, 132–37.

147 Sallares 2002:18–19.

148 Sallares 2002:192–200.

149 See Kuhnke 1993:456; Armstrong et al. 1998.

150 See Brown 1993; Grmek 1989/1991:210–44.

ZOONOSES

A few references to zoonoses or links between human activity and animal disease are tantalizing:

> August 13. If in any way it should thunder, there will be plague upon the bodies of both humans and dumb animals.
> November 5. If it thunders, it signifies storm for the state, and disease for humans and dumb animals alike.

Borgognini Tarli (1992) has suggested that the high incidence of porotic hyperostoses in skeletons of the Bronze Age inhabitants of Italy was a consequence of cohabitation with domestic animals. Enteric diseases contracted from eating meat contaminated with bacteria or from handling infected animal products could compromise digestion and absorption of nutrients, and lead to anemia or other deficiency problems. Anecdotal references in Classical literature occasionally show an awareness of zoonosis. For instance, Livy (3.6.2–4), describing Rome's war with the Aequi in 463 BC, says it was "a year of sickness in city and country, affecting sheep just as much as men" (*forte annus pestilens urbi agrisque nec hominibus magis quam pecori*). The suffering was exacerbated when flocks were brought for safety into the confines of the city. Perhaps the most immediate zoonosis reference is the description by Herodotus (1.167) of the affliction that crippled equally "sheep, draft animals or men" (ὁμοίως πρόβατα καὶ ὑποζύγια καὶ ἄνθρωποι) after the Caeretans stoned the captives from Alalia.

"Dumb animals" (ἄλογοι) or "the four-footed ones" (τετραπόδοις, at August 8) indicate domesticated species, such as flocks and herds, that could harbor parasites, anthrax, brucellosis, or bovine tuberculosis.[151] In the modern world, these diseases are spread by long-distance trade in livestock. Iron Age Sardinian boat models imply the long-distance transport of animals by water from an early date, as do histories of Phoenician and Greek colonization,[152] and it is generally assumed that Iron Age Etruria imported equine breeding stock from the Levant.[153] Whether or not the horses that came with Phoenician or Cypriot chariots harbored the rhinoviruses responsible for the common cold, it is possible that other diseases, like rinderpest, could have circulated as early as the eighth century BC.[154]

A propensity to zoonoses is further supported by circumstantial evidence from the Campanian discoveries in the village at Nola-Croce del Papa (ca. 1780 BC – see Chapter 5). Animal pens (with pregnant goats and a guard dog) were found in the village among the huts, and thousands of footprints of humans and animals (cattle – some inbred – sheep, goats, pigs) were preserved in the

[151] See in general Miller 1987; Krauss et al. 2003; T-W-Fiennes 1978; Acha and Szyfres 2003.
[152] Camporeale 1967:138–40, pl. 34, and 1969:94, pl. 32.4.
[153] Emiliozzi 1997:*repertorio*, 311–35 et passim; Azzaroli 1972.
[154] Cf. T-W-Fiennes 1978:18–20.

rising volcanic ash.[155] Although the mass escape was an unusual situation, it is clear that humans and their livestock ate, slept, and gave birth in close proximity to each other – prime conditions for propagating zoonoses.

AVIAN ZOONOSIS: NEWCASTLE DISEASE AND MUMPS?

At February 23, "If it thunders, it threatens deformity for men, but destruction for birds (ὀρνέοις)." What diseases are shared by men and birds? Etruscan art abounds with birds, both wild and domesticated, living in close proximity to humans, and also as subjects for augury. Expressions such as "winged ones" (τὰ πτηνά) seem close to the meaning of Archaic Italic texts, such as the Umbrian *Iguvine Tables*, which refer to birds as the "messengers of the gods." Might the augural perception of birds reflect some understanding of them as graphic harbingers of adverse conditions, like the proverbial canary in the coal mine?

Today, influenza strains do affect domestic fowl (and/or swine) before humans, but that is due to mutation of a virus in dense and stressed flocks. Ancient Etruria did not have a large enough population of fowl living under "factory" conditions and in close proximity to large numbers of humans, as is today unfortunately the case.[156] Influenza has been traced by some scholars to Classical antiquity through the Hippokratic account of the epidemic of Perinthos.[157] Because its main symptoms are respiratory, with fever, however, influenza does not seem an obvious cause of "deformity for men." Psittacosis/chlamydiosis (parrot fever) is a serious disease of birds, but it is not transmitted in such a way that it would be perceived as an epidemic.[158]

The avian paramyxoviruses, however, could have caused the full set of symptoms predicted in the *Brontoscopic Calendar*, because they include Newcastle disease, which kills birds,[159] and *Rubula* virus, which results in

[155] Mastrolorenzo et al. 2006; Albore Livadie 2002.

[156] T-W-Fiennes 1978:37–45, references 186–91; Aufderheide and Rodríguez-Martín 1998:210–12; Crosby 1993B. For a good bibliography of recent literature on zoonoses, sorted by animal species, see the electronic database maintained by D. Shapiro at http://medicine.bu.edu/dshapiro/zooref.htm (accessed July 13, 2001). Some basic works are S. R. Palmer, Lord Soulsby, and D. I. H. Simpson, eds., *Zoonoses: Biology, Clinical Practice, and Public Health Control* (Oxford: Oxford University Press, 1998). P. Acha and B. Szyfres, *Zoonoses and Communicable Diseases Common to Man and Animals*, 2nd ed. (Washington, DC: Pan American Health Organization, 1987). J. Diamond, *Guns, Germs, and Steel: The Fates of Human Societies* (New York: W. W. Norton, 1998), 195–214; and A. Karlen, *Man and Microbes: Disease and Plagues in History and Modern Times* (New York: Simon & Schuster, 1996), 29–63, offer lively discussions of the density dependent diseases of zoonotic origin.

[157] *Epidemics* IV; see Grmek 1989/1991:305–39.

[158] Yung and Grayson 1988.

[159] Newcastle disease can directly affect modern workers in close proximity with infected birds; see D. G. Trott and R. Pilsworth, "Outbreaks of Conjunctivitis Due to the Newcastle Disease Virus among Workers in Chicken-broiler Factories," *British Medical Journal* 5477 (1965):1514–17. Other disfiguring infections may be linked to avian diseases, but, again, the conditions of modern factory-farms that promote them simply did not apply to ancient

parotitis – mumps – in humans. Mumps would certainly account for the description of "deformity" at February 23 and at January 12, when "men shall be damaged in their faces," likewise at February 4, when "men shall be troubled not only in visage but also in their very minds." If this last comment really is a reference to communicable disease, then perhaps the fever of mumps accounts for the troubling of minds. Kraus and colleagues note that although the two disparate diseases are not caused by the *same* virus, other pathogenic animal viruses exist that are closely related to human paramyxoviruses.[160] Newcastle disease, "atypical fowl plague," is endemic worldwide, and is lethal to poultry. It causes symptoms in the respiratory and gastrointestinal tracts and central nervous system, and it can affect people who work with and slaughter domestic fowl, such as chickens and turkeys. Symptoms can appear within one to four days of contact with infected birds, so it would have been easy for an ancient observer to make the association between the two species.[161]

The Hippokratic *Epidemics* (1.1) unequivocally describes an outbreak of mumps in a dry early spring on the island of Thasos in the fifth century BC.[162] The Thasian victims had slight or no fevers and the classic manifestation of mumps, swellings on one or both sides of the ears. Some suffered dry coughs, and young men, but not women, experienced more discomfort, and in some the testicles also became inflamed. (In subsequent seasons, the Thasian survivors suffered other, severe infections of various kinds.) It was noted that most people showed no serious fever and little distress apart from their mumpy appearance, which does sound like "deformity for men" or "men ... damaged in their faces."

In temperate climates, mumps clusters in cool seasons, which fits the dates of February 4 and February 23 in the *Brontoscopic Calendar*. The incubation period is two to three weeks.[163] Now a childhood disease, mumps appearing for the first time in a naive adult population would have been much more severe. This has been noted especially during wartime, when outbreaks have disabled armies of conscripts.[164] Today, the only reservoir of mumps is the human population, but the same climatic conditions associated with seasonal

farming practices. Cf. A. A. Mutalib, J. M. King, and P. L. McDonough, "Erisypelas in Caged Laying Chickens and Suspected Erysipeloid in Animal Caretakers," *Journal of Veterinary Diagnostic Investigation* 5(2) (1993):198–201. Cf. A. G. Carmichael, "Erysipelas," in Kiple 1993, 720. T-W-Fiennes 1978:37–45. I thank Dr. Robert Moffatt, VMD, for discussion of Newcastle disease.

160 Kraus et al. 2003:123.

161 Kraus et al. 2003:124–25.

162 Cf. Grmek 1989/91:336. See text in Jones 1923/1995:146–49.

163 I am grateful to Dr. Jonathan King, VMD, for discussions of zoonoses, and to Dr. Deborah Goldberg, MD, for suggesting mumps for these entries.

164 For the most recent in an extensive literature on mumps, see Holmes 1997; Kim-Farley 1993 notes that the droplet-borne pathogen needs a large human population as reservoir.

outbreaks of mumps could have provoked the related scourge of Newcastle disease, fatal to birds and contagious to their human handlers.

"*THE REPTILES*"

Two more predictions are tantalizing and involve τὰ ἑρπετά, potentially any "creeping things," but almost surely here, "the reptiles." The exact identification of these "reptiles" is problematic, but Mesopotamian omen literature does refer especially to (poisonous) snakes, and at July 13 and 28, and February 14, they are called "poisonous." At August 11, "the downfall of reptiles and harm to men," and at February 24, destruction is predicted for fish and reptiles. Tortoises and turtles, snakes, lizards, and frogs are all portrayed in Etruscan art, beginning in the Early Orientalizing period. They are occasionally attested in archaeological finds of bones, often from meals, and, of course, they are still present in Italy. Most species were casual rather than intimate members of Etruscan households, like the frogs and lizards found in the thatch of Bronze Age Croce del Papa. And although vipers may have been feared for their poison, as they are in Tuscany today, many other snakes would have been appreciated for their pest-controlling habits.

Because of obvious differences in metabolism, body temperature, preferred environment, and the like, humans share fewer afflictions with reptiles than they do with birds or mammals. Unfortunately, the modern pet industry has made us aware of the dangers of infection with the gastroenteritis of *Campylobacter*; in reptiles it can be even more severe. *Salmonella* is also associated with reptiles and birds, thanks to the inhumane practices of modern commercial breeding, though today it is more harmful for humans than for the turtles and other small pets forced to live in filth.[165] Another possible disease group shared by humans and reptiles is tuberculosis, recognized in different types afflicting cold-blooded animals, birds, or cattle, and also humans.

Probably environmental changes account for the differential prediction of February 24: "If it thunders, it threatens good health for men, but destruction for both fish and reptiles." The aftermath of winter floods bearing silt or salt might have caused observers to see fish kills or dead lizards. In fact, at April 12 and May 22, destruction or ruin of fish is noted in conjunction with heavy rains – when water runoff might lead to algal blooms and the production of lethal toxins. Some bacteria of humans, like *E. coli*, can be pathogenic to reptiles, and a period of good harvest might create additional sewage or runoff

[165] There is an extensive literature on this topic; for instance, H. P. Minette, "Epidemiologic Aspects of Salmonellosis in Reptiles, Amphibians, Mollusks and Crustaceans," *International Journal of Zoonoses* 11(1) (1984):95–104; J. Y. D'Aoust et al., "Pet Turtles: A Continuing International Threat to Public Health," *American Journal of Epidemiology* 132(2) (1990):233–38. Cf. T-W-Fiennes 1978:146.

from the processing of foodstuffs.[166] At December 10, the opposite situation is invoked, when thunder "threatens slaughter for men from diseases, but the fish shall be abundant." We should probably read the omens for fish and reptiles as expressing concern for food supplies, not for the creatures themselves.

SOCIAL ASPECTS OF DISEASE: "THE FORTUNES OF SLAVES"

A few entries in the *Brontoscopic Calendar* seem to acknowledge the social aspects of disease in Etruria:

> July 6. If in any way it should thunder, it threatens death-bearing diseases to the fortunes of slaves.
> September 7. . . ., it threatens a disease and out of it, a disaster for the servile class.
> December 26. If it thunders, it threatens diseases for the slaves.
> January 7. . . ., there will be a slave revolt and recurring illness.

In the other predictions, the victims of afflictions are generic, "men" or "people," and childbirth for women is simply a natural association, but one class is singled out for specific health warnings. At three or four dates, disease is predicted specifically for slaves – something that one might expect under the conditions of the Roman *ergasteria* (agricultural slave dungeons) of the Late Republic and Empire, but not necessarily among the serfs or bondsmen of earliest Etruria. We might reasonably expect disease to be a result of problems related to malnutrition, but we lack archaeological evidence of chattel slavery in Iron Age and Archaic Etruria. In fact, apart from rare references in the Classical authors, the strongest evidence of a servile class comes from onomastic studies that seem to identify slaves' names as signatures in some well-to-do ceramic workshops, where they were hardly a desperate and oppressed population (see Chapter 7). If "slaves" referred to prisoners of war, or some other unprivileged class, then perhaps conditions of crowding or the expected deaths from wounds incurred in battle might explain this problem. Although later, Roman slaves were said to get plainer rations, mainly cereals, there is no indication through the Archaic and Classical periods that one stratum of society, even slaves, ate a radically different diet than their masters. For instance, studies of skeletons by Bartoli and colleagues showed that the Paleovenetic population of Este (8th to 6th c. BC) and Piovego (6th to 4th c. BC) enjoyed essentially the same diet, whether affluent or not (as gauged by the number of grave goods

[166] See T-W-Fiennes 1978:146; see Oceanic Studies Board 1999:43–70, on infectious diseases and algal blooms.

in their tombs).[167] The later evidence of anatomical votive models includes at least one dedication by a freed slave, a class growing in economic and social power by the third century BC.[168] Slaves in the *Prophecy of Vegoia* are portrayed as the extension of their masters, yet also as actors independent enough to attract divine anger to themselves.

The selective affliction of slaves is a perplexing situation, perhaps to be understood as the gods' willingness to communicate helpful messages to them and to their aristocratic masters. Because we do not know the Etruscan word used for "slave" in the *Brontoscopic Calendar*, we cannot be sure that the original text did not refer to a geographically or ethnically separate group, such as rural serfs. The entries on slavery in the *Brontoscopic Calendar* must have been written by Etruscans in Etruria, for these passages have no parallels in Mesopotamian texts. Their viewpoint is similar to that implicit in the other categories on health and society, relatively matter-of-fact but spoken for the benefit of the persons it names. The chattel slaves of later Rome, undoubtedly suffering diseases and injuries, are not the subject of these omens.[169] Iron Age slavery must have been a different situation, where captives' sufferings would have been apparent and immediate to their masters. (See the next chapter for more on social issues in Etruria.)

CONCLUSION

The worldview and understanding of health implicit in the predictions of the *Brontoscopic Calendar* differ radically from the academic philosophy expressed by the authors of the Greek Hippokratic *Corpus*. Its tone also contrasts with historical accounts of prodigies by writers like Livy. Mesopotamian omens on disease and physical afflictions are wary in tone: such suffering is sent from the gods as punishment to sinful mortals. This belief seems to have been common across the entire Near East and has been analyzed for Mesopotamian, Levantine, and Egyptian texts. Bardinet (1988) noted the strong influence of Canaanite/Semitic texts and vocabulary in the Egyptian medical papyri and other writings, from the New Kingdom (and before) through the Ptolemaic and Coptic periods. Although the Levantine papyri themselves are not extant, their influence may be clearly discerned. To the "audience" of the Etruscan *Brontoscopic Calendar*, misfortunes and illness probably seemed to fall

[167] Bartoli et al. 1999:220.

[168] Vel Tiples: *ET* Ta 3.5; Turfa 2004:363, no. 301; Colonna 1966:pl. 51.

[169] The Herculaneum victims may have given us graphic illustration of the physical insults of slavery, such as a man whose childhood injuries had to heal after he returned to back-breaking labor, a frail man who worked hard every day, a little boy (9–10 years old) who performed heavy labor most days of his short life, a girl caring for her master's infant, and another pregnant too soon with her own; see Capasso 2001: individuals Erc 49, Erc 27, E 92, Erc 10, and Erc 110.

unexpectedly from the sky, but the actual symptoms predicted by thunder had to have meaning in their world – in other words, the text's creators did not simply copy unfamiliar terms from Mesopotamian models. Rather, the crafters of the calendar created many original health predictions without correlates in other literature, but with definite relevance to life in Iron Age Italy.

I have suggested only a partial assessment of the most obvious health-related entries in the calendar, in which the gods were expected to give warning of occasional outbreaks of disease among both humans and animals. The underlying assumption is that reasonably good health is normal, but noticeably *good* health, prosperity, and well-being are cause for special consideration, and bad health appears to occur almost always at random, but at the will of the gods, and is only rarely linked to situations over which man might have some control, like war.

In a time relatively free from modern scourges like cholera or yellow fever, Etruscans did expect to experience digestive disorders, parasites, respiratory ailments, skin diseases, and what to them were plagues – perhaps some were virulent diseases now demoted to childhood nuisances, like mumps. Other infections such as brucellosis and tuberculosis were brought on by their pastoral and agricultural lifestyle, with malaria so endemic a problem that it lurked almost unremarked in the background, as if waiting for the polyglot masses of Imperial Rome to attain its critical mass.

Many diseases that we might expect, from a reading of the Classical medical authors, such as the tertian and quartan fevers, seem not to be cited – probably frequent bouts with fever marked an integral part of Etruscan life. The skeleton of the unusual boy buried at Tarquinia Pian di Civita tells a story of frequent, periodic bouts of childhood fever and/or hunger that is echoed in many other communities, down to the residents of Flavian Pompeii and Herculaneum. Perhaps a more cautionary interpretation is implied in the prophecy of May 9, "If it thunders, it signals plague, but not especially life-threatening." Any modern reader perceives plagues as unthinkable, uncommon catastrophes, but the ancient populations of central Italy, living with the constant threats of bad harvests, flood, storm, or insects may have come to expect periodic, plague-like bouts of infectious disease. They contented themselves with the assurance that at least some of the "plagues," though causing suffering, would spare them their lives. Even if we see the "spotted diseases" and the nonlethal "plagues" in a modern light, as the result of microbes and not thunderbolts, these references do offer hints of the health and vital concerns of the Etruscan populace.

CHAPTER 7

THE SOCIETY OF THE *BRONTOSCOPIC CALENDAR*: "THE WOMEN AND THE SLAVES WILL CARRY OUT ASSASSINATIONS"

HUMAN SOCIETY: POWER AND DIVINATION

The third realm among the *Brontoscopic Calendar*'s predictions, human society, is of equal importance – to judge by the number of citations – with land and food supply, and with the health of humans and animals. The calendar's omens encompass familial events, such as the birth of children, the social status and empowerment of women, and the friction of a society in which social strata, slavery, and stigmas have emerged and can sometimes lead to unrest. There are also the relations (war and peace, harmony and discord) of this society with other groups, such as allies and enemies. I have suggested that the initial recording of the *Brontoscopic Calendar* took place during the Iron Age, perhaps very roughly around 700 BC. (This is based on the depth of its evident Mesopotamian influence, the date of the adoption of literacy in Etruria, and the historical context of the written *etrusca disciplina.*) Circumstantial evidence strongly suggests that it could not have been a first-century BC or later forgery. If it was composed as an Etruscan document, at the beginning of the seventh century during the formalization of the great city governments and cults, then it can only describe (serendipitously) the character of society in Etruria of the ninth through seventh centuries BC. I will survey what is now known of that culture (and its *sequellae*) to show that, just as the agriculture and health of Iron Age Italy were reflected in the calendar's predictions, so too all the elements of a complex and stratified society, including its strife, both foreign and domestic, were already in operation by that early date. In fact, some excavators have presented archaeological evidence that could place the stratification of society, and the precursors of urban planning, as far back as the ninth century.[1] Put simply, Etruscan society was ab initio radically different from the societies of the Near East, and Etruscan ancestors, as their cities

[1] See Bartoloni 2006; Damiani et al. 2006, Pacciarelli 1998, and more in this chapter's notes.

developed, had to craft a new subset of predictions to cover their own human and daily concerns beyond the omens culled from Mesopotamian texts.

The palace societies of Mesopotamia would have seemed alien to the people of Etruria: the crown princes, usurpers and standing armies of Babylonian omens do not find counterparts in the Etruscan *Brontoscopic* document. For that matter, Etruria adapted the tangible elements of Near Eastern society – chariots, parasols, fans, thrones, and footstools – to its own purposes. These symbols appear in numerous tombs, including those of women – whereas, in the Near East, only the Great King and crown prince would be seen riding in a chariot beneath a parasol.[2] By the early sixth century, all sorts of Etruscan women were riding in this fashion, and even driving themselves. We can view this as "broadening the market base," except that the original chariots were so valuable that they were probably never "sold" by simple merchants, but rather must have been bestowed (along with horses, grooms, and chariot repairmen) as diplomatic gifts upon rulers whose permission was sought by outsiders for more canonical forms of trade, like access to metals. In all the formal expressions of regal status that they acquired from the East, the Etruscans (or those catering to them) adapted the usage and the meaning of these royal goods to native Italian or Etruscan objectives, and the circumstances of the *Brontoscopic Calendar* are analogous.

The fields in which the Etruscan or Italic social and political background are most evident are the status of women, the delicate and often explosive balance between different urban social/political classes, the ever-present concerns for and about slaves, and the waging of war, which implies close contact over the long term with different ethnic or regional groups. All the conditions noted in the *Brontoscopic* document were already present to some degree by the Late Iron Age in Etruria. They were all still fomenting throughout the Archaic period, even though our current and woefully limited knowledge of Etruscan history makes us more familiar with similar events and situations that occurred in the fourth century and later.

The underlying perception of society in the calendar's predictions is of a city population, still living close to the land and identifying with the land for its life. At least eight apodoses refer specifically to a city (or "the" city), and several more speak of "the state" or "the commonwealth" (οἳ κοινοί, ἡ πολιτεία). Sometimes the "affairs of state" (πράγματα τῆς πολιτείας) are just "affairs" (πράγματα). The city is ἡ πόλις. At June 29, the improved affairs are of the "queenly city" (τά τῆς βασιλίδος πόλεως), which in *Byzantine* parlance would mean Constantinople, and in earlier literature it referred to Rome. If there was an original Etruscan line closing the *Brontoscopic Calendar*, where the Figulus text now ends: "This brontoscopic almanac Nigidius claimed was not universal, but was only for Rome," such a "colophon" was probably similar

[2] Cf. Turfa 1999.

to the closings of the other documents of that time, the Zagreb *Liber linteus* and Capua tile, which assert their authority with a statement about the writing of the text. If so, the Etruscan brontoscopy must have recorded the Etruscan city for, or in, which it was created.

The urban setting:

> June 5. If in any way it should thunder, ill-omened for the countryside. Those responsible for villages or towns will be thrown into a state of disorder.
> July 22. …, it signals good things for the affairs of state, but for the bodies, diseases around the head.
> November 5. If it thunders, it signifies storm for the state, and disease for humans and dumb animals alike.
> November 24. …, the watch post shall complete for the state good service against enemy tricks.
> December 16. …, newfangled affairs for the state.

Mention is made, without prejudice, of both "country folk and city-dwellers," or citizens (ἀστειοτέρι); the people can also be simply the δῆμος. By the end of the ninth century BC, the nuclei of the great cities were in place, on the sites of Caere, Tarquinii, Veii, Vulci, and others. Each proto-city already manifested traits of a distinctive regional character, in pottery and metalwork, and in funerary customs.[3] Across the settlement's territory, we see what Torelli has called strong bonds of economic and social community, which, like infrastructure, are criteria for definition of the city.[4] Some basic developments in Etruscan society and political organization may be illustrated by a diachronic survey of selected archaeological sites.

LARGE, WELL-ORGANIZED COMMUNITIES: BRONZE AGE TERREMARE

The archaeological evidence establishes that an organized society of large settlements linked by cults and commerce was fully operative by the Middle Bronze Age in the regions ancestral to Etruscan culture.[5] As early as the mid-second millennium BC, the so-called Terremare culture, famous from early discoveries in northern Italy, showed the hallmarks of a sophisticated society, with careful attention to communal planning and preparation for various forms of danger. Terremare villages were laid out according to a highly

[3] Cf. Torelli 1987:40.

[4] Torelli 1987:41.

[5] For background on the Terremare and *palafitte* societies, see Gleba 2008:9–10; Bernabò Brea and Mutti 1994; Bernabò Brea et al. 1997. For a concise survey of recent research into protohistoric processes, see Bietti Sestieri 2005 and 2010. I thank Anna Maria Bietti Sestieri for her new book, which appeared after my manuscript was prepared.

formalized plan; although there is no disparity in burial rituals or offerings to indicate social stratification, family and community values were expressed in other aspects of mortuary practice. The number of people living within a walled settlement must have led to occasional problems and the need to obey some central authority. Traditional "women's work" in agriculture, gathering, food storage, and textile production would have been recognized as indispensable to the community and family. Shared cultic celebrations were conducive to harmony.

ORGANIZATION AND SPECIALIZATION OF LABOR AND CREATION OF INFRASTRUCTURE: *PALAFITTE*

By the Late Bronze Age, a number of settlements in northern and central Italy held large populations, and the infrastructure must have been maintained by a central authority. The systems employed in *palafitte* ("platform") villages of the Late Bronze and Early Iron ages efficiently performed these duties, with carefully constructed drainage ditches, well-maintained wells, and central ponds or cisterns for water supply. Even more noticeable are the specially constructed bulwarks or caisson-type flood ramparts. In the Po valley, flood surges may only have occurred once in ten years, but they would have overwhelmed any town that was unprepared – so we may assume that annual or more frequent labor was expended on the building and upkeep of these works, just as the structures of the platforms themselves had to be constantly renewed. An elaborate variety of details and the purposeful adaptation of the finely cut and joined timbers attest to the growth of groups of craftsmen who could specialize in these techniques, as do discoveries of their specialized tools.[6] Reliance on people with specialized skills betrays the seeds of a stratified society.

The *palafitte* of the North, from Switzerland to the Po delta, represent several different ethnic or linguistic groups;[7] the cold, wet period of European climate (ca. 800 BC) prompted another flourishing of this art of construction. Near Trente, the site of Fiavé, inhabited since the third millennium and flourishing in the Early Iron Age, had aerial platforms supported on mortised pine pilings preserved to 10 m high in a peat bog.[8] More complicated carpentry was used in water cisterns at Les Grandes Corvées (near Metz, early 11th c. BC).[9] Similar wooden structures provided support, defense, and water supply for centuries at Ledro, Barche di Solferino, and Banche; builders used standardized plans, tools, and uniform styles of size and shape for pilings.[10]

[6] Pellegrini 1998:28.

[7] Balista and Leonardi 1996. See Säflund 1939:219–24, 238–39.

[8] Marzatico 1992.

[9] Boura et al. 1990/1992:75, fig. 6.

[10] See Balista and Leonardi 1996.

Negroni Catacchio and Miari (1995) have discerned the social and political stratification that resulted in a hierarchy of village types and of construction techniques and materials.[11]

A TIME OF GROWTH AND CONSOLIDATION: THE PROTOVILLANOVAN PERIOD

Etruscan urban society can be said to have formed during the final stage of the Bronze Age, when sophisticated Italian metal technology came to be associated with a new version of the funerary practices (the change from inhumation to cremation) and pottery types developed in transalpine Europe, in the so-called Urnfields culture. (What prompted such dramatic change in the deeply held beliefs of families concerning death and burial? There is no sign of European aggression or invasion, and one possibility is religious reform spread perhaps by itinerant prophets.) New radiocarbon dating techniques have pushed the date back to circa 1000 BC for the inception of the new outlook in peninsular Italy.[12] By the dawn of the Protovillanovan era (ca. 1200 BC), material culture with distinctive flanged axes and cremation burial using biconical urns were relatively homogeneous across the entire peninsula. But by the beginning of the Iron Age (traditionally ca. 900 BC, now understood as ca. 1000 BC), all the different regional cultures had emerged, as seen in the archaeological and typological record.

Some excavated evidence shows a sophisticated hierarchy of tomb types in the Final Bronze Age in the region of Caere-Tarquinia, Veii, the Fiora Valley (Vulci), and Narce, indicating a precocious beginning for social stratification and economic hierarchies.[13] By the beginning of the ninth century, small children interred with valuables must betoken inherited wealth,[14] marking the difference between "haves" and "have-nots" in the Final Bronze society of the southern Etruscan mining regions.

According to Bietti Sestieri, this visible diversity points to large communities linked to others in political association with networks crossing the peninsula.[15] She associates this with the need for stability in the control of long-distance routes for access to metal sources and increased trade in raw materials, farm surpluses, and manufactured goods. A shift to easily defended plateau sites allowed some communities to control long-distance traffic. Territorial control

[11] Turfa and Steinmayer 1996. See Bietti Sestieri 1973:387–88, and 1986:6, no. 18; and Turfa "The Technology of Wooden Structures: Etruscan Temples and Shipbuilding," *Etruscan Studies* 7, 2000:113–15. Malnati and Manfredi 1991:64, pl. 20.1, also comment on the emergence of technical specialists during the 8th c.

[12] See Bietti Sestieri 1997:378; Turfa 2006F; Bartoloni and Delpino 2005.

[13] See Pacciarelli 1998.

[14] Damiani et al. 1998.

[15] Bietti Sestieri 1997:378.

makes it imperative that leaders maintain a hierarchical and thorough system of organization; intensified political structure follows, and, given human nature, this is sure to breed some discord. Ethnic identities became further defined as many people traveled, traded, or otherwise interacted with individuals from different regions: Tyrrhenian Etruria, Umbria, Latium and the Sabine territory, and Puglia, Basilicata, and Calabria in the south. The Adriatic/Po region appears as a visibly Etruscan enclave beginning in the Protovillanovan phase.

Bietti Sestieri characterizes the Protovillanovan communities as chiefdoms, or kinship-based communities with stratified relationships within and between settlements, symbolized in the display of prestige objects.[16] In other words, the early "states" began to form during the Protovillanovan period, with Etruscan culture exerting a technological and artistic hegemony over much of the peninsula. Regular trade to transalpine Europe and the eastern Mediterranean was recalled by ancient authors (Cato, in Servius *Ad Aen.* XI, 567, and Livy 5.33.7–8), preserving the perception that Etruscans had "ruled" most of Italy, from the Tyrrhenian to the Adriatic.[17]

The sociopolitical complexity and power plays we understand for the Archaic and later periods, then, may also be applied to the Iron Age society of the time the *Brontoscopic Calendar* was developed in Etruria. Specialist studies in the typology of metal artifacts show the wide circulation of people across Italy and into Europe, in part surely the result of aristocratic intermarriage, but in other cases reflecting commercial routes.[18]

LARGE, ORGANIZED COMMUNITIES WITH SHARED FACILITIES TO COORDINATE

Many isolated villages disappeared, and organized clusters of huts were erected at the sites of the future cities: the forces behind such mass relocations and new foundations must have loomed as crises for those involved. During the Protovillanovan period the hierarchy of settlements is illustrated by the site of Sorgenti della Nova in what would become the territory of the city of Vulci.[19] It appears that agricultural goods and metals were brought to centers for marketing or tribute, and leaders there saw to it that roads, bridges, drainage, and water channels were maintained. At Sorgenti della Nova, the large settlement encompassed an estimated 300 closely situated huts, home to 1,000–1,500 people during the eleventh to tenth centuries BC, supported by a hierarchy of smaller outlying settlements.[20]

[16] Bietti Sestieri 1997:379.
[17] See Bietti Sestieri 1997:399.
[18] Documented in the series *Prähistorische Bronzefunde*; also Camporeale 2001; Pallottino et al. 1992. See now Bietti Sestieri 2010.
[19] For concise background, see Moser 1996.
[20] Negroni Catacchio 1995; 1982:18–19; Negroni Catacchio and Miari 1995.

Bartoloni (2006 and 2005) has furnished evidence, confirmed by radiocarbon analysis, that speaks to the formal foundation of a city at Veii as early as the ninth century BC. On Piazza d'Armi, the future acropolis, a tomb without offerings was covered by a Villanovan hut, and the sacred(?) space later was marked with walls. The burial (Torelli once proposed it could be a *bidental*, the mark of a person killed by lightning) was the focus of a cult through the seventh century. The orientation of the tomb and elliptical hut were respected by the orthogonal street plan of subsequent, Archaic structures, and there seems to have been a regular plan to the village that preceded it. In the seventh century, there may even have been a traditional foundation ritual, as noted for Rome, of ceremonially plowing the circuit of the city boundaries, for parallel furrows have been found beneath the later *decumanus*, a main street running through the city of Veii.[21]

On Lake Bolsena, Gran Carro shows a carefully planned, densely constructed habitation area, by the ninth century already at some disadvantage for lack of green space between its huts.[22] Apart from the intensive social and probably legal arrangements that must have been needed to regulate such close contact between neighbors, the town plan has the earmarks of the forced movement and implantation of a settlement. One possibly relevant historical parallel for the deliberate abandonment and refoundation of a town is the Roman *Parilia*, the festival commemorating Romulus' torching his shepherds' old huts to force them to move into his founded city.[23] Another possible scenario is a religiously inspired lustration or departure from an area of ill-fortune, as has been suggested for cases of heavy metal pollution of soils in mining areas.[24] A civically guided relocation into houses with tiled roofs occurred fairly abruptly in many seventh-century towns and must have caused ill-feeling when accomplished through the inducements or constraints of centralized authority. Early housing developments at San Giovenale, Veii Piazza d'Armi, and elsewhere show that the move from huts to houses was a concerted effort among Etruscans.[25] What social processes are reflected in the last ghetto huts that remained on the fringes of the elaborately tiled houses of Acquarossa?

During the seventh to sixth centuries BC, habitation sites continued to show the inexorable stratification of society, with some families distinguished by wealth and privilege. At the settlement at Lago dell'Accesa in the region of Massa Marittima and Vetulonia, social position was expressed in a hierarchy of dwelling types. The sprawling town also reflected society's commitment to

[21] See Bartoloni 2007:26.

[22] Tamburini 1995:209–21, 359–62.

[23] April 21; Scullard 1981:103–5; cf. Torelli 1987:87–89.

[24] See Harrison et al. 2009.

[25] It undoubtedly benefited most residents; see the suggestion on fire codes by Örjan Wikander 1988 and 1992; also Steingräber 2000; Donati 2000:319–23.

the risks of surplus production (of metals) for commerce. Mining and metallurgy require a firm support base and political directive, and the prosperity of Accesa seems to indicate their efficiency.[26] Founded in the ninth century,[27] Accesa too was abandoned at the end of the sixth century BC. Camporeale suggests an organization of the site into *quartieri*, each with its own specialties, such as mining.[28]

Some quarters, like District A, were laid out with a single, multiroom house for the "aristocrat" in charge, flanked by several small or one-room dwellings. But all households were run by freeborn matrons, identified by the remnants of their looms and spinning equipment; and inscribed objects left behind illustrate a relatively high level of literacy that we would not expect in a servile or poor population. Further, the grave goods show an affluent, if frugal, population, and several houses still held fragments of imported Attic pottery once used for domestic events.[29]

One metalworking plant was built over an old necropolis: this must mean that the families who built the tombs had died out or left the town. Such harsh changes in the use of land probably indicate social distress or discord. Other metallurgical centers also hold hints of social imperfections. The late sixth-century abandonment of Poggio Civitate (Murlo, near Siena) may represent the final stages of the process by which Etruscan polities established control over metallurgical resources, asserting it in innovative architectural engineering and design.[30] Studies of craft and the industrial production of surplus goods, as indicated by Nijboer (1998), Berkin (2003), Gleba (2000 and 2008), and others, also point out the seventh-century results of an "urban" process of technical and commercial growth that contributed significantly to the stratification of society and the uneven division of wealth in Etruscan cities.

IRON AGE SOCIAL STRATIFICATION: FUNERARY EVIDENCE

By the Villanovan period, the institutions associated with urban life were manifested across Etruria. Large populations, well-run villages, interregional trade, and formalized intermarriage were well under way. Although the graves look egalitarian in the ninth century, by the eighth century, funerary offerings differentiate between persons. Some mature adults were buried with almost no permanent offerings; yet in other families, some *children* buried with wealth were the offspring of ruling classes for whom land ownership had become hereditary. In a study of nearly 100 burials of the eighth century in

[26] Camporeale 1997.
[27] According to the evidence of tombs of elaborate type; see Doro Levi 1933.
[28] Camporeale 1997:418.
[29] Camporeale and Giuntoli 2000:23, 30, 65, 68, 76.
[30] See Turfa and Steinmayer 2002.

the Quattro Fontanili necropolis at Veii, Teegen (1995) noted this phenomenon for children under fifteen years. It has been suggested that in Etruria, as in Rome, the token of a *bulla*, the hollow amulet, was a signal of the wearer's citizenship.[31] If this can be extrapolated back to the earliest finds of *bullae*, in the Late Villanovan burials of Veii, Narce, and others, where elite women, children, and men were buried with bronze lentoid pendants, then we might have additional, tangible emblems of the stratification of the great cities. Zifferrero has suggested that these symbols of sun and moon were already worn to signal hereditary rank among the elite of the later Villanovan period.[32] Incidentally, the source of this astral iconography and ornament type is indisputably the Phoenician Levant, yet another case of borrowing Eastern symbols of power.[33] We should expect that the growing, complex communities were also establishing and maintaining formal religious cults, these too enriched by borrowings from foreign cultures.

ETRURIA AS CULTURALLY DISTINCT FROM NEIGHBORING CULTURES BY THE NINTH CENTURY BC

Wide-ranging regular contacts through exchange and diplomacy are underlying traits of the society portrayed in the *Brontoscopic Calendar.* Specialist studies in the typology of metal artifacts show the wide circulation of people across Italy and into Europe, with northern types of metal artifacts appearing regularly in southern Etruria and beyond;[34] in part, this was the result of diplomacy and aristocratic intermarriage, but it also encompassed organized, commercial routes of exchange.

During the second half of the eighth century, ties between Latium and Etruria, especially Veii, intensified, as an aristocratic class strove in Latium to consolidate its hold over the land and turned increasingly to bonds with the warrior-rulers of Etruria. At Osteria dell'Osa, an Etruscan "prince" who had become a member of the ruling class was buried in Tomb 600. His bronze-crested helmet, armor, iron sword, wheeled censer, and ornate vessels are all of Villanovan type, and they would have been displayed as symbols of his power within the community.[35] The phenomenon of an outsider becoming a town's leader had already occurred in Etruria, but during the eighth century, such adventurism was imposed upon the more egalitarian society of Latium, no doubt because of wars. The Romans credited Etruria with the creation of

[31] See Palmer 1996; Zifferero 2004:38–329.

[32] Zifferrero 2004:330. See Zifferero 1995 and 2004; Dolfini 2004 and 2005; Turfa 2005:133, no. 88, 177–78, no. 169; cf. Warden 1983.

[33] See Culican 1973; Zifferrero 2004:329.

[34] Camporeale 2001; Pallottino et al. 1992.

[35] De Santis 1995.

the symbols of power, which they adopted for the Roman state, including the magistrate's stool or throne, the purple *toga picta* of a *triumphator* (and the triumph ceremony with it), the *lituus*, and the *fasces*, and it is clear that with the symbols came the behaviors, laws, offices, and such already developed within the Etruscan states.[36] Because ornamental *fasces* or similar axes and other regalia can be identified in the burials of "princes" of the late eighth century in Tyrrhenian and Adriatic Etruria (Veii, Tarquinia, Vetulonia, Populonia, Bologna, Verucchio), we should attribute the appropriate legal, civil, and military systems to them by the beginning of the Orientalizing period (ca. 730 BC), if not before.[37] There are even hints of an enduring and close association between Etruscans and their Italic neighbors in vocabulary, and even syntax, shared across cultures, as seen, for instance, in the Etruscan term *macstre*, seen in the name Mastarna, related to Latin *magister*, the Etruscan use of the enclitic *-c*, and words like *nefts*, "nephew," and *vinum*, "wine."[38] The integration of men and women of different ethnic backgrounds must have enriched the communities of Iron Age Etruria and Latium, but disparities – an Etruscan giving orders to Latin soldiers, a homesick bride, or dissatisfied village blacksmith – must at times have fostered unrest or poor morale among fellow citizens, a situation recreated in the *Brontoscopic Calendar*.

FOREIGN CONTACT IN THE EIGHTH CENTURY AND COSMOPOLITAN CITIES

The Tyrrhenian coast had engaged in exchange with other parts of the Italian archipelago since the Neolithic period, when trade for Lipari obsidian flourished. Throughout the Bronze Age, trade with Sicily and Sardinia continued, and during the Iron Age interaction increased, especially with Nuragic Sardinia.[39] Imported goods in burials and settlements of both cultures, and other evidence of seafaring, illustrate close contact, including regular exogamy between the two groups.[40] This conscious diversity of overseas contacts became a sign of status in some Etruscan communities. Gilda Bartoloni (1991 and 2009) has portrayed Etruscan Populonia as the most important producer of metals and a leader in commerce with Sardinia.[41] Not only did Populonia trade with Nuragic communities, but it also shows deeper, multigenerational cultural engagement with Sardinia, first in valuable imports (bronzes, including boat models) and then in the imitation of Sardinian goods in Populonia.[42]

[36] Cf. Tassi Scandone 2001; Jannot 1993; Schäfer 1989.
[37] See Delpino and Bartoloni 2000:237–43.
[38] Camporeale 2000:201.
[39] Cf. d'Agostino 1990.
[40] See Lo Schiavo 2002.
[41] Bartoloni 1991:105–8.
[42] Lo Schiavo 2002.

At Vetulonia, imitations abound in such profusion that Camporeale has postulated concerted immigration by Sardinians, who brought with them a ceremony that served some special beverage from Nuragic-style jugs.[43] In a second level of association, there is evidence at Populonia of the marriage of Phoenician-trained Sardinian craftsmen into Etruscan families, in the form of unique craftsmen's emblems (miniature pendants) buried in the tombs of their Etruscan (we assume) wives.[44] The Sardinian connection also became prevalent at maritime Tarquinia, and had repercussions even in the *Ager Faliscus*.[45] Also, Populonia, like several other major settlements, shows strong commercial ties to distant Felsina-Bologna, because it served as middleman in commerce between Bologna and Nuragic Sardinia. Meanwhile, in Bologna, a single workshop seems to have been exporting (maybe along with their female owners, in marriage) its unique products, such as fine bronze distaffs.[46]

Early Etruscan contacts with the eastern Mediterranean are exemplified in artifacts dedicated abroad, and in interaction and intermarriage with the earliest of the colonists from the eastern Mediterranean. Pithekoussai provides unequivocal evidence of this in distinctive offerings that proclaim the ethnic origins of those buried in the eighth-century tombs. Vases, elements of costume, and some evidence of ritual (Levantine unguent flasks) show that some Etruscans (and/or Latins and Campanians) along with other Italians (Daunians) had married into Pithekoussan families of mixed Levantine and Greek genealogy.[47] Ship models in ninth-century Tarquinian tombs demonstrate that the early Etruscans were already a technically adept and well-traveled people and that some shipowners represented by models were powerful women.[48] (In fact, evidence is growing for Villanovan-period contacts between Etruria and the western Mediterranean, including Iberia.)[49]

One category of offerings at the great sanctuaries of the eastern Mediterranean are the distinctive Villanovan fibulae, many of "leech" or other types that belonged with the clothing of women.[50] They are probably

[43] Camporeale 2000:37; see Delpino 2002.

[44] Babbi 2002; Bartoloni 1991:109–13; Bartoloni et al. 2000:238–40.

[45] Cf. Turfa 2005:129–30, no. 78.

[46] Now recognized as distaffs: Gleba 2008:110–13, type A1; cf. Turfa 2005:82–83, no. 3; Turfa 2009:529, no. 18.

[47] See Ridgway 2000, and 1992:52–54, 67–77, 111–18.

[48] Turfa and Steinmayer 1999 and 2001; Turfa 2001.

[49] See Docter 2006; Docter et al. 2004; Milcent 2006; cf. Ridgway 2000.

[50] For Etruscan goods abroad, see Turfa 1986:69n48; Close-Brooks 1967. For additional Etruscan finds, see Bietti Sestieri 1986:19, no. 20; see Naso 2000B; von Hase 1981 and 1997; von Hase 1997:315, figs. 21–22. There is now evidence, from the 7th c. on, of Etruscans involved in the cults of Gaul and Iberia; see Gran-Aymerich 2006 and 2009, and Colonna 2006. For fibulae in Greek sanctuaries, see von Hase 1997:297–301.

remnants of actual sets of colorful clothing – Etruscan national costume, so to speak, fastened with the nonperishable fibulae. Women from Italy may have been traveling to the shrines or sending token sets of garments for dedication, perhaps because they had married into families of the Aegean or Near Eastern elite. Dynastic marriages among Greeks and Levantines would not surprise us, and mixed marriages were frequent in the new world of Geometric Pithekoussai.[51] Perhaps one of the Etruscan women adorned with leech fibulae was actually married to a "king of the East," like that cited in the *Brontoscopic Calendar* at July 16 (if said king did not merely rule an east-central Italian territory).

As early as 700–690 BC, epigraphic evidence shows important Etruscans as special guests at Latian funerals. At the Cumae necropolis of Fondo Maiorano, Tomb 17 held an Early Protocorinthian *lekythos* inscribed with an *abecedarium* and the phrase *hisamenetinnuna.* This is read by Colonna as *Hisa Tinnuna mene[ce]* or "Hisa Tinnuna has given/donated."[52] As David Ridgway points out, "it follows that Hisa Tinnuna is a person of high rank: though a foreigner at Cumae, he is allowed to make a funerary gift there; and, even in this solemn context, he is allowed to indicate the fact epigraphically in his own language. In the Greek terms appropriate to the time and place, he is thus a *xenos*, an honoured visitor from the north rather than a permanent and fully integrated resident."[53]

SOCIAL STRATIFICATION: THE EVIDENCE OF NOMENCLATURE AND LITERACY

Social complexity and stratification in Late Villanovan/Orientalizing Etruria have also been demonstrated in studies of Etruscan epigraphy and nomenclature.[54] The earliest inscriptions, owners' or donors' names, appeared around 700–675 BC. Rex Wallace suggests that two-part names may have begun just before 700 BC among the Faliscans and quickly disseminated through Etruria to match the social networks being developed in the new cities.[55] In cases where a double name is used, we may reconstruct a complex society in which clans and classes are in effect, and by 650 BC three-part aristocratic names appear in force across Etruria and Latium. Coincidentally, the phenomenon of "commemorative gifts" inscribed with the names of donor and recipient, begun in the seventh century, seems to end with the sixth century, as if further

[51] Coldstream 1993.

[52] Colonna 1995:332–38.

[53] Ridgway 1997:241–43.

[54] See Heurgon 1970; Colonna 1977; Marchesini 1997:116–66, see esp. 154–59 and 117n3 for previous scholarship; also Bagnasco Gianni 1996.

[55] Wallace 2008:79–80.

social changes have occurred that render the elite gift-giving ceremony less important or no longer in practice.[56]

In Caeretan inscriptions, both men and women carried gentilicial titles. On occasion the name indicates foreign (Greek) origins in a form that implies servile background.[57] Beyond the borders of conventional Etruria, around 700 BC, the *lekythos* offered in Cumae Fondo Maiorano Tomb 17 furnishes the gentilicial title of an Etruscan man, Hisa Tinnuna. Older still is the inscribed oil flask at Osteria dell'Osa, from an elderly woman's tomb (ca. 775 BC, perhaps even 9th c.). It is now read as Etruscan *ni lue*, "do not steal." The sentiment is seconded by the unabashed tone of a vase buried at Cumae (ca. 675–650 BC), which warns that whoever steals this *lekythos* belonging to an Italic girl, Tataie, "may he go blind!"[58] It might not indicate class struggles, but certainly shows a stressful social environment for some people.

Other inscriptions use an ethnic or the name of a city, which must be the result of an even earlier network of stratified, interactive urban systems. Examples of foreigners integrated into the highest levels of Etruscan society are inscribed on their various belongings, gifts, or offerings: for Latins, Kalatur, Sabellic Peticina, and Vestirikina, and for Greeks, Telicles and Rutile Hipucrates. Rutile's first name betrays Italic inheritance and also Greek. Although his life was concluded as an Etruscan citizen in seventh-century Tarquinia, his heritage began elsewhere, and his Greek single name was converted into an Etruscan *gentilizio*.[59]

By 650 BC, Greek immigrants were integrated into Etruscan society, as seen on a Caeretan *bucchero aryballos* at New York University inscribed *mi larthaia telicles lechtumuza*, "I am the little lekythos of Larthai, daughter of Telikles."[60] Her Greek father, Telikles, had been previously adopted into an Etruscan family. Larthai's valuable vase speaks also to the early literacy and power of women, and the association of some of them (in her mother's generation at least) with foreigners and slaves or ex-slaves. At the end of the seventh century, adoptive citizenship had been extended to another Latin ethnic, Tite Latine, given a rich tomb in the Picazzano necropolis of Veii, an indication that the stratified society was still open to augmentation.[61] Later epigraphic evidence for slave nomenclature includes Arnth Praxias, the fifth-century vase painter,[62] and Metru, who inscribed a fine red-figured kylix found in

[56] Wallace 2008:178–79.

[57] Marchesini 1997:119, 130; Colonna 1975 and 1977; de Simone 1981 and 1993.

[58] See Colonna 2004 and 1995:331–32, 342, also for 8th-c. letters on bronze hoards from Ardea and Bologna.

[59] *ET* Ta 6.1: Bagnasco Gianni 1996:no. 156; De Santis 1995:372; Ampolo 1976–77; Bartoloni 1987:52–53; cf. Torelli 1987:46; Colonna 1977 and 1975:189.

[60] See Bonfante 2006; *ET* OA 2.2; Marchesini 1997:119, 130, no. 42; Bagnasco Gianni 1996:no. 313; Colonna 1970.

[61] Tomb 17, Bagnasco Gianni 1996:no. 113; Palm 1952:56–57.

[62] See Colonna 1975:188–91.

Populonia: his single name, lacking a patron's gentilicial, must indicate ongoing servile status.[63]

ISSUES REFLECTED IN THE *BRONTOSCOPIC CALENDAR*: SOCIAL STRIFE IN THE CITIES

There is simply no way to have hundreds or thousands of people sharing agricultural land or living in close proximity in urban settings without fostering some degree of dissatisfaction and discord. It seems only logical that the great Protovillanovan settlements, such as Sorgenti della Nova, would have had to develop means of coping with and quelling factional strife and assuring compliance with a town's growing needs for conformity and public cooperation. But what are the archaeological correlates for expressions of discord if it stops short of warfare, destruction, or executions? Although a few instances have been documented of conflagrations in Etruscan and Italic villages or houses, as at Fidenae and Gran Carro, or the fire that sent men and women running for their lives at Murlo near the end of the seventh century, all of these could be ascribed to simple domestic or industrial accidents rather than urban or factional warfare.[64]

One tantalizing piece of epigraphic evidence may have been found in the text of the Capua tile inscribed in the early fifth century BC. Line 30 holds the word *natinusnal*, linked by Pallottino to the gloss that Festus (166, 167), on *natinatio, natinatores*, likened to the concept of Latin *discordia*.[65] It is curious, if this is a correct connotation, to find another calendar-formatted religious document citing civil discord and attesting to an Etruscan term for it as early as the Archaic period.

Many scholars have remarked upon the similarities of the calendar's description of social conflict and slavery, with historical events recorded from the fourth through first centuries BC. Although we do know of servile revolts at Arezzo (*seditiones Arretinorum*, 302 BC, Livy 10.5), and coups by the formerly disenfranchised freedmen of Volsinii (264–263 BC), these were undoubtedly not the first such situations, but obviously we cannot find written evidence of such troubles during the Iron Age.[66] The authors of the Late Republic and Principate had focused on the travails by which Rome managed to acquire these Etruscan cities, that is, under cover of urban warfare between Etruscan *domini* (or *principes*) and *servi*.[67] Torelli sees (in naming conventions, etc.) the evolution of a "'ceto medio' contadino," a sort of "middle-class

[63] Colonna 1975:190–91.

[64] Murlo disaster: Nielsen 1987:92–93; Berkin 2003:7–26.

[65] Pallottino 1948–49:170 = 1979:599.

[66] For the situation in Etruria during the 2nd c. BC, see Domenici 2001:87–90, *coniuratio servorum* (196 BC); Livy 33.36.

[67] Torelli 1987:90–92.

gentry," that, during the second century BC, lived in rural settlements around Chiusi, Perugia, and Volterra,[68] there is no reflection of any dichotomy of city and country in the *Brontoscopic* predictions – as in Classical Greece, *polis* and *chora* are inseparable. We may envision a hierarchical social system that included patrons and clients, because this is clearly in operation in the later epigraphic sources, but we do not know its exact legal character or the historical means by which it came about.[69]

Social stratification is further apparent in the epigraphic testimony of the class labeled *etera* in several Etruscan inscriptions, but consensus is lacking on the exact meaning or civil identity of this group: the term was symbolic of some respected social condition such that it was used in funerary epitaphs, and some cities had special magistrates to administer it (*zilaθ* or *camθi eterau*, etc.), but we still do not know for certain what the word meant. Steinbauer suggested it described a citizen class or public function.[70] Facchetti (2002) has surveyed all the epigraphic evidence and concluded that the term *etera* designated persons of an inferior social class corresponding to either "plebeian" or "client" in Roman parallels, thus a class of free men whose legal and social standing is better than that of the freedmen, the *lautni*, some of whom gained the intermediate status of *lautneteri* or *cliens libertus*. What is certain is that the *etera* were invariably affiliated with a *dominus* or head of family, through whom they achieved their civil and legal rights. Such a specialized term does not have a correlate in the *Brontoscopic* predictions, nor is it epigraphically attested before the Late Period, but the existence and legal control of such a class by the *gentes* implies its survival from an earlier day. In surveying all the titles of office found in funerary *cursus honorum*, especially among the tombs of Tarquinia and Orvieto, Maggiani (1998) found that, though most epitaphs are Hellenistic, the diversity of offices and their geographical distribution show a complexity that must have taken centuries to achieve. Benelli reads *etera* as "member of the *iuventus*" but cautions that there are females named in this class, so it is not a direct match for the Roman *iuventus*.[71]

One other correlate has been sought to the social categories designated in the *Brontoscopic Calendar*, a class determined by (relative) age. Just two entries name a discrete group known as "youths" or "wellborn youths":

> July 25. If in any way it should thunder, it will go badly for a band of youth and also for the crops along with them. It will be a disease-bearing time.
> August 24. …, it threatens the loss of wellborn youths.

[68] Torelli 1987:93.

[69] Torelli 1987:58–60.

[70] Steinbauer 1999:418; see Rix 1963:371; Maggiani 1998:119–20.

[71] Benelli 2003:219.

This could represent some special category of the heirs of noble families whose *paterfamilias* is still alive, but it may simply recall the flower of the youth going off to war.[72] (Conflicts involving the *iuvenes* at Caere in 354–353 BC, and Faliscan *iuvenes* ca. 357 BC show them as a distinct social category; Livy 7.16.2; Diodorus 16.31.7.)

One archaeological find has been proposed as evidence of some special category for elite youth, in the unusual Tomb 6337 ("Dei Due Giovinetti," ca. 700 BC) in the Monterozzi necropolis of Tarquinia.[73] Two young men buried side by side wearing identical silver anklets had iron-tipped spears placed by their sides and an array of offerings at their feet, mostly vases for banquet and libation; a bronze basin and iron knife represented sacrifice and serving meat. This may have been an idiosyncratic funeral for two very rich youths who died together, but the ritual shows that Tarquinian society deliberately commemorated their distinctive association.

In the *Brontoscopic Calendar*, every month includes some predictions of civil troubles, with seven such omens in the months of August, September, November, and January; June and December have the fewest (two each). The events range from coups or power plays – some successful, some tragic – by a popular/rich/powerful man to dissension or discord among the "common people." Potential Mesopotamian prototypes show far fewer references to common people and virtually none to revolt and civil discord; likewise, they may refer to a usurper or king but not to the array of possible tyrants that appear in the *Brontoscopic Calendar.* Again, the Etruscan document has many specially designed apodoses relevant to the different culture of Italy. The main term for civil unrest is στάσις, as it would have been, for instance, for all the Greek cities, where civil strife often led to emigration if not revolt. Omens include: "war, and a prosperous man shall disappear from public life" (June 16), "abundance and at the same time murders to the people" (June 17), "the underlings of the wellborn will foment revolution in the state" (Sept. 11), and "the lower classes will have the upper hand over their betters, and the mildness of the air will be healthy" (Oct. 4).

Carmine Ampolo suggests that the Latin authors misunderstood the original Etruscan designation of the lower social order(s), and that they were not slaves but simply a lower class of society without a clear-cut analogue in the Greek cities or Rome.[74] A similar confusion is probably also behind the Greek account (Pseudo-Aristotle *De mirabilibus auscultationibus* 94) of Oinarea, where power fell to a group of slaves and one became tyrant.

[72] Cf. Massa-Pairault 1986.

[73] Leighton 2004:59–60; Cataldi 2001.

[74] Ampolo 1990–1991:193–94; for analyses, see Harris 1971; Torelli 1981:257–58; M. R. Torelli 1978:257–61.

The case of the Pyrgi plaques, which seem to illustrate uncomfortably close links between the "king" of Caere, Thefarie Velianas, and the state (the military?) of Carthage, is a hint that Rome was not the only city to experience social unrest and a coup toward the end of the sixth century BC. As Livy's tale of the would-be king of Veii at the *fanum Voltumnae* shows, most Etruscan statesmen by the fifth century regarded the installation of a king as a giant step backwards in civic affairs.[75] The implication is that most Etruscan states had evolved more open governments, presumably with councils and elected magistrates. At Pyrgi, Temple A, the choice of the Seven Against Thebes for the gable *columen* plaque has been read as another expression of concern over civil war and conflicts that must have grown apace during the Archaic period.[76]

"THE WOMEN"

Women are cited at several points in the calendar, and evidence of early Etruscan female drivers, landowners, and proprietors of businesses is increasing with recent mortuary and epigraphic studies.[77] The evocation of a servile revolt in which women also participate, at August 19, could have been a very real concern of Etruscan urbanites, even as early as the seventh century. As for the identities of the slaves, we do not know the terms that were used in the original Etruscan calendar, but, from the mid-seventh century on, funerary inscriptions do name many persons whose origins were foreign and some who very likely arrived in Italy as slaves.

As early as the Villanovan period, the equality of burial ritual for men and women at Tarquinia, Veii, and Vulci commemorated the elevated and permanent position of the Etruscan matron. Boccolini (2004) observed that the ubiquity of textile equipment in female burials seemed like an egalitarian marker – but some tombs hold more valuables and more complex rituals than others. Presumably some differences served to set priestesses apart from other women, and their common bond, symbolized with weaving, was never forgotten (cf. Fig. 22). Early figural art served to record their status in the simple drawings of a seated couple that appear on hundreds of eighth-century biconical urns used for both male and female interment.[78] The scene is sometimes read as a *hieros gamos* (sacred marriage ritual) or simply an aristocratic couple seated honorifically.[79] Urns at Chiusi, Fermo, and Bologna show the same imagery as at Tarquinia: the concepts of social values and the equality of the married couple were elements basic for the whole of Etruria.

[75] Livy 5.1; see Cornell 1995:312.

[76] Cf. Ridgway 1991.

[77] Rallo 1989; Galeotti 1986–88.

[78] See Donati 2005, for examples and references.

[79] Donati 2005:378.

The composition is a sort of funerary banquet or fantasy scene of the reunited couple. Donati suspects that the disappearance of this favored motif during the eighth century accompanied the changeover from small villages to the great city of Tarquinia, a time when the ties to land became tenuous, as did the link to the ancestors, whose continued cultivation through worship guaranteed the survival and success of the rural family.[80]

Thus it would appear that wrenching changes affected the early cities from the beginning, and we should assume that unrest and adjustment troubles were constant. Land "reforms" requiring farmers to move into the city or become clients of aristocratic landlords, and the deliberate enhancement of city populations and amenities, would have been paralyzing to farm families separated from their land. A growing class of landless persons whose skills as artisans and builders led them to sell their labor in the cities would have reacted to any threats to their well-being: soon after the founding of Rome, such persons established craftsmen's guilds there. The divide between social classes, at Rome better known in the fifth century and later as "patricians versus plebeians," was surely inherent in Etruria in the preceding centuries.[81]

Women of substance were emerging in the eighth century alongside their warrior husbands, as we see in the numerous extravagant female "princely" burials,[82] preceded by the ninth-century Tarquinian female owners of longships.[83] Longships were undoubtedly used for long-distance trade – but also for raiding – and we should count these "princesses" among those responsible for the epithet of pirate that the Greek and Roman historians leveled at Etruscans (even though it was equally true of many Greek and Italic seafarers). In turn, "princesses" of the mid-seventh century, such as Larthia, wife (possibly daughter) of Velthur, of the Caeretan Regolini Galassi Tomb,[84] or Vetusia of the Bernardini Tomb at Praeneste, have lent names to the next generation of personages of this kind.

At June 2 and February 11, easy childbirth is predicted, and positive comments accrue at August 5, November 19, and March 29; September 6 notes, "there shall be power among the women greater than [what is] appropriate to their nature." The *Brontoscopic Calendar* seems to cover women of a certain social status, able to be recognized for their skills or power in society. They may plot with slaves to carry out assassinations (August 19), but there are none of the (rare) Mesopotamian omens of court intrigue, such as *Šumma izbu*, tablet 1, omen 8, "If a woman gives birth to a pig – a woman will seize the throne."[85]

[80] Donati 2005:379.
[81] Cf. Momigliano 2005:178–79; Eder 2005: 250–52.
[82] Cf. Rathje 2000; Bartoloni 2000.
[83] Hencken 1968:30–31, 48, 136, pls. 76–77.
[84] See Buranelli and Sannibale 2005.
[85] Leichty 1970:32.

The power of women is not the sort of thing that Romans would have invented – but it is difficult to prove such conditions without epigraphic or literary sources. Hints of women's status in Iron Age Etruria have always been there – in the parity of funerary rituals, in the symbolic respect accorded women's work, as seen in their ornamental and valuable textile equipment, and the use of those textiles in regal display.[86] The ability of women to own and dispose of property, found tombs, and own and drive chariots, like ninth- to eighth-century ship ownership, implies civic status and economic power.[87] The control of a household and its servants, which could often constitute a small workshop for clothing, foodstuffs, pottery, and more, conferred real social and political power on affluent women.

The extension of women's role in the aristocratic household to include industrial production and the supervising of slaves and servants is a logical division of labor when men in the household may be away on trading, diplomatic, or military ventures. Artists' signatures of the seventh and early sixth centuries have furnished intriguing data on the status of Etruria's slaves and even linked some of them with female owners. Giovanni Colonna has established the name of the painter of a Caeretan white-on-red pyxis (Louvre D 151) for the *Pittore della nascita di Menerva*, who apparently signed his work *kvsnailise*: he was the slave (or just possibly the client) of a woman of the Cusna family.[88] Colonna suggests she may have been the widow of the *paterfamilias*.[89]

A number of Hellenistic vases are marked with the signatures/stamps of foreign-born artists working in Etruria, whose names betray their servile condition: Lethe, Pultuce, Thanse, Creice, and so on; and at least two stamps (*vel numnal* and *θanseca numnal acil*) name the slave-artisans belonging to a woman, Numnei.[90] Colonna has pointed out another piece of evidence in the oldest known tile-stamp from Etruria, on an early sixth-century funerary tile found in a small necropolis near Nepi, inscribed *mi cusul puiunal*, "I am [the work of] Cusna [slave] of Puiuna."[91] There are other servile signatures of Archaic date, too, but they are not associated with female patrons. They name Aranth slave of Hercnas, who painted the scene of a man defecating in the *Tomba dei Giocolieri* at Tarquinia (ca. 530 BC), and Kape Mukathesa (Cape [slave of] Mucathe), who painted a vase in the workshop of the Micali Painter, circa 500 BC.[92] The ability of a slave to place his or her name on goods should

[86] As at Verucchio; cf. von Eles 2002:passim.

[87] See Emiliozzi 1997:passim; Galeotti 1986–88.

[88] Colonna 1993; Bagnasco Gianni 1996:no. 67. This vase was skillfully overpainted in the 19th c. and merits further scrutiny before theories can be built upon it; see Geppert and Gaultier 1999:217–18; Ridgway 2003:465n12.

[89] Colonna 1993:66.

[90] Colonna 1993:66n34.

[91] Colonna 1993:66–67.

[92] Colonna 1975:184–88.

be a warning that this condition in Etruria was not identical with the chattel slavery of contemporary Greece or Rome. With women and their *familia* of slaves engaged in joint occupations in the large urban household, occasions could well arise for them to make common cause, but we may be surprised to find them engaged in bloody revolt against the government.[93]

SLAVERY

Material evidence of slavery in past cultures is extremely difficult to identify because the difference between poor commoners and slaves is hard to differentiate in the archaeological record, and burials might not exist. The attempt to reconstitute the Etruscan Late Iron Age has often been overshadowed by our relatively greater knowledge of the Late Period, the fourth through first centuries BC, for which some literary and epigraphic references are available.[94] Greek and Roman authors portrayed Etruscans as degenerate because of their reliance on slaves. And, of course, the *familia rustica* could be armed and led into battle by its Etruscan master, a fact recognized by Roman opponents.[95]

It is nonetheless true that one aspect of the accumulation of wealth in seventh-century Etruria must have been the acquisition of slaves or servants for the affluent household. Equipment such as long-handled fans that must be wielded by an attendant had already appeared in eighth-century burials of elite men and women.[96] Artistic representations of persons being served at banquet begin during the seventh century with the Montescudaio urn.[97] If we count both equipment and imagery, the custom of being served, whether by a chattel slave of foreign birth or a free household member of lesser social standing, extends throughout Etruria, perhaps indicative of a homogeneous, stratified social system.

Mario Torelli links slavery to the emergence out of a subsistence economy in the mid-eighth century: the phenomenon created numbers of people who wholly or partially lacked civil rights.[98] In such systems, the institution of slavery may still have been marginal: the usefulness of persons captured in local raids would have been limited. With the advent of industrial production, as stimulated by the work of foreign craftsmen and the need to supply metals and other products commercially, likewise the need to feed a growing urban

[93] For slavery and gynaecocracy in Greek literature, see Vidal-Naquet 1970:75; Roman festivals of note for slaves and women include *Saturnalia* and *Matronalia* (Macrobius *Saturnalia* 1.12.7; Lydus *De mensibus* 3.22). Issues of intermarriage across classes must have arisen in early, as in Hellenistic, Etruria. Consider Rome's legislative and sacerdotal constraints in the *confarreatio* traditions enforced for the *flamines maiores*, as explored by Linderski 2005.

[94] E.g., Rix 1994.

[95] Livy 9.36; cf. Camporeale 2000:174.

[96] Veii, Narce, elsewhere, cf. Turfa 2005:123–24, no. 67.

[97] For the Montescudaio urn, see Camporeale 2000:174, pl. 5; Haynes 2000:111–12, fig. 92.

[98] Torelli 1987:87, 88.

population and the availability of laborers without civil rights, must have been perceived as worthwhile.

One other situation was presented by Weidemann, a possible analogy with slavery in precolonial Africa: as an "open" process, slavery was a means of bringing an outsider into a kinship group – then later there were mechanisms for making him or her a full-fledged member of that community.[99] Under Roman law, manumission granted a slave not only freedom but some gradation of citizenship. Is it possible that in the newly formed cities of maritime Etruria, such mechanisms were developed in order to reorganize the imported aliens with talents needed by the city? If the institution of manumission was operating by the seventh century BC, when we encounter the nomenclature of freedpersons or their children, then slavery must have been common at least one generation before that, thus during the eighth century. In fact, the proprietary inscriptions of the families of freedmen attest to their ownership of valuables (like Larthai, daughter of Telicles) and/or their ability to make donations to public cults, good indicators of adoption into society.[100] A compelling example is a seventh-century jar given to the "beautiful" *Raqunntu Tipeia* at Veii and found, presumably in the vicinity of her aristocratic house, on the Piazza d'Armi.[101] Colonna indicates that her father's name betrays servile and foreign origins: the name Τίβειος or Τίβιος is Anatolian (Paphlagonian) and was common in the Classical period among slaves and ex-slaves residing in Athens.[102] If the new *gentilizio* was derived from a Greek servile name, then Tipei or his father probably entered Etruscan society as a slave of foreign (Eastern) origin (cf. the *Brontoscopic Calendar* for March 26, "acquisition of imported slaves").

The excavators of the Pithekoussan necropolis have suggested servile condition for some cases where an adult inhumation burial (sparing of costly fuel) was made without any offerings in a "family plot" that held affluent persons, including children.[103] Likewise, Bietti Sestieri has suggested that the Iron Age burials at Latin Osteria dell'Osa that lack grave goods may represent those of clients of families in the stratifying society of Gabii.[104]

At eighth-century Bologna, in the necropolis of San Vitale, Giardino and colleagues detected "the expression of a ranked and complex society, in which numerous roles existed."[105] There may even be evidence of the sacrifice of women in funerals of high-ranking men, but this is not entirely clear – infectious disease might account for multiple deaths of family members. Another part of

[99] Weidemann 1981:45.
[100] Bonfante and Bonfante 2002:138–40, no. 11.
[101] Wallace 2008:160, no. 36.
[102] Colonna 2002:355.
[103] See Ridgway 1992:52.
[104] Bietti Sestieri 1992A:209.
[105] Giardino et al. 1991:12.

the necropolis, however, holds unmarked inhumations. Because all but two have no grave goods whatever, these "may represent a social class linked to the individuals buried with cremation rite by servitude."[106] The situation of a society in transition from the more egalitarian mode of the Final Bronze Age to a highly formalized and stratified structure would surely have fostered socially dynamic communities with rulers, commoners, slaves, and occasional periods of discord, just as in the *Brontoscopic Calendar.*

Most commentary on slavery in Etruria has been prompted by references of Classical authors, illustrated from Hellenistic historical events; to them, Etruscan society was exotic, alien, and undoubtedly repressive. Concerted analysis began with the 1957 articles of Heurgon and Mazzarino, and it continues today.[107] Thérèse Frankfort (1959), in an article to which we may add merely a few new epigraphic finds, distinguished classes of slaves, emancipated slaves, and *clientes*, but she cautioned that the available sources begin only in the second century BC, and mostly cover only the third-century event of the Volsinii revolt and downfall. Etruscan epigraphic documents are found in the north of Hellenistic Etruria, especially at Perugia and Chiusi, and a few images from tombs at Tarquinia and Orvieto seem to depict servants.

Such phenomena as the name of the Roman king *Servius* Tullius, or the Greek writers' assertion that Etruscan slaves dressed as well as their masters (Poseidonios, in Diodorus Siculus 5.40.3), demonstrate that the institution of slavery was not the same as under the Roman Empire. One additional situation to ponder is found in the *Twelve Tables* of Rome, recorded in the fifth century, but containing older material, some of it attributed to the Regal period. Weidemann pointed out that the *Twelve Tables* allow the practice of enslavement for debt, but they stipulate that Roman citizens so enslaved must be sold "across the water," that is, on the Etruscan side of the Tiber, rather than enslaved to their fellow citizens and neighbors.[108] One religiously motivated reform of Numa Pompilius was his prohibition of fathers selling off legally married sons – their wives should not be subjected to a lesser status than they had contracted for (Plutarch *Numa* 17.4).

The Figulus document, in Lydus' translation of course, only employs a narrow set of terms for slaves: at December 26, "servants, caretakers" (θεράποντες); at August 19, it is "the women and the servile class" (τὸ δουλικόν) who will dare to carry out assassinations. The two different terms should warn us that the original Etruscan may have preserved distinctions among the unfree, or otherwise indicated that "servants" were distinct from "slaves." The well-dressed, named servitors of fourth-century tomb paintings seem quite different from

106 Giardino et al. 1991:12, 17.

107 See Benelli 1996 for full discussion.

108 Weidemann 1981:39–40.

chattel slaves, and closer to those affluent "slaves" mentioned by Poseidonios, cited by Diodorus 5.40.3–4.[109]

The *Brontoscopic Calendar* offers several predictions relating to slaves, scattered throughout the year. Disease affecting slaves comes at July 6, December 26, and January 7. September 7 predicts "a disaster for the servile class," probably also disease, and "slave revolt" is cited at January 7, 15, and 25, and assumed at August 19. Why does January have so many predictions of unrest? This would be the quietest time of the agricultural year, when there was less work in the fields and people were more constantly indoors together. Likewise, why are slaves singled out for diseases? Late Republican and Imperial slave farms must have harbored such problems, but it is hard to imagine infectious disease touching some members of a household and not others, except perhaps among new arrivals to Italy, who were not immune to its diseases, such as the local strains of malaria. (This was the case in the Imperial period, when the population of foreign slaves and immigrants was constantly augmented. Malaria has also been postulated for the high infant mortality rate of immigrants at Pithekoussai.) The affluence of urban Etruria endowed some of its domestic servants with an unusually high standard of living, disturbing to Greek authors.[110] These Etruscan slaves may have been skilled craftsmen permitted to run businesses as investments for their masters, so their segregation hardly seems to put them at risk. The Mesopotamian predecessors of the *Brontoscopic Calendar* make almost no reference to slaves (though legal documents abound to show the trafficking in human lives in those societies). A few omens predict the escape or capture of a runaway, even a "famous slave," but they are portrayed as mere economic items, quite different from the Etruscan references.[111]

Where cultures differed, the historians had to use their own approximations of technical terms, which must fail to capture all the original societal nuances. For instance, Classical Greek authors describing the palace of the Persian king used the term δούλος for all persons belonging to the royal household, even though not all were chattel slaves.[112] Greek and Latin designations of slaves

[109] Several 5th- and 4th-c. tombs (Tarquinia Tomba delle Bighe, dei Leopardi, Orvieto Tomba Golini) seem to depict servants or slaves as important participants in funerary banquets (Steingräber 1986:289–91, 319, 278–79). The sarcophagus of *Ramtha Viśnai* of Vulci (now in Boston) shows husband and wife attended by male and female servants, all except one man quite well clad (Haynes 2000:289, fig. 232b); but in the Tomb of the Shields, also mid-4th c., the girl who fans her well-clad mistress wears black and red leather boots identical to hers, and a similar white dress; both have blond hair, but the servant's is cut in a short bob (Haynes 2000:310, fig. 249; Steingräber 1986:341–43). None of these examples are old enough to tell us much about slavery in the Iron Age, but they should serve as a warning that the condition of servitude in Etruria was not identical to that in Greece, Rome, or modern states.

[110] Poseidonios, cited by Diodorus 5.40.3–4; see Benelli 1996:341.

[111] See Koch-Westerholz 2000:339.

[112] Weidemann 1981:40.

or other disenfranchised persons only partially match known Etruscan situations: Οἰκέται is used by Diodorus (5.40), Pseudo-Aristotle (*De mirabilibus auscultationibus* 94), John of Antioch (Fr. 50 = *FGH* IV, 557), and Zonaras (8.7 P.I. 380), indicating the *household* slaves whose takeover went so far off course at Volsinii. Roman authors simply say *servi*: these are Livy (5.1.5, 9.36.6); Valerius Maximus (9.1.2); Florus (*Epitome* 1.16; *De viris illustribus* 36); Orosius (4.5.3–5). The last two works additionally have the term *libertini*, "emancipated slaves."

Only Dionysios of Halikarnassos (9.5.4) brings up the term πενέσται: they must be the "associated rural population" summoned by the Etruscan *principes* to fight Rome in 310 BC: Livy (9.36.12) calls them *agrestium Etruscorum cohortes*. The original πενέσται were the native peasants of Thessaly, whose servile condition had ethnic overtones and smoldered with threats of revolt.[113] The unrest was exploited by enemies of the Thessalian state (Xenophon *Hellenica* 2.3.36), but Benelli warns against a narrow definition of the term.[114]

There is some background in the historians for slave uprisings in Etruria, but nothing certain before the fourth century BC.[115] An insurrection at Arezzo in 302 BC (Livy 10.3.2) seems to have been a revolt against the *gens Cilnia*. The slaves in the fourth-century and later histories were, according to Torelli, essentially a different sort from the earlier variety attested in the seventh- and sixth-century signatures of potters and painters. The Late Period slaves were those unfortunates also known from Roman social troubles, such as foreign prisoners of war who were being captured or bought in increasing numbers for export to large farms, such as the second- and first-century BC *latifundia*, organized to produce huge surpluses for export.[116]

The revolution and subsequent destruction of Volsinii in 263 BC may have arisen out of long-standing inequities in the hierarchy of the Etruscan city, where former slaves resented their disenfranchisement, and probably sought civil rights reforms. Could a situation in which "the women and the slaves" made common cause to seek advances like recognition of marriage between free citizens and freedmen actually have been at the root of the Volsinian revolution of 264? There is not enough evidence to extrapolate as far back as the eighth to seventh centuries BC, but logically we may assume that large populations, rigid class systems, and growing economic inequality, exacerbated by wars with other groups, must have predisposed Etruscan society to all the ills later cataloged by the Classical authors.

Terminology recorded in the ultimate social status of persons given a traditional burial may illustrate additional gradations in Etruscan society. Some

[113] Athenaeus 6.259f–265c; Benelli 1996:339; Aristotle *Politics* 1269a37.

[114] Benelli 1996:339–41.

[115] Capozza's (1966) study covered the earliest historical events, revolts affecting Republican Rome in the years 501–184 BC.

[116] Torelli 1987:94; Plutarch *Tiberius Gracchus* 8.5.7.

philologists have proposed the word *lethe* or name *Lethaie* as indicating a slave/former slave.[117] *Lethaies* as a cognomen is attested at Veii circa 570–560 BC in an Etrusco-Corinthian vase dedicated at the Portonaccio sanctuary.[118] If *Lethaies* derived his name from an ancestor of servile origin, then that person must have been a slave during the seventh century if not earlier. The terms *lautni* and *lautniθa* ("of the family" or "freedman/woman") are, in some bilingual epitaphs, equivalent to *libertus* and *liberta*. The terms appear almost exclusively in the north and interior (Chiusi, Perugia) in the Hellenistic period.[119] Rix pointed out that before around 90 BC, freedmen apparently did not receive full rights of Roman citizenship, for it is only after that time that we see them creating family names for themselves, often out of their former slave or ethnic designations.[120]

Fridolf Kudlien, writing on the psychosocial associations between slavery and divination, did not approach the *Brontoscopic Calendar* as an Etruscan document, but he has pointed out some interesting aspects of the relevant literature:[121] most slaveowners tended to fear slaves' flight, theft, seduction of family members, insurrection, and assassination. Writing of the uprisings in Sicily during the Republican period, Diodorus Siculus noted that inhumane treatment had triggered the initial revolt.[122] The later *Tonitrualia* and *Brontoscopia* do not evince any interest in the slaves as persons, but, as Kudlien discerned, the Figulus text does in a certain sense (and this is another reason to separate it from the others).[123] Kudlien notes that the *Tonitrualia* evince a universal or political viewpoint and treat slaves only as a social category.[124]

Kudlien interprets the entry of March 26 as an advisory to buyers that, if it thunders, the market will be selling freshly imported slaves, in a similar tone to the tablet predictions for the Babylonian bourse.[125] The next day, March 27, predicts "abundance" imported from abroad, and April 27 and May 3 likewise show imported food, and I would read this as part of the bigger picture: the ships will return from war abroad, bringing booty in the form of slaves (but those ships might as easily be Villanovan longships as Republican Roman

[117] Originally advanced by Emil Vetter in 1948; see full references in Briquel 2009:57–58; Morandi 1989:581–85.

[118] *mini muluvanice laris lethaies mi(ni) zinace ve[thur a]cinies*; Briquel 2009:58n43–45; Frankfort 1959:8; *ET* Ve 3.44 = *CIE* 6449; Morandi 1991:46, 88–94, fig. 7, pl. 7.1; cf. the 6th- to 5th-c. tomb of another Lethaes (*ET* Vs 1.142); Rix 1963:371, and others, however (Steinbauer 1999:436), do not concur.

[119] See Steinbauer 1999:435; Rix 1963:356–72.

[120] Rix 1963:369–72. I suspect that the practice may have begun somewhat before 90 BC in Etruria, but apparently not as early as the Archaic period; cf. Turfa 2005:141–42, no. 103. For 4th-c. issues, see Turfa 2011.

[121] Kudlien 1991:50–52.

[122] See Urbainczyk 2008:81–90.

[123] Kudlien 1991:109–10.

[124] Kudlien 1991:102–4.

[125] Kudlien 1991:110.

naves onerariae). In the Greek of Lydus, we see different kinds of threats, from outright "slave war" (δουλομαχία) to a "slave uprising" (στάσις δουλική) – and στάσις is used also for civil discord among the commons or between free classes.

RULERS

A king, ruler, or autocrat is cited at seventeen dates; some are good, others clearly evil, but all imply an entrenched, central authority focused on a single man.[126] The most citations of rulers are for the months of July and September (six entries each); a king is cited four times in January, once "hated by many" (Jan. 21); the 16th of July, November, and January each bear a royal reference, but the most tantalizing is the "king of the East," who at July 16 will make war – but the text is damaged at that entry. Most early commentators have seen in the allusions to coups and tyrants veiled references to the Roman Civil War and Julius Caesar, which would imply tampering with the text either by Figulus or by someone soon after his demise, but there is never sufficient detail or correlation of dates to support this thesis: the Ides of March cites hot spells, mice, and fish, not assassination. The prediction of March 12, close to the Ides, that "a powerful man in politics or a general is endangered; on his behalf, battles will be waged, and the wild beasts shall fall upon man" has seemed to some commentators to be an insertion after the assassination of Caesar, doctored for allegorical impact. The precedent in the *etrusca disciplina* for the warning given by Spurinna and rebuffed by Julius Caesar, however, came from haruspicy, not thunder.[127]

Gilda Bartoloni (2009) has demonstrated from archaeological evidence of Iron Age kings and their courts that they were just as active in the north as in the cities of southern Tyrrhenian Etruria. The great necropoleis held burials of rulers, but now their palaces and banquets are coming to light. At Populonia (Falda della Guardiola), an eighth-century city fortification has been excavated, and similar walls are being analyzed at Veii and Vulci, in a distinct reversal of the old picture of unwalled settlements.

Bartoloni notes this "increasingly systematic occupation of rural areas must be attributed to centralized political organizations," based on the family, *gens*, *curia*, or *populus*.[128] An eighth-century rectangular building excavated at Populonia Poggio del Telegrafo has been identified as the palace, albeit under a thatched roof. It was used for some decades before being abandoned in the

[126] See Torelli 1997; Stoddart 1989; for other studies of power and society, see Herring et al. 1991. For illustrations of rulers, symbolic artifacts, and evidence of conflicts from the Neolithic on, see Marzatico and Gleirscher 2004.

[127] The sacrificial victim had no heart; Cicero *De divinatione* 1.119, cf. Haack 2006:110–12, no. 75.

[128] Bartoloni 2009:162.

first quarter of the seventh century, when a special de-foundation deposit of over eighty identical cups (*kyathoi*) was placed in a pit cut into the holes left by its posts. This must be the remnant of a symbolic toast, a communal act by 100 citizens commemorating the peaceable dismantling of the ruler's headquarters.[129] An early porticoed structure at Casalvecchio-Casale Marittimo may have been a ruler's house with meeting area.[130] At Veii Quattro Fontanili Tomb AA1, a cremated warrior buried in a bronze urn with his crested helmet, arms, armor, and a scepter has been unabashedly called a king by Francesco Boitani.[131] Several other candidates for this title come to mind among the buried populace of Volterra, Veii, Vetulonia, Caere, Tarquinia, Bisenzio, Narce, and others.

Romans recognized the tradition that the insignia of civic power had descended from Etruscan forms: the ivory throne, purple robe, *fasces*, scepter, and *lituus*.[132] Ancient authors preserve the term *lucumo*, and an Etruscan root *lauchm-* certainly existed, but it remains enigmatic for lack of pertinent epigraphic sources. The term *zilaθ* (*zilaχ*) instead is well documented and seems to have applied to the highest magistrate, by the sixth century, and probably by the late seventh century, to judge from an inscription on an Orientalizing *cippus* from Rubiera in the north.[133]

An impasto chalice in the Louvre, inscribed in the territory of Caere early in the seventh century for Laucie Mezentie, places in the Iron Age the name of a family linked to the kingship of Caere, but not nearly as early as the Late Bronze Age king Mezentius in books 7 to 10 of the *Aeneid*.[134] Royal adventurism of the early sixth century is attested at Veii in the Portonaccio sanctuary with the name of the (in)famous *condottiero*, Aulus (Avile) Vibenna (comrade of Servius Tullius), inscribed (dare we imagine by his own hand? probably not) on a showy *bucchero* vessel.[135]

Ending the Archaic period are the Pyrgi plaques, which perhaps betray a Carthaginian-assisted coup: they dramatically name Thefarie Velianas *zilath* of Caere: yet he is known in the Punic version as *mlk 'l Kisry'*, *king* of Caere (*ET* Cr 4.4). The presence – and power – of kings in the Etruscan cities of

[129] Bartoloni 2009:162–67.

[130] Esposito 2001:25–27, fig. 11. See also the Middle Orientalizing round house at Roselle, and communal libation rituals are attested at the Roman Regia, Latin Ficana, and elsewhere; for background, see Torelli 2000; Bartoloni 2009.

[131] Boitani 2004; see review by F. R. S. Ridgway 2005:468.

[132] Dionysios of Halikarnassos *Ant. Rom.* 3.59–62; Jannot 1993; see Tassi Scandone 2001:201–31, on the Tarquinia Civita deposit. For the silver and gold floral scepter in the Barberini Tomb, see Delpino and Bartoloni 2000:224.

[133] Its funerary inscription names a woman, *Kuvei Puleisnai*, presumably the wife or relative of the *zilaθ*; Camporeale 2000:62, fig. 15, 152, 200; Amann 2000:102; Bagnasco Gianni 1996:no. 291.

[134] Also Cato *Origines* 1.9–12; see Gaultier et al. 1989–90; Bagnasco Gianni 1996:108, no. 84; Steinbauer 1999:444; de Grummond 2006:205–7; Wallace 2008:15, no. 32.

[135] *ET* Ve 3.11 = *CIE* 6456, see Briquel 2009:52.

the seventh to sixth centuries is evident in the histories of figures like Lars Porsenna of Clusium, the would-be nemesis of the Roman Republic. Veii had kings into the fifth century, some from the Tulumne/Tolumnius family.[136]

The *Elogia Tarquiniensia*, analyzed by Torelli, also give evidence of monarchy in latter days, with the *elogium* of Aulus Spurinna, whose adventures on behalf of Tarquinia during the fourth century seem to have involved deposing a "king of the Caeretans" and dealing with a slave war in Arezzo.[137] Inscriptions of the Hellenistic/Late Republican period attest various kinds of *zilaθ*, just when most Etruscan cities had lost their autonomy to Rome. These range from the *zilaθ meχl rasnal* ("of the Etruscan commonwealth") to *zilaθs* labeled *ceχaneri, eterai/eteraias, marunuχva, parχis, spurana, purtsvana.*[138] Some terms have been linked to social groups such as plebeians, freedmen, clients, foreigners, and so on, but there is insufficient evidence to satisfy all scholars.[139] If the *camθi eterau* is read as the head of a college of youth, it might mirror the "band of youth" cited in the *Brontoscopic Calendar* at July 25 and August 24. In the Roman system, where the *paterfamilias* wielded power of life and death over grown sons with families of their own, junior family members might well have sought outside alliances.

At the necropolis of Quattro Fontanili at Veii, arms appear suddenly in Veii phase II, with numbers of spears, swords, and helmets increasing during phase IIB – as if more men now trained for ceremonial dueling among aristocrats as a means of conflict resolution.[140] One can only imagine the human dramas that must have played out while shifts of power from one family to another were being accomplished.

FORMALIZED RITUAL AND CULT: THE FOUNDATION DEPOSIT AT TARQUINIA PIAN DI CIVITA

Symbolic rituals conducted in Villanovan sacred areas also reflected societies' newly adjusted values focused on the warrior-priest. A rare example of such a ritual has been identified at the site of Tarquinia Pian di Civita.[141] During the Protovillanovan period (late 10th c.), remains of animal sacrifices and ritual

136 Livy 4.17.1–8. Cf. Edlund-Berry 1991; *ET* Ve 3.2 and 3.6; Briquel 2009:54, 62–63; Morandi 1989:592–94.

137 Torelli 1975:39–42, no. 4; Camporeale 2000:152.

138 See Rix 1984. I wonder if a magistrate for "Etruscan culture" held jurisdiction for cases in which Etruscan law was applied, on loose analogy to the situation in Ptolemaic Egypt, under which mixed families could in some cases choose whether to submit a case to Greek or Egyptian adjudication. Women usually opted for the better treatment afforded by Egyptian law; see Chauveau 2000:96–98, 162.

139 See Camporeale 2000:153.

140 De Santis 2005:629.

141 See Chiaramonte Treré 1988; for state formation at Tarquinia, see Leighton 2004:59–85; for the latest on this site, see Bonghi Jovino 2010.

meals deposited in a trench probably represent worship. Then late in the ninth century, a special boy was buried, wearing a cloak and bronze pendant. The brain aneurism of which he died would have caused him to see and hear things that no one else could: perhaps a sign of a prophet, like the *puer senex* of the Tages legend?

Then, early in the seventh century, a rectilinear building was erected with Phoenician-style masonry, with a special altar draining into the boy's grave. Use of Levantine construction technique for an Etruscan cult (ca. 680 BC) constitutes some of the earliest evidence of Near Eastern influences here. The threshold of the cult building covered a foundation deposit in two pits, each with different artifacts and bones from cattle, sheep/goat, and pig. One pit held a sacrificial ax of antique type, a musical *lituus* horn, and a bronze parade shield, all deliberately destroyed. The other pit held the remains of a ritual meal, including ten plates incised with an "X" and two cups, all deliberately smashed. Cristina Chiaramonte Treré attributes this to a sacrificial banquet, recalling historic instances, such as the sacrifices established by Romulus for various *curiae* in his new city of Rome (Dion. Hal. 2.23).[142] The bronzes may represent a civic deposit of royal symbols, illustrating complex society and intricate rituals in Etruria.[143] The Pian di Civita cult buildings disappeared during Classical-era remodeling, no doubt indicative of a shift in government and society away from the forms of the Iron Age.[144]

The legend of the king Numa Pompilius falls in the same period, the early seventh century; his reign was credited with the institution of religious ceremonies and public banquets as a means of "taming" the restive, newly agglomerated citizens of Rome (Plutarch *Numa* 8.3 and 15.1–2). Numa is even associated with a charm to protect from the effects of lightning and thunder (Plutarch *Numa* 15.4–5). Divination played a role also, because he apparently could interpret "apparitions of supernatural beings and disquieting voices" from the heavens (φάσματα δαιμόνων ἀλλόκοτα καὶ φωνὰς οὐκ εὐμενεῖς; Plutarch *Numa* 8.3). His civic developments include the adjustment of the Roman calendar from a set 360 days by the addition of intercalary days every other year (Plutarch *Numa* 18).

WAR OR RUMORS OF WAR?

The *Brontoscopic Calendar* offers just four predictions of peace, scattered over the year (July 8, Aug. 16, Jan. 4, April 25), reflecting cessation of war or urban warfare. Predictions of war and combat are more numerous, but no

[142] Chiaramonte Treré 1988:586.

[143] Bonghi Jovino and Chiaramonte Treré 1987:77.

[144] Background: Haynes 2000:25–29; Turfa 2006A, 2006B; de Grummond 2006:202–7; Briquel 1996.

month holds more than four war apodoses. Outright "civil war" (as opposed to unrest or factional strife) appears five times, once in August, April, and November, and twice in December. Two of the war omens sound more like the Mesopotamian texts, promising "eastern war" at May 14, and "northern war" at May 25. Many a state could lie to the east or north of the major Tyrrhenian cities, and eighth-century fortifications at Populonia, Veii, and Vulci mean that defense was a citywide endeavor. Several war omens note abundance of food or crops, but the tone at December 15 is more realistic: "If it thunders, many will set out for war but few shall return."

ARMS AND ARMOR

With rulers and uneasy classes come wars, but as yet no one has proven warfare to be the source of the fires that occasionally left debris on many seventh- to sixth-century sites. Pallottino (1987) commented on the combat depicted in the François Tomb, no doubt in honor of Vel Saties' ancestors' part in some early sixth-century intercity warfare. Rawlings has found that "aspirations and activities of powerful individuals, and not clans, are the most significant dynamic in these early societies."[145] Thus far, our best evidence of warfare is the apparent preparation for it.[146]

One clear indication of war or raiding as an accepted means of living is the manufacture, exchange, and burial of arms and armor. This begins in the ninth century in Etruria: the same types may have been in use sooner and simply too precious to bury.[147] Perhaps the Levantine and Greek visitors who reached the Tyrrhenian coast early in the eighth century really only brought rumors of wars – enough, however, to cause the Villanovan elite to arm themselves with additional Near Eastern maces and chariots. Villanovan smiths were already producing bronze armor and swords, carried only for the purpose of fighting men, and displayed in warriors' funerals. The Bronze Age winged or flanged ax became emblematic of the warrior class, and continued into the sixth century in funerary iconography.

THE TARQUINIA PIAN DI CIVITA WARRIOR

One of the most telling proofs of Iron Age/Villanovan warfare – and its tragic aftermath – is the eighth-century burial of a sacrificed warrior discovered in the heart of the presumed cult place at Tarquinia Pian di Civita. His bones show a long history of combat, and his sacrifice may mark him as an enemy

[145] Rawlings 1999:99.

[146] See Stary 1980A, 1981; Delpino and Bartoloni 2000:225–27.

[147] For tangential and theoretical material to consider in evaluating social aspects of Iron Age warfare, see Harding 1999 and Kristiansen 1999.

in conditions of intercity war. The thirty-five- to forty-year-old warrior's skeleton bears sword wounds from recent battles – and a shattering blow to his skull made by a winged ax; the forensic evidence indicates violence like that seen in animal sacrifice.[148] The deed was performed in a public place, where he was then ritually interred without benefit of a traditional cremation funeral, evidence of a ritualized act, perhaps taken on behalf of the entire community.

This ritual killing occurred at about the time of the historical Romulus, and it links such activity to the state cult of a full-fledged city, conjuring up the tale of the killing of Remus on the walls of Rome by the Etruscan "Celer." If there is a germ of truth in Remus' seemingly needless death at the wall of Rome, perhaps it lies either in a battlefield duel or in some ritual sacrifice of a dedicated warrior or noble adversary taken as a prisoner of war. And just possibly, the "shovel" that killed Remus, as it is called in the literature, was actually an Etruscan flanged ax, its original, Etruscan name forgotten by the time Diodorus Siculus (8.6), Ovid (*Fasti* 4.837–53), and Plutarch (*Romulus* 10.2) recorded this tale. Perhaps even their sources, Fabius Pictor and Ennius, lacked this information when they wrote at the end of the third century and beginning of the second century BC.[149]

NEAR EASTERN INFLUENCES IN WARFARE

Another indication of a climate of warfare or socially sanctioned armed combat is the presence of weapons in everyday life. Some members of Villanovan societies were acquiring new forms of military equipment – and presumably learning the accompanying battle tactics – from the Near East, in the form of chariots, now attested as early as the ninth century at the Poggio Impiccato (Tarquinia) necropolis.[150] The hardware used for the chariots, and the breed of horses, show that the sources of this paraphernalia lay in the Levant, Cyprus, and Near East.[151] Archaic Etruscan representations of a military-style *triga* race, with the third horse carrying armor, show the serious, military background of this vehicle.[152] The commitment to use and maintain chariots and horses imposes a strain on owners, but by the Villanovan period, many considered it a necessity.[153] Ultimately this led instead to the use of wheeled vehicles and paved roads within and between cities, a development that Rome

[148] Bonghi Jovino, Malegni, and Usai 1997.

[149] For general discussion of violence detected in the archaeological record, see Robb 1997 on the Pontecagnano burials, and Carman 1997.

[150] For Tomb 39, Tarquinia IB, see Delpino 2005.

[151] See Azzaroli 1972.

[152] Turfa and Steinmayer 1993.

[153] For early chariots disseminated by Phoenicians and Cypro-Phoenicians, see Emiliozzi 1997:passim, and 2001.

would continue. As in the fields of divination and architecture, only close association and trust between Etruscans and Levantine or Cypro-Phoenician metics could have fostered such a major borrowing.

Hand-to-hand combat, too, was initially influenced by Near Eastern paradigms of warfare; such honorific Near Eastern weapons as maces appeared in Villanovan tombs.[154] By the mid-eighth century, the *kardiophylax*, a disc worn to protect the heart, became warrior's equipment. Tracing it to Assyrian armorers of the period of Tiglath-Pileser III, Sargon II, Sennacherib, and Ashurbanipal (750–630 BC), Peter Stary notes the significance of these comparisons: not only technological expertise, but actual patterns of behavior are being shared between Near East and Etruria.[155]

IMAGES OF SOCIETY AND WAR

A Late Archaic relief sarcophagus from Perugia (Fig. 24) may preserve the aftermath of warfare: a procession of men, women, and animals, including spear carriers and some mature (bearded) men tied with a rope around their necks. Is this a return from a raid or perhaps a triumph celebrated by the family buried in the Sperandio Tomb?

Early works of art express the unique character of Etruscan society, including warriors, women of high status, farmers and hunters, and even children. A fine bronze sword scabbard from Vulci (late 8th c.) has a man and woman of equal size, presumably the warrior and his wife, symbolic of war and fertility, flanking the hilt.[156] A wheeled censer stand, from a woman's tomb at Bisenzio (Fig. 17), though badly corroded, holds thematic groups of figures on its struts: birds and monkeys, a stag, wolves (one with a lamb in his jaws), a goat, and a ram; two hunters with dog, bow, and javelin attack a boar, and a farmer plows with a team of oxen. The family appears as a warrior in crested helmet, his wife (of equal stature), and a smaller figure, probably their son, holds a shield.[157] A sword duel between two warriors must symbolize war. Another composition illustrates the popular Orientalizing fable of "the woman at the well" depicting the woman as the purveyor of the necessities of life, symbolized by her water jar.[158] Michele Cupitò has suggested a complex spatial-theoretical interpretation linking the design of the piece with the world of Iron Age Italy: agricultural land/forest/hereditary lands/sacred grove, and so on.[159] Could the couple's son who proffers a shield recall

[154] Maces appear in Assyrian palace reliefs of Shalmaneser II through the reign of Ashurbanipal, ca. 858–631 BC; Stary 1980B:32. See Stary 1980B, 1981, and 2001.

[155] Stary 1980B:29, 31–32; Stary 2000.

[156] See Haynes 2000:21, fig. 20; Amann 2000:123.

[157] 750–700 BC; Haynes 2000:21–24; Amann 2000:122–23, pl. 9b.

[158] Richardson 1984.

[159] Cupitò 2005; see also Menichetti 1994:21ff.; cf. Torelli 1997:38–46.

the "band of noble youth" in training for their future role as defenders, as cited in the *Brontoscopic Calendar*?[160]

Other favored images show elite matrons engaged in textile production, such as the Bologna *tintinnabulum* (Fig. 22), made circa 600 BC in the form of a sacrificial ax, thus compounding the lady's identity with that of a priestess.[161] Human figures in association with celestial objects appear in a composition on a Villanovan woman's bronze belt found in Vulci Tomb 42F (Fig. 18, ca. 680 BC).[162] Bosses with zig-zag or sickle-shaped sets of parallel "rays" streaming from one side may be celestial bodies; another is edged with a border of triangles like a primitive drawing of the sun (compare Fig. 1). Immediately to the left (in Etruscan art to be read as "after") of the sun are two stick figures with arms upraised as if dancing. A set of dots with streamers (a comet?), lines, and triangles may be other celestial bodies or elements of landscape. A number of late eighth- and seventh-century gold ornaments from Iron Age Etruscan burials depict stars, suns, or crescent moons patterned on motives found in Phoenician jewelry.

The Plikaśna *situla* (Fig. 19) is a silver-gilt vase from a tomb at Chiusi (ca. 650 BC), labeled with its owner's name and decorated with two incised friezes that illustrate rural and urban life among the Etruscan elite, but its drawing style is strongly influenced by Phoenician silver bowls.[163] Two processions of men and women bringing offerings (sheep, pig, boxes) converge on a pair of men, each with young attendant, who are boxing before a large cauldron; musicians, armed warriors in a dance, and a horseman with attendant warrior bring up the rear in each file. The second frieze depicts a herd of domestic pigs led by an armed boy with a flute and a dog.[164] Although it may depict funerary games, races and the prizes, or sacrifices to be offered, the vase surely mirrors the characters and chattels of Orientalizing Etruscan cities and the countryside, down to the introduction of sports and the luxuries of a ruling class. (The women wear bobbed hair and short dresses, possibly indicative of servile status – we do not see anyone who is portrayed as the ruler/recipient of this festival.)

One other artistic document, the Tragliatella *oinochoe* (Fig. 20), made in the region of Caere (ca. 630–600 BC), refers to urban society and its leaders, both male and female (and their heirs, in this case female). The vase in Etrusco-Corinthian style is incised with narrative images, probably for the funeral of

[160] Finds of children buried in rich, adult-type rituals, such as the adolescent boy whose remains were placed in the Calabrese house urn hint at such status; see Becker 1998.

[161] Morigi Govi 1971. I am indebted to Cristiana Morigi Govi for the gift of this article, and many others.

[162] Dohan 1942:94–95, pl. 50, no. 23; Turfa 2005:101–2, no. 31.

[163] See Haynes 2000:108–10, fig. 90.

[164] Cf. Barbieri 1987.

the woman depicted on it and labeled "I am Thesathei."[165] Some parts of it are obscure: two scenes of intercourse, two horsemen exiting a circular maze labeled *truia*; birds, a boat, and a nude man with a goat on a lead illustrate ownership of land, livestock, and transport; a hare and hound evoke the hunt, and a long snake could have various connotations. On the neck of the vase, framed by the goat and boat images, a man grasps a woman's hand, indicating that they are a married couple, surely the same pair featured on the main scene. There, a militia of seven soldiers with identical round shields march to the left followed by a civilian, a nude man with a staff. At their head are a man, "Mamarce," holding the hand of a little girl, "Velelia," as they converse with a plaid-robed, veiled woman (Thesathei); they exchange an egg(?) and a fruit, and the woman (in a second image) turns and proceeds into a rock-framed doorway, no doubt her new tomb. The prominence of little Velelia in the family farewell scene may be the female counterpart of the boy who holds his father's shield on the Bisenzio incense burner.

CONCLUSION

We are much more familiar with the social systems of Hellenistic or Late Republican Italy than those of the Early Iron Age, but that is no reason not to assign an Iron Age date to the Etruscan *Brontoscopic Calendar*. All the elements that stand out as distinctly Etruscan in the text of Lydus can be illustrated from protohistoric sites and archaeological finds: elite, powerful women who own valuable land, slaves, or property, a deeply hierarchical society, both within the city and across its agricultural hinterland, strife among those classes, particularly commons versus elite, slaves, serfs or perhaps freed slaves versus freeborn or property holders, the governed versus corrupt autocrats. War, too, whether a slave revolt or a campaign against some neighboring state that might bring in riches and "slaves" is a constant concern in the predictions of the calendar. And as in real life, war often accompanies famine, and servile conditions foster disease.

[165] Haynes 2000:97–99, fig. 83; Martelli et al. 1987:102, 270–72, no. 49; Camporeale 2000:174, pl. 76; Amann 2000:130, pl. 15b.

PART IV

SOURCES AND SUCCESSORS OF THE ETRUSCAN *BRONTOSCOPIC CALENDAR*

CHAPTER 8

MESOPOTAMIAN INFLUENCES AND NEAR EASTERN PREDECESSORS OF THE ETRUSCAN *BRONTOSCOPIC CALENDAR*

INTRODUCTION: THE CASE FOR NEAR EASTERN PROTOTYPES

Increasingly, scholars examining the early culture of the Aegean have had recourse to the adage *ex oriente lux*. As our access has improved to Near Eastern art and texts, yet more Near Eastern prototypes for many facets of Greek culture have been identified. The phenomenon of the *Brontoscopic Calendar* has reopened the issue of the interdependence of early Italy with the states of the Levant and beyond. This chapter surveys the background of Mesopotamian divination and considers its potential influence on Etruscan cultic literature by analysis of texts with similar subject matter and expressions and by estimating the likelihood and modes of transmission of documents from Near Eastern cults to the Etruscan culture of the Iron Age. The small sample of texts in translation offered in Appendix B illustrates topics and comments significant of Mesopotamian influence in this unequivocally Etruscan document.

TECHNICAL FRAMEWORK: TYPES OF DIVINATION

Many modern scholars follow the formal distinction expressed by Cicero (*De divinatione* 1.11 and 2.26) juxtaposing "artificial" and "natural" divination, and they equally apply this to the corpus of first-millennium Near Eastern divinatory literature. "Natural" divination refers to direct contact initiated by the gods, in the form of visions, dreams, or oracles. Such activities, under the heading of prophecy, were well known in the Near East and Levant and figure large in biblical literature.[1]

[1] For basic background on Mesopotamian divination, see Koch-Westenholz 1995; Rochberg 2004:5ff; Hunger and Pingree 1999:5–26; Rochberg-Halton 1988:8–17. A clear, brief survey of the texts appears in Maul 2009.

The bulk of ancient Near Eastern texts, however, constitute "artificial divination," requiring deliberate observation and interpretation. Artificial divination can take one of two forms: provoked or unprovoked. Provoked divination is brought about by some deliberate action of the seer; for example, in lecanomancy (from Greek, *lekane*, bowl), the diviner stirs oil on a bowl of water to reveal messages, and in extispicy, the priest removes or handles the entrails of a slaughtered victim. Such provoked omens occur when the diviner seeks signs in answer to specific questions, and perhaps even asks them directly of the gods in the form of a prayer.[2]

Unprovoked divination results from phenomena seen in the sky or otherwise experienced, such as eclipses, cloud patterns, storms, the behavior of animals, or even the actions of people in city settings. In Neo-Assyrian scholarly circles, the diviner was called the *tupšarru*, "scribe," or occasionally *ummânu*, "scholar," but quite different terms were applied to prophets or "natural" diviners of both sexes (*mahhû*, "ecstatic," *šabrû*, "seer," or *raggimu*-/*raggintu*, "prophet (-ess)."[3] Prophecy in Assyria was inspired by the goddess Ishtar, who might even speak through slaves. Some prophets were even enrolled in the temple food-rationing system.[4] It appears, from names cited in the texts, that most Assyrian "prophets" were women, but most, if not all, "scholars" were men.[5]

The tone and content of scholars' letters to the kings show that they took an active role as advisers on the running of the Assyrian state.[6] Ultimately, some prestigious scholars were called *tupšarru Enūma Anu Enlil*, "scholars expert in the *Enūma Anu Enlil*," the major compendium of celestial omens, consulted throughout the Near East by the first millennium BC.[7] This title continued through the Hellenistic period for some officials attached to the temple Esagila in Babylon.[8]

ORIGIN AND BACKGROUND OF FIRST-MILLENNIUM DIVINATION TEXTS

The recording, reworking, and dissemination of written versions of *omina* was a constant process in Neo-Assyrian and Neo-Babylonian courts and temples.[9] The Assyrians said that Ea, the god of wisdom, had himself created the famous series of texts that included the *Enūma Anu Enlil*, and many

[2] Cf. Starr 1990:xxxvi–lv.
[3] Rochberg 2004:48; see Parpola 1997:xlv–xlvii; Nissinen 2000:90–95.
[4] Nissinen 2000:95–97; for full background on prophetic cult, see Parpola 1997:xviii–xliv.
[5] Nissinen 2000:109.
[6] Especially the haruspices; see Starr 1990:xxx.
[7] Koch-Westenholz 1995: 56; Pongratz-Leisten 1999:18–23.
[8] Boiy 2004:256–57, 297–98.
[9] See Leichty 1993 for a concise explanation of Mesopotamian scholarship.

scholars have detected the roots of first-millennium divination in the scraps of texts from early second-millennium archives in Mesopotamia and nearby kingdoms, such as Mari. In fact, divination was probably institutionalized by the third millennium.[10] Literary texts with bearing on interpretation of the *Brontoscopic Calendar*, however, are really a phenomenon of the Neo-Assyrian, particularly the Sargonid, state.

Knowledge of divination was prized as a special gift of the gods, kept secret among an elite, but not necessarily affluent, group of scribes and scholars. A text found at Nineveh says that the gods Shamash and Adad (symbolizing sun and weather) revealed to the legendary antediluvian scribe (and king of Sippar) Enmeduranki the techniques for divination by oil on water and by liver examination (haruspicy). It notes that he handed down the knowledge to the men of Nippur, Sippar, and Babylon.[11] A catalog of authors and tablets made at Nineveh claims that Enmeduranki was the author of *Bārûtu* and *Kalūtu*, texts on "haruspicy" and the "art of lamentation"; he also copied a tablet set of the *Enūma Anu Enlil*.[12] Whatever the date of the historical Enmeduranki, it may have been his efforts that determined the canonical form of the Assyrian texts eventually assembled in the eighth- to seventh-century royal archives.[13] Paolo Xella (1999) has noted that the "biography" of Enmeduranki illustrates the close link, throughout Mesopotamian history, between institutionalized divination and the power of the state.

The third-millennium, Early Dynastic period in Mesopotamia already partook of institutionalized divination. Peter Huber (1987) has suggested that some lunar eclipse accounts refer to events that must have occurred in the second millennium, recorded under some already established system of observation.[14] According to Francesca Rochberg, an Akkadian tradition of written omens may be traced to the early second millennium BC, including the *Šumma izbu*, the *Alamdimmû*, celestial omens, and *Bārûtu*.[15] All these texts hold parallels for predictions found in the Etruscan *Brontoscopic Calendar*. By the eighth century BC, scholars working in Akkadian would have considered them part of an ancient, divinely inspired tradition.

Inscribed liver models and tablets found at Mari in Syria, including an assortment left in a room of King Zimrilim's palace circa 1775 BC, attest to the institutionalization of divination as early as the nineteenth century BC. Inscribed in Mari dialect mixed with Early Old Babylonian terms, the sub-

[10] Koch-Westenholz 1995:32.

[11] Koch-Westenholz 1995:74, with additional references.

[12] Lambert 1962:64, text 1:1–4.

[13] For the Mesopotamian perception of religious texts, canonical and otherwise, see the concise explanation of Pearce 2004.

[14] See Rochberg 2004:65.

[15] Rochberg 2004:63.

jects of the divinatory tablets are so close to Mesopotamian types that they are believed to have been copied in Mari from Old Babylonian originals.[16]

THE DEVELOPMENT OF THE *ENŪMA ANU ENLIL* SERIES

The best-known, standardized work on divination, the *Enūma Anu Enlil*, already existed in a fairly standardized condition as early as the fourteenth century.[17] This constitutes clear evidence that the tradition of copying and redacting divine literature was widespread across Middle and Late Bronze Age Mesopotamia. It was also familiar in the neighboring regions, long before it was exploited by the Assyrian court. Over the centuries the large tablet series was handed on by a steady stream of human editors, commentators, and copyists, and some Assyrian texts offer names for the persons who began to organize this omen corpus. One was the legendary antediluvian *apkallu* ("sage") Adapa,[18] but another figure was historical, the scholar Esagil-kīn-apli, who worked under Assyrian king Adad-aplu-iddina (1068–1047 BC).[19] A colophon on a tablet from Nimrud shows that he edited the famous series of medical and physiognomic omen texts, the SA.GIG and *Alamdimmû*. The label claims that he studied the many copies of *Enūma Anu Enlil* and assembled a correct edition out of a plethora of variants.[20]

We come closer to the historical Esagil-kīn-apli with tablet 20 of the "Babylonian" recension of the *Enūma Anu Enlil*. A subscript on the tablet says that it was written according to a writing board inscribed in the eleventh year of Adad-aplu-iddina (ca. 1057 BC). The timing is such that this tablet actually could be a copy of the famous scribe's own edition. In any case, we have confirmation, even before 1000 BC, of the existence of a complete set of the *Enūma Anu Enlil* texts and of an ongoing scholarly tradition.

The eleventh century had a special *Drang* for compiling and codifying the venerable texts of the second millennium. The famous astronomical text the MUL.APIN ("The Plow Star") seems to have been crystalized in the form we

[16] Koch-Westenholz 1995:32.

[17] Koch-Westenholz 1995:50; see Reiner 1999 for celestial texts.

[18] The Seven Sages, the *apkallu*, are known from the Hellenistic and Roman accounts of Berossos and Philo (channeling the text of Sanchuniathon allegedly written ca. 1000 BC; see Dalley 1998:16). They were immortal beings sent by the god Ea from the underground fresh water of Apsu to enlighten men; each sage was associated with a particular city and one of its historical kings. In teaching men the arts of civilization, such as agriculture, metallurgy, music, and kingship and how to form cities, inevitably a sage would offend the gods, and then be banished back again into the Apsu. Because of their association with living waters, the *apkallu* were sometimes represented in Assyrian reliefs as heroic humanoids wearing large fish costumes.

[19] Koch-Westenholz 1995:42–43, 74; Finkel 1988.

[20] Liebermann 1990:note 182. See Finkel 1988:149, lines 18′–21′, 24′–28′.

know around 1000 BC, and cross-fertilization had already occurred with the *Enūma Anu Enlil*.[21] It is evident that canonical omen-text series were already consulted by the Assyrian kings by the mid-eleventh century BC.

The best source of our knowledge of Mesopotamian divination literature comes from the so-called library of Ashurbanipal found in the Northwest Palace, and also in rooms 40 and 41 of Sennacherib's Southwest Palace, and in the Temple of Nabû. Most tablets in the library had been inscribed during the reigns of Esarhaddon and Ashurbanipal and were preserved by the conflagration that destroyed Nineveh in 612 BC. They are now in the British Museum, designated by inventory numbers beginning K- for Kuyunjik, the mound in which Layard excavated them.

MESOPOTAMIAN TEXT SERIES FORMING BACKGROUND FOR THE *BRONTOSCOPIC CALENDAR*

By the first millennium BC, the Mesopotamian corpus of works on divination covered a wide array of topics and situations.[22] In addition to the *Enūma Anu Enlil* ("When Anu and Enlil") listing thousands of celestial phenomena, the frequently consulted works (also titled by modern analysts after their incipits "opening words") included the MUL.APIN, a venerable star catalog with some omen predictions, the *Šumma ālu* ("If a city"), citing terrestrial events, like construction projects, as protases; the *Ziqīqu* ("Dream god"); *Alamdimmû* ("The form"), analyzing human physiognomy; the *Šumma izbu* ("If the anomalous newborn"), listing teratological omens or birth defects; and the SA.GIG ("Symptoms"), a medical diagnostic compendium. Related tablets include the *namburbû* texts ("The omen's undoing"), outlining rituals that could be performed to turn away a predicted catastrophe (for kings this sometimes involved installing a scapegoat).[23] Even the Sargonid "Great Star List" includes omens such as "Venus is for thriving cattle."[24]

The sampling in Appendix B illustrates the tone and topics of the sort of Mesopotamian omens that influenced Etruscan scriptures; I have found no single text or section of a text that I can say was duplicated (in translation of course) to create the document that is the Etruscan *Brontoscopic Calendar*. Single predictions found, often in more than one series of Mesopotamian tablets, crop up in no particular order in the *Brontoscopic Calendar*, as if a broad selection of Mesopotamian divination literature was known to its creators.

[21] Koch-Westenholz 1995:76–78; Hunger and Pingree 1989:12–18.

[22] See Rochberg 2004:49–50, 78–97; Hunger and Pingree 1999:12–20; Koch-Westenholz 1995:74–92; Reiner 1995:83–88; Pongratz-Leisten 1999:19–23.

[23] See Reiner 1995:81–96.

[24] See Koch-Westenholz 1995:187–205.

FORMATTING OF MESOPOTAMIAN OMEN TEXTS

All the cuneiform omen texts were formulated as conditional clauses. Each omen is named as the protasis (condition) and then an apodosis (outcome) or series of possible outcomes is stated, with each entry marked off by an incised line. Instead of indicating "item" at the beginning of each new statement, as in the well-known tablet lists of Sumerian ideograms and Akkadian vocabulary words, a vertical wedge was used to mark the end of each entry. The new beginning was usually read "if" (the Akkadian word *šumma*), and eventually the various omen texts were called by scholars "The ifs" and marked by a pseudologogram, ŠUM.MA.ME. In comparison, note that the Etruscan *Brontoscopic Calendar* scrupulously adheres to a similar formula, using either, "If it thunders" (εἰ βροντήσῃ, September 17 to May 30) or, "If in any way it should thunder" (ἐὰν βροντήσῃ, June 1 to September 16).

The "binary" formulae that also characterize the well-known cuneiform lists of vocabulary terms are used in the Mesopotamian omens for analyzing phenomena to be observed;[25] for instance, celestial phenomena show formal juxtapositions of dark-light, right-left, or labeling, such as the color of the dog that urinates on the questioner.[26]

THE STANDARD MESOPOTAMIAN CALENDAR (*SEE* APPENDIX B, TABLE B.1)

If weather phenomena are to be tied into a standard scheme, they must be dated in a universally recognized fashion. The calendrical scheme used in the Etruscan *Brontoscopic Calendar* is different from the early Latin/Italic calendars that scholars have postulated for Iron Age Italy in that it comprises a theoretical set of twelve months, all containing thirty days. Obviously, like all Classical, pre-Metonic, and pre-Julian calendars, any document with months of an equal number of days is going to have to be interpreted by experts, who decide when to call intercalary periods and such. This does resemble the so-called Standard Mesopotamian Calendar, which by the first millennium BC was used all over the Near East as the civic calendar, on which cultic or festival calendars often were also based.

Mark Cohen calls it a "manufactured" calendar compiled with material from several different cultures to serve as a unifying civil and commercial framework in the restless and newly unified Babylonian "empire" of second-millennium Mesopotamia.[27] Its base was a second-millennium or older Sumerian calendar, but its month names betray assorted origins. For instance,

[25] Cf. Guinan 2002:9–11.

[26] Cf. *Šumma izbu*, tablet 23; Leichty 1970:193–94.

[27] Cohen 1993:302–4.

the "eighth month," Arahsamnu, derives its name from the Old Persian month, Markašan, and other months (Nisannu, Ajaru) are linked to Ugaritic/Syrian forms.

It appears that the switch to the "standard" calendar was ordered by king Samsuiluna of Babylon (1749–1712 BC), "who sensed the urgency of fostering a sense of nationhood among the cities of his empire which had recently rebelled against him."[28] By around 1100 BC, the uniform system of this calendar was in use throughout the Near East. Centuries later, Tiglath-Pileser I (1114–1076 BC) mandated the adoption of the Standard Mesopotamian Calendar by the Assyrian kingdom, and it spread and survived for civil administration and cults, also flourishing in commerce and legal affairs.[29] It was the obvious template when the famous divination texts the *Enūma Anu Enlil* and the MUL.APIN were codified (ca. 1100–1000 BC). This format also appeared in various "hemerologies" ("day-to-day calendars"),[30] such as the *Enbu bêl arhim*, the royal version of a calendar of auspicious times (*uttuku*): for each day of a month, it notes whether it will be favorable, unfavorable, or outright dangerous for certain activities.[31]

The Standard Mesopotamian Calendar was the only form of civil or cultic calendar that would have been available for dissemination to the West during the first millennium BC. Some version of it would have been used by merchants trading out of the Levant and eastern Mediterranean, and it must have been known to all the bureaucrats and businessmen within the network of Phoenician colonies that stretched all the way to the Atlantic in the ninth and eighth centuries BC. It remained in use for a long time – and incidentally became the Judaean calendar after the Babylonian Captivity.[32]

DIVERGENCE BETWEEN MESOPOTAMIAN SCHEMATA AND ETRUSCAN

The Mesopotamian New Year begins in the spring with the beginning of the first lunar cycle after the spring equinox, thus, the month of Nisan. (For comparison of calendars, see Appendix B, Table 1.) The *Brontoscopic Calendar* begins just slightly later, with the month of June, so it is also perhaps keyed to the equinox. After the sixteen-day *akitu* (New Year) festival, the first major event was the spring flood, expected in Ajaru (April–May), after which grain would be harvested. Summer would see drought, and then the planting season would take place in autumn. Clearly, the Tyrrhenian Italian seasons and watersheds behaved differently, and the *Brontoscopic Calendar* reflects this in

[28] Cohen 1993:297, 305.

[29] See Cohen 1993:299–301.

[30] See Labat 1972–75; Livingstone 1993–97.

[31] For this text, see Livingstone 1993:98–99, 101; Landsberger 1915:103–5.

[32] Cohen 1993:299–300.

seasonal weather patterns and the human activities they prompted, but it does adhere to the same theoretical schema of twelve, thirty-day months, requiring scholarly or priestly interpretation for odd days.

THE *ENŪMA ANU ENLIL*

The *Enūma Anu Enlil* is a long document known from several copies found in the library of Ashurbanipal at Nineveh; of these, one nearly complete version was recognized in a set of approximately seventy tablets now in the British Museum. There has never been a complete translation and publication of this document, and the well-studied portions already published mostly cover astronomical/astrological observations, such as lunar eclipses or conjunctions of planets. Hunger and Pingree point out that composition of this document, part of the Babylonian "canon," was already in progress by the second half of the second millennium.[33]

The *Enūma Anu Enlil* is customarily divided into four sections named for the appropriate gods, following the 1905 study of Ch. Virolleaud: lunar (Sin), solar (Shamash), weather omens (Adad), and omens from stars and planets (Ishtar). Some tablets at Nineveh had been copied by the famous eighth-century scribe Nabû-zuqup-kēnu of Kalḫu (Nimrud).[34] An ivory, sixteen-board wax polyptych excavated at Nimrud may have held this scholar's royal copy of the *Enūma Anu Enlil*. Other tablets of the *Enūma Anu Enlil* were taken as booty in 647 BC by Ashurbanipal from his brother Shamash-shum-ukin, according to their labels. Such religious documents were widely disseminated as part of the activities of the state, the monarch, and would-be rulers.

The *Enūma Anu Enlil*, comprising almost 7,000 omens, was significant for the interpretation of all celestial phenomena, weather included. There developed around it a large series of commentaries, guides, and individual communications, in the form of letters and analytical reports furnished to the Assyrian kings by scholars, who often claim to have checked multiple texts and commentaries. Many cities had schools of scholars in residence, and collections of these omen tablets have been identified at Babylon, Aššur, Borsippa, Kalhu, Uruk, and Nineveh.[35]

Unfortunately, the meteorological tablets, numbers 40 through 49/50, that contain references to thunder and related phenomena are not yet accessible in translation; they include conditions of lunar eclipse, lightning strikes, cloud formations, rain, the rainbow, wind, and earthquakes, sometimes presaging war, death, fertility, and so on. They also designate some geographic topics,

[33] Hunger and Pingree 1999:12. For basic background on the *Enūma Anu Enlil*, see Rochberg 2004:66–78; for critique of the sources and studies of the *Enūma Anu Enlil*, see Hunger and Pingree 1999:12–23.

[34] See Guinan 2002 on his library and activity.

[35] Koch-Westenholz 1995:82–93; Weidner 1941:181.

such as relations with kings, enemies, or peoples of the north/south/east/west.[36]

THUNDER-OMENS: *ENŪMA ANU ENLIL*, TABLETS 40 TO 49/50

The meteorological omen series seems especially relevant for comparison with the Etruscan divination text. Ernst Weidner discussed very briefly the tablets of this series, where thunder is specified in tablets 44 through 46; tablet 43 begins with a lightning bolt striking in the south.[37] Within the other omen series of the *Enūma Anu Enlil*, where the protases are astronomical events, predictions also show many similarities to the *Brontoscopic Calendar*'s apodoses. An ancient scholar's commentary written on *Enūma Anu Enlil*, tablet 39+/-, which discusses solar and meteorological phenomena, calls them "*omina* of Shamash and Adad" (thus the sun and weather gods).[38] Other texts linking weather or thunder with earthquake were found in the Babylonian Temple of Nabû,[39] and at Nuzi.[40]

Weidner (1968–69) published a list of collated texts and relevant topics in the *Enūma Anu Enlil*, but his work was unfinished. The original publication of transcriptions of tablets, facsimiles, and such, Ch. Virolleaud's *L'astrologie chaldéene, le livre intitulé <enuma (Anu) ^{ilu}Bêl>* (Paris 1905–1912), is now outdated, and it never included a translation of the so-called Adad passages.[41] Erlend Gehlken, working especially with the British Museum tablets from the library of Ashurbanipal, and collating other exemplars in the series, has begun the publication of the Adad tablets, with tablets 42 and 43 (the latter poorly preserved) in press and subsequent tablets to appear soon in the *Cuneiform Monograph* series (Leiden). He has estimated that approximately 500 tablets and fragments are known from the Adad series.[42] Other works, especially the *Iqqur īpuš*, and some of the lunar eclipse tablets from the *Enūma Anu Enlil*, in their present state have more relevance as comparanda for the Etruscan *Brontoscopic Calendar.* Gehlken has pointed out that these texts influenced Aristotle's *Meteorology*, Euktemon's *Peri semeion*, and the Roman-era *Brontologion* attributed to Hermes Trismegistus (see Chapter 9); he even finds strong echoes in the *Bauernregeln*, farmers' proverbs of early modern Europe.[43]

36 Cf. Koch-Westenholz 1995:56, 109, 115.
37 Weidner 1968–69:70–74.
38 Koch-Westenholz 1995:55n2.
39 Wiseman and Black 1996:no. 18.
40 See Hunger and Pingree 1999:11; Lacheman 1937; Virolleaud 1905–1912: Adad 20:35–48.
41 For references to other publications of these sections, see Weidner 1968–69.
42 Gehlken 2005:238. I am grateful to Erlend Gehlken for enlightening me on these tablets.
43 Gehlken 2005:235–36, 238.

The topics, as determined mainly from tablets and colophons excavated at Nineveh and Aššur, are summarized in Appendix B, Table 3. Within the *Enūma Anu Enlil* there do not appear to have been any series of *omina* organized, like the Etruscan document, solely by month and day, and I can find no series in the translated corpus that can be shown to have been duplicated in its entirety, or even in serial order, within the *Brontoscopic Calendar* preserved by Lydus.

LUNAR ECLIPSE OMENS

A brief survey of the remainder of the published tablets of the *Enūma Anu Enlil* series furnishes many additional parallels for elements of the *Brontoscopic Calendar*. Taken in numerical order of the tablets, the topics cover lunar observations, especially eclipses, solar phenomena, and planetary and constellation observations. Some are better preserved (and more thoroughly analyzed by scholars) and serve to give an impression of what the meteorological texts were like.

The *Enūma Anu Enlil* lunar eclipse omens furnish predictions that are more detailed and lengthy,[44] including Old Babylonian schemata for equating eclipse directions with countries of the Mesopotamian sphere[45] (south=Akkad, north=Subartu/Guttium, east=Elam, and west=Amurru). In times of political stability (read Assyrian hegemony), Assyrians could interpret omens pertaining to both Akkad and Subartu as referring to themselves, at least in times when the Assyrian king really was the overlord of Akkad (Babylonia) along with Aššur.[46]

THE MUL.APIN: CELESTIAL DIVINATION

Another work with distinctly celestial omens is the astronomical text MUL.APIN ("The Plow Star"); famous in antiquity, its lore was reprised by cultures well beyond Mesopotamia. It had arrived at its present form by 1000 BC.[47] Hunger and Pingree noted the high degree of uniformity in different copies of this document; naturally, its best preserved exemplar comes from the library of Ashurbanipal. Its protases cite the risings of stars and constellations, occasionally accompanied by weather phenomena, and the apodoses show many parallels to other Mesopotamian divination texts – and also to the

44 I thank Erlend Gehlken for pointing this out to me (pers. comm., February 24, 2008).

45 Rochberg-Halton 1988:53–55.

46 Koch-Westenholz 1995:98; for full details, see chap. 5, 97–136.

47 See Koch-Westenholz 1995:76–78; Hunger and Pingree 1989:9, 12–18. Two tablets from different sources express the dates of their copying, one in 687 BC and the other during the Seleucid period.

Brontoscopic Calendar. See Appendix B for sample entries and Table B.4 for topics and cross-references.

OMENS OF URBAN LIFE: THE *ŠUMMA ALU INA MĒLÊ ŠAKIN*

Another important Mesopotamian text series is known by the incipit "If a city is built on a hill ..." (*Šumma alu ina mēlê šakin*). Sets of unusual events are neatly listed in numbered series by topic, on tablets found in mid-seventh-century contexts at Ashur and Nineveh. Some of the earliest tablets also came from the library of Nabû-zuqup-kēnu, moved from Kalḫu to Nineveh after his death. Although thematically different from the other texts, many lines result in similar predictions of devastation, plague, and "violent deaths," and some apodoses simply note "the same" (as does the *Brontoscopic Calendar*). The *Šumma ālu* is one of the few Mesopotamian divination texts to mention servants (though not slaves). Ann Guinan (1989, 1990, and 2002) has analyzed many of these omens, and she notes the underlying perceptions of human structures vis-à-vis the kosmos, and the beginnings of psychological analyses, in the behavioral omens. Commentaries on the *Šumma ālu* continued to be written into the Hellenistic period at Babylon.[48] It also contains passages similar to other divination texts: *Šumma ālu*, tablet 60, with omens "if the earth exudes honey...," has parallels in *Enūma Anu Enlil*, tablet 43, which incorporates thunder-omens.[49] In other words, much Mesopotamian divination literature was fluid, or shared across disciplines, a situation that would have been conducive to cross-cultural sharing also. (See Appendix B.)

THE *ŠUMMA IZBU* TERATOLOGICAL OMEN SERIES

Another field of enquiry for Mesopotamian diviners was the observation of abnormal births. As with the haruspical texts, the *Šumma izbu* ("If a malformed newborn") lists anomalies in humans and animals by body part and type and then gives different apodoses for each malformation or oddity. Here again, although some predictions are relevant to the animal's owner or his household, like "that ox-fold will prosper," as with haruspicy, most outcomes relate to the state, city, or population as a whole. The tradition of *izbu* divination is believed to have been orally disseminated until around 1600 BC, when a version of the *Šumma izbu* was first recorded; Ugaritic versions were created at Ras Shamra about 1450 BC, and Hittite and Hurrian translations were copied at Hattuša in the mid-fourteenth century. Erle Leichty noted that standardized versions circa 1000 BC were disseminated throughout Assyria

[48] Boiy 2004:27.
[49] Gehlken 2005:241.

and southern Mesopotamia, and they were also found as far east as Susa. An important collection comes, as usual, from the library of Ashurbanipal, but the series continued to be copied and used until around 100 BC in the scribal schools of Babylonia.[50]

Some *Šumma izbu* omens were excerpted and repeated in the *Šumma ālu* series.[51] Leichty postulated the transfer of *Šumma izbu* and haruspical lore to Etruria and thence to the Roman state, but at that time he could only suggest that it was "the tradition, if not the texts" that was disseminated.[52] A few protases describe unusual behavior or conditions of adult animals, such as wild pigs entering towns, oxen breaking a horn, dogs kissing other dogs or sitting on people's furniture, but most describe malformed newborns. Just a few apodoses are specific, such as, "If a woman gives birth and (the child) has no hands – the enemy will conquer the city (where the) birth (took place)."[53] Only a few predictions are offered for the mother, such as "that woman will experience misfortune/trouble experience happiness."[54] See Appendix B for Table 5 and sample entries.

Lightning is predicted in tablet 20, omen 41′, but the text is too damaged to tell more. The apodoses for the generalized *izbu* omens of tablets 6–17 very much resemble those of celestial phenomena and such: the king will defeat his enemy, the land will be laid waste, the nobles will rebel against the king, the crown prince will kill the king, reign of a despotic king, famine/hunger, prosperity, sickness/plague, war/peace, fire, flood, weather, and so on.

OTHER RELEVANT MESOPOTAMIAN TEXTS: *IQQUR ĪPUŠ*

Other relevant texts in the canonical works on divination include hemerologies and menologies, documents for divination by the day or month of the year. Essentially the texts simply warned rulers of days that were auspicious or dangerous for certain activities, or perhaps for any activity at all – on some days, a king was warned not to even leave the palace.[55] Most significant of the hemerologies is the *Iqqur īpuš* ("He tore down and rebuilt"), essentially a set of dated omen predictions, tagging whole months for specific activities and noting day by day the details of dangerous or favorable activities, sometimes qualified by astronomical, natural, or catastrophic phenomena, such as fire or storms, using a theoretical year of twelve, thirty-day months.[56] Later

[50] Leichty 1970:21.
[51] Leichty 1970:26.
[52] Leichty 1970:21, 14–16; 1960:82–85.
[53] Tablet 3, omen 50; Leichty 1970:59.
[54] Tablet 3, omen 69; Leichty 1970:61.
[55] Survey: Labat 1972–75. Cf. Livingstone 1993–97; Landsberger 1915.
[56] See Reiner 1995:81, 88–91.

astrologers would use this document to draw up lists of auspicious days for their clients. The same omens were arranged in two series, first by the subject of the protasis, and then by the month in which an omen could occur. Some tablets are actually formulated as tables with cells for omen types cross-referenced to days of the month; some cells are blank or else marked with the ideogram "favorable."[57] A few linguistic clues imply origins in the reign of Tiglath-Pileser I (1115–1047 BC) or even earlier.[58]

There was much excerpting and copying of omens between the *Iqqur īpuš* and other texts, including the *Šumma ālu* (human/city activities) and the *Enūma Anu Enlil* (celestial phenomena). The first sixty-six sections of the *Iqqur īpuš* cover daily human life, the last third natural phenomena.[59] Standard readings are on the order of omen 43: "If a man digs a well," … in the month of Ajar, then he will be in want of grain; … in the month of Tebet, he will have a long life."[60] And, "If in the month of Kislev, [the king (as opposed to a commoner)] plants a garden of palm trees: he will die in the course of the year;"[61] or, "If [a fire occurs] on the 10th day [of any month]: the gods, by a revolt, will depose the king."[62]

Appendix B gives a sampling of the omen lists from the *Iqqur īpuš*, based on the readings of Labat (1965). There seem to be more predictions about crops, food, and weather – everyman's concerns – in this document, another feature similar to the *Brontoscopic Calendar*. The epithet of the thunder god Adad is of interest: "The month of Shebat is (the month) of Adad, the guardian of the waters of heaven and of the earth."[63] The omen protases of the *Iqqur īpuš* do offer a few meteorological situations, including the rumbling of thunder, always described as "when Adad hurls his voice."

The thunder-omens in many cases seem to be associated with political developments, the reigns – and downfall – of kings and princes. Presumably, the status of Adad, the divine source of thunderstorms, blatant expressions of power, is a rationalization of the control of the state. Earthquake, which is often associated with a sound like thunder, occurs in sections 100 and 101, predicting calamity for the entire kingdom; omen 8 of section 101 cites "pillaging and assassinations in the country."[64] It seems that the *Iqqur īpuš* offers the largest number of omens, and the most extensive series of lines for comparison with the Etruscan *Brontoscopic Calendar*, but there is nowhere a serial, one-to-one correspondence between the two texts. Also, though fragmentary, the *Iqqur*

57 See Labat 1965:12, and pl. 1.
58 Labat 1965:19–20.
59 See Labat 1939, 1965; Weidner 1968–69:72n25.
60 See Labat 1965:111.
61 Labat 1965:115, section 46, omen 10.
62 Labat 1965:136–39.
63 Labat 1965:196–97.
64 Labat 1965:191.

īpuš may be judged as lacking several important omens that are present in other Mesopotamian literature – and are also reflected in the *Brontoscopic Calendar.*

Sections 86–87, describing phenomena of Venus and the "light of the zodiac," also offer predictions that sound familiar in the context of the Etruscan document, citing harvests, happiness, famine, the king, the juxtaposition of wheat and barley "futures," the linking of rain predictions with harvests or the fate of "the people," the advent of locusts, revolts, and more. The south wind or storms that will sweep away the trees sound allegorical for devastation in human political or military events, and both resonate within the Etruscan document also. Thunder is the condition that actually determines the omens in sections 88–94, which are sampled in Appendix B.

SCHOLARS' RESEARCH REPORTS TO THE ASSYRIAN KINGS

From libraries at Nineveh and elsewhere come many reports and letters, part of the vast correspondence between the palace and a network of scholars writing from various cities in the Assyrian realm. Many were generated in the years around 670 BC,[65] or perhaps tablets of those precarious times were more often saved from the daily correspondence that must have inundated the court. Although even the Chief Scribe might complain that he was impoverished and starving, the palace seems to have employed a huge workforce of literate persons. Some corresponding scholars with prestigious reputations could even criticize the intervention of "those who read to the king" for twisting their predictions.[66] All the better, then, that Prince Ashurbanipal was trained by the top palace scholar, Balasî, and was able to do his own research and fact-checking!

The large body of royal reports to Esarhaddon and Ashurbanipal is a good indicator of the strong link between divination and the state. Most reported omens (except some one-time predictions based on a baby's birth),[67] and some *Šumma ālu* items, were pronounced on behalf of the king or the state. It was only after the change in ideology with the Achaemenid Persian rulers that such techniques as celestial divination could be banalized for private curiosity. After all, if just anyone could predict the downfall of the king …

The royal reports' expository, highly personal style contrasts with the Etruscan *Brontoscopic Calendar*; its inspirational sources were more formal literature. But in the royal reports, omens concerning political/military

[65] Chadwick 1989; Oppenheim 1969.

[66] See Parpola 1993A:xxiv–xxvii.

[67] Another feature of Etruscan belief was the recording of a person's lifespan, especially preserved in epitaphs inscribed at Tarquinia and Volterra: noble families assiduously recorded the exact day and time of a child's birth.

situations are mingled with general concerns, such as trade and food supplies/ agriculture; they serve to illustrate the living character of divination literature and how it was used for consultation. Many private libraries of standardized works existed in diverse locations, and trained professionals eagerly exercised their prerogatives of interpreting and disseminating such information – always within the approved, royal, circles. And the thunder-omens found in several reports probably take us closer to the "Adad omens" of the *Enūma Anu Enlil.*[68]

The reports for a single day show a high degree of similarity, whether in Babylonian or Assyrian dialect, presumably because the scholars were honestly and accurately consulting the same divination texts for their interpretations.[69] In a report sent from Borsippa by Nabû-iqiša, for the date June 11, 679 BC, a conjunction of Jupiter and Mars is interpreted as "the fall of a great army, evil in the land, death of the cattle, death of the king of Akkad, but a prospering crop … this is an evil for all lands," and the adviser urges the king to undertake an apotropaic ritual to avert the predictions.[70] (See Appendix B.)

HARUSPICY: A PARALLEL EXAMPLE FOR MESOPOTAMIAN INFLUENCE IN ETRUSCAN RELIGION

The adaptation of the discipline of haruspicy in Etruria forms an interesting potential parallel case to the development of the Etruscan *Brontoscopic Calendar.* Enough details are present to demonstrate some deliberate borrowing from Mesopotamia, yet the end result (compare Fig. 12) is quite different from any Near Eastern versions. It can only have resulted from a long period of Etruscan intellectual activity and concerted interaction with Near Eastern worshipers or priests.

There is a rational basis for the use of sheep livers in divination, for situations in which the health of flocks or humans, and possibly the state of the food supply, may be predicted.[71] Ancestral Etruscan communities had been sheepherders for centuries, probably millennia, by the Iron Age, and so some native tradition of scrutiny of carcasses and viscera probably was already in operation. Melding that approach with the more formal formatting of the discipline that is seen in Near Eastern models and texts would have been a reasonable

[68] In many cases, the works these scholars were consulting must have included the thunder-omen series from the *Enūma Anu Enlil.* For reports on storms, thunder, and earthquake omens, cf. Thompson 1900:lxxviii–lxxxiv, 89–98; for thunder-omen reports, lxxix–lxxxi, 92–95.

[69] Koch-Westenholz 1995:180–85.

[70] *SAA* 8 288 (State Archives of Assyria), quoting Koch-Westenholz 1995:122.

[71] Turfa and Gettys 2009.

extension of valuable knowledge. Comparing the labels of Mesopotamian and Etruscan livers shows two quite different systems of belief, however.

Ulla Koch (2005) has translated portions of a tablet series on haruspicy found in the Ashurbanipal library at Nineveh; Jan-Waalke Meyer (1987) published a large selection of model livers, some of which were inscribed. In first-millennium Mesopotamia, haruspicy was practiced in many levels of government and society, with specific questions asked in prayers and the sacrificed animal then scrutinized for answers. One of the major entities to avail itself of haruspicy was the army, along with campaigning rulers, and thus a great number of the responses recorded in state archives refer to war.[72]

Neo-Assyrian haruspical documents partake of a tradition reaching far back into the second millennium. By 750 BC, a canonical text was in use across the entire Near East; it continued essentially unaltered into the Seleucid period.[73] (A short sample of haruspical omens appears in Appendix B.) Most omens interpret aspects of the appearance of the fresh liver, according to its size, shape, and other variables like color (with white/red/green/black spots, white filaments, holes, blood, etc.). Religious expressions also appear, such as "reconciliation of god with man." The language and topics of Mesopotamian haruspicy and its gods were also entwined with the matter of meteorological divination. For instance, on an Old Babylonian liver model, one of the sectors states *ina warkat šatti adad i-ša-si-ma*, "Adad will thunder in the autumn."[74] Many *Greek* anatomical terms used in haruspicy appear to be direct translations from the technical jargon of Akkadian liver texts: the parts of the liver are named "the gate, the head, the path, the river."[75] So someone of Near Eastern background must have *explained* the anatomical terms to the Greeks, probably on multiple occasions.

Naturally, Etruscan haruspicy has been linked to the Near East, and at last the evidence is growing to support this, but the old connection suggested by Alfred Boissier in 1901 between the Sumerian ideogram HAR and the *har-* in the Latin term *haruspicina* must be false, because this sign would have been read as the Akkadian word *amutu*, and neither word resembles the appropriate Etruscan terms *netśvis, netśrac*. The use of three-dimensional models of sheep livers, many of them inscribed with texts, is a distinctive Mesopotamian development. The sharing of this system of information must have occurred in Etruria by the seventh century BC, yet it is not a wholesale borrowing but a well-integrated sort of system upgrade. Although the stylized configuration of the liver models, with prominent gall bladder and starkly geometric *processus pyramidalis*, is very close to Mesopotamian versions, the written matter

[72] For background on Mesopotamian interpretation of liver omens and the literature it generated, see Koch-Westenholz 2000:11–37.

[73] Koch-Westerholz 2000:27.

[74] CT 6 2:37, *CAD* Š II [1992] p. 151, 1g.

[75] Cf. Liedermann 1990; Mayer 1987:53–74; Koch-Westenholz 2000:38–70.

is utterly different. Instead of geographic terms or nicknames like "the gate," "the path," the Piacenza liver is marked with names of Etruscan gods, many of them rather obscure in terms of art or literary references.

Burkert maintained that "Some more complicated rules bring the Etruscan and Mesopotamian discipline even closer together … to make an art out of this and a system, using models for demonstration, this is peculiar to the corridor leading from Mesopotamia to Etruria. There are no liver models of this type in Greece … since the Etruscan discipline clearly was fully developed by the 6th century, the foreign influence will go back to the 7th, which is the fully Orientalizing period just for Etruria."[76] In fact, subsequent discoveries have put the inception of Orientalizing, and of literacy, into the eighth century. The telltale is that "it can even be shown that both the Assyrian and the Etruscan models diverge from nature in a similar way: that is, they are derived not directly from observation but from common traditional lore."[77]

Thus far, the earliest liver model in Italy is probably the uninscribed Falerii terracotta (ca. 300 BC); its anatomical details are remarkably similar to the Mesopotamian scheme, including a model found at Levantine Megiddo.[78] An irregular lead, fifth-century plaque from Magliano, perhaps in the outline of a liver, is covered with a spiral inscription naming the *aiser*, "gods," and specifically Cavtha, Maris, Thanr, Calu, Śuri, Tin, and Lur; it notes rituals to be performed "Eighty times," "every month," "yearly," and so on.[79] No one could argue for a second-millennium, Babylonian *origin* for extispicy in Etruria, but the Neo-Assyrian models and the wealth of first-millennium texts analyzed by Meyer (1987) and others do support the notion of some sort of knowledgeable interchange between Mesopotamian *Bārûtu* and Etruscan *netśrac*.

The Archaic costume of tall cap, fringed cloak, and fibula venerated Etruscan shepherd ancestors and their cult.[80] Hellenistic funerary images and epitaphs of men who served as haruspices show that several Etruscan cities harbored noble families expert in this discipline: Lars Pulenas of Tarquinii actually *ziχ netśrac*, "wrote on haruspicy";[81] at Chiusi Nae (Gnaeus) Cicu is described as a *netśvis*;[82] the bilingual epitaph of Lars Cafates/Lucius Cafatius (Fig. 6; see Chapter 3) proclaims him as *netśvis* and *trutnut frontac* at Pesaro in the late first century BC;[83] and Aule Lecu of Volterra on the lid of his urn holds a model(?) sheep liver.[84] At Chiusi, Arnth Remzna's urn[85] and two sarcophagi in

[76] Burkert 1985:117, order of sentences changed; cf. Burkert 1992:46–48.

[77] Burkert 1992:48.

[78] Pfiffig 1975:117; van der Meer 1987:153–54, fig. 71, Villa Giulia inv. 3786.

[79] *ET* AV 4.1; van der Meer 1987:169–71; Capdeville 1996; Colonna 1993[1994]; Morandi 1988.

[80] Roncalli 1981 and 2010; Ambos and Krauskopf 2010; Maggiani 1989; Steuernagel 2005.

[81] *ET* Ta 1.17 = *CIE* 5430.

[82] *ET* Cl 1.1036 = *CIE* 978.

[83] *ET* Um 1.7.

[84] *ET* Vt 1.128, 1.129 = *CIE* 81, 92.

[85] *ET* Cl 1.166 = *CIE* 1192. Turfa 2005:263–65, no. 295.

the Tarquinian *Alveθna* family tomb[86] represent the men with distinctive *apex* headgear; a mural in the earlier, mid-fourth-century Tomba Golini at Orvieto shows at least one member of the Vercna-Cnezus family similarly attired. He may have been Laris Vercnas, who was a *zilaθ meχl rasnal* (*praetor Etruriae*), thus combining priestly calling with high government office.[87] The tradition of haruspicy was also carefully guarded in Etruria, handed on in families in many cases, even into the Imperial period.[88] It would not have required many practitioners to effect the adaptation of Mesopotamian teachings and to pass on the results; a few priests would have sufficed to formulate and disseminate the new system.[89]

SURVIVAL OF MESOPOTAMIAN TEXTS AFTER 600 BC, AND SUBSEQUENT TRADITIONS

Use and dissemination of documents like the *Enūma Anu Enlil* actually continued through the Achaemenid and Hellenistic periods in the Near East, when, as late as the second century BC, scholars making celestial observations were still titled "Scribe (*tupšar*) of the *Enūma Anu Enlil*."[90] But after the Achaemenid Persian and Hellenistic conquests, the field of personal horoscopes burgeoned, rather than state-centered divination.[91] The character of these later documents sets them apart: there is no chance that a Hellenistic divination text written, say, during the Seleucid empire, could have been mistaken for an older Etruscan text.

The literature exerted an influence of long duration across an immense area: "Echoes of classical Babylonian omen astrology may be found in Egyptian, Greek, Latin and Indian astrology and even in Aramaic texts of the Byzantine period. The parallels between Babylonian omens and Hellenistic popular astrology pertaining to earthquakes and thunder are striking," states Koch-Westenholz.[92] As in the Sargonid period, rulers still acknowledged the usefulness of formal divination to their realms. Even through the Seleucid

[86] *ET* Ta 1.142.

[87] *ET* Vs 7.20 and 7.27 = *CIE* 5101, 5108. Steingräber 1986:279, no. 33. Third-century bronze votive figurines of worshipers with livers or haruspical costume (see Figs. 2–3) give additional family names and locations, such as Arezzo, but it is uncertain where or whether the dedicants held haruspical office; see Turfa 2006A:106, fig. VI.17, and 2006C.

[88] See Haack 2006:passim.

[89] Burkert (1992:48–49) discusses the burgeoning of hepatoscopy in Greek religion, where it does not occur in the "older strata" of the Homeric epics, but occurs in the final version of ca. 700 BC, with the story of the seer Kalchas (*Iliad* 1.69); Attic vases show scenes of hepatoscopy beginning ca. 530 BC. For Plato, hepatoscopy was more prestigious than bird augury (*Phaedrus* 244C).

[90] Hunger and Pingree 1999:14.

[91] See Rochberg 2004:chaps. 3–5; Koch-Westenholz 1995:162–79.

[92] Koch-Westenholz 1995:53, citing Bezold and Boll 1911:45–54, explication and commentary on *Adad Tablet* 20.

period, Hellenistic Greek rulers still maintained cuneiform libraries of divination texts, updating them with their own recorded queries on historical events.[93]

Other Hellenistic divinatory works in Greek, according to Pingree, are "late reworkings of texts composed in Hellenistic Egypt."[94] They contain material from different sources, some quite unrelated to Babylonian divination; modern scholarship at times links these to "Petosiris," but this is problematic, and the texts bear little similarity to the Etruscan *Brontoscopic Calendar*. From the Roman Empire, Ptolemy's *Almagest* also cites Chaldean astronomical data, and it is clear that already ancient Mesopotamian documents somehow were still accessible to scholars whose own works were written in Greek. The extensive Hermetic literature of Egypt only began in the second century BC, and it continued to be produced or edited into the Roman period. It was attributed to a (fictitious) savant, Hermes Trismegistus, and actually did include a text on divination by thunder, a so-called *brontologion*.[95] Most of the texts in the Hermetic corpus, as we know it, are of Roman Imperial or even Byzantine date.

Again, this scholarship appeared too late to have influenced the Etruscan *Brontoscopic* text, but the Near Eastern sources of its inspiration may have included some documents seminal to both traditions. Franz Boll, who edited the Berlin manuscript of the *brontologion* attributed to Trismegistus, noted that it shared the same source as another divination text, attributed to Fonteius Capito (see Appendix C, Text 1).[96] The Egyptian astrological imprint is evident, in geographical references to the Nile, and in use of the zodiacal signs as markers for observed events. The calendar of the Trismegistus text begins in January, another feature that distinguishes it from the Etruscan *Brontoscopic Calendar*, and also from the Mesopotamian texts.

David Pingree noted that

> there exist in various languages in the Middle East texts containing a curious mixture of astral omens of ultimately Babylonian origin with Hellenistic and Roman astrology. To the Greek and Arabic versions of books of Hermes can be added the Apocalypse of Daniel in both of these tongues and in Syriac, the Syriac Book of Medicines, the Mandaean Book of the Zodiac, and numerous brontologia, seismologia, and similar tracts in Greek, Arabic, and Syriac. The interrelations of these texts to each other, and the channels through which the Babylonian material reached each, have yet to be investigated.[97]

[93] Rochberg (2004:9) cites a diary entry generated about the victory of Alexander at Gaugamela.

[94] Pingree 1978:620–21.

[95] See Rochberg-Halton 1988:12n25, with references.

[96] *CCAG*:vol. 7 [1908], 226–30 = Berlin codex 26, F. 152^{v}.

[97] Pingree 1978:620.

A scrap of an Aramaic *brontologion* was found in Qumran Cave 4, and it is likely an example of this sort of literature (see Chapter 9).

INTELLECTUAL OR ARCHIVAL SOURCES FOR THE PROTOTYPES OF THE *BRONTOSCOPIC CALENDAR*

Koch-Westenholz suggested that the Assyrian duplication and self-conscious preservation of divination texts was (for Mesopotamia) a late development, and that "there is little evidence that any Assyrian kings before Sargon II (722–705 BC) took interest in astrology. Our major source for the so-called canonical works are the archives of the time of Sargon II to the death of Ashurbanipal in 627 BC."[98] It is interesting that this period of renewed engagement of scholars and kings with the literature of divination in Assyria should have coincided with the beginning of the Orientalizing period in Italy.[99]

SOCIAL, POLITICAL, AND PRACTICAL BACKGROUND OF THE MESOPOTAMIAN DIVINATION TEXTS

In the Neo-Assyrian and Neo-Babylonian periods, divination was enmeshed in palace politics and the social relationships of the court. It seems likely that any scribe or scholar trained in divination and sent abroad or introduced to foreigners would have been associated with the state or king. The *bārû* (haruspices) not only consulted for the palace but also served with the army on military campaigns.[100] Correspondence in royal archives shows that, in the Sargonid era, and probably before, many people were trained to perform several different types of divination, including quite disparate types, such as teratomancy (*izbu*), analysis of physiognomy, medical diagnosis, and so on.[101] In the eighteenth-century Syrian kingdom of Mari, diviners swore a loyalty oath to the king, a very clear indication of divination perceived as serving the state.

Royal Assyrian historical inscriptions, especially frequent under Sennacherib, Esarhaddon, and Ashurbanipal, cite specific *omina* as supporting the king, confirming his policies, or heaping trouble upon his enemies.[102] The phenomenon shows that the Assyrian kings appreciated the value of divination as a means of stimulating support for their power – a situation that any ruler might find valuable. The scholars derived their status from royal favor and believed they were, by their interpretative efforts,

[98] Koch-Westenholz 1995:51.
[99] See Bartoloni and Delpino 2005.
[100] Starr 1990:xxx.
[101] Rochberg 2004:5, 76; see also Parpola 1993B:51–52.
[102] Koch-Westenholz 1995:152–61.

protecting the king from adverse fate.[103] David Brown argues that Sargon II (721–705 BC) began his dynasty's exploitation of the divination cult to maintain control over a restive southern Mesopotamia, noting the distribution of families of scholars over a wide region from Mesopotamia to Anatolia.[104]

In addition to private libraries, a collection was uncovered in the Temple of Nabû at Nimrud.[105] Dated colophons (814?, 787, and 676 BC) show that this archive was formed no later than the eighth century BC; it continued in use through the seventh century. In addition to epics, medical texts, historical documents, and lexical works, it held the *Enūma Anu Enlil*,[106] *Šumma ālu*, *Iqqur īpuš*, *Šamaš bēl dīnim* (extispicy), physiognomic omen texts, and *Namburbû* rituals. If any travelers departed for foreign lands during the ninth to eighth centuries, or any visitors came thence, the rich literature of divination would certainly have been available to them, with the help of Mesopotamian or perhaps Levantine interpreters.

INTERPRETATION IN MESOPOTAMIA

Simo Parpola noted that, as a sort of god or divine representative on earth, the Assyrian king had much more stringent constraints on his behavior than mere mortals, and thus he needed to surround himself with the most expert advisers in order not to misstep when imperfection could mean disaster for his state.[107] In fact, the notion of celestial signals showing approval or disapproval and warning the king sounds somewhat like the Latin authors' descriptions of Jupiter's first two *manubiae*, the lightning bolts that warn gently or scorch but offer prospect for improvement – the doctrine makes the most sense if taken in the context of royal, "divine right" kingship. The king's moral state is reflected in the omens, and, in some cases, punishment for his sinning will be meted out to his whole kingdom.

SCRIBES, SCHOLARS, AND THE DISSEMINATION OF MESOPOTAMIAN LORE

During the second millennium, and increasingly in the first-millennium Near East, scribes also became translators, already trained in the written and spoken forms of Sumerian and Akkadian. The Mari tablets cite the presence of translators in the commercial caravans from Mesopotamia, and Middle Babylonian translators traveled with the Mitannian army. There is

[103] Cf. Brown 2000:33–47.

[104] Brown 2000:51–52.

[105] See Wiseman and Black 1996.

[106] See Wiseman and Black 1996:11, no. 18, pls. 16–17.

[107] Parpola 1993B:52–53.

ample evidence for the use of Assyrian/Mesopotamian translator-scribes in exchanges with the Hittite kingdom and Egypt, as the Amarna tablets attest. At court, an official title was "head of the translators."[108] A small number of literate individuals could account for the introduction and continued elaboration of the divinatory texts that have been found in the peripheral kingdoms of the Near East – even though the impact of these documents was great.

At the royal level, an ivory polyptych perhaps inscribed by the scholar Nabû-zuqup-kēnu and found at Nimrud places a much-read copy of the *Enūma Anu Enlil* in the royal household.[109] It had been copied at the end of the eighth century for Sargon's palace at Khorsabad.[110] Colophons on other copies state that a number of clay-tablet omen series were copied from waxed tablets, suggesting such media as the means by which Akkadian literature was disseminated to the West.[111] Writing tablets of fine woods or ivory have been found in other elite or royal contexts – for instance, the Canaanite Uluburun ship (ca. 1300 BC), a seventh-century princely tomb at Etruscan Marsiliana d'Albegna (Circolo degli Avori and the Isola del Giglio wrecked ship [ca. 580 BC]).

Rulers and scribes took pride in Mesopotamia's religious heritage: Ashurbanipal said,

> I learnt the craft of the sage Adapa, the esoteric secret of the entire scribal tradition; I have observed and discussed celestial and terrestrial signs in the meetings of scholars; I ponder with expert diviners the liver, the image of heaven; I can solve complicated mathematical problems lacking solution; I read sophisticated texts written in obscure Sumerian and hardly understandable Akkadian; I have studied stone inscriptions from [the] time before the flood.[112]

Surely the texts that reached the kingdoms of Anatolia, Syria, and others were sent or loaned through diplomatic channels and accompanied by scribes. (In Archaic Etruria, scribes enjoyed high status, according to the Roman tale of the *scriba cum rege sedens*, the "scribe sitting with the king," who was mistaken by a Roman assassin for King Porsenna himself.)[113]

[108] See Pearce 1995A:2274.

[109] See Robson 2008:145–46.

[110] See Wiseman 1955:7, who noted that at Nineveh the palace library had copies possibly written by the same famous scribe.

[111] Wiseman 1955:13.

[112] Lambert 1962:72, tablet VI 15; see Parpola 1993B:59n30, citing Streck 1916:252, vol. 1:13–18.

[113] Colonna 1976.

ADOPTION OF BABYLONIAN DIVINATION TECHNIQUES BEYOND MESOPOTAMIA

Mesopotamia shared with other cultures both canonical texts like the *Enūma Anu Enlil* and also reworked documents, many of them translated into the local languages.[114] The palace finds of nineteenth- to eighteenth-century Mari show the early spread of uniquely Babylonian forms of divination, such as haruspicy and astrology. Burkert proposed "migrant charismatics" as vectors for the transmission of Mesopotamian culture to the West during the Late Bronze and Iron Ages; a similar phenomenon appeared during the Hellenistic-Roman period, when they became known as Magi.[115] The rulers of Mesopotamia, Ugarit, the Hittite kingdom, and Alashiya-Cyprus made specific requests for scholars, including diviners, to be sent from the courts of Egypt and Babylon, and the practice continued into the Iron Age, as seen in Assyrian palatial archives. Documents show the existence of an Assyrian *karum* (trading post) on the Egyptian border, and similar commercial outposts in Tyre, Sidon, and Arwad that could have furnished diplomats and scholars for expeditions west.[116]

ANATOLIA

Before its demise at the close of the second millennium, the Hittite court at Hattuša (Boghazköy) developed rich parallels for the phenomenon of Near Eastern diplomatic and scholarly exchange with Mediterranean cultures. *Karum Hattuša* saw the archiving of dozens of cuneiform documents, augmented with hymns, Akkadian translations of Hittite historical texts, liver oracles, medical texts, wisdom literature, and even the Gilgamesh epic and other myths copied and housed in the palace archives.[117] The *Enūma Anu Enlil* is represented in a Hittite translation of its introduction, and there is also evidence of the *Prayer to the Gods of the Night*, used in haruspicy rituals.[118] Beckman's list of long-term Mesopotamian residents in Hatti includes scribes, priests, diplomats, and women sent to marry into the royal family (like Tawananna, third queen of Šuppiluliuma I).[119] Two "medical missions," including various *ašipu* ("exorcist/magician") and *ašu* ("physician"), sent

[114] See Koch-Westenholz 1995:44–51; Hunger and Pingree 1999:8–12; Rochberg-Halton 1988:30–35.

[115] See Burkert 1992:42–46; 1985.

[116] Dalley 1998:28.

[117] See Beckman 1983:101, fig. 2, for a chronological list of Mesopotamian literary material found at Boghazköy.

[118] Cf. Pongratz-Leisten 1999:22–23.

[119] Beckman 1983:108, fig. 4. Dynastic marriages transferred numbers of skilled and multilingual persons (some slaves) in the retinues and dowries of princesses shipped off to foreign courts; see Bryce 2003:chaps. 5–6.

from Babylon in the time of Muwatalli and Hattušili III, brought professional men who married local women and raised families in Hattuša; one effectively established a scribal school in the capital.[120]

BORDERS OF MESOPOTAMIA AND SYRIA

The *Enūma Anu Enlil* by the fourteenth century BC was studied in some parts of Syria, particularly the lunar eclipse passages arranged by date.[121] At Nuzi, tablets in Nuzi script, discussing earthquake omens, parallel *Enūma Anu Enlil*, tablets 22 and 43, and section 100 of the Babylonian *Iqqur īpuš* omen text. Although in local language, another omen betrays Mesopotamian origins by its use of Babylonian month names. In the kingdom of Elam, celestial divination texts in Akkadian and Elamite were used during the mid-second millennium BC; a tablet found at Susa parallels part of *Enūma Anu Enlil*, tablet 22.[122] A pre-Achaemenid text in Elamite is arranged by months, and on its back surface are inscribed parts of the *Iqqur īpuš*.

This phenomenon must have continued among the Iron Age successors to the Syrian kingdoms, although the archaeological situation in the Levant, and the Phoenicians' preference for papyrus have not been conducive to the preservation of texts. At Emar, a large group of thirteenth-century Akkadian texts, excavated in the library of Temple M1, contain about seventy-five divinatory documents, with about fifteen celestial omen texts. Among them is a fragment of the *Enūma Anu Enlil* in Akkadian. A fragmentary second-millennium Akkadian text with parallels to the *Enūma Anu Enlil* was found at Qatna; its spelling shows that it was copied by local scribes, who themselves spoke a West Semitic language.

LEVANT, UGARIT

At Ugarit on the north Syrian coast, several divinatory texts in Ugaritic script and language were in circulation when the city was destroyed (ca. 1190 BC). In the view of Paolo Xella, they result from a deft fusion of Babylonian traditions with West Semitic ritual practices, particularly the model livers, but there are also celestial and birth omens.[123] The vigor with which Ugaritic scholars embraced and recast Mesopotamian tradition may be a parallel for the later Etruscan integration of information from Babylonian texts – they were not copied wholesale but are very thoroughly considered and reworked. As Xella notes, this was an active process with adaptations and new creations appearing

[120] Beckman 1983:107–8; Rochberg 2004:93–96. For vassals asking to borrow skilled professionals, see Bryce 2003:121–28.

[121] Koch-Westenholz 1995:50.

[122] Rochberg-Halton 1988:31.

[123] Xella 1982:328.

at the same time.[124] An Ugaritic version of the teratological series, the *Šumma izbu*, was found with twenty-one model sheep livers, four inscribed with interpretational texts.

Neo-Assyrian, first-millennium contexts have also been identified.[125] In addition to Boghazköy, Mari, Emar, Mumbaqat, and Hamrin, Bronze Age liver models were found at Hazor, Megiddo, Ebla, Tell B'ia (on the Syrian Euphrates), and Ugarit. (Liver omen tablets have also been identified at Susa.) Meyer also cites an uninscribed terracotta liver model with Ugaritic affinities found in Cyprus, in the settlement of twelfth- to eleventh-century Enkomi.[126] They place the literature of Mesopotamian divination inside the palaces of many cultures and demonstrate that standardized Old Babylonian writings were readily translated into the local languages.

POSSIBLE SOURCES: CANAANITE AND PHOENICIAN DIVINATION LITERATURE

Sporadic similarities are evident between the Etruscan *Brontoscopic* text and divination texts from the Late Bronze Age Levant. Whether this is an epigraphic coincidence or a token of common descent from genuine, older Mesopotamian cultic literature, or whether the cults in both places simply experienced parallel evolution, is not clear.[127]

The only surviving Ugaritic astrological omen is the highly fragmentary KTU 1.163, which has been compared to the *Enūma Anu Enlil*. A section is quoted here:

> 1. [If ...], the cattle will die.
> 2. [If ...] on the new moon, the people [...] will be slaughtered.
> 5–6. If for three times it is seen month after month, [(there will be) storm clou]ds, rain will fall.
> 12–13. If the moon in its ascension becomes red, he will lack good influence.
> 14–15. If the moon in its ascension becomes yellow, [...] the cattle will die.[128]

124 Xella 1982:323.

125 At Assur and elsewhere; see Meyer 1987:11 (chronological list); Rochberg-Halton 1988:30–35.

126 Meyer 1987:273. For Enkomi, see Caubet and Courtois 1986.

127 Del Olmo Lete warns of semantic and cultural differences in these documents; see his 1984:7–16, for methods of dealing with fragmentary cultic texts: because the brontoscopic text exists only in the Byzantine version, it is not appropriate here to attempt to analyze in detail its syntax or vocabulary.

128 After the translation of Del Olmo Lete 2004:352.

In other Canaanite texts, some of the predictions divined by haruspicy sound just like apodoses for Mesopotamian series – and also resemble Etruscan predictions. In documents on teratomancy, various defects of human or animal fetuses are linked to outcomes such as:

many people will fall in the land
the young of his small cattle will be weak
there will be hunger in the land
the land will be destroyed
the king will take the land of his enemy
the gods will forsake the land[129]

Likewise, Canaanite and Phoenician documents name species of animals and types of food or materials to be sacrificed to the gods or the dead on certain dates, such as cattle, birds, fish, and such also named in omens.[130] Del Olmo Lete notes that Canaanite ritual texts (especially KTU 1.87) indicate the midpoint of a lunar month for ritual purification of the king[131] – and several months in the *Brontoscopic Calendar* do seem to speak (negatively) of a ruler on either the 13th or 14th (of June, July, Oct., Jan.).

MESOPOTAMIAN INFLUENCE IN MEDITERRANEAN CULTURES: BY WAY OF THE LEVANT

There was extensive interaction between the Near East and Greek settlements/rulers of the Late Bronze Age Aegean, the most visible evidence are the finds of exotic imports in Mycenaean tombs. Geometric-period interaction perhaps affected more levels of society than before, including the transmission of Near Eastern myth and ritual.[132] Sarah Morris outlines a plausible scenario of Late Bronze Age interplay, with the Levant and Cyprus as intermediate stages in the westward dissemination of cultural influences.[133]

The Uluburun ship with its royal cargo was possibly attended by Aegean diplomats (recognized by their shaving equipment); it also carried persons, perhaps mercenaries, from farther afield – one owned a sword made in the Italian archipelago.[134] In technological development, the famous cargoes of metal were accompanied by proper tools for processing it according to

[129] Quoted after Del Olmo Lete 2004:353–59 (KTU 1.03+1.45, 1.140).
[130] Del Olmo Lete 2004:40–42.
[131] Del Olmo Lete 2004:125–26.
[132] West 1997:586–624.
[133] Morris 1992:104–5, and chap. 5.
[134] Pulak 2001:45–47.

the latest standards of Levantine and Cypriot industry.[135] Skilled craftsmen undoubtedly traveled with such cargoes.[136]

For the Late Bronze Age, Canaanite Ugarit appears as a paradigm of the international mixing pot. Tablets in aristocratic private houses and palace archives show that records and literature were kept in both vernacular (Ugaritic) and international languages (Akkadian and Hurrian). Imported Aegean goods and locally produced imitations also attest the high value placed on foreign cultures. Owners of the cultic texts were also equipped to implement the rituals they recorded: the private library of the priest Agaptari at Ugarit, for instance, stored utensils for divination.

Nineveh remains very far removed from Tarquinia, however, but some attempt to establish commercial (i.e., diplomatic) relations with the West might not have been beyond the dreams of such rulers as Esarhaddon, who maintained control of Cyprus and Phoenicia and took Phoenician ships into the Mediterranean to hunt sea monsters. The scholars who have recently explored the Near Eastern stimuli to early Greek high culture have looked to the commercial kingdoms of the Levant as possible intermediaries in the exchange of information, and this may be the most likely scenario for Etruria also.

As a probable venue for such exchange, the settlement of Pithekoussai in the Bay of Naples, flourishing by around 775 BC, housed many multilingual families, with children learning to apply one parent's Semitic alphabet to the other's Greek or Italian language. West makes the case for Phoenician as a sort of first-millennium *lingua franca* in circulation across the Near East, as we know from epigraphic and other evidence from Lydia (Sardis), Cilicia, and Cappadocia. The Iron Age Phoenician colonies scattered across the shores of the central and western Mediterranean, Atlantic Iberia, and North Africa by the late eighth century BC would all have been bilingual (at least).

Such seafaring cultures undoubtedly had methods for coping with alien trading partners. The account of Hanno's voyage along Atlantic Africa around 500 BC makes reference to his Carthaginian colonists taking on some Libyans at their old colony of Lixus who could interpret for them as they ventured down the African coast.[137] Slavery, too, was conducive to multilingualism, and the Homeric epics describe Sicilian or Phoenician women kept as nursemaids in Greek households.

[135] Cypriot tools have been identified in Bronze Age Sardinian sites, and the Gelidonya wreck carrying both ingots and scrap metal also held tools for metalworking. See Vagnetti 2000:313–17; Kassianidou 2001; Lo Schiavo 2001; Macnamara 2001; Bass 1967:102.

[136] Compare the Telchines of Strabo 10.3.7; 14.2.7; Pausanias 9.19.1; see Morris 1992:105.

[137] *Periplus* 35; Lipinski 2004:439, 457.

LANGUAGE, FROM BRONZE AGE TO IRON AGE

By the time of the Neo-Assyrian empire (9th–7th c.), Aramaic was supplanting Akkadian as the international language – and it was at this time that Greeks were being taught to write by Phoenicians. "Parchment scribes" who wrote in Aramaic used these perishable materials for the faster note-taking their language permitted, the reason why they were appreciated in the Assyrian state. They would also have been at least bilingual – and by the seventh century, most of the literate population of Assyria must have been colloquially bilingual in Aramaic and Akkadian.[138] Scribes inked Aramaic colophons and notes on tablets in the Assyrian palaces, where royal reliefs of foreign conquests depicted them, in the company of cuneiform (Akkadian) scribes, inventorying the loot pillaged under Tiglath-Pileser III (744–727 BC) and Sennacherib.

It seems likely that the transmission of the information reprised in the *Brontoscopic Calendar* came through official channels, because the divination cult was so closely linked to the palace. One more means of transmission is suggested by the mechanisms used to consolidate conquests of the East: one Assyrian tactic for subduing a land taken by the sword was deportation. We know that the later, Persian control of Asia Minor, when met by the Greek revolt, triggered not only Persian deportations but also deliberate, voluntary evacuations of Greeks, in some cases to Greek-speaking colonies in the central and western Mediterranean.

An earlier period of forced migrations in the Near East and Levant could have sent a core group of elite or royal relations fleeing to Italy, perhaps, like the Corinthian royal Demaratus (ca. 650 BC), following trade routes that were already familiar to the sailors in their entourage. Assyrian expansion to the west began during the ninth century as Ashurnasirpal (883–859) and Shalmaneser III (858–824) made incursions into the Levant, collecting tribute from cities in Phoenicia and Cilicia.[139] Alternately, this was a likely time for a maritime diplomatic venture by the Assyrian state. Under Tiglath-Pileser III, Assyrian and Babylonian settlers were moved into Phoenicia. Assyria conquered Israel about 722, and by 700 BC Damascus and Cyprus had capitulated. Sidon fell to Assyria in 672, giving a foothold on the coast and access to the wider Mediterranean. Ashurbanipal replaced the Israelites deported by Sargon with people from Babylonia and Elam; similar exchanges occurred with the subjugation of Carchemish (717), Hamath (720), and Cilicia. As late as the seventh century, the eponymous king (680–669 BC) established Kar-Esarhaddon in the territory of conquered Sidon, so there would even have been permanent

[138] Pearce (1995B) notes the term *kuššaru* as possibly meaning "parchment scribes." For bilingualism, see Steiner 2009.

[139] See West 1997:614–16.

settlements in the Levant, outliers of the Assyrian government, as potential repositories of scholarly information to be taken west.

BACKGROUND: THE PHOENICIAN PRESENCE IN THE CENTRAL MEDITERRANEAN

The Phoenicians are the obvious candidates for the dissemination of Mesopotamian divination literature, cult practices, and more to the central and western Mediterranean, because their tracks are so easily identified in the trade goods and colonies scattered across the Mediterranean. For the crucial eighth century BC, Phoenician contact with Etruria could have fostered Etruscan adoption of Near Eastern state rituals. Much depends upon the social level at which Phoenicians interacted with Etruscans. It might take only a single, peripatetic prophet to spark a religious movement, but if the prophet(ess) was speaking Babylonian, Assyrian, or Aramaic, no one could have understood it, so a support group is a given.

There is some evidence for the movement of royal kinsmen across the Phoenician Mediterranean network. The ninth- and eighth-century Phoenician foundations in Iberia, from Atlantic Cadiz to the Andalusian commercial towns at Toscanos and Malaga, were full-fledged cities with all the governmental infrastructure one might find in the metropoleis.[140] Some of the tomb structures and funerary offerings at necropolis sites such as Almuñecar and Trayamar, appear to be of royal caliber.[141] Stone or wooden constructions, offerings, and sarcophagi in some cases are only paralleled in the royal necropoleis of Phoenicia, or (in different ethnic contexts) in the so-called princely tombs of the kingdom of Salamis in Cyprus. Several Andalusian burials held Egyptian alabaster jars marked with pharaonic cartouches: most were probably royal gifts to Phoenician kings who then perhaps regifted the vessels to relatives and/or officials, who carried them on their migration to the "new world" of Iberia. In the East, certainly, royal hosts expected to be visited by embassies of suitably flattering rank, and Easterners had every reason to believe this is what the chieftains of the West, like the king of Tartessos, would expect.

In the subsequent period, some important fifth-century BC residents of Cadiz were buried in anthropoid sarcophagi of a type known from the Sidonian royal necropolis. Simpler versions are found in the Phoenician homeland, namely, at Arwad; the stone sarcophagi are linked to royal families and also have been found in the Cannita necropolis of Phoenician Panormos/Palermo.[142] Surely a king would have dispatched as viceroys persons related

[140] See Ruiz Mata 2002; Niemeyer 2002.

[141] See Pellicer Catalán 2002.

[142] See Moscati 1999.

to the palace by blood, marriage, or social rank. And with such personages would have traveled an entourage of skilled diplomats, scribes, and priests.[143]

In the *Etruscan* coastal towns, evidence of resident aliens is sparse prior to about 580 BC, when foreigners, both Levantine and Greek, are attested in ports such as Pyrgi and Graviscae, where their settlements would have been foreign enclaves with their own sanctuaries and cults. However, at the Tarquinia Pian di Civita sanctuary, Phoenician residents in Italy participated in the local cults and, in addition to making valuable offerings, actually influenced the rituals and structures of native cults. The first examples of foreign influence are the rectilinear cult rooms (*edificio beta*, ca. 680 BC), with interior *penetralia* on the order of Levantine temples and built in what looks like Phoenician pier-and-panel masonry (*muri a pilastri*).[144] This masonry style is highly distinctive, recognized in the homeland (Sarepta), and a hallmark of the early colonies in Spain, North Africa, Sardinia, and Sicily.[145] Certainly Ciafaloni views the complex at Tarquinia Civita as firm evidence of the influx into Etruria not only of ideas from the Levant, but of a significant number of people, and not only of material goods, but of an ideological complex of far-reaching traditions.[146]

BACKGROUND: ETRUSCAN RELATIONS WITH NEAR EASTERN CULTURES

A persuasive case can be made for concerted interaction with the eastern Mediterranean cultures during the Late Bronze Age, when Sardinian bronze technology and copper production show Cypriot contact. Two major Aegean symbols, the figure-eight shield and the sun boat, appear together on native Italian objects, such as razors.[147] A number of ceramic types in Italy, such as tripod vases, *calefattoi*, and miniature cauldrons, have been traced to Aegean or Near Eastern prototypes, and Renato Peroni has indicated some Near Eastern imports in the bronze hoards of Piediluco and Contigliano.[148]

[143] On the Orientalizing phenomenon here, see Aubet Semmler 2002.

[144] See Bonghi Jovino 2010; Prayon 2001:335–40; Bonghi Jovino 1991.

[145] Ciafaloni 2006; Bonghi Jovino 1991; cf. Elayi 1980.

[146] Ciafaloni 2006:149–50. Some scholars even see the ritual sacrifice of *prodigia* in the (very fragmentary) burials, over the course of a century, of three or four infants in the Tarquinia Civita precinct (Fornaciari and Mallegni 1997), and associate them with the Phoenico-Punic practice of infant sacrifice. There are problems, however, in matching these burials to Punic infant sacrifice: the genuine ritual only used cremation. Further, Pian di Civita did not have a precinct like all known examples of the Punic "tophet" burial ground for sacrifices.

[147] Bettelli 2004. Colonna (1991) linked miniature shields in Latian cremation burials of Final Bronze and First Iron Age to the ceremonial *ancillae* of the Salii, the Roman priesthood founded by Numa Pompilius and graced with a shield that fell from the sky (Plutarch *Numa* 13).

[148] Peroni 1989:390–91.

ASSYRIA AND ADJACENT REALMS, NINTH TO SEVENTH CENTURIES BC

The reign of Ashurbanipal (668–631 BC) was characterized by peace, stability, and improved welfare for the populace. His reign is symbolized in the so-called garden relief from the palace at Nineveh, in which Ashurbanipal reclines on a couch drinking as the queen sits enthroned beside him.[149] These luxury traditions were rapidly embraced by other cultures – the reclining banquet with all its accoutrements (couches, gold, silver, or bronze utensils, wine, servants, fans) appears overnight in Greece and Italy, particularly Etruria and Latium. Easterners had much to offer in desirable goods and ideas; the upheavals in the Near East sent many skilled craftsmen abroad and also caused many officials to seek stable foreign contacts.

GOODS AND INFLUENCES

Seals (scarabs) circulated widely; at Pithekoussai nearly 100 seals were buried in eighth-century tombs, especially with infants.[150] Eighth-century Phoenician metal bowls were offered in the sanctuaries on Cyprus, and in Athens, Olympia, and Delphi, and buried with Etruscan and Italic rulers during the seventh century (Pontecagnano, Praeneste, Falerii, Caere, and other Etruscan sites). The bowl at Falerii bears a cuneiform inscription, and one each at Olympia, Pontecagnano, and Praeneste is inscribed in Phoenician/Aramaic, but perhaps this was performed before their disposition abroad.

Not only the goods but the customs in which they were used must have been exchanged, and this implies intensive interaction, that is, cohabitation. The use of Near Eastern incense was instituted during the Early Orientalizing period, as illustrated in a growing number of finds in Etruscan tombs.[151] The discovery at Protovillanovan Sorgenti della Nova of donkey bones (in a hut and a ritual deposit) shows the introduction into Italy, by the end of the second millennium, of this species, perhaps via the Levant, and ultimately from Egypt where it had been domesticated.[152]

The princely goods deposited in Etruscan tombs of the mid-seventh century must have been formally presented to the rulers of Tarquinia, Caere, and others.[153] Fluted bronze (or gold or silver) bowls surely came to Etruria tagged with the aura of royalty – in the Assyrian reliefs, for instance, as with chariots, parasols, thrones, and footstools, fluted or hemispherical bowls are the

[149] Illustrated by Curtis and Reade 1995:122.

[150] Burkert 1992:15, with references; Ridgway 1992:51–52, 65–67.

[151] Casale Marittimo, 7th-c.; Esposito 1999:90–92; Burkert 1992:20.

[152] See de Grossi Mazzorin 1998:173–74.

[153] For most recent bibliography, see Fletcher 2007; for range of goods and phenomena, see Naso 2000C.

prerogative of the king and his family. In Etruria they must have broadcast the owner's lofty (albeit relatively recently established) status within the city, like the senator's ivory stool or triumphator's purple robe. Because neither ivory nor purple dye is indigenous to first-millennium Italy (but always were hallmarks of Phoenician material culture), these symbols must have been part of the "package" of urban statecraft that arrived from the Near East during the eighth century BC. (In the course of the seventh century, additional emphasis of Near Eastern concepts was instilled through monumental sculpture.)

The mobility of skilled persons (and their families) in the Near East was legendary; in the Assyrian/Aramaic "novel" *Ahiqar*, the king of Egypt sends for an architect from Nineveh.[154] Hiram of Tyre sent Phoenician artisans to Jerusalem for the construction of Solomon's Temple. Resident foreign artisans were the source of the Orientalizing art developed in Etruria, for when it first appears, it is always bound with new technologies, for instance, wheel-made pottery, glass, goldsmithing techniques, and armor types.[155] The last imply an early interaction with Near Eastern "military advisers."[156] Maurizio Sannibale (2008, 2006) has shown that such goldwork as the pectoral of Larthia from the Regolini-Galassi Tomb (Caere) drew inspiration ultimately from Egyptian concepts of the universe and the gods. Such intellectual property implies a long and intimate interaction with people versed not only in the technology but in Eastern ideology.

THE NEAR EASTERN ORIGINS OF ETRUSCAN REGALIA AND KINGSHIP PATTERNS

Luciana Aigner-Foresti (2000) has succinctly outlined the rich visual evidence for a Near Eastern version of kingly symbols, implements, and actual political patterns having been adopted during the eighth century in Etruria. They are most obvious to us in the extravagant status symbols found in the princely tombs of the early and mid-seventh century, but the origins of both statecraft and its regalia have their roots as early as the revolution in settlement begun in the tenth century. She offers a long list of items, including the double ax bound with rods (*fasces*), traced by Roman scholars to Vetulonia, where by chance a fine example, in iron, has been identified in the "Tomb of the Lictor."

In the Tomba delle Statue at Ceri, a stone-cut chamber tomb of circa 660 BC, thrones were integral to the monumental sculpture. One of the confronted, bearded, robed men holds a scepter ornamented with a Phoenician

[154] See Burkert 1992:22–24.

[155] Cf. Canciani 1981:54, fig. 1; Martelli et al. 1987:66–80, 242–54, nos. 1–25; Martelli 1994; Turfa 1986:67n16; Canciani 1981:55, fig. 2.

[156] See Stary 1979:181, 190; 1980B; and 1981:passim.

palmette, which, like the style of carving, and distinctive furniture, betrays the Levantine sources of this new artistic technology.[157] In a painstakingly documented article, Jodi Magness (2001) enumerates several elements of Etruscan material culture, most convincingly funerary architecture, furniture forms, techniques of divination, and hydraulic engineering, with many correspondences to traditions in the Near East. Certain regal furniture found in princely chamber tombs, especially the roll-footstool,[158] is so closely paralleled in tomb furnishings of Palestine and the Archaic Phoenician world that its prototype must once have been an actual footstool presented to an Etruscan ruler. Multiple examples show that many families in several cities had close associations with Eastern people, causing them to adopt new traditions from the Near East. As Aigner-Foresti notes, however, the throne used with this Levantine symbolic stool is the native Etruscan cylindrical throne.[159] I suggest that the comfortable mixture of oriental and native furniture implies a long process of borrowing and development of Near Eastern customs.

The cross-cultural aspects of using imported vessels in native ceremonies to express political or social dominance have been analyzed by Ferdinando Sciacca with respect to Assyrian exchange with Etruria. He maintains that ever since the expansion begun by Tiglath-Pileser III, the Assyrian court, via the efforts of Phoenician merchants, stimulated the Orientalizing phenomenon throughout the central and western Mediterranean.[160] Pertinent to this argument is the ribbed or fluted drinking bowl held by Ashurbanipal's queen on the garden relief in his North Palace at Nineveh, in which he celebrates consolidation of his power.[161] In Assyria and environs, ribbed metal bowls were part of royal tradition since the mid-ninth century, restricted to the hands of the king and his consorts and used specifically for libations and purification ceremonies. By the 730s BC, the vessels appeared in central Italy, in tombs at Latin Castel di Decima, Osteria dell'Osa, Etruscan Veii, Vetulonia (Circolo del Tridente), Pontecagnano, and Faliscan Narce.[162]

During the seventh century, more than 300 ribbed bowls were in central Italy, some imports, but many certainly local imitations. The imitation response is just as important as the presence of proven imports, for it demonstrates the profound effect that contact with Easterners had on the societies of Etruria, Latium, and Campania, causing native industries to spring up to produce new types of goods for new modes of behavior. Assyrian imports include bronze cauldrons and the wonderful lion-head *situla* found at Veii in Casale

157 In addition to Aigner-Foresti 2000, see Prayon 1977 and Colonna and von Hase 1986.

158 See Jurgeit 2000.

159 Aigner-Foresti 2000:276–77; cf. MacIntosh 1974.

160 Sciacca 2005:782.

161 Delpino 2000:194.

162 Sciacca 2005:782–83; for Narce, woman's tomb, see Turfa 2005:16–20, 109–32; for ribbed bowl, see Dohan 1942:37, no. 10, pl. 18.

del Fosso Tomb 871 – buried in the last years of the eighth century, it has fine parallels in the great tomb of Tumulus MM at Phrygian Gordion and representations of the era of Sargon II (722–705 BC; see Fig. 16).[163] Animal-head *situlae* were symbols of Assyrian ritual banquets and religious ceremonies, exchanged as presentation pieces between the Assyrian kings and foreign vassals or allies. This is assumed to be the context of the vessel in the royal tomb at Gordion, so why not a similar scenario for the one buried with a Veientine prince?

RECEPTION OF FOREIGN INFLUENCES AND EARLY LITERACY IN ITALY

The great reliance upon texts throughout the Near East is mirrored in the Etruscan veneration for the documents of its *disciplina*. In Mesopotamia, the tradition of revelation by a god, or some semidivine personage like the Apkallu sages, imbued cultic literature with divine authority, and it was renewed as fresh texts were disseminated, as seen in the cuneiform *Catalogue of Texts and Authors* known from the library of Ashurbanipal.[164] In fact, the notion of inscribing a document with the mark of its authority begins with the Mesopotamian tablets. The attribution of the Etruscan *Brontoscopic Calendar* to a divinely inspired Tages also sounds very much like the Mesopotamian literary milieu with its attributions to the antediluvian sages.

Some of the very basics of Western literacy have been traced to Near Eastern roots: written contracts and the practice of "subscription," inscribing the author's name and title of a work at the end of the book (as seen in the Capua tile and Zagreb *Liber linteus* calendars).[165] Although no cuneiform tablets have yet been found in Italy, a bronze bowl from a mid-seventh-century tomb at Falerii does bear a Neo-Babylonian official's proprietary cuneiform inscription.[166] Several objects from princely tombs had been inscribed in Phoenician or Aramaic.[167] The Western extent of cuneiform clay tablets to date reaches only to Tarsus and Cyprus, but they are texts on hepatoscopy (liver divination).

PITHEKOUSSAI AS A LITERATE MELTING POT

For exchange of customs and traditions, people must live and work side by side, and the multiethnic community of Pithekoussai is a good paradigm for such a place. It also shows a high degree of literacy and respect for "literature,"

[163] Marchesi 2000; Ebbinghaus 2008; Sciacca 2003.
[164] Lambert 1962:72–73.
[165] For background, see Sassatelli 2000.
[166] "Belonging to Nabû-iddin, son of Banya, the qīpu"; Cristofani and Fronzaroli 1971.
[167] Pontecagnano and Praeneste: Markoe 1985:72; Magness 2001:83.

even in its first generation.[168] In fact, the excavators postulated eighth-century written documents as the legal basis of this settlement, suggesting that the presence of Corinthians in addition to Euboeans may have been mandated by treaties that guaranteed passage over the Isthmus and use of Corinthian trade routes to the West – and the Levantine contingent among the colonists, with a background of Near Eastern-style diplomacy, would certainly have maintained documentation of land and trade permits, and probably many more contracts.[169]

Pithekoussai by mid-eighth century shows skilled commoner-citizens of Levantine origin on the same footing as Greeks and Italians; funerary ritual and some possible offerings from the acropolis reflect the activity of Phoenician cults.[170] An egalitarian haven for underappreciated immigrants with commercial skills from Greece and the Levant, Pithekoussai was also the home of many Italians, whether of Etruscan or Italic ethnicity. These eighth-century citizens are recognized by their Etruscan fibulae, of types indicating male and female costume, and by native metalwork and pottery.[171]

CONCLUSION

A review of the evidence on Mesopotamian literature and archives, and the dissemination of Mesopotamian information to the wider world, supports the notion that some of its lore could and did reach Etruria, at the highest levels of Etruscan society. Major Mesopotamian works that contain predictions like those found in the *Brontoscopic Calendar* were codified by 1000 BC. In Sargonid Assyria, copies of all such works were archived in the palace and with scholars dispersed across the kingdom.

Hepatoscopy, the examination of the entrails, especially livers, of sacrificial victims, is a field worthy of further scrutiny in Etruscan religion, because of distinct parallels with Mesopotamian practices and texts, including inscribed models of sheep livers. Stefan Maul has noted, "Around the turn from the second to the first millennium BC, the written sources about sacrificial divination underwent a process of large-scale systematization, most likely in response to the increasingly powerful role of Mesopotamian kings, who demanded a comprehensive and reliable system of divination."[172] A compilation of sacrificial omens was codified in a series of 100 tablets known to us as the *iškar baruti*, preserved in the library of Ashurbanipal, who actually inscribed a set

[168] See Ridgway 1992:94, fig. 26, 55–57. Burkert (2004:20) surmised that whoever inscribed the cup alluding, in verse, to Nestor of the *Iliad*, must have seen that epic, or similar poetry, in hard copy, because he or she formatted the inscription with a new line for each verse.

[169] See Ridgway 1992:107–20.

[170] Ridgway 1992:111–14, 117–18.

[171] Ridgway 1992:65, 121–44; Colonna 1995.

[172] Maul 2009:369.

himself.[173] The omens focus on the king, royal family, and state, especially disasters and epidemics, harvests, and wars, and were not for popular consumption, but the people would have been aware that such archives were in the rulers' custody and were continually consulted.

Mesopotamian celestial divination was invariably bound up with the monarch and the state. A major emphasis on such pursuits seems to have developed under the Sargonid dynasty, especially its later kings (Esarhaddon, Ashurbanipal) – just the period when Iron Age Etruria was entertaining an influx of foreign traders and diplomats from the East. The earliest surviving version of thunder-omens in the tradition of the Etruscan *Brontoscopic Calendar*, namely, the Qumran papyrus (4Q318), offers predictions for the whole state or community, with the same viewpoint as the Figulus document – but this is a different perception than the subsequent *brontoscopia*, which often include Hellenistic material akin to the personal horoscopes of the Achaemenid, Ptolemaic, and Seleucid archives. No single-protasis *Brontoscopic Calendar* per se has yet been identified among the thousands of cuneiform texts: their ample thunder-omens were cataloged under celestial gods, weather, and such. The notion of creating a document linked solely to thunder seems to have originated with the author(s) of the text that was translated by Figulus, and its omens all show close correlation with the environment and society of Iron Age Etruria/Italy.

I suggest that the *Brontoscopic Calendar* was in the hands of families who held power in the burgeoning Etruscan "states," and it was used as a priestly reference or training tool. But the usefulness of the text depended upon the rest of the population knowing that they held it (and assuming that their ancestors had received it through divine intervention). Citizens must have been familiar enough with its lore (or its perceived accuracy) to respect it. Spawned in a time of extra storms and stormy social adjustments, it probably seemed especially inspired to its first users, and thus it would have carried that cachet among later generations. Quite possibly, the haruspices called to Rome in 56 BC consulted it to identify the noise heard in the skies of Latium.

The concepts in the calendar are mostly simple and straightforward: the person who formed the interface between Mesopotamian religion and Etruscan believers would have been able to refer to the recipients' familiarity with the agricultural year – and with basics of livestock, crops, weather, and human existence. It would be easy to find correlates for such experiences in another culture, and it is evident that the Etruscan document is free of most of the complexities of urban life in an ancient, intensely stratified Near Eastern kingdom, even though, by the end of the eighth century, the major settlements of Etruria (like Caere, Tarquinii, Vulci, Veii) were already large, complex, and in the process of urbanization.

[173] Koch-Westenholz 2000:27–31; Jeyes 1997.

Could this hold the key to the Etruscan development of such forms of divination as the *Brontoscopic Calendar*? The timing is reasonable: although we think of Ashurbanipal as the first seriously educated, scholarly prince (certainly the first, and only, able to write in cuneiform), his predecessor Esarhaddon (680–669 BC) clearly supported scholarly endeavors. And the reigns of *his* predecessors, Sargon II (722–705) and Sennacherib (704–681), fall just when Etruria began to urbanize under strong cultural influence from the eastern Mediterranean and Levant, when Assyria made a bid for the Levantine Mediterranean coast, and when Near Eastern luxury gifts appear in the Mediterranean.

CHAPTER 9

OTHER *BRONTOSCOPIA* IN THE CLASSICAL TRADITION

INTRODUCTION

Classical antiquity knew other documents of brontoscopy besides the Mesopotamian texts and the Etruscan *Brontoscopic Calendar*. The Assyrian, Ugaritic, and Babylonian texts preceded Figulus' document by many centuries, and all the other extant brontologies may be said to derive more or less from the Near Eastern literature, though by a more roundabout route than the Etruscan *Brontoscopic Calendar*. They arose in the tradition of esotericism that began with the Persian conquest of the Near East – this political development brought with it a radical shift in ideology under a different regime. Combined with the Persians' uniting of a vast area that comprised scores of different ethnic groups, their more laissez-faire attitude permitted the burgeoning of horoscopy and astrology along with scientific, observational, and mathematical astronomy. The stage was set during the period of Achaemenid domination, and subsequently the economic and social conditions fostered in the Hellenistic East and Ptolemaic kingdom would bring all these schools of thought to fruition.

THE ROLE OF MESOPOTAMIAN DIVINATION IN THE HELLENISTIC, ROMAN, AND BYZANTINE *BRONTOLOGIA*

As was the case for the astrological literature of the first centuries BC to AD, the ultimate sources of the Hellenistic and Roman, and also Byzantine, brontoscopic texts are the same Mesopotamian divination series outlined for the *Brontoscopic Calendar*. They continued to be processed and copied by scholars under Persian, Seleucid, and Ptolemaic institutions. Pingree noted the existence of an Egyptian Demotic papyrus (2nd–3rd c. AD) that preserves eclipse omens originally written in the Achaemenid period and extracted from the *Enūma Anu Enlil*, but these were thoroughly reworked to make them applicable for Egypt.[1] Another Ptolemaic set of omens was recorded in Greek, based on parts of the Sin, Shamash, and Adad series of the *Enūma Anu Enlil*, but again this was not a mere translation of the Akkadian texts. Pingree discusses the time of its likely creation, either Achaemenid or Hellenistic, but not earlier; a different, mid-second-century BC Greek treatise attributed to "Petosiris" is the oldest example of the tradition.[2] Its astral omens appear to have been extensively altered from the Mesopotamian tablet traditions, probably via Greek versions of the second century BC. Clearly there were a number of scholars writing in this tradition, few of whom are represented in any surviving texts.

The Roman second-century AD author known as "Claudius" Ptolemy included some related omens in such works as his *Apotelesmatika* (*Effects of Astral Phenomena*) 2.9.18; 2.14. The same body of information was replicated in Byzantine works, including an *Apotelesmatika* written by Hephaistion of Thebes (ca. 415 AD, see below), and also in Lydus' Περί σημαίων (*On Signs*, composed ca. 550 AD). As with the later *brontologia* cited below, the protases of these omens are lunar and solar eclipses, comets, winds, rain, thunder, and lightning. Pingree notes that the information on comets is Greek and much more detailed than any found in Mesopotamian omen literature.[3] Over the next centuries an immense amount of literature was derived from Ptolemaic Greek sources and disseminated in Greek, Latin, Aramaic, Syriac, Hebrew, Arabic, and Persian. It appears that Ptolemaic omen literature spawned scores of variant texts, most claiming illustrious antecedents; yet none credited the genuine Assyrian or Babylonian tablet series from which they actually (by a long process) derived. As Pingree cautions, "Much remains to be done to clarify the interrelationships of all these texts, which are written in virtually every language of Europe and the Near East."[4]

As for the Persian-Seleucid literature whence these Late Hellenistic, Roman, and Byzantine documents finally evolved, Koch-Westenholz has summarized:

> The development is still not too well understood since many sources are in poor condition and many remain unpublished. New types of omen texts were generated by combining old and new ideas or rearranging old material. They existed side by side with the traditional texts of EAE [*Enūma Anu Enlil*]

[1] Pingree 1998:132. In the Late Achaemenid period (mid-4th c. BC), Egyptian astronomers developed their own twenty-five-year intercalation cycle, again influenced by, but not directly copying, the Mesopotamian calendar cycle.

[2] Pingree 1998:132–34; see Pingree 1974.

[3] Pingree 1998:133.

[4] Pingree 1998:134.

which continued to be copied throughout the closing centuries of the first millennium.[5]

Many of the scholars responsible worked in the cities of Uruk, Babylon, and Sippar. A few family archives excavated at Uruk, such as those of the Šangu-Ninurta family (6th–5th c. BC), the Hunzû family (3rd c.), and the Sîn-lēqi-unninni family (Seleucid period), illustrate the character of private scholarly collections.[6] Along with lamentations, school exercises, and medical works, all owned some religious texts, including parts of the *Šumma ālu* and *Šumma izbu*, the *Enūma Anu Enlil*, and hepatoscopic *Bārûtu* tablets. The later brontological texts *could* easily have begun as treatises created in such environments as the homes or temples where such Seleucid-era scholars studied.

The older Mesopotamian system of celestial observation used a series of seventeen constellations in the "Path of the Moon," but in the late fifth century BC, an Egyptian-influenced zodiac of twelve signs (essentially our own zodiac) became common. This is a telltale for interpretation and dating of the various brontological manuscripts, because most of them use the zodiac as a calendar by which to identify the thunder-omens – instead of listing each day of a month, they cite the zodiacal sign, thus a period of several days. Apart from the Etruscan *Brontoscopic Calendar*, those other documents that eschew the zodiac very clearly betray a later date in their reliance on the Julian calendar; in many cases, they even date by the Roman Kalends, Nones, and Ides (the 1st, 9th, 13th (or 15th) day of a month).

Further, Koch-Westenholz pointed out that even the names used for the signs of the zodiac in the Persian- and Seleucid-era texts betray their late reception: different cultures see different images in the stars.[7] The common zodiacal symbols used in the manuscripts always depict the exact Greco-Roman name for a constellation; for instance, Aries is symbolized by a ram, but the Akkadian/Babylonian name for the constellation was *lú-hun-gá*, "the Hireling."

ARAMAIC: PAPYRUS FRAGMENTS QUMRAN 4Q318

Interest in Mesopotamian-style divination and literature continued in Palestine, in some (to us) surprising venues, as attested by the finds in Qumran Cave 4, conventionally dated to the first century AD. Scholars assume that this earliest of the extant *brontologia* was deposited there during the first century AD, probably at the time of the Jewish Revolt. It gives the signs of the zodiac for the days of each month and then predictions based on thunder-omens for

[5] Koch-Westenholz 1995:162.
[6] Robson 2008:227–37, 240–42, 244–53.
[7] Koch-Westenholz 1995:164.

that day.[8] Fragments of three columns preserve a list in Aramaic with entries beginning with a month and then zodiacal symbols as dates, with apodoses completing the lines. After the lines of zodiacal dates, column VIII contains four lines with thunder-omens.

The dating mode, using Hellenistic/Ptolemaic zodiacal signs rather than days of the month from some civil or cultic calendar, is part of the Hellenistic, post-Babylonian tradition in divination. The predictions as preserved, however, seem to relate to the state, ruler, or community as a whole and are not the personal material of horoscopes, such as were popular by that time.[9]

The best preserved section is column VIII, lines 6–9:[10]

> [If in Taurus] it thunders (there will be) *msbt* [*sic*] against [= [and] affliction for the province, and a sword [in the cou]rt of the king, and in the province, [ ] will be. And to the Arabs [ ], hunger, and they will plunder each oth[er vac]at If in Gemini it thunders, (there will be) fear and sickness from the foreigners and m[

Geller emphasizes the Akkadian predecessors of this text, noting that "the omina resulting from thunder predict 'toil in the city' (*'ml bmdynt'*), destruction in the royal court (*hr[b bd]rt(?) mlk'*), hunger (*kpn*), and 'one plundering another' (*lhwwn bzzyn 'ln b'[ln]*)."[11] Because of the zodiacal tags, Geller makes the association with the MUL.APIN, which offers some similar apodoses, such as plague and winds. The zodiacal signs of the Qumran Aramaic document match those of the Akkadian MUL.APIN, as does the starting point of Taurus (mid-May – Greek zodiacs would have begun at Aries). Geller reminds us that "as in the Aramaic '*brontologion*,' the omina in MUL.APIN refer to phenomena affecting the entire land, rather than individuals."[12]

Wise has indicated a vast reservoir of astrological, hemerological, and brontoscopic literature that seems to have been consulted in Palestine during the Second Temple period, even by the Jewish faithful, in spite of its nonbiblical outlook.[13] Only the fortunate preservation of papyri in the Qumran caves allowed scholars to acknowledge this literature. Under the Byzantine rule of Palestine, texts continued to be generated in the fields of *brontologia* and *selindromia* (distinguishing lunar phases). These latter texts, in Greek, give the protasis, "If the moon," followed by apodoses that sound alarmingly close to the, by then, ancient language of the Babylonian texts of the

[8] Greenfield and Sokoloff 1989:202n12; translation in Greenfield and Sokoloff 2005.
[9] Cf. Geller 1998:224–27.
[10] From Greenfield and Sokoloff 2005; also transcribed and translated by Wise 1994:17–20.
[11] Geller 1998:224–27.
[12] Geller 1998:226.
[13] Wise 1994:13–15.

first millennium BC.[14] Indeed, Bezold and Boll (writing before the Qumran discovery) identified a long list of predictions and expressions in the Greek texts that are closely related to apodoses or astronomical schemes employed in the *Enūma Anu Enlil*, such as the downfall of men in power, miscarriages of women, hostilities with enemy states, attacks by dangerous animals, plenty or famine, and so on.[15] Table 9.1 indicates main known or extant brontological and related works.

ROMAN AUTHORS, CONTEMPORARY WITH FIGULUS AND/OR EMBEDDED IN LYDUS' *DE OSTENTIS*

The text of Lydus' *De ostentis* names or quotes several Roman authors whose works touched upon brontoscopy, and others were cited by authors ranging from Cicero and Pliny to Lactantius and Ammianus Marcellinus. Roman scholars include the Late Republican Caecina, Tarquitius Priscus, and Fonteius (Capito?), the possibly Augustan Claudius Tuscus, and the Late Imperial (3rd or 4th c. AD) Labeo and "Vigellius." Mingled with the lore of thunder are other types of omens, especially earthquakes and lightning; all the extant examples show the influence of zodiacal astrology that does not appear in the Figulus document. Several of these Roman authors are represented among the antiquarian treatises embedded, in Greek translation, in Lydus' *De ostentis*.

A table of contents for *De ostentis* would be:

1–8. Introduction by Lydus, on origin and development of divination among the Romans; the myth of Tages (2–3)

9–10. On solar phenomena

11–16. Excerpts from "Campestrio" on comets[16]

17–20. On phenomena of the moon, through twelve signs, including ancient treatises

21–26. On thunder, including Egyptian teachings on thunder

27–38. *Brontoscopic Calendar* according to P. Nigidius Figulus, translated from Tagetic writings

39–41. Observations on thunder drawn from writings of Fonteius

42. General remarks on omens according to phases of the moon, translated from the books of Labeo

43–46. A commentary on the fulgural disciplina of the ancients, based on Latin authors

47–52. Book by Labeo on lightning strikes

[14] See Greenfield and Sokoloff 1989:213; cf. Bezold and Boll 1911:25–37.

[15] Bezold and Boll 1911:14–24.

[16] Campester was a 3rd–4th-c. AD Roman author; see Dominici and Maderna 2007:63–88, 146n73.

TABLE 9.1. *Other* brontologia *and related texts in the classical tradition*

Item	F	T	Length	Author	Begins/ Format	Qualifiers	Other/Apodosis topics	Cf. Etruscan?
P. Nigidius Figulus 1st c. BC	Y	Y	Long	"Tages"	Mesop. calendar June 1 omen per day	None *Brontoscopic Calendar*	Topics Mesopotamian and Etruscan	Etruscan
Tarquitius Priscus Late Republican			- -	Translated Etruscan texts	- -	No works preserved (wrote an *Ostentarium arborarium*)	- -	- -
Fonteius Capito* 1st c. BC		Y	Long	Fonteius	Lunar/ zodiac Capricorn	Day/night	Hellenistic-Roman regions Nile, Persians, Asia, Europe	Some topics same
Claudius Tuscus Augustan			Long prose	Claudius Tuscus	Roman Day system January	Latin poem: Stars, winds, weather	"from the sacred writings of the Etruscans, word for word" (πρὸς λέξιν)	Different topics
Labeo 3rd c. AD			Very long prose	Labeo	Lunar/ zodiac (Aries)	Lightning, other phenomena	Mentions M. Cato, Apuleius and *etrusca disciplina* Hellenistic-Roman regions	Clearly later than Figulus
Vigellius Imperial			Long	"Tages"	Solar/ zodiac "sun in Aries"	EARTH-QUAKES – not a *brontologion*	"Following verses of Tages"– notes Apuleius Hellenistic-Roman regions	Different material
Hephaistion of Thebes *Apoteles-matika*		Y	Long prose	"Nechepso" "Petosiris" (7th c. BC & later)	Lunar/ zodiac/ celestial/ weather (Aries)	Time of day, many qualifiers	Mentions Hellenistic peoples	Many same topics
Geoponica 1.10*		Y	Short prose	"Zoro-aster"	Lunar/ zodiac Aries	Annual: first thunder No other qualifiers	Mentions Arabs Byzantine + older docs	Many same topics
CCAG:								
Milan 19 F. 315		Y	Long		October to "End"	Weather, astronomical	July extra-long	Many same topics
Bologna 18 F. 275				David the Prophet	?	Listed *CCAG* 4:41	- -	?
Bologna 18 F. 275				Hermes Tris-megistus	?	Listed *CCAG* 4:41	- -	?

(continued)

TABLE 9.1 *(continued)*

Item	F	T	Length	Author	Begins/ Format	Qualifiers	Other/Apodosis topics	Cf. Etruscan?
Bologna 18 F. 327v		Y			Zodiac+ month Aries/ March	Direction of thunder, earthquakes, weather	Mentions Egypt	Many similar topics
Naples 20 F. 125v		Y	Very short		Month January	Gives number of days, hours Daytime, earthquakes, lunar eclipses	"lands across the sea" breaks at August	Many topics similar
Berlin 25 F. 12					Lunar+ solar Month+ zodiac April- Aries	None	Hellenistic-Roman regions named Cites Eudoxus Astrological tradition	Most different
Berlin 26 F. 152v			Long	Hermes Trismegistus	January	Day, night, weather	Hellenistic-Roman regions named See Fonteius, Vigellius in Lydus, *De ostentis*	Most different
Paris 28 F. 108				"Heraclii"	Zodiac + month October	Listed *CCAG* 8.3:4	(not published)	?
Paris 35 F. 77					Month: January to "Telos"	Earthquake # days, hours	Condensed from Bologna 18 and duplicates some other MSS	Many similar topics
Paris 43 F. 277		Y	Short		Lunar/ zodiac Aries	None	Hellenistic-Roman regions named	Some topics similar
Paris 44 F. 325v		Y	Very short	David the Prophet	Roman months January # days, Hours	Gives number of days, hours per diem earthquakes	April-December also in Paris 53	Most topics similar
Paris 44 F. 326v			Short		Zodiac+ month Aries/ March	Lunar phases	Apodoses simplified	Some topics similar
Paris 44 F. 441			Long		Lunar calendar Roman system January	Earthquakes		Many topics similar

Item	F	T	Length	Author	Begins/ Format	Qualifiers	Other/Apodosis topics	Cf. Etruscan?
Paris 61 F. 42v		y	fr.		Zodiac+ sign Aries	Astrological notes	Some Hellenistic topics: barbarians, Babylon, Egypt. Arabia, Eudoxus	Some topics similar
Naples 20 F. 123v		Y			Roman month-zodiac April/ Taurus	Day/night, plus earthquake omens		Many topics similar
London 47 F. 148					Zodiac Aries –	First thunder in the year	Roman month names	Many topics similar
London 48 F. 46		Y			Month March	Constellations, quakes, lunar phases, day/ night, lightning strikes	Linked to London 47 No glaring Hellenistic references	Most topics similar
Athens 2 F. 185		Y	Very short	Esdras the prophet	Roman month September	Earthquake	Notes unlucky days Mentions barbarians	Many topics similar
Athens 5 F. 161v		Y			Lunar-zodiac Virgo	Plus earthquake omens	Two texts melded?	Many topics similar
Athens 10 F. 34v		Y	Long	Heraclius the sage	Month+ zodiac Sept./ Libra	None	Astrological tags Cites The City, Arabs	Most topics similar
Athens 11 F. 15v		Y	Short		Greek numbers, zodiac Virgo	Earthquakes	Shortened apodoses also cite Egyptians, The City	Many similar topics
Escorial 1 F. 161		Y	Very short		Roman Month September	Plus earthquake omens	Condensed from some other text	Most similar topics
Escorial 1 F. 165*		Y	Short		Zodiac only Aries	Direction of thunder	Very close to Mesopotamian	Most similar

Note: F = Format similar to Etruscan? T = Topics in apodoses similar to Etruscan? Y = Yes

See *CCAG*.

* See translation in Appendix C.

53–54. On earthquakes
55–58. On earthquakes, according to the 12 signs, from Vigellius, following the carmina of Tages
59–70. A zodiacal weather calendar, from the writings of Claudius Tuscus
71. Conclusion/astrology by geographic regions

It is surprising that of all the different texts that Lydus translated and embedded in *De ostentis*, none are genuine treatises – they do not furnish analyses, commentaries, or an author's own insights into the material. They all seem to be copied from cryptic, "scriptural" religious texts. Labeo, for instance, was well known for his Neoplatonic philosophy, yet his long *fulgurale* text is similar to many of the others – and not overtly philosophical in approach.

AUTHORS OF THE LATE REPUBLIC AND EMPIRE LINKED TO ETRUSCAN DIVINATION LITERATURE

Most of the authors cited or quoted by Lydus have names or backstories that demonstrate their connections to Etruscan culture, often through family ties. A few other authors noted here are cited in Classical sources as having written treatises on the *etrusca disciplina* or having promoted/preserved its documents. Most of the Late Republican authors, like Nigidius Figulus, were active members of Roman society and government who took part in political and military activities and also religious cults and scholarship. They must have regarded their priestly and scholarly duties as important to their state careers, and more than one acted with integrity upon his strong political convictions – usually to his personal detriment.

AULUS CAECINA

It is surprising that it should today be so hard to identify so well known a personage as Caecina, whom his contemporaries considered seminal to the preservation and diffusion of the *etrusca disciplina*.[17] The Etruscan name *ceicna*, with a variety of *praenomina*, appears, usually in funerary epitaphs, at Volterra, Orvieto, Orte, Chiusi, and Vulci. A grave stele in Bologna depicts a fifth-century Vel Caicna as perhaps a *trierarch*, posed on the deck of his warship.[18] Cicero defended an Aulus Caecina of Volterra in a property case in 69 BC and later corresponded with his son.[19] Cicero (*Ad fam.* 6.6.3) implies they are a distinguished and wealthy equestrian family, skilled in the *etrusca*

[17] Cf. Weinstock 1950.

[18] For full discussion of the Caecina family, see Hall 1984:204–24, no. 18, with references; *RE* III, col. 1237 no. 6 "Aulus Caecina"; Torelli 1969:297–98; for the Bologna stele, see Basch 1987:409, fig. 873.

[19] Cf. *Pro Caecina* 18; *Ad fam.* 6.5–9, 13.66; see Rawson 1978/1991:297–99.

disciplina, on which Aulus, the son of his client, wrote a treatise. Some later family members were associated with Augustus' rise to power, and another was apparently a friend of Claudius.

Aulus Caecina the author was cited by various Latin authors, including Cicero (*Ad fam.* 6.7.1), Pliny (*Nat. Hist.* 2, index), Seneca (*Quest. Nat.* 2.39, 49, 56), and Suetonius (*Julius* 75.5) – Caecina wrote a political pamphlet criticizing Caesar during the civil wars, and later wrote hoping to have his resulting exile reversed. Hall notes the scholarly debate as to whether these two Caecinae, father and son, were Cicero's only correspondents of that name.[20] Because none of the works of Caecina the author(s) survive, the question remains moot.

TARQUITIUS PRISCUS

This Late Republican authority on the *etrusca disciplina* was mentioned by Pliny as a source for book 2 of the *Naturalis Historia* and was cited or named by several later authors.[21] Pliny said, in the index to book 2 (*Nat. Hist.* 2.1b.1), *Caecina qui de etrusca disciplina, Tarquitio qui item*, "Caecina who [wrote] on Etruscan scripture, Tarquitius who [did] the same." Tarquitius would presumably have been a contemporary of Aulus Caecina, thus Late Republican, or first century BC; he could be the C. Tarquitius Priscus who was distinguished by his military career in the 80s BC. The name seems to derive from *tarchna*, that is, Tarquinia; another variant would be Tarquinius.[22] Hall places the author in the Late Republic or Early Imperial period, suggesting that he was a son or grandson of the famous Tarquitius Priscus who had served as legate to Sertorius in Spain in 72 BC, along with other disaffected Etruscans who defected after the defeat of Marius.[23] Epigraphic evidence, mostly late epitaphs, locates a *gens Tarquitia* at Caere and Veii.[24] Macrobius (*Saturnalia* 3.7.2) cited Tarquitius Priscus as a source for the commonplace notion that a ram of unusual coloring was an omen meaning happiness for the king and firmly associated him with a "book transcribed from Etruscan divination scripture" (*liber Tarquitii transcriptus ex ostentario tusco*).[25] Apparently one of his specialties was a treatise on tree portents, and according to *Saturnalia* 3.20.3, he named the trees that were *infelices*, associated with the deities of the Underworld.[26]

[20] Hall 1984:445–46n49.
[21] See Guillaumont 1986:122; Heurgon 1953; Kroll 1932.
[22] See full discussion and references in Hall 1984:382–89.
[23] Hall 1984:387–88, no. 87.6.
[24] Hall 1984:383–84.
[25] Cf. Briquel 1999B.
[26] See Guittard 1990; cf. Harrison et al. 2009:§ 8.

The identification of Tarquitius as a translator of Etruscan texts comes from Lactantius (*Div. Inst.* 1.10.2) and Ammianus Marcellinus, who show that his learned treatises were used by contemporary Etruscan haruspices studying a meteor omen: they "referred to the books of Tarquitius titled *On Divine Matters*," *etrusci haruspices ex Tarquitianis libris in titulo de rebus divinis id relatum esse monstrantes* (Ammianus Marcellinus 25.2.7 – J 363). Lydus is believed to have cited his work (cf. *De ostentis* 2). A Latin inscription at Tarquinia, part of the refurbished Roman monument of the *Elogia Tarquiniensia* that commemorated famous Etruscan citizens (*CIL* XI.3370 + 7566) may even refer to Tarquitius' "publication of hymns [in translation]" (*carminibus edidit*):

]uitio m[
]m comiti[a?
[n]umina Arusan[ia?
[signumque I]ovis et Iustitiae et[
[consec]crandum; discipul[inam]
]i carminibus edidit[
]ta annis ampliu[s]

The inscriptions are fragmentary and not the original Etruscan versions, and they have been variously interpreted. Heurgon (1953) had suggested a Claudian date for this inscription, but Torelli's studies on the *Elogia Tarquiniensia* again placed its subject in the first century BC.[27] The most recent critical appraisal by Haack, however, is skeptical, noting the many lacunae in the text and dating this particular member of the *Ordo LX Haruspicum* to the period of Augustus or Tiberius.[28] Although we have no *brontoscopia* or other treatise remaining from him, Tarquitius must have exerted a formidable influence on the literature of divination.

FONTEIUS CAPITO

The author on religious topics cited by Lydus simply as "the Roman Fonteius" is probably the same whose *tonitruale* Lydus translated (*De ostentis* 39–41; see Appendix C.1).[29] According to Lydus (*De mensibus* 2.2; 4.80), a Fonteius said that the identity of the god Janus was a sort of allegory for the concept of time, and that the month of May was named for the goddess Maia/Gaia. He further noted that Fonteius had recorded an oracle attributed to Romulus (*De magistratibus* 2.12; 3.42 = *De mensibus* Fr. 7). It may well be that this author is the

[27] Torelli 1975:96, 105–6, 110–15, 192. First half of 1st c.; cf. Rawson 1978/1991:301n61.
[28] Haack 2006:143–44, no. 118; cf. Rawson 1978/1991:301n61.
[29] See Graf 1998; Domenici and Maderna 2007:14, 99–101.

same C. Fonteius Capito who was a supporter of Antony and accompanied Maecenas to Tarentum for the abortive conference between Octavian and Antony in 37 BC (also the occasion for Horace's grumpy trip to Brundisium, *Satire* 1.5.32). Caius Fonteius Capito also escorted Cleopatra to Antony during the latter's ill-fated sojourn in Syria in 37–36 BC (Plutarch *Antony* 36.1), and he was *pontifex maximus* and *consul suffectus* in 33 BC.

The *Brontoscopia* of Fonteius, "translated word for word" into Greek by Lydus, is a short, prose *brontologion*, with many apodoses adapted ultimately from the Mesopotamian literature. It occupies only three short sections (39–41) of *De ostentis*, and it is arranged in paragraphs in order of zodiacal periods, qualified by whether the thunder is heard during the day or night, beginning with Capricorn and ending in Sagittarius. Its predictions show the heavy overlay of Hellenistic-Roman tastes, where omens are tagged for regions significant to the Roman state or the Hellenistic Greek kingdoms. Its reference to the Nile might be anticipated in the voluminous hermetic and astrological literature of the Ptolemaic milieu whence it also derived.

Many of Fonteius' predictions match closely the topics of the Figulan/Etruscan document, but the format (by zodiac) and many expressions differ. Among the topics are the fortunes of a tyrant, a lack of foodstuffs, the Nile flood, strife between parents and children, peace, war, enemies, the downfall of famous men, winds and floods, locusts, fish, wild beasts, herds, parasites, deadly reptiles, grapes, storms, women in labor, disease, and deaths. Terms like "the four-footed," and soft versus hard "fruits" or crops (which mean fruits and grains respectively), or the "subject allies" do match the terms of the *Brontoscopic Calendar* as they stand in the Greek translation of Lydus.

In contrast, however, predictions for good and bad markets, and the omen that evil spirits ("apparitions," φάσματα) will afflict bodies with disease, do arise from the Mesopotamian tradition, but they are not found in the Figulus text. Likewise, Fonteius offers more motifs, such as children rebelling against parents and references to a royal court, which did figure in Mesopotamian literature, but these do not match the Figulan brontoscopy. There is mention of Persians, western Europe or Asia, Arabs, and incursions into Roman territory by "the enemy," all topics that set this text firmly in the milieu of Hellenistic and Roman astrology. The Figulus text appears to be older and simpler in its lack of astrological references, its pedantic format, and simpler society that lacks markets, royal courts, and far-flung enemies.

CLAUDIUS (CLODIUS?) TUSCUS

Tuscus was the author of a Latin poem, in calendar format, on weather signs intended to evoke the nostalgia of peasant weather lore; Wissowa (1900) maintained that the version furnished by Lydus (*De ostentis* 59–70) must be a poor Greek translation of the original. Tuscus is possibly the same author mentioned by Aulus Gellius (*Noctes Atticae* 5.20.2) as a poet of the Augustan period. Servius (*Ad Aen.* 12.657) implies that he also wrote a grammatical treatise. Ovid (*Tristia ex Ponto* 4.6.20) mentions a Tuscus, and some scholars have implied that his calendar poem was used as a base for the *Fasti*, though this is quite unlikely. The name Claudius, coupled with the adjective Tuscus (i.e., Etruscan), might have been invented to contrast with that of Claudius Ptolemy, the astronomer.[30]

The document translated by Lydus is relatively long, about the length of the *Brontoscopic Calendar* but quite different in format. It concludes with Lydus' comment that this information "Claudius has from the sacred writings of the Etruscans, word for word" (ἐκ τῶν παρὰ Θούσκοις ἱερῶν πρὸς λέξιν – first line of section 71), but he goes on to name several Greeks whose calendars were supposedly of the same sort.[31] Beginning in January, it notes days by the Roman system of Kalends, Nones, and Ides. Most days are characterized only by the prediction of winds or weather conditions; some include celestial phenomena, such as constellations, zodiacal risings, or the occasional note on seasonal fruits ripening. The similarities in content with Columella's *De re rustica*, book 11, have long been noted.[32] The *Ephemeris* resembles the Figulus document only in having a calendrical template paired with an item, not really an omen, for each day of the year. Days number thirty-one for most months, but thirty for April, June, September, and November; February has twenty-eight, another telltale of post-Julian editing. There is no reason to doubt that a Clodius Tuscus had been familiar with some Etruscan cultic calendar, but if this is the same thing, it seems necessary to agree with Wissowa that the text preserved by Lydus is a denatured and much-edited version of the original.

CORNELIUS (?) LABEO

Little is known of the author Labeo, whose works provoked polemics from early Christian authors, such as Arnobius and Augustine; his works were much discussed in nineteenth- to twentieth-century *Quellenforschung*. According to Briquel, Labeo may now be placed securely in the second half of the third-century AD.[33] Paolo Mastandrea (1979) has characterized him as a Neoplatonist. His works on religion were said to include *De diis Penatibus* (*On the Penates*), *De diis animalibus* (*On the Spirit Gods*), and *On the Oracle of Klaros Apollo*. Labeo was one of the few Roman (as opposed to Greek) voices of the pagan

[30] See Domenici and Maderna 2007:151n120, with additional references.

[31] A new Italian translation of the text is furnished by Domenici and Maderna 2007:117–34; and a Greek text and a new English translation have been published by Lehoux (2007:162–63, 343–75).

[32] Lehoux 2007:80; Domenici and Maderna 2007:151n120.

[33] See Eck 1997; Ramelli 2003:118–21; Briquel 1997:119–37, and 1999A:51n1, for past bibliography; Domenici and Maderna 2007:14, 102–3, 107–10.

opposition to Christianity. Fulgentius says that he dedicated fifteen books to the *etrusca disciplina* and cited the prophet Tages as his authority.[34] One of his fields of expertise was haruspicy; he also wrote on the *dii animales* funerary sacrifice ritual for making a dead human spirit into a god in the afterlife – which occasioned Augustine's denunciation of the use of "animal victims," *hostiae animales* (*De civitate dei* 22.28).[35]

Two works of Labeo are embedded, in Greek translation, in *De ostentis* (42, 47–58): a short text on zodiac and weather, and a very lengthy, prosaic treatise covering multiple categories of observed omens, including lightning bolts and other phenomena.[36] This fulgural text follows a lunar dating format beginning with the summer solstice as the moon in Aries has just made the transition from Cancer. The text then cites the sun in Aries, and this lunar-cum-solar zodiacal format continues for the rest of the work. Many omens refer to regions relevant to the Hellenistic Greek kingdoms that still made sense for the Roman world.

Lydus says (*De ostentis* 42) that he has translated the text of Labeo into Greek "word for word," and many Greek phrases here are similar to those used in the *Brontoscopic Calendar*, but the content differs. The work begins with general weather observations of clouds, thunder, lightning, trees struck down, and plague-bearing heat. The apodoses cited either lack or plenty of storms, drought, wine, oil, grain, and so forth; and the next sections cover lightning bolts (47–52) and earthquakes (53–58). There is a kind of sympathetic logic to such predictions as, "If in Aries, a thunderbolt ... hits a vine, it means a decrease in wine, if it hits a fruit tree it predicts a decrease in the harvest of that fruit." The subjects of the outcomes are determined by the object or being that the lightning strikes: danger for enemies, flocks, river floods, fires, serpents, and so on. In Virgo, a thunderbolt will predict omens affecting women (note similar logic in Fonteius' *Brontoscopia*, Appendix C.1); sometimes the omen is for the owner of the tree that is blasted, a condition similar to many in old Mesopotamian documents, like the *Šumma ālu*. The tone and many predictions are quite different from those of the Etruscan *Brontoscopic Calendar*. Like the earthquake omens (53–54) linked to regions of the Hellenistic world (Egypt, Scythia), there are many telltales of a Ptolemaic or later source.

VIGELLIUS

Lydus cites Vigellius as the author of a treatise on earthquake omens, claiming that he developed them from the "verses (στίχων) of Tages," "which Apuleius discussed."[37] This is the only attestation of a "Vigellius," who cannot be more

[34] *Sermo antiqua* 4 = Labeo fragment 9 Mastandrea.
[35] See Briquel 1997:128–37.
[36] On its attribution, cf. Briquel 1997:53n7.
[37] *De ostentis* 55–58 = Domenici and Maderna 2007:113–16.

closely identified and is very probably Late Roman.[38] The lengthy document begins, like many Ptolemaic-Roman astrological texts, with the sun in Aries. Omens are listed by zodiacal sign and geographic region (Europe, Sicily, Italy, Asia, etc.); the author also notes whether the sign is male or female, a trait of Hellenistic astrological lore but far from the outlook of the Etruscan calendar. Other omens relate to "men living in the Near East." The discursive text has few points of contact with the material of the Figulus brontoscopy and is notable here only because of Lydus' interest. Clearly, with the progression of time, these late texts diverged ever farther from the tone and focus of the *Brontoscopic Calendar.*

HEPHAISTION OF THEBES, APOTELESMATIKA*: FRAGMENTS OF NECHEPSO AND PETOSIRIS?*

The scholar Hephaistion of Thebes claimed, in his three-book astrological treatise, the *Apotelesmatika*, published circa 381 AD, to have translated or quoted "literally" long passages from much older authors, the Egyptians Nechepso (an unlikely allusion to the 26th Dynasty pharaoh of 677–672 BC) and Petosiris (supposedly his high priest and spokesman). Hephaistion also claimed among his sources (and this is evident), Ptolemy's *Tetrabiblos* and a poetic work by Dorotheus of Sidon, both of Imperial date.[39] Four different *Epitomes* exist for this work. A Hellenistic Egyptian author using the name Nechepso did write a verse astrological work, titled *Astrologoumena*, probably around 150–120 BC, but no passages can be attributed to him here.[40]

The excerpted sample presented as Appendix C.2 is taken from fragment 6 (ed. Riess) of Hephaistion's *Apotelesmatika* 1.21, quoting from the modern translation of Naphtali Lewis.[41] The long work incorporated lunar eclipses qualified by zodiacal seasons, celestial phenomena, and weather conditions such as winds; the predictions sound very much like Mesopotamian omens, with an overlay of Hellenistic peoples and places (Egypt, Syria, Cyprus, Media, Cilicia, Rome, Armenia, Cappadocia, Euboea, etc.) It shows the bipolar schemes of Mesopotamian texts, with outcomes keyed to colors, time of day, or length of phenomenon, and uses opposites and repetitions ("the same as preceding") in similar fashion. Some traits can only have developed within Roman-era scholarship that used Ptolemaic astrology and perhaps Seleucid-Babylonian divination texts: reliance on the zodiac, a somewhat Egyptocentric viewpoint, and a historical geography that includes some peoples who can only have signified during the Roman Empire.

[38] See Domenici and Maderna 2007:140n32.

[39] See Engelbrecht 1887; Pingree 1973:v–xviii; Gundel and Gundel 1966:241–44.

[40] See Pingree 1974; Gundel and Gundel 1966:27–35; Riess 1892.

[41] Lewis 1996:139–43. Complete passage in Pingree 1973:52–63; Lewis 1996:139–44.

GREEK PAPYRI OR MANUSCRIPTS

It is probable that the Qumran Aramaic *brontologion* belonged to a large body of astrological, hemerological, and brontoscopic literature that circulated across the Seleucid and Ptolemaic kingdoms and in Palestine during the Second Temple period and thereafter.[42] Under Byzantine rule, this region continued to produce texts of *brontologia* and *selindromia*, usually in Greek. They normally begin, "If the moon," and then offer predictions of statewide events that resemble the Neo-Assyrian and Neo-Babylonian texts of the first millennium BC.[43] Other texts include *seismologia*, which are similar in format and predictions to the *brontoscopia* but refer to earthquakes for omens. Many of these short texts were written or copied/edited in the Byzantine period, and they now bear tags that attribute them (erroneously) to biblical prophets or other historical figures. Other documents of more evident Classical tradition were also embedded in the writings of Lydus; these, too, use weather or celestial phenomena as portents and derive from the tradition of Ptolemaic-, Seleucid- and Roman-period erudition that was itself rooted in Mesopotamian divination literature. There is a high degree of cross-fertilization throughout the corpus, with several texts that duplicate each other or contain excerpted lines from another text. Most show traces of the old Mesopotamian literature, but none share the formatting of the *Brontoscopic Calendar* or the (preserved fragments of) Republican Roman authors, and most betray Hellenistic terms in their geographic citations. Many use the language or data of the Greek agricultural *parapegmata*, stone markers, often artistically embellished, which note the dates of seasonal phenomena and agricultural activities. Lehoux reminds us that the *parapegmata* themselves parallel Mesopotamian tradition, but, unlike the astrological texts, they do not actually derive from it.[44] Lydus also furnished additional texts similar to *parapegmata* in his *De mensibus*.[45]

BRONTOLOGICAL TEXTS KNOWN FROM CLASSICAL AND BYZANTINE SOURCES

See Table 9.1 for a list of major documents. Most of the Byzantine texts are known only by their manuscript designation in the corpus compiled by Cumont, Boll et al. (1898–1953) (*CCAG*), but several announce their "pedigrees," as it were, by an attribution to some famous prophet or author, including "the prophet David," "the prophet Esdras," Hermes Trismegistus, or Heraclius [the emperor?]. A précis of each text is offered below.

[42] Wise 1994:13–15.

[43] See Greenfield and Sokoloff 1989:213; also discussed by Bezold and Boll 1911:25–44.

[44] Lehoux 2007: 115.

[45] See Lehoux 2007:163–64.

MILAN CODEX 19 F. 315: περὶ βροντῶν ("CONCERNING THUNDERS").[46] Beginning with October (written as number 1, in Greek, α'), this gives several lines per month beginning, "If the sound of thunder is heard"; finishing with September, it states "End" (Τέλος). Within each month, the prediction is qualified by weather and astronomical conditions, such as the behavior of sun, moon, or stars. Many apodoses are similar to those of the Etruscan document, such as "there will be a lack of wine/food," or the downfall/death of a great man, a bad uprising, vicious wild animals, and so on. The discussion of the month of July is nearly three times longer than others, which may reflect a prototype developed during the Early Empire, when Julian allusions would have been intended to flatter an emperor.

BOLOGNA CODEX 18 (FF. 327V–329): *TONITRUALE IGNOTI AUCTORIS* ("THUNDER LORE BY AN UNKNOWN AUTHOR").[47] This Byzantine manuscript is considered very corrupted and seems to have combined two texts, one arranged by zodiac, the other by Roman months; the editors have pointed out its many similarities to documents copied by Lydus in *De ostentis*. As it stands now, it begins with Aries/March and ends at Pisces/February, and some entries include qualifiers, such as earthquakes, "If it thunders in the East," or great winds blowing. The apodoses are similar to those of the *Brontoscopic Calendar*, including "dearth of wine, grain, and all crops," "disease for the countryside," "war," or "death of the greatest men."

NAPLES CODEX 20 F. 123V–125V: Ἡ ἑρμενεία τοῦ σεισμολογίου διὰ τῶν δώδεκα ζῳδίων ("THE INTERPRETATION OF EARTHQUAKE OMENS ACCORDING TO THE TWELVE SIGNS OF THE ZODIAC").[48] Although titled a *seismologion*, this text begins each entry with a long set of omens: "If it should thunder," compiled by month linked to a zodiacal sign, starting at April/Taurus. Bassi and Martini, the editors of these two fragments, noted that the manuscript broke off in mid-January (Pisces). Weinstock has indicated close affinities between this text and Harleian (London) Codex 47 (below) and others. Boudreaux, editing Paris Codex 61 (below), also linked it to several other Byzantine texts. Each month/sign begins with the standard ἐὰν βροντήσῃ ("If in any way it should thunder") and then lists a number of grim predictions, including famine, crop failures, grain shortages, harm to animals or people, occasionally abundance and good fortune, and also predictions on treachery or other situations in the royal court. Foreign relations are implied in April with the Ἄραψιν λιμός, the Arab *limes* (frontier), another

[46] *CCAG*:vol. 3; *Codices Mediolanenses*, eds. A. Martini and D. Bassi (1901), 50–52. Also edited by R. Wünsch in *Byzantinische Zeitschrift* (1896):V, 416–18.

[47] *CCAG*:vol. 4; *Codices Italicos*, eds. D. Bassi, F. Cumont, A. Martini and A. Olivieri (1903), 128–31.

[48] *CCAG*:vol. 4; *Codices Italicos*, 170–72.

Hellenistic element. Similarities with many other texts are evidence of the pseudoscholarly milieu of Hellenistic-Roman astrology, in which a given document probably had numerous sources, which themselves were not directly descended from cuneiform texts, but must have been compilations created by scholars of the Persian or Hellenistic period in the East.

NAPLES CODEX 20 F. 125V–127: Ἑρμενεία τοῦ βροντολογίου τοῦ μηνιαίου ("KEY TO THUNDER INTERPRETATION BY THE MONTH").[49] Bassi and Martini associated this short manuscript with others through the so-called Melampous manuscript;[50] a portion of it was copied in Paris Codex 5 (see discussion later in this chapter). It is formatted with the month name "Month January" in nominative case, then cites the number of days and hours of daylight and night. "If it thunders" is qualified by "during the day" or "if there is an earthquake," or by lunar eclipse. Many apodoses resemble lines of the Etruscan *Brontoscopic Calendar*: destruction of the four-footed, abundance of wheat, war, destruction of men in power, and so on. "Lands across the sea" imply a Hellenistic background to the text, and notes on lengths of daylight and night link it to agricultural calendars. The manuscript breaks off in August.

BERLIN CODEX 25 F. 12: *TONITRUALE IGNOTI AUCTORIS* ("THUNDER LORE BY AN UNKNOWN AUTHOR," CF. *GEOPONICA*).[51] This text uses both lunar and solar formats, with lunar months likened by its editor Boll, an expert on ancient astronomy, to the system found in the Byzantine compilation known as *Geoponica* 1.10 (see Appendix C.4).[52] He also noted affinities with the *tonitruale* Naples Codex 20 F. 123v–125v (above). Boll makes a strong case for its author having been influenced by the text of Eudoxus, the famous fourth-century astronomer, and having borrowed from the *seismologion* attributed to "the Roman Vigellius" in *De ostentis*.[53]

Beginning in April, "with the Moon being in Aries," and then "with the Sun being in Aries," the text alternates loosely between the same two formulae seen in the Figulus document, "If it thunders" or "If in any way it should thunder" – εἰ βροντήσῃ or ἐάν βροντήσῃ. Assuming that Lydus faithfully recorded differences in the original language of this document and the *Brontoscopic Calendar*, then one of the oldest of the Berlin codex's prototypes could have been the *Brontoscopic Calendar*. Berlin 25 F. 12 ends under the sun in Pisces, with a prediction for the crops in nearby Lydia, Cilicia, and Pamphilia, "but

[49] *CCAG*:vol. 4; *Codices Italicos*, 172–73.
[50] See *CCAG*:vol. 4, 110–111n1, on Modena Codex 11.
[51] *CCAG*:vol. 7; *Codices Germanicos*, ed. F. Boll (1908), 163–67.
[52] *CCAG*:vol.7,163.
[53] Wachsmuth 1897 = 57ff, and prolegomena at xxviii. Vigellius cannot be identified with extant evidence; see Domenici and Maderna 2007:14, Italian translation of his text, 113–16.

the females of living creatures both those who speak [humans] and those that are dumb [animals] will look angry/sad, even the female constellations of the zodiac" (*CCAG*, p. 167 = F. 12 lines 21ff.).

BERLIN CODEX 26 F. 152V: "*BRONTOLOGION* OF HERMES TRISMEGISTUS."[54] This text, allegedly by Hermes Trismegistus, is slightly longer than most, and Boll found that it owes much to the text attributed to Fonteius by Lydus, *De ostentis* 39–41. It began at January and incorporated in the February entry material apparently copied from a text attributed by Lydus (*De ostentis* 55–58) to Vigellius (see discussion earlier in this chapter) – the first six lines are identical. Mentions of the Nile, the Persian king, Babylon, barbarian peoples, and such place the text in the milieu of Ptolemaic-Roman astrological Egypt, as of course does its attribution to Hermes Trismegistus, a favorite for hermetic/magical lore.[55] The predictions, "If it should thunder or there should be lightning," are mostly qualified by conditions such as "in the daytime," "at night," or by weather conditions, and so on that are not factors in the Figulus brontoscopic text.

PARIS CODEX 35 (PARIS GR. 2118 P. 77): Ἑρμενεία τοῦ βροντολογίου τῶν δώδεκα μηνῶνη ("INTERPRETATION OF THUNDER SIGNS FOR THE TWELVE MONTHS").[56] Part of the same document is preserved in Naples Codex 20 F. 125v–127 (discussed earlier in this chapter). It also has been associated with Paris Codex 44 F. 427 and Paris Codex 53 F. 163v. The editor, P. Boudreaux, suspected it was a condensed version of Bologna Codex 18 (discussed earlier in this chapter), and also linked it to the content of Paris Codex 44 (discussed later in this chapter); Boudreaux offered Paris Codex 44 FF. 441–447 as an archetypical text for this sort of divination (cf. discussion later in this chapter).

Paris 35 begins in January and notes the number of days and hours of day and night, with qualifiers for the thunder-omens. In January, for example, thunder can signal abundance of grain, but earthquake signals war. It uses many of the same terms as other documents, including the Etruscan *Brontoscopic Calendar*, such as σημαίνει or δηλοῖ – so that thunder "signals" or "indicates," and many apodosis events match either those of the Figulus text or common Mesopotamian omens: an abundance or lack of food/grain/livestock, flight of the rulers, war, and so on. Like several other manuscripts, it ends with the note Τέλος, "End." It has clearly derived from a multiplicity of other documents, including agricultural "literature."

[54] *CCAG*:vol. 7; *Codices Germanicos*, 226–30.

[55] For background, see Fowden 1993; Copenhaver 1995.

[56] *CCAG*:vol. 7, pt. 3; *Codicum Parisinorum*, ed. P. Boudreaux (1912), 123–25.

PARIS CODEX 43 (PARIS GR. 2315 F. 277–278): Περὶ βροντολόγιον τῶν ιβ′ ζῳδίων ("CONCERNING THUNDER LORE OF THE TWELVE SIGNS OF THE ZODIAC").[57] Boudreaux indicated that this short manuscript was more recent than some (perhaps 9th c. AD). It is formulated only according to zodiacal signs, beginning in Aries, with a single sentiment for each sign, prognosticating about food supplies and rulers, naming regions (India, Egypt, the land beyond the sea) considered of interest to Hellenistic kingdoms or the Roman Empire. Further evidence of a late composition may be the closing prediction: "good for men and great peace for the world (*kosmos*)."

PARIS CODEX 44 (PARIS GR. 2316 F. 325V): βροντολόγιον Δαβὶδ τοῦ προφήτου ("THUNDER TREATISE OF DAVID THE PROPHET").[58] More than one *brontologion* was attributed, extremely posthumously, to the "Prophet David" (see Table 9.1).[59] This is a very short text. The entries for April through December were linked by Boudreaux to Paris Codex 53 F. 163v, which is practically identical.[60] Like some others, it gives the number of hours and days, beginning in January and ending in December, and combines earthquake omens with the thunder predictions.

PARIS CODEX 44 (PARIS GR. 2316 F. 326V–327V): Ἕτερον βροντολόγιον τοῦ σεληνιακοῦ ("A DIFFERENT THUNDER TREATISE ACCORDING TO THE LUNAR CALENDAR").[61] The short text begins, "With the Moon being in Aries, in the month of March," and ends in Pisces, with rather simple predictions, such as "for the people" ... "a good period of time." After "the month of April," entries are recorded in peculiar format: "month #3. With the moon in Gemini. In the month of May," and so on. If thunder occurs during the waxing phase of the moon, good conditions usually ensue. There is little similarity to the Figulus document in either format or apodoses.

PARIS CODEX 44 F. 441–447: *LUNARIUM ANONYMUM*: Σεληνοδρόμιον, σεισμολόγιον, βροντολόγιον ("ANONYMOUS LUNAR CALENDAR: PHASES OF THE MOON, INTERPRETATION OF EARTHQUAKE AND OF THUNDER-OMENS").[62] Another anonymous, and quite long, document combines lunar phases, zodiacal appearances, earthquake omens, and thunder-omens in lengthy commentaries for each month, beginning with January and ending with December. Some, but not all, dates are recorded in the Roman system of Kalends,

57 *CCAG*:vol. 8, pt. 3; *Codicum Parisinorum*, 166–67.
58 *CCAG*:vol. 8, pt. 3; *Codicum Parisinorum*, 168–69.
59 In addition to this, see *CCAG*, 4:41, part of Codex Bologna 18 F. 275; and *CCAG*, 4:50, Codex Milan 19 F. 315, above.
60 *CCAG*:vol. 8, pt. 3, 75.
61 *CCAG*:vol. 8, pt. 3; *Codicum Parisinorum*, 169–71.
62 *CCAG*:vol. 8, pt. 3; *Codicum Parisinorum*, 181–92.

Nones, and Ides. Many apodoses are standard, identical, or very close to some in the Etruscan *Brontoscopic Calendar*, including good/bad harvests, famine, disease, disorder among enemies, the kingdom, and so on. This multitopic text does not harbor the telltale Hellenistic or Roman regional references found in many late *brontologia*, and it may have derived from (a) different (combination of) source-documents. Boudreaux noted close affinities to material in Lydus' *De ostentis* 17–20, an unattributed, general discussion of lunar phenomena.[63]

PARIS CODEX 61 (PARIS S. GR. 1191 F. 42V–F. 47): βροντοσκόπιον τῶν δώδεκα ζῳδίων καθ' ἑκάστην νύκτα <κατὰ> τὸν τῆς Σελήνης δρόμον ("THUNDER INTERPRETATION BY THE TWELVE SIGNS OF THE ZODIAC FOR EACH NIGHT ACCORDING TO THE COURSE OF THE MOON").[64] This text is incomplete, breaking off in Capricorn; Boudreaux linked it to several other manuscripts, including Naples Codex 20 F. 123v–124v (discussed earlier in this chapter), which is titled a *seismologion* but begins each entry with a set of omens, "If it should thunder." Another relative is the omen text attributed to the Emperor Heraclius.[65] The first section comprises astrological notes on constellations and months; next, beginning in Aries, it offers a paragraph marked by a horoscopic symbol with brontological omens. Apodoses include standard predictions of famine or plenty, activities of enemies and kings, and so on. Others use the jargon of Hellenistic/Roman astrologers, citing an onslaught of "barbarians" on the cities, causing citizens to flee for their lives; exotic regions include Babylon, Egypt, and Arabia. The famous fourth-century BC astronomer Eudoxus is cited several times, implying that the document was produced by Greek-speaking, Hellenistic scholars in the tradition of late Ptolemaic and Roman astrology.

LONDON CODEX HARLEIAN 47 (HARLEI. 6295 F. 148): βροντολόγιον ("TREATISE ON THUNDER-DIVINATION").[66] Weinstock indicates the close affinities between this text and codices Naples 20 (discussed earlier in this chapter), Berlin Codex 25 F. 12 (lunar section, discussed earlier in this chapter), and others. Predictions cover the whole year, determined by the sign of the zodiac, beginning in Aries. Weinstock pointed out that the extensive editing of so many of these manuscripts to use Roman month names instead of the terminology of the Hellenistic Greek or Ptolemaic calendars makes them much later than their presumed original, Hellenistic Greek sources. Stock apodoses do match many items in the *Brontoscopic Calendar* and Mesopotamian literature, but they seem to have been used piecemeal:

[63] See *CCAG*:vol. 8, pt. 3, 180–81, and 181n1 for references to other, related brontological manuscripts.

[64] *CCAG*:vol. 8, pt. 3; *Codicum Parisinorum*, 193–97.

[65] For Codex Paris 28 F. 108v, see *CCAG*:vol. 8, pt. 3, 4.

[66] *CCAG*:vol. 9, pt. 2; *Codices Britannicos*, ed. S. Weinstock (1953), 120–23.

crops (pulses, etc.), want, disease, kings, weather, creatures. Hellenistic elements include exotic peoples, such as Arabs.

LONDON CODEX REGIUS 48 (REG. 16 C. II (F. 46): Ἀρχὴ σὺν Θεῷ βροντολογίου καὶ Σεισμολογίου τῶν ιβ′ μηνῶν ("BASIC TREATISE UNDER GOD ON THUNDER- AND EARTHQUAKE-DIVINATION ACCORDING TO THE TWELVE MONTHS").[67] This text is definitely linked with the preceding (London Harleian 47 F. 148) by Weinstock, who noted its descent from Mesopotamian divination literature. It makes additional references to other constellations than the standard zodiac and begins with March, qualifying the time of thunder by the phases of the moon, whether it is day or night, and also whether lightning strikes occur. Apodoses are standardized: dearth of grain/wine, disorder among the people(s), and good times for men. It does not contain the exotic regional references of obvious Ptolemaic-Roman documents and so may have earlier sources, but it is firmly grounded in the Hellenistic astrological tradition.

ATHENS CODEX 2 (BIB. PUBL. 462 F. 185): Ἀρχὴ σὺν Θεῷ βροντολογίου καὶ Σεισμολογίου τῶν ιβ+′ μηνῶν ("BASIC TREATISE UNDER GOD ON THUNDER- AND EARTHQUAKE-DIVINATION BY THE TWELVE MONTHS").[68] A very short text once labeled *ex libello Esdrae prophetae* ("from the booklet of the prophet Esdras," to whom God supposedly revealed these omens) begins "in the month of September" and is phrased somewhat conversationally – "If it thunders, it does not signal good," "If an earthquake happens, wise men will die and there will be many wars." Other apodoses include disease, trouble for rulers, famine, and social unrest. In August, "If it thunders, the salvation of the world" (σωτηρία τοῦ κόσμου) is surely a Christian sentiment. Each month ends with a note of which days will be "heavy" (meaning either unlucky, as in the Mesopotamian hemerologies, or simply overcast in terms of weather).

ATHENS CODEX 5 (BIB. PUBL. 778 F. 161V–164V): Ἀρχὴ σὺν Θεῷ βροντολογίου καὶ σεισμολογίου κατὰ τὴν ἡμέραν τοῦ μηνός ("BASIC TREATISE UNDER GOD ON THUNDER- AND EARTHQUAKE-DIVINATION ACCORDING TO THE DAY OF THE MONTH").[69] Dated by the position of the moon in relation to a zodiacal sign, the text begins in Virgo. Many apodoses are stock Mesopotamian types similar to those in the Figulus document: abundance or dearth of pulses, grain, fish, weather or wind conditions. Delatte suggested that the text melded two different sources, one a treatise on earthquakes. He linked it to Athens

[67] *CCAG*:vol. 9, pt. 2; *Codices Britannicos*, 124–26.
[68] *CCAG*:vol. 10; *Codices Athenienses*, ed. A. Delatte (1924), 58–59.
[69] *CCAG*:vol. 10; *Codices Athenienses*, 60–62.

Codices 11 and 10 F. 34v (discussed next). Apodoses citing the "downfall of the four-footed and wild animals" are very close to the Etruscan *Brontoscopic Calendar* and its Mesopotamian sources, but the formatting and melding of multiple texts betrays a much later creation.

ATHENS CODEX 10 (BIBL. PUBL. 1275 F. 34V–37): Βροντολόγιον τῶν δώδεκα μηνῶν ("THUNDER-DIVINATION FOR THE TWELVE MONTHS").[70] A lengthy *brontologion* formatted by months linked to the zodiac begins with September/Libra and was attributed to "Heraclius the sage."[71] Also, like several other texts, it uses the term ἔνδοξος, normally "glorious," to modify χειμῶν, "winter"; this may have originally been a reference to the fourth-century astronomer Eudoxus.[72] Each month is captioned, as in "Month September has zodiac sign Libra"; the first word of each entry simply repeats that sign. Then, "If in any way it should thunder" introduces omens of the same stock phrases as other Byzantine texts and the Etruscan *Brontoscopic Calendar*, with the verbs δηλοῖ or σημαίνει for predictions.

Most apodoses are standard: war, slaughter, an onslaught of disease, rashes, famine, wild beasts, failed harvests in the mountains or plains, reversals, river floods, death of men and women, misfortunes, destruction of the four-footed, or of birds, cattle, reptiles, or wild beasts, fruits, crops, pulses, barley, wheat, attack by locusts, harsh/mild/late/changeable winter, winds, peril on the sea, or disease in the royal palace. "The City," presumably Constantinople, is cited, along with ruling men, generals, and kings, including the "king of the East," and "destruction of Arabs" betrays the document's roots in the Hellenistic and Roman-period Near East.

ATHENS CODEX 11 (BIB. PUBL. 1350 F. 15V): βροντολόγιον καὶ σεισμολόγιον ("THUNDER- AND EARTHQUAKE-DIVINATION").[73] A short text headed with the Greek letters used as numbers 1 through 12 and by zodiacal signs beginning with Virgo (as in the preceding text); each section ends with an additional prediction for cases of earthquake (e.g., "If an earthquake occurs, many diseases and deaths"). Its apodoses are not complete sentences, but their subject matter resembles topics of the Etruscan *Brontoscopic Calendar* and/or Mesopotamian omen series: danger for ships, abundance of pulses, barley, and fish, winds unstable, winter late, grain in short supply. "Wars among the Egyptians" is a telltale, however, for a Ptolemaic or Roman-period origin. Several predictions for "the City" might indicate Constantinople,

[70] *CCAG*:vol. 10; *Codices Athenienses*, 129–32.

[71] See Delatte's comment; *Codices Athenienses*, 129.

[72] See Boll's note in *CCAG*, 4:163.

[73] *CCAG*:vol. 10; *Codices Athenienses*, 140–42.

Alexandria, or Antioch, but there are Near Eastern precedents for it meaning the current capital of the empire, or the city in which thunder is heard.

SPAIN, EL ESCORIAL CODEX 1 (F. 161): βροντολόγιον ἕτερον ("A DIFFERENT TREATISE ON THUNDER-DIVINATION").[74] A very short text, probably condensed from some other document, gives one sentence of omens, "If it should thunder," and one for, "If an earthquake happens" beginning "in the month of September." Stock apodoses are of familiar types: death, famine, happiness for the upright, sickness, war.[75]

SPAIN, EL ESCORIAL CODEX 1 (F. 165): βροντολόγιον ("TREATISE ON THUNDER-DIVINATION").[76] (See translation in Appendix C, Text 3.) Another short text that begins in Aries, each sign concluding with an earthquake prediction. Some thunder-omens are qualified by locale – for example, "if thunder is in the East." The apodoses are close to the stock Mesopotamian types also found in the Etruscan *Brontoscopic Calendar*: the fortunes of the king, poor harvests, dearth of food, various diseases, fire in the city, disease among the cattle/sheep. Parallels with omens in the Figulus document are striking – yet the format, by zodiacal sign only, is a demonstrably later development. Certainly, the text's originators partook of the same Mesopotamian divination literature, but they must have done so through a Seleucid or other late filter.

GEOPONICA 1.10: *BRONTOLOGION* ATTRIBUTED TO ZOROASTER.[77] (See translation in Appendix C, Text 4.) The *Geoponica* is a compendium of fragments of numerous ancient Greek works on agriculture and related topics, sponsored (not authored) by the Byzantine emperor Constantine VII Porphyrogenitus (945–59 AD), a sponsor of other such compendia. The basic work was said to have been a collection of manuscripts made in the sixth or seventh century AD by Cassianus Bassus, who himself credited two earlier authors, Anatolius Vindonius of Beirut and Didymus of Alexandria. It does contain fragments of genuine, more ancient works, but the embedded, short brontological text (§ 1.10) attributed to the Persian prophet Zoroaster (variously dated either ca. 1000 BC or 6th c. BC) cannot have been a Persian original. It does, however, show the stock apodoses common to the Mesopotamian tablet texts and to the Etruscan *Brontoscopic Calendar*. It differs radically in that it is keyed to the first observed occurrence of thunder in the year, begun

[74] *CCAG*:vol. 11, pt. 1; *Codices Hispanienses*, ed. C. O. Zuretti (1932), 145–46.

[75] See Zuretti's notes; *Codices Hispanienses*, 145.

[76] *CCAG*:vol. 11, pt. 1; *Codices Hispanienses*, 155–57.

[77] *Geoponica sive Cassiani Bassi scholastici de re rustica eclogae*, ed. H. Beckh (Stuttgart and Leipzig: Teubner, original edition 1895, reprinted 1994), 19–20.

in Aries – thus, this text would only be invoked once in a year, and it seems unlikely that the last months would ever see the year's first electrical storm.

DISCUSSION OF GREEK ASTROLOGICAL TEXTS

Several of the Byzantine manuscripts incorporate *seismologia*, earthquake omens, but some modern editors make clear their suspicions that a manuscript copyist or author had simply melded two disparate texts. Some of the *brontologia*, like the *seismologia*, furnish observational categories absent from the Etruscan *Brontoscopic Calendar*: zodiacal signs, celestial phenomena, such as lunar phases or eclipses, weather or wind signs, or hours of daylight and darkness. Many apodoses bear remarkable resemblance to those of the Figulus document and to Mesopotamian literature. They must have been culled from the same sources, presumably texts preserved in the region of Syria through the Persian and Seleucid periods; frequent *contaminatio* with Egyptian references indicates other modes of transmission through the scholarly milieu of Ptolemaic Egypt, most probably Alexandria.

The manuscript texts that show the closest similarities to the Etruscan *Brontoscopic Calendar* are the Greek texts of El Escorial Codex 1 F. 165 and Athens Codex 11 F. 15v (both above). El Escorial Codex 1 F. 165 is furnished in Appendix C.3 in Greek, with my English translation. These manuscript texts originated and were copied in the Byzantine tradition but were based on or copied from documents of late Hellenistic-Roman astrological lore. The Roman astrological texts themselves were derived from Neo-Assyrian and Neo-Babylonian divination literature, as further developed in the cultures of the Achaemenid, Seleucid, and Ptolemaic Near East and Egypt. In contrast to the majority of late *brontologia*, these two examples, Escorial 1. F. 165 and Athens 11. F. 15v, although not formatted like the *Brontoscopic Calendar* of Figulus, and not nearly as long or complex, have so many apodoses that are similar or even identical to the Etruscan series that I am tempted to suggest that their master texts derived from the same set of texts that were used to formulate the Etruscan *Brontoscopic Calendar*. This could have been a compendium of excerpts from several famous documents, such as the *Enūma Anu Enlil, Iqqur īpuš, Šumma ālu*, and *Šumma izbu*: no single text contains all the apodoses that were collected for the Etruscan work. Perhaps one day, a Late Assyrian or Babylonian compendium with the same selections will be identified, or perhaps the person(s) who shared the Mesopotamian information with foreigners were skilled in all the major texts. And, of course, the Etruscan *Brontoscopic Calendar*, in its preserved form, has several dramatic apodoses that do not appear in any Near Eastern literature, *or* in the Greek brontological manuscripts: the band of noble youth, the women and the slaves, and certain diseases.

The Etruscan *Brontoscopic Calendar* of Figulus is longer and better developed than any of the Byzantine texts of thunder-omens, with an entry for each day of the theoretical year. Many of its apodoses clearly were created to fit the situations of Etruscan society and environment. It also seems to fit Seneca's criticism of Etruscan thunder-omens, namely, that the thunder lore is not carefully cataloged according to variables for ease of interpretation. In contrast, many of the Byzantine texts *are* qualified with conditions, like day versus night or geographical region, overlaid on the thunder observation. The effects of Greek and Hellenistic philosophical enquiry and writing have made their mark on the later *brontologia*. The complete absence, in the Figulus document, of regional references that would betray Hellenistic links, such as "the Arabs" or "the Egyptians," and of Oriental-Hellenistic situations relating to the "royal court" with crown princes and usurpers (as opposed to a mere "king"), are additional indications of the earlier derivation of the Etruscan document and more direct influence from Mesopotamian divination literature.

CHAPTER 10

CONCLUSION: ASSESSING THE ETRUSCAN *BRONTOSCOPIC CALENDAR* AND ITS HERITAGE

Much more could be revealed about the character of early Etruscan society – if we could only be certain of the date and details of the Etruscan original from which Figulus made his translation of the *Brontoscopic Calendar*. Because we can only work with the Byzantine Greek translation of Lydus, even though it is probably as accurate a rendering (via Figulus' Latin) as was possible in the sixth century AD, we cannot draw many firm conclusions about such things as diseases, animals, social unrest, and the like. The omens with moral overtones or calls for the expiation of misdeeds (e.g., May 16) all seem in keeping with what little we know of the *etrusca disciplina*, and no lines seem to be at odds with extant (albeit sparse) Etruscan religious texts.

One of the main discoveries that has emerged from my study is the astonishingly close correlation of approximately half of the *Brontoscopic Calendar*'s predictions with the sentiments and topics of the great works of Mesopotamian divination. No Akkadian or Babylonian tablet series features thunder alone as the triggering omen, nor do any give the same kind of day-by-day template for the predictions, but several of the great works, such as the *Enūma Anu Enlil*, the MUL.APIN, and *Šumma ālu*, do encompass dozens of the same apodoses. The *Iqqur īpuš*, a compendium of excerpted omens loosely formatted as a calendar, holds many of these predictions, but it cannot be the sole source of the Etruscan borrowings. Further, in no case does an "Etruscan" omen's outcome slavishly copy a Babylonian item, which would have had little meaning in Italy. In other words, there are no date palms, crown princes, or sesame seeds in the Etruscan *Brontoscopic Calendar*. And at the same time, several apodoses have no analogues in the cuneiform documents: the Etruscan authors have added a number of diseases, the notable status of women, and the more egalitarian situation of civil unrest (as opposed to king, court, and faceless masses in Mesopotamia), and also the perception of a servile class with its own tribulations – and opportunities for revolt.

It appears that the Standard Mesopotamian Calendar system and the literary works closest in subject and details to the *Brontoscopic Calendar* all were in use in Near Eastern court and private archives around the time of the Sargonid Neo-Assyrian empire. Certainly, the best sources for modern scholars for most of the relevant texts are the so-called library of Ashurbanipal at Nineveh, destroyed at the end of the seventh century BC, and similar troves found in other Assyrian and Babylonian excavation sites. Yet by what modes were cuneiform tablet texts accessible to Iron Age Etruscans? There is growing evidence – as archaeologists begin to excavate relevant early levels of tombs and settlements, and as we improve in the ability to identify foreign imports – of either direct contact between Etruscans and Assyrians or close, regular contact through intermediaries from an elite level of Levantine society. These were not simple traders but, rather, shipowners, diplomats, or other officials. These people deliberately cultivated their "fellow" elite (the rulers of central Italy), understood each other socially if not always linguistically, and communicated respectfully at the appropriate level of contact. The Near Eastern visitors were not proffering cheap baubles to naive indigenes, but rather they were purveying the details of a regal lifestyle, with chariots and parasols, thrones and ivory couches, to their perceived cousins in power in the West.

We have evidence from Roman history that the kings, beginning with Romulus, used divination (did Romulus cheat? many Romans thought so) to establish and then consolidate their control over the eighth-century city they founded on the shoulders of an unruly, heterogeneous population. (Whether or not we dispute the historical identity of Romulus, it is clear that the Romans themselves found perfectly plausible the notion of autocratic rulers or chieftains, the early exploitation of public divination, and the manipulation of formal religion by protohistoric states.) Some old Roman families probably even had archives and keepsakes going back to the Regal period. Kings like Numa Pompilius were known, to authors like Livy, to have created certain religious and civic institutions as frameworks for assuring and rewarding the cooperation of their subjects of various classes. They would have accomplished this by establishing formal priesthoods and religious shrines, public ceremonies and sacrifices ending with banquets, and by using lightning and thunder among other varieties of divination.

This "founder's strategy" is now attributable to eighth- to seventh-century Etruria in archaeological finds made at several cities. For instance, at Tarquinia Pian di Civita, permanent structures, cult buildings in foreign-style masonry, attest to a formalization of religious institutions at just the crucial period circa 690–680 BC. The public ceremonials are even preserved for us in a "foundation deposit" buried under the threshold, comprising the symbols carried in the inauguration ceremonies – sacrificial ax, ruler-warrior's parade shield, priest's musical *lituus* – and the banquet vessels and leftover consecrated food, all buried in sacred ground. At Veii Piazza d'Armi, there is the remnant of a formal foundation ceremony for the city in the form of a ritually plowed furrow, and other sites, from Como and Pisa to Vetulonia and

Marsiliana d'Albegna, hold possible *bidentalia* that look like tridents or forks, deposited in the burials of men who may have been divination priests and interpreted lightning and thunder.

Although Mesopotamian literature certainly encompasses lightning and thunder (the storm god Adad's "mighty voice") among other celestial omens, no actual thunder calendar per se has been found among the libraries of the first-millennium Near East. The first such fragments, found in one of the Qumran Caves, are of Roman date (1st c. AD), thus later than Figulus' Latin translation of the Etruscan *Brontoscopic Calendar.* Even though the Ashurbanipal library has yet to be fully published, it does not seem likely that an Akkadian or Neo-Babylonian *Brontoscopion* will be identified today.

How the Etruscan *Brontoscopic Calendar* was created was much more complex than simple translation or the borrowing of a complete (i.e., multi-tablet) text. Under Sargonid rule, the Neo-Assyrian court extended its control and influence firmly onto the Levantine coast, at a time just slightly preceding the Phoenicians' well-known initiative of colonization into the western Mediterranean. The cultures of East and West saw many intimate ties of statecraft and state cult, functioning at a high level of intensity just as the Phoenicians made their commercial excursion into central Italy, in such phenomena as the shared settling of Pithekoussai and the distribution of royal luxury goods to Etruria and Latium. It was particularly ironic that the climate of Italy and Mediterranean Europe had just entered, in the late ninth century, into a stormy phase that would last through the period of urbanization and codification of the cultic texts, increasing the likelihood of thunder on any given day. The Hallstatt atmospheric minimum almost guaranteed that Etruscans would pay attention to thunder-omens.

Of course, the Etruscans (and their Italic neighbors) already had a fully evolved religious system, even if, during the ninth to eighth centuries, they did not yet erect permanent monumental structures to commemorate it. Surely the indigenous belief systems already incorporated thunder and lightning lore as part of their mythology, just as they must have benefited from divination by hepatoscopy, to take advantage of the sacrifice of sheep from the flocks that had been bred since the Neolithic period. Yet the details of the theoretical "plan" of the Piacenza bronze and other Etruscan liver models simply must have been influenced by the rich "geographical" terminology and schematics of Mesopotamian liver divination. Unfortunately, it seems that the only way we might prove an abiding link to the Near East would be through texts and vocabulary, which for Etruria are conspicuously lacking.

Another field ripe for comparative investigation is the study of Mesopotamian traditions of mensuration and boundary commemoration with stone-carved *kudurrus*, which assert the patronage of the gods – and especially goddesses – over the allotting, recording, and marking of owners' lands. A number of carved *kudurrus* are known, resembling *cippi* carved with images of rulers, gods, symbols, and inscriptions – they often warn of punishment for those who disrespect the boundary. These monuments represent the last stage in a long process of allotment and legal transactions, and they replicate documents deposited with the legal or religious authorities. The *kudurru* tradition extended from the fourteenth through seventh centuries BC, but it seems to have flourished especially in the period ca. 1300–1050 BC, in the region of northern Babylonia (Babylon, Sippar, the Diyala Valley). Many *kudurrus* ended up in temple precincts for safekeeping, but they may have originally been erected on the actual boundaries they delineated. There might not be a direct correlation between the *Prophecy of Vegoia*, the Perugia *cippus*, or the *Tabula Cortonensis* and the Mesopotamian *kudurrus*, but there is surely a related outlook in the concept of divine mensuration and patronage of land division and physical boundaries along with the close involvement of the government/ruler in the procedure.[1] For instance, one famous *kudurru* shows Nebuchadnezzar I himself (ca. 1125–1104 BC) bestowing a land grant on a favorite official. Such institutions would undoubtedly be useful in the ordering of newly urbanized territories.

An additional instance of the link between gods – actually goddesses – and measurement is the Mesopotamian myth of the Descent of Inanna-Ishtar: as she descends into the Underworld, she is stripped of her robes and regalia but retains, almost until the end, her measuring rod and tape. This story is at least third millennium in origin and defines the goddess's link to justice;[2] it may even be behind the imagery of such Etruscan treasures as the Cannicella goddess. According to a novel interpretation presented by Nancy de Grummond in the Orvieto Workshop of the 2009 Annual Meetings of the Archaeological Institute of America, the nude and much-handled statuette from a shrine in the Orvietan necropolis may have been dressed and undressed in the course of rituals of Eastern inspiration.[3] Once again, though, the rich Sumerian and Akkadian literature of goddesses who measure and mete justice contrasts with our total loss of comparable Etruscan texts.

Contact with and exposure to Near Eastern texts and divination practices happened for Etruria at just the right moment, when such things as ceremonies, calendars, religious scriptures, and symbolic implements, long used by Assyrian rulers to consolidate their power over a diverse populace, could be adapted and put to good use in the establishment of the Etruscan cities. In the case of secular emblems and power, many cities must have picked up on

1 For full references and description on *kudurrus*, see Robson 2008:166–76; for the Cortona mortgage bronze plaque, see Agostiniani and Nicosia 2000, Facchetti 2000:59–88.

2 See Robson 2008: 115–23.

3 Compare the Old Babylonian Burney Relief in the British Museum, nicknamed "The Queen of the Night," in which Inanna, nude and morphing into a bird of the Underworld, still brandishes her measuring lines; see Robson 2008:cover illustration; von der Osten-Sacken 2002, ca. 1800 BC, otherwise identified as Ereshkigal.

these developments at the same time, for we see similar symbolic riches and ceremonies across the peninsula, at Adriatic Verucchio just as at Tyrrhenian Vetulonia and Tarquinia, and inland Chiusi and Orvieto. With the symbols must have come the appropriate traditions – not just a throne and footstool, for instance, but a code of regal behavior. We see traces of lengthy ceremonials in the details of seventh-century foundation deposits and sixth-century images of processions, and in funeral rites. What exactly did those rulers at Verucchio and Caere, Tarquinia and Murlo actually believe and say, and with whom did they confer when developing the organization of their people and land?

In contrast, we may not need to postulate multiple sites and parallel "evolution" for the dissemination of Near Eastern religious teachings in Etruria. The brontoscopic tradition, maybe even the entire original recension of the *etrusca disciplina*, could have descended through a single priest and his family, just as the extant myth more imaginatively records a single supernatural seer, Tages, and a single, noble recorder, Tarchon. Somewhere, some Etruscan must have sat down – at length – with some Near Eastern emissary (a pious Phoenician diplomat? an itinerant Babylonian prophet or a scholarly translator such as a *tupšaru Enūma Anu Enlil*?) and learned the gist of some of the great Mesopotamian literature. Already known to our Etruscan author were folkloric proverbs or even cultic refrains linking thunder to certain environmental events, such as drought, famine, or disease. Eventually, early in the seventh century BC, he or she wrote down something uniquely Etruscan, combining both traditions. In contrast, to directly copy either a complete text or long excerpts from several works would have taken a lifetime of learning. To gain an understanding of the concepts behind some passages in the famous literature, with Mesopotamian scholarly help, and then create a unique blend of Italian environmental lore and Near Eastern formality would have been eminently feasible. Many of the recently postulated points of comparison between the Etruscan *Brontoscopic Calendar* and the Mesopotamian texts could only have been adopted as the result of lengthy and cordial association of literate and devoted individuals (similar situations brought the alphabet and monumental architecture from the Levant to the Mediterranean world).

In addition to Near Eastern modes of sacrifice and divination, it appears that Etruria may have derived at least a portion of its expertise in engineering and hydraulics from the East, perhaps even through the exchange of craftsmen, as is well known in Mesopotamia and the Levant (King Hiram of Tyre's loan of architects to Solomon for the building of the Temple comes to mind). If, as Jodi Magness (2001) has suggested, some eighth-century Levantine immigrants actually joined the Etruscan elite, through marriage or adoption, we have a logical explanation for the avidity and rapidity with which some families seem to have adopted the funeral traditions of the East. These appeared full-blown as large chamber tombs with gabled interiors ornamented with imitation beams, pillars, and stone-carved tomb furniture, including beds/benches with semicircular headrests. Enthroned statues of ancestors with Levantine floral scepters and Etruscan *litui* sat with their feet on Judaean-style roll-footstools; the previous generation had known nothing at all comparable to this. Thulin,[4] Burkert, West, Morris, and others have suggested that various details of religious observance adopted in Greece and the Mediterranean cultures began with cults in the Near East, among them the use of incense, liturgy, animal sacrifice, temple building, and certain myths. Although we may contest particular aspects of the process, it is clear that prolonged and good-willed human interaction across cultures did create an Iron Age amalgam of cultic and civic traditions.

We do not really know if the *etrusca disciplina* was a universal sort of phenomenon in the Iron Age – it might have been unique to one city like Tarquinia and may have spread only much later (all the surviving literary references to it are 4th c. or later). It does seem reasonable, though, to extrapolate from epigraphic texts, such as the early fifth-century Capua tile, and predicate an Archaic, seventh- to sixth-century BC character for the formal discipline. On the other hand, noting how rapidly and thoroughly the *civic* symbols spread across central Italy, it would not be surprising to find an early acceptance of shared (Etruscan) cultic texts. We simply lack the actual texts, whether recorded on organic media or valuable materials, such as bronze or ivory. If, as I suspect, the uniquely Etruscan additions to the repertoire of brontoscopic omens derive from some sort of oral, folk tradition, then it should not be surprising to find this form of divination all over Italy. The early *bidentalia*, if they are indeed symbols of this, as they were in Hittite, Urartian, Assyrian, and other Eastern cultures, do show a wide and early distribution in Italy, beginning with one buried in an eighth-century Golaseccan chief's tomb near Como, and represented by fine examples deposited in elaborate funeral rituals at seventh-century Vetulonia, Pisa, and Marsiliana d'Albegna.

No single text known today holds examples of *all* the Mesopotamian apodoses that were adapted for the Etruscan *Brontoscopic Calendar*. So it is evident that the process must have been more complex than a simple borrowing, even if it did not involve some Etruscan(s) actually learning cuneiform Akkadian, or alphabetic Aramaic. It is possible that some discrete number of Italians spent time at some Near Eastern or Levantine court. Such finds as the Assyrian metal bowls and lion *situlae* (Fig. 16) show that understanding and gift-giving certainly occurred on royal levels (even if the Etruscan chieftains may have reciprocated with shiploads of iron ore). The deeper meanings of regal scepters, ivory furniture, chariots, thrones and footstools, parasols and fans, were understood by many Eastern cultures, and certainly these implements – in parallel, though slightly earlier, sets – reached Cyprus and Spain

[4] Thulin 1906:vol. 1, x–xv.

by Phoenician agency. Again, the concepts and specifics of regal behavior, formal law codes, state religious cults, and such were also shared. We may see, in the masonry at Tarquinia Pian di Civita, the material proof that some perceptive Etruscan rulers recognized and grasped Near Eastern models for support in the new cities they were launching. I think it is a logical extension of this phenomenon to see a parallel early engagement with the written texts that formed integral parts of the regimes of any of Etruria's Eastern visitors.

It has been easier to track actual objects, such as Assyrian imports and their local imitations, than to track the complex concepts that must have been shared simultaneously across cultures at the onset of the Orientalizing period. The lore of divination is surely an example of the process of stimulus diffusion, by which a concept, rather than a physical object, is copied across cultures. We know that this was also happening – at the same times and places – with the material building blocks of the Etruscan cities, the terracotta tiled roofs. We can examine the first roof tiles in Greece, found in the seventh-century Corinthia; tiles next appear in mid-seventh-century Etruria at Acquarossa and Poggio Civitate (Murlo). The Greek examples, the original invention, represent a wonderful breakthrough in architecture, aesthetics, and safety. But they were huge and ungainly and could only be handled or fitted by two men: their awkwardness was not a problem in Greece, for they were used only on monumental structures, especially temples, where the work of worshipers was a part of their devotional contribution, and a time span of many years was expected to complete the project. In Etruria, however, the roof tiles are from the very beginning smaller, simpler, more modular – and they are found, almost overnight, on common people's homes and on monumental buildings. It is clear that what Etruria derived from the Corinthian builders was the concept, not the material details. No large compound tiles were imported and slavishly copied in 650 BC, but rather, some smart workers got together with their opposite numbers from an alien culture and talked about the new invention, perhaps even working side by side on an actual building site.

Only in this way would the new builders gain sufficient confidence in the endeavor. This has even been associated with the legend of Demaratus, a royal Corinthian exile who married into an elite Tarquinian family and fathered a future king of Rome.[5] This diffusion of tiled roofing implies great familiarity and trust – first, in the use of some common means of communication, and second, in the matter of safety issues (why won't this heavy thing fall down on my head and kill me?), and also in the committing of skilled labor to a newfangled and risky proposition. (If this building does not work, my crew could starve.)[6]

[5] See Ridgway and Ridgway 1994; Williams 1980; Ampolo 1976–77.

[6] For references on tiled roofing, engineering, and safety, see Turfa and Steinmayer 1996; Wikander 1988, 1992.

A similar process must have operated in the case of the *Brontoscopic Calendar*, but on a smaller scale than with tiled roofing. Many different families had to adopt, simultaneously, the new housing technology. In contrast, the composition of the Etruscan *Brontoscopic Calendar* could have been the work of only one priestly family, perhaps that of a ruler-priest, as exemplified by the noble warrior buried at Verucchio in the Tomb of the Throne. Such a family would have interfaced over a period of time with elite foreign residents, metics, or diplomats; they may even have sent a child to an Eastern court to be educated. The ultimate textual product would have been formulated from the Etruscan priest's own stock of native folklore and divination expertise, compounded with a Near Eastern standard calendar template (also useful as a civic tool), and a mixture of apodoses as known from the *Enūma Anu Enlil* (possibly from tablets 40–50, but not necessarily), the MUL.APIN, *Šumma alu ina Mēlê Šakin*, and *Šumma izbu*. The text we know for the great compendium cast in calendar format, the *Iqqur īpuš*, contains some but not all of the relevant omen apodoses that occur in the *Brontoscopic Calendar*. Some similar anthological text, composed of carefully indexed excerpts from the great works, would be the simplest source for the inspiration of the Etruscan document – but so far none has been found, and so we must regard the creation of the *Brontoscopic Calendar* as a more lengthy and complex process, perhaps some anonymous priest's lifework, encouraged by one or more Mesopotamian or Levantine scholars.

A few of the Late Roman-Byzantine *brontologia* cataloged in the *CCAG* seem, on superficial reading, to contain similar material to the text of Figulus. Apart from the first-century AD Qumran papyrus fragment, the earliest extant texts are the short pieces attributed to Fonteius Capito (1st c. BC) and Claudius Tuscus (Augustan/1st c. AD). The Capito document (Appendix C, Text 1) shares a few omen apodoses with the Etruscan *Brontoscopic Calendar*, but it uses the zodiac and territorial designations from Ptolemaic or later scholarship. Although it claims to be "from the sacred writings of the Etruscans, word for word," the Latin environmental poem of Claudius Tuscus uses a different format, different apodoses, and the Roman day system of Kalends, Nones, and Ides, beginning in the month of January like Roman Imperial calendars. Subsequent texts diverge ever more widely from the details and format of the Figulus text (see Table 9.1).

It seems possible that one or two of the late texts (see Appendix C) were derived in part from the Figulus document, but there is nothing compelling to indicate that the Etruscan *Brontoscopic Calendar* itself was derived from Hellenistic, Ptolemaic, or Seleucid versions of the old Mesopotamian tablet series.

A knowledge of the great Mesopotamian tablet texts, lost for almost two millennia and excavated only in the nineteenth century, can allow us to understand the composition and roots of the Etruscan *Brontoscopic Calendar*.

It can only be fully appreciated, however, if we interpret it in the light of recent archaeological discoveries of the life of Iron Age settlements and their ecology. They illustrate the vivid and dynamic environment in which Etruria was formed. It does not take very much imagination to envision the society of the Iron Age as portrayed in the *Brontoscopic Calendar*. It was a time when anything could – and often did – happen: a frail child might prophesy and disappear, a whole community could move overnight from a cluster of huts to a row of tiled, cozy houses in a new city like Acquarossa or Veii. (Those left behind in huts or otherwise disadvantaged by the new urban customs would contribute to a new atmosphere of civil unrest.) For each day that the city assured a regular supply of food and safe, clean water, there might be another day when war raged, or riots, or diseases appeared that had not been known before. Already, ecological diversity was being lost, and no one cared to change that. Now, only the elite could spare the time or land for the hunting of deer and boar; although wine was now within reach of most people, at least occasionally, and grains were being made into daily bread, many of the wild foods, fruits, and nuts were disappearing, and many more pests were caught damaging crops and food stores. And where before everyone seemed to be on equal footing in their hardships, now society was stratifying, and in order for there to be haves, a group of have-nots would be fostered, which later authors could only describe as slaves. Although that condition may not have been as abject as it was in the Roman Late Republic and Empire, such disadvantaged persons, whether natives who had given up land rights or prisoners of war from nearby Italic settlements, must have been a constant concern for those who sought to control them. Sometimes the thunder in the Etruscan sky must have been the rumblings of a guilty collective conscience. (We know of an event in the 3rd c. BC, at Volsinii for instance, where the situation erupted into a revolt of the freed and enfranchised slaves, terminating in the Roman take-over and evacuation of Orvieto – but the slavery problem, like all the other issues of urban life, may be traced back to the Iron Age also.)

Furthermore, the tradition of the Etruscan *Brontoscopic Calendar* may have exerted a greater influence on Roman and Western society than has previously been acknowledged. That rare window into Republican Roman cult practices, Cicero's speech *De haruspicum responso*, enables us to suspect that Etruscan liturgy remained a mainstay of Roman state religion down to the end of the Republic, at least. And the tone, dating formats, and concerns expressed in the predictions of the later brontological texts owe much to the *etrusca disciplina*, even though the documents used for the core of their scholarly composition were undoubtedly the products of Ptolemaic or Seleucid astrological literature. The *Brontoscopic Calendar*'s focus on issues such as class friction, social unrest and revolt, the changing status of women, and the institution of slavery, once read as the expression of fourth-century or even first-century BC social troubles and land reforms, may now, thanks to archaeological research, be understood as the underlying concerns of the founders of the Etruscan cities of the Late Villanovan period. They turned often to the Near East to borrow symbols and solutions to the previously unknown travails of an urban society growing complex and troubled – and rich. One of the tools they utilized was the *Brontoscopic Calendar*, the product of millennia of Mesopotamian scholarship, forged by the creative genius of some anonymous Etruscan man – or woman.

APPENDIX A

TEXTS RELATING TO THE STUDY OF ETRUSCAN RELIGION

Unless noted, English translations are my own.

TEXT 1. CICERO *DE DIVINATIONE* 2.50–51: THE STORY OF TAGES AND THE *ETRUSCA DISCIPLINA* (LATIN TEXT AFTER FALCONER 1923/1996)

Sed quid plura? ortum videamus haruspicinae; sic facillime quid habeat auctoritatis iudicabimus. Tages quidam dicitur in agro Tarquiniensi, cum terra araretur et sulcus altius esset impressus, extitisse repente et eum affatus esse, qui arabat. Is autem Tages, ut in libris est Etruscorum, puerili specie dicitur visus, sed senili fuisse prudentia. Eius aspectu cum obstupuisset bubulcus clamoremque maiorem cum admiratione edidisset, concursum esse factum, totamque brevi tempore in eum locum Etruriam convenisse; tum illum plura locutum multis audientibus, qui omnia verba eius exceperint litterisque mandarint; omnem autem orationem fuisse eam, qua haruspicinae disciplina contineretur; eam postea crevisse rebus novis cognoscendis et ad eadem illa principia referendis.

Haec accepimus ab ipsis, haec scripta conservant, hunc fontem habent disciplinae. Num ergo opus est ad haec refellenda Carneade? Num Epicuro? Estne quisquam ita desipiens, qui credat exaratum esse – deum dicam an hominem? Si deus, cur se contra naturam in terram abdiderat, ut patefactus aratro lucem aspiceret? Quid? Idem nonne poterat deus hominibus disciplinam superiore e loco tradere? Si autem homo ille Tages fuit, quonam modo potuit terra oppressus vivere? Unde porro illa potuit, quae docebat alios, ipse didicisse? Sed ego insipientior quam illi ipsi, qui ista credunt, qui quidem contra eos tam diu disputem.

But what more? Let us examine the origin of haruspicy; thus we shall very easily judge its authority. It is said that, once upon a time, in the district of Tarquinii, while the earth was being plowed, the plowshare plunged deeper and a certain Tages suddenly sprang forth and spoke to the man who was plowing. Now this Tages, according to the books of the Etruscans, is said to have had the appearance of a boy, but his wisdom was that of an old man. The peasant was shocked by his apparition, and marveling raised a great call; a crowd gathered at that place in a short time, and the whole of Etruria met on the spot. Tages then spoke at length to his many listeners, who received all his words and committed them to writing. His whole speech, however, was on the field of *haruspicina* (divination). Afterward, as new facts were learned and tested against these principles, they were added to the original knowledge.

This we have received from the Etruscans themselves, this their writings preserve, this they hold is the source of the *disciplina*. Now is it necessary for a Carneades or an Epicurus to refute this? Is there anyone so stupid that he believes that a god – or should I say a man – was ever plowed up? If a god, why did he, contrary to his nature, hide himself in the ground so that he could be exposed to the light of day by a plow? What? Wasn't this god able to pass on this learning to men from a loftier vantage point? If, however, Tages was a mortal, then by what means could he have survived beneath the earth? And whence was he himself able to learn that which he taught to others? But I am stupider than those people who actually believe this stuff, since I have wasted so much time debating them.

TEXT 2. *PROPHECY OF VEGOIA*

The text is found in the *Gromatici veteres/Corpus agrimensorum* from Lachmann and A. Rudorff 1848:350.17–351.11. The basic reference is Valvo 1988:3; see also Heurgon 1959. A recent English translation is by Celia Schultz (2006:222–23); she reminds us that around line 7 the text is corrupted and words must be missing. I quote below the translation of de Grummond and Simon 2006:191–92, to emphasize that the language is similar in style to that of such terse religious documents as the Etruscan *Brontoscopic Calendar.*

Idem Vegoiae Arrunti Veltymno.

Scias mare ex aethera remotum. Cum autem Iuppiter terram Aetruriae sibi vindicavit, constituit iussitque metiri campos signarique agros. Sciens hominum avaritiam vel terrenum cupidinem, terminis omnia scita esse voluit. Quos quandoque quis ob avaritiam prope novissimi octavi saeculi data sibi [corrupt text] homines malo dolo violabunt contingentque atque movebunt. Sed qui contigerit moveritque, possessionem promovendo suam, alterius minuendo, ob hoc scelus damnabitur a diis. Si servi faciant, dominio mutabuntur in deterius. Sed si conscientia dominica fiet, caelerius domus extirpabitur, gensque eius omnis interiet. Motores autem pessimis morbis et vulne-

ribus efficientur membrisque suis debilitabuntur. Tum etiam terra a tempestatibus vel turbinibus plerumque labe movebitur. Fructus saepe ledentur decutienturque imbribus atque grandine, caniculis interient, robigine occidentur. Multae dissensiones in populo. Fieri haec scitote, cum talia scelera committuntur. Propterea neque fallax neque bilingius sis. Disciplinam pone in corde tuo.

(Prophecy) of Vegoia, to Arruns Veltymnus

Know that the sea was separated from the sky. But when Jupiter claimed the land of Aetruria for himself, he established and ordered that the fields be measured and the croplands delimited. Knowing the greed of men and their lust for land, he wanted everything proper concerning boundaries. And at some time, around the end of the eighth *saeculum*, someone will violate them on account of greed by means of evil trickery and will touch them and move them. [...] But whoever shall have touched and moved them, increasing his own property and diminishing that of another, on account of this crime he will be damned by the gods. If slaves should do it, there will be a change for the worse in status. But if the deed is done with the master's consent, very quickly the master will be uprooted and all of his family will perish. The ones who move [the boundaries] will be afflicted by the worst diseases and wounds, and they will feel a weakness in their limbs. Then also the earth will be moved by storms and whirlwinds with frequent destruction, crops often will be injured and will be knocked down by rain and hail, they will perish in the summer heat, they will be felled by mildew. There will be much dissension among people. Know that these things will be done when such crimes are committed. Wherefore be not false or double-tongued. Keep this teaching in your heart.

TEXT 3. QUOTED LINES FROM THE HARUSPICES' RESPONSES OF 56 BC, CITED BY CICERO *DE HARUSPICUM RESPONSO* (LATIN AFTER LOEB TEXT, WATTS 1923/1961)

§ 9 *loca sacra et religiosa profana haberi* – "sacred and hallowed sites were being turned to secular purposes"

§ 20 *quod in agro Latiniensi auditus est strepitus cum fremitu* – "inasmuch as a rumbling and a noise have been heard in Latin lands"

exauditus in agro propinquo et suburbano est strepitus quidam reconditus et horribilis fremitus armorum – "in the immediate outskirts of the city there has been heard a subterranean rumbling and an awful noise as of arms"

postiliones esse Iovi, Saturno, Neptuno, Telluri, dis caelestibus – "arrears of sacrifice are due to Jupiter, Saturn, Neptune, Earth and the gods of heaven"

§ 21 *ludos minus diligenter factos pollutosque* – "the games have been performed with laxity, and have been desecrated"

§ 34 *oratores contra ius fasque interfectos* – "ambassadors have been slain contrary to law human and divine"

§ 36 *fidem iusque iurandum neglectum* – "fealty and oath has been neglected"

§ 37 *sacrificia vetusta occultaque minus diligenter facta pollutaque* – "ancient and secret sacrifices have been performed with laxity, and have been desecrated"

§ 40 *ne per optimatium discordiam dissensionemque patribus principibusque caedes periculaque creentur, auxilioque divinitus deficiantur, qua re ad unum imperium provinciae redeant exercitusque deminutioque accedat* – "let not death and danger be wrought for the fathers [senators] and for statesmen by reason of the discord and division of the nobles, and let them not be bereft of divine power; and let the state lapse not to the rule of one" [the last lines are corrupted in the MSS]

§ 55 *ne occultis consiliis res publica laedatur* – "that the republic be not harmed by secret designs"

§ 56 *ne deterioribus repulsisque honos augeatur* – "let not honor be increased for the baser and the rejected"

§ 60 *ne rei publicae status commutetur* – "change not the condition of the republic"

TEXT 4. DIODORUS SICULUS *THE LIBRARY OF HISTORY*, 5.40.1–2: ETRUSCAN CULTURE

(Text: Diodori Bibliotheca Historica, Vols. 1–2. Diodorus Siculus. Immanel Bekker. Ludwig Dindorf. Friedrich Vogel. in aedibus B. G. Teubneri. Leipzig. 1888–1890)

[1] λείπεται δ᾽ ἡμῖν εἰπεῖν περὶ τῶν Τυρρηνῶνυ οὗτοι γὰρ τὸ μὲν παλαιὸναἀνδρείᾳ διενεγκόντες χώραν πολλὴν κατεκτήσαντο καὶ πόλεις ἀξιολόγους καὶ πολλὰς ἔκτισανκόὁμοίως δὲ καὶ ναυτικαῖς δυνάμεσιν ἰσχύσαντες καὶ πολλοὺς χρόνους θαλαττοκρατήσαντες τὸ μὲν παρὰ τὴνaἸταλίαν πέλαγος ἀφ᾽ ἑαυτῶνυἐποίησαν Τυρρηνικὸν προσαγορευθῆναισ τὰ δὲ κατὰ τὰς πεζὰς δυνάμεις ἐκπονήσαντες τήν τε σάλπιγγα λεγομένην ἐξεῦρονο εὔχρηστοτάτην μὲν εἰς τοὺς πολέμους, ἀπ᾽ ἐκείνων δ᾽ ὀνομασθεῖσαν Τυρρηνήν, τό τε περὶ τοὺς ἡγουμένους στρατηγοὺς ἀξίωμα κατεσκεύασαν, περιθέντες τοῖς ἡγουμένοις ῥαβδούχους καὶ δίφρον ἐλεφάντινον καὶ περιπόρφυρον τήβενναν, ἔν τε ταῖς οἰκίαις τὰ περίστῳα πρὸς τὰς τῶν

θεραπευόντων ὄχλων ταραχὰς ἐξεῦρον εὐχρηστίαν: ὧν τὰ πλεῖστασ̓Ρωμαῖοι μιμησάμενοι καὶ πρὸς τὸ κάλλιον αὐξήσαντες μετήνεγκαν ἐπὶ τὴνεἰδίαν πολιτείαν.

[2] γράμματα δὲ ράὶ φυσιολογίαν καὶ θεολογίαν ἐξεπόνησαν ἐπὶ πλέον, καὶ αὰ πλέὶ λὴν κεραυνοσκοπίαν μάλιστα πάντων ἀνθρώπων ἐξειργάσαντο: διὸ ινὶ μέχρι τῶν νῦν χρόνων οἱ ῆς οἰκουμένης σχεδὸνχὅλης ἡγούμενοι θαυμάζουσί τε τοὺς ἄνδρας καὶ αραὰ αὰς ἐν τοῖς κεραυνοῖς διοσημείας τούτοις ἐξηγηταῖς χρῶνταιτ

(40.1) It remains for us now to speak of the Tyrrhenians. This people, excelling as they did in manly vigor, in ancient times possessed great territory and founded many notable cities. Likewise, because they also availed themselves of powerful naval forces and were masters of the sea over a long period, they caused the sea along Italy to be named Tyrrhenian after them; and because they also perfected the organization of land forces, they were the inventors of the salpinx, as it is called, a discovery of the greatest usefulness for war and named after them the "Tyrrhenian trumpet." They were also the authors of that dignity which surrounds rulers, providing their rulers with lictors and an ivory stool and a toga with a purple band; and in connection with their houses they invented the peristyle, a useful device for avoiding the confusion connected with the attending throngs; and these things were adopted for the most part by the Romans, who added to their embellishment and transferred to them their own political institutions.

(40.2) Letters, and the teaching about Nature and the gods they also brought to greater perfection, and they elaborated the art of divination by thunder and lightning more than all other men; and it is for this reason that the people who rule practically the entire inhabited world show honor to these men even to this day and employ them as interpreters of the omens of Zeus as they appear in thunder and lightning. (Diodorus/Oldfather 1939:207–9)

TEXT 5. LUCRETIUS *DE RERUM NATURA* 6.164–73: PERCEPTION OF THUNDER (TEXT AND TRANSLATION: MELVILLE 1999:184, 190)

Sed tonitrum fit ut post auribus accipiamus,
fulgere quam cernant oculi, quia semper ad auris
tardius adveniunt quam visum quae moveant res.
id licet hinc etiam cognoscere, caedere si quem
ancipiti videas ferro procul arboris auctum,
ante fit ut cernas ictum quam plaga per auris
det sonitum, sic fulgorem quoque cernimus ante
quam tonitrum accipimus, pariter qui mittitur igni
e simili causa, concursu natus eodem.

Hoc est igniferi naturam fulminis ipsam
perspicere et qua vi faciat rem quamque videre,
non Tyrrhena retro volventem carmina frustra
indicia occultae divum perquirere mentis,
unde volans ignis pervenerit aut in utram se
verterit hinc partem, quo pacto per loca saepta
insinuarit, et hinc dominatus ut extulerit se,
quidve nocere queat de caelo fulminis ictus.

Our ears receive the sound of thunder later
Than our eyes see the lightning, for this reason:
Things always come more slowly to the ears
Than to the eyes; as this example shows:
If in the distance you observe a man
Felling a tall tree with twin-bladed axe
You see the stroke before the sound of it
Reaches your ears; so also we see lightning
Before we hear the thunder, which is produced
At the same time as the fire, and by the same cause,
Born of the same collision of the clouds.

Thus the true nature of the thunderbolt
Can now be understood, and how it works;
Not by unrolling scrolls of Tuscan charms
To search in vain the hidden minds of gods
And ask them whence the flying fire has come
Or to what other quarter of the sky
It went, and in what way it penetrated
Through walls of buildings, and having worked its will
Inside, made its way out again, or ask what harm
The stroke of a thunderbolt from heaven can do.

TEXT 6. CICERO *DE DIVINATIONE* 2.42: DIVINATION BY LIGHTNING AND THE STATE (LATIN TEXT AFTER FALCONER 1923/1996)

Atque hac extispicum divinatione sublata omnis haruspicina sublata est. Ostenta enim sequuntur et fulgura. Valet autem in fulguribus observatio diuturna, in ostentis ratio plerumque coniecturaque adhibetur. Quid est igitur, quod observatum sit in fulgure? Caelum in sedecim partis diviserunt Etrusci. Facile id quidem fuit, quattuor, quas nos habemus, duplicare, post idem iterum facere, ut ex eo dicerent, fulmen qua ex parte venisset. Primum id quid interest?

deinde quid significat? Nonne perspicuum est ex prima admiratione hominum, quod tonitrua iactusque fulminum extimuissent, credidisse ea efficere rerum omnium praepotentem Iovem? Itaque in nostris commentariis scriptum habemus: "Iove tonante, fulgurante comitia populi habere nefas." Hoc fortasse rei publicae causa constitutum est; comitiorum enim non habendorum causas esse voluerunt. Itaque comitiorum solum vitium est fulmen, quod idem omnibus rebus optimum auspicium habemus, si sinistrum fuit.

And so, in dismissing divination by extispicy, we have dismissed haruspicy. Now will follow portents and lightning. (You say that) lengthy observation is required for lightning (divination), and reasoning and discussion are usually applied for portents. So what is it then that is observed in the case of lightning? The Etruscans divided the sky into sixteen parts. Of course it was easy to double the four quarters, which we use, and then double them again, and then to tell which sector a lightning bolt had come from. First of all, what difference does that make? And then, what does it mean? Isn't it evident that, out of the primitive wonder of men, who feared thunder and lightning strikes, they came to believe that these things were brought about by all-powerful Jupiter? And so we find it written in our [augural] commentaries: "With Jupiter thundering, sending lightning, impious to hold an election of the people." Perhaps this was established for political purposes; they may have had reasons for not wanting to hold meetings. And so lightning is only undesirable in case of an election; in all other affairs, we consider it the best omen, if it appears on the left.

TEXT 7. PLINY *NATURALIS HISTORIA* 2.137–38: ANALYSIS OF THUNDERBOLTS, INCLUDING ETRUSCAN DIVINATION (LATIN TEXT AFTER RACKHAM 1938/1991)

Fulminum ipsorum plura genera traduntur. quae sicca veniunt non adurunt sed dissipant, quae fumida non urunt sed infuscant. tertium est quod clarum vocant, mirificae maxime naturae, quo dolia exhauriuntur intactis operimentis nulloque alio vestigio relicto, aurum et aes et argentum liquatur intus, sacculis ipsis nullo modo ambustis ac ne confuso quidem signo cerae. Marcia femina princeps Romanarum icta gravida partu exanimato ipsa citra ullum aliud incommodum vixit. in Catilinariis prodigiis Pompeiano ex municipio M. Herennius decurio sereno die fulmine ictus est.

Tuscorum litterae novem deos emittere fulmina existimant, eaque esse undecim generum, Iovem enim trina iaculari. Romani duo tantum ex iis servavere, diurna attribuentes Iovi, nocturna Summano, rariora sane eadem de causa frigidioris caeli. Etruria erumpere terra quoque arbitratur, quae infera appellat, brumali tempore facta saeva maxime et exsecrabilia, cum sint omnia quae terrena existimant non illa generalia nee a sideribus venientia sed ex proxima atque turbidiore natura: argumenturn evidens, quod omnia e superiore caelo decidentia obliquos habent ictus, haec autem quae vocant terrena rectos. et quae ex propiore materia cadunt ideo creduntur e terra exire quoniam ex repulsu nulla vestigia edunt, cum sit illa ratio non inferi ictus sed aversi. a Saturni ea sidere proficisci subtilius ista consectati putant, sicut cremantia a Martis, qualiter cum Volsinii, oppidum Tuscorum opulentissimum, totum concrematum est fulmine. vocant et familiaria, in totam vitam fatidica, quae prima fiunt familiam suam cui que indepto. ceterum existimant non ultra decem annos portendere privata praeterquam aut primo matrimonio facta aut natali die, publica non ultra tricesimum annum praeterquam in deductione oppidi.

Of actual thunderbolts several varieties are reported. Those that come with a dry flash do not cause fire but explode; those that are smoky do not burn but blacken. The third kind is what they call "bright," of an exceptionally amazing nature, that drains storage jars dry without damaging their lids and without leaving any other trace, that melts gold and copper and silver in their bags without even singeing the bags themselves, and even without melting the wax seal. Marcia, a Roman noblewoman, was pregnant when struck by lightning; her child was killed but she herself survived without any other injury. Among the portents relating to the Catilinarian conspirators, a decurion of the Pompeian town-council, Marcus Herennius, was struck by lightning on a clear day.

The writings of the Etruscans claim that nine gods can send thunderbolts of eleven kinds, with Jupiter hurling the other three. Only two of them have been retained by the Romans, who attribute daytime thunderbolts to Jupiter and nighttime ones to Summanus, since these are rare because of the colder sky at night. Etruria thinks that some also erupt from the earth, and calls these underworld (bolts), made exceptionally violent and accursed in the winter season, because they believe that all the earthly bolts are not the ordinary ones coming from the stars but from the closer and more turbulent element: a clear proof of this is that all those coming from the upper heaven deliver slanting blows, however these which they call earthly strike directly. And those that fall from the nearer elements are believed to issue from the earth because they leave no traces (on the ground), although that is the action not of a downward blow but of a slanting one. Those who pursue these enquiries in greater detail think that these bolts come from the planet Saturn, just as the fiery ones come from Mars, as, for instance, when Volsinii, the richest town of the Etruscans, was entirely consumed by fire from a thunderbolt. Also the first ones that occur after a man sets up a household (*familia*) for himself are called *familiaria* as they foretelling his entire family life. Other people think that private portents, except for those that occur either at a man's first marriage or on his birthday, do not prophecy more than ten years, nor public portents beyond the thirtieth year, except those occurring at the colonization of a town.

TEXT 8. PLINY *NATURALIS HISTORIA* 2.140: HISTORICAL LIGHTNING INCIDENTS (LATIN TEXT AFTER RACKHAM 1938/1991)

Exstat annalium memoria sacris quibusdam et precationibus vel cogi fulmina vel inpetrari. vetus fama Etruriae est inpetratum, Volsinios urbem depopulatis agris subeunte monstro quod vocavere Oltam, evocatum a Porsina suo rege. et ante eum a Numa saepius hoc factitatum in primo annalium suorum tradit L. Piso, gravis auctor, quod imitatum parum rite Tullum Hostilium ictum fulmine.

There exists a record in the annals of thunderbolts being either caused or summoned by certain sacred rites or prayers. There is an old story in Etruria, when the monster which they called Olta came to the city of Volsinii, when its territory had been devastated; it was called forth by its king Porsina. And before his time, as is recorded by the reliable author Lucius Piso in the first book of his Annals, this was often performed by Numa, but when Tullus Hostilius performed the same ritual improperly he was struck by lightning.

TEXT 9. PLINY *NATURALIS HISTORIA* 2.143–144: ON QUARTERING AUGURAL PRECINCTS FOR OBSERVATION OF NATURAL PHENOMENA (LATIN TEXT AFTER RACKHAM 1938/1991)

In sedecim partes caelum in eo spectu divisere Tusci: prima est a septemtrionibus ad aequinoctialem exortum, secunda ad meridiem, tertia ad aequinoctialem occasum, quarta obtinet quod reliquum est ab occasu ad septemtriones; has iterum in quaternas divisere partes, ex quibus octo ab exortu sinistras, totidem e contrario appellavere dextras. ex his maxime dirae quae septemtriones ab occasu attingunt. itaque plurimum refert unde venerint fulmina et quo concesserint. optimum est in exortivas redire partes. ideo cum a prima caeli parte venerint et in eandem concesserint, summa felicitas portendetur, quale Sullae dictatori ostentum datum accepimus; cetera ad ipsius mundi portionem minus prospera aut dira. quidam fulgura enuntiare non putant fas nec audire, praeterquam si hospiti indicentur aut parenti.

The Etruscans in making these observations divided the heaven into sixteen parts: the first quarter is from the North to the equinoctial sunrise [East], the second to the South, the third to the equinoctial sunset [West], and the fourth occupies what remains from West to North; these quarters they again divided into four parts, of which the eight on the East they called the left-hand regions and those opposite the right-hand regions. Of these the most significant extend from West to North. And so it is most important to know where the lightning bolts are coming from and going to. It is best if they return to the sectors of sunrise. Accordingly, if they come from the first part of the sky and retire to the same place it will be an omen of supreme happiness, which we learn was portended to the dictator Sulla; but all the other directions are less fortunate or actually unlucky, in relation to their exact sector of the world. Some think it is not right to announce or even listen to omens of thunderbolts, except if they are divulged to a guest or a parent.

TEXT 10. FESTUS, UNDER *MANUBIAE* (114.5–14 L) ESSENTIALLY PARAPHRASES SENECA (*QUAEST. NAT.* 2.41.1–2) ON TYPES OF LIGHTNING BOLT

Manubiae Iovis tres creduntur esse, quarum unae sint minimae, quae moneant placataeque sint. Alterae quae maiores sint, acveniant cum fragore, discutiantque aut divellant, quae a Iove sint, et consilio deorum mitti existimentur. Tertiae his ampliores, quae cum igne veniant; et quanquam nullum sine igne fulgur sit, hae propriam differentiam habeant quod aut adurant, aut fuligine deforment, aut accendant; quae statum mutent deorum consilio superiorum.

There are believed to be the three *manubiae* of Jupiter, of which one kind is negligible, that warn or may be placated. Another that are more important, come with damage, and they shatter or explode and are supposed to be from Jupiter but are thought to be sent with the consent of the gods. The third kind are more powerful, that come with fire; and since a lightning bolt is hardly anything without fire, these have a particular distinction because they either consume by burning, or scorch with flame, or set on fire; these change conditions by the agreement of the higher gods.

TEXT 11. SENECA *QUAESTIONES NATURALES* 2.32.2: ETRUSCAN PERCEPTION OF THE UNIVERSE (LATIN TEXT AFTER CORCORAN 1971/1999)

Hoc inter nos et Tuscos, quibus summa est fulgurum persequendorum scientia, interest: nos putamus, quia nubes collisae sunt, fulmina emitti; ipsi existimant nubes collidi ut fulmina emittantur. Nam cum omnia ad deum referent, in ea opinione sunt, tamquam non, quia facta sunt, significant, sed quia significatura sunt, fiant.

This is the difference between us and the Etruscans, with whom resides the utmost learning for interpreting lightning: We believe that lightning is caused by clouds colliding, whereas they believe that clouds collide in order to create lightning. Since they attribute everything to the divine, they are led to believe not that events have a meaning because they have happened, but that they happen in order to express a meaning.

TEXT 12. SENECA *QUAESTIONES NATURALES* 2.41.1–2: TYPES OF LIGHTNING BOLT (CF. TEXT 10) (LATIN TEXT AFTER CORCORAN 1971/1999)

Haec adhuc Etruscis philosophisque communia sunt. In illo dissentiunt quod fulmina a Iove dicunt mitti et tres illi manubias dant. Prima, ut aiunt, monet et placata est et ipsius Iovis consilio mittitur. Secundam mittit quidem Jupiter, sed ex consilii sententia, duodecim enim deos advocat; hoc fulmen boni aliquid aliquando facit, sed tunc quoque non aliter quam ut noceat; ne prodest quidem impune. Tertiam manubiam idem Jupiter mittit, sed adhibitis in consilium diis quos superiores et involutos vocant, quia vastat in quae incidit et utique mutat statum privatum et publicum quem invenit; ignis enim nihil esse quod fuit patitur.

These things up to this point are held in common by Etruscans and philosophers. In this they disagree, because (the Etruscans) say that lightning bolts are sent by Jupiter and they attribute to him three *manubiae*. The first, as they say, warns and is gentle and is sent by the decision of Jupiter himself. The second Jupiter also sends, but following the advice of a council, for he summons twelve gods; this lightning bolt sometimes accomplishes some good, but even then it also brings damage; it does not help without harming. The third *manubia* Jupiter also sends, but he summons to council the gods whom the Etruscans call the Superior, or Veiled Gods (*Dii superiores et Involuti*), because the lightning destroys whatever it strikes and everywhere alters the state of private or public affairs that it encounters, for fire allows nothing to remain as it was.

TEXT 13. SENECA *QUAESTIONES NATURALES* 2.47: CLASSIFICATION OF THUNDERBOLTS (LATIN TEXT AFTER CORCORAN 1971/1999)

Huic illorum divisioni non accedo. Aiunt aut perpetua esse fulmina, aut finita, aut prorogativa. Perpetua, quorum significatio in totam pertinet vitam nec unam rem denuntiat sed contextum rerum per omnem deinceps aetatem futurarum complectitur; haec sunt fulmina quae prima accepto patrimonio et in novo hominis aut urbis statu fiunt. Finita ad diem utique respondent. Prorogativa sunt quorum minae differri possunt, averti tollique non possunt.

I do not agree with their [Etruscan] classification. They say that lightning bolts are "perpetual," "limited," or "deferred." The perpetual ones pertain to the entire life, nor do they forecast a single event but embrace the complex of events which will happen throughout the future lifetime; these are the lightning bolts which first occur when someone has received an inheritance or when a new phase begins for a man or a city. Limited ones correspond to a single day. Deferred are those whose threats can be postponed but cannot be averted and lifted.

APPENDIX B

SAMPLE MESOPOTAMIAN DOCUMENTS AND ADDITIONAL DATA

This sampler of Mesopotamian divination literature is not undertaken as a comprehensive study of the Near Eastern texts, but rather to give an impression of the types of documents upon which the creators of the Etruscan *Brontoscopic Calendar* could have drawn. Not all works are fully translated and published. Likewise, modern translations can differ, and we lack both the original Etruscan and Figulus' Latin *Brontoscopic Calendar*, but in the Near Eastern texts cited, a similar tone and outlook characterize Mesopotamian and especially Assyrian, cultic literature. It yet remains possible that a document even closer in type to the Etruscan *Brontoscopic Calendar* will one day be identified in a Neo-Assyrian library. It should be evident, however, that nearly all the topics and the format of omen texts can be identified within the extant, published corpus that includes the *Enūma Anu Enlil*, MUL.APIN, the *Iqqur īpuš*, *Šumma ālu*, *Šumma izbu*, haruspicy texts, and scholars' research reports to the Assyrian kings.

REVELATIONS FOR THE KINGDOM OF NIPPUR

Some recorded exchanges in the category of prophecy also employ the tone, language, and topics of celestial omens. An example was excavated at Nippur and published by Robert Biggs (1967). It records revelations to the kingdom of Nippur from the god Enlil; Adad's thunder is one of the omens. The text includes the murder of a king of Babylon, revolt, plague, famine, and attack by enemies. Biggs noted the points of similarity with other Mesopotamian texts, and a sample here will show the common themes and expressions seen in all the divination texts:

> The shrines of the great gods will be obliterated, there will be a reduction of the inhabitants of Nippur by slaughter ... A son of the king who is not mentioned by the people will seize the throne ... there will be death among the domestic animals in the land, the mighty weapon of Erra (i.e., plague) will be in the land

and the people of the land, the children [of the king(?)] who have experienced famine will experience good times whose duration(?) will be indicated to you as seven years; end of the reign, complaint of the land, downfall of the king together with his family. The counsel of the land will change; the entire country will take up arms, and then one throne will overthrow another.[1]

TEXT SAMPLES: EXPLANATION OF SAMPLING AND FORMULATION

For the sake of clarity in reading and comparison to the Etruscan document, I have excerpted relevant sections from recent publications of the Mesopotamian tablets. I have noted elision with ellipses (…) and lacunae with brackets [ ], rather than retaining the meticulous markings set by the translators to indicate the exact condition of specific cuneiform originals, to simplify comparison with the English *Brontoscopic Calendar.* Most of the texts presented are known from multiple copies. I have also regularized spellings to a single style, thus always using Ululu for the month that some authors translate as Elul.

CALENDARS AS A FORMAT FOR OMEN TEXTS

In reading the table, note that Greece, among other regions, saw a multiplicity of calendars, often unique to each city; the Athenian calendar is noted here for comparison.

[1] Excerpted from the translation of Biggs 1967:123, §§ 18, 21–26.

TABLE B.1. *Comparison of calendars of the Iron Age world*

	Logo-Grams	Babylonian months ITI, *arhu*	Ugaritic (likely)	Hebrew months	Gregorian months-equivalents	Etruscan@	Latin (Julian) months	Greek (Athenian only)
1	bar	Nisannu	yrh ib'lt	Nisan	Mar-April	Velcitanus (velcitna)	Martius	Elaphebolion
2	gu_4	Aiaru/Ajar	yrh hiyaru	Iyar	April-May	Cabreas (capre)	Aprilis	Mounichyon
3	sig	Simanu	yrh hiyaru	Sivan	May-June	Ampiles	Maius	Thargelion
4	šu	Du'uzu	yrh gn	Tammuz	June-July	Aclus (acale)	Junius	Skirophorion
5	izi	Abu	yrh itb	Av	July-Aug	(Traneus)	Quinctilis	Hekatombaion
6	kin	Ululu	yrh ittbnm	Elul	Aug-Sept	[H]Ermius	Sextilis	Metageitnion
7	du_6	Teshritu	yrh riš yn	Tishri	Sept-Oct	Celius (celi)	September	Boedromion
8	apin	Arahsamnu	yrh nql	Heshvan	Oct-Nov	Chosfer	October	Pyanepsion
9	gan	Kislimu	yrh magmaru	Kislev	Nov-Dec		November	Mainakterion
10	ab	Tebetu	yrh pgrm	Tevet	Dec-Jan		December	Poseideon#
11	zíz	Shabatu	yrh šm[ ]/išigu	Shevat	Jan-Feb		Januarius	Gamelion
12	še	Addaru	yrh šm[ ]/išigu	Adar	Feb-Mar		Februarius	Anthesterion
13		Addaru *arkû*		*	intercalary		**	

Addaru *arkû* and Ululu potential intercalary periods (cf. Rochberg-Halton 1988:37).
@ Etruscan months Masan (Pyrgi plaque) and Thucte (perhaps August, *Liber Linteus* VIII.1, see van der Meer 2007:121) have not yet been correlated with Latin months.
* Adar Shenia for leap years.
** *Mensis intercalaris* for leap years, nicknamed Mercedonius in Rome
Hekatombaion was the first month of the Athenian year, and included the Panathenaia festival.
Athenian Poseideon II a potential intercalary month.
The Etruscan *Brontoscopic Calendar* begins in June.
Sources: See Cohen 1993; Labat 1965:25; Michels 1967; Edlund 1992; Bonfante and Bonfante 2002:186–191, 224.

THE *ENŪMA ANU ENLIL*: TOPICS AND SAMPLE OF TEXTS

TABLE B.2. *Cross-references between* Enūma Anu Enlil *and other texts*

Tablet	Topics	Cross-references with other texts
1–14	Lunar phenomena	MUL.APIN II
15–22/23	Lunar eclipses	*Iqqur īpuš* 69–81
23/24–39/40 (22)	Solar omens	*Iqqur īpuš* 67–68 *Iqqur īpuš* 100
40–49/50 -43	Meteorological phenomena	*Iqqur īpuš* 88–102 Nuzi tablet, etc. *Šumma ālu*, tablet 60: "if the earth exudes honey" (Gehlken 2005:241)
50–70+ (53) (56) (59–63) (64–65)	Planetary, stellar appearances (Pleiades) (Venus) (Jupiter)	*Iqqur īpuš* 87 MUL.APIN II *Iqqur īpuš* 82–86

TEXT SAMPLES: THE ENŪMA ANU ENLIL: *LUNAR ECLIPSE TABLETS, CITED FROM ENGLISH TRANSLATIONS*

Tablet 15, col. iii, sections 6′ and 7′ (Rochberg-Halton 1988:74–75). "If an eclipse begins in the south and the east wind blows: Destruction of Elam and Guti; … the king of Akkad will die; Adad will trample; there will be pestilence." … "If an eclipse begins in the north and the north wind blows: There will be a depletion of barley in the storage baskets of Akkad; the king of the enemy land will die."

Tablet 16, omens 5–6 (Rochberg-Halton 1988:84–85). "If an eclipse occurs and the weather becomes dark: Hard times will beset the king; regarding the people of the land, (they) will experience terrible famine. If an eclipse occurs and Adad thunders: The gods will order the destruction of the land."

Tablet 16, Source E, section III, omens 4 and 5 (reconstructed by Rochberg-Halton 1988:90). "If an eclipse occurs on the 14th of Simanu and begins and clears in the west: The king will come to shame in the eyes of his courtiers; there will be hostility; there will be an epidemic in the land. If an eclipse occurs on the 14th of Simanu and a meteor falls: One who has no right will seize the throne."

ENŪMA ANU ENLIL: *LUNAR ECLIPSE, TABLET 22 (AFTER ROCHBERG-HALTON 1988:251–272)*

Tablet 22: part I, section V, paragraphs 2–5. "If [in the fifth month] the 15th day: Nergal will ravage; there will be plague in the enemy country.

If the 16th day: pregnant women will miscarry.

If the 20th day: lions will run wild and block the roads.

If the 21st day: the fish spawn, the produce of the sea, will disappear."

Tablet 22: part II, section III, paragraph 2. [In the month of Simanu] "If Adad thunders: the barley and flax on the threshing floor will decrease; famine."

Tablet 22: part II, section IX, paragraphs 1–2. "[If an eclipse occurs in Kislimu from the 1st to the 30th day: Dogs will rage;…

"If Adad thunders: in a storm there will be the death of the herds."

ENŪMA ANU ENLIL: *SAMPLE SOLAR OMENS*

Tablet 28(29), lines 75–76. "If the sun is seen in a black web: the people of the country will tell lies, 'heaviness' will disappear from the country. If the sun is seen in a yellow web: the country will fall into anarchy by its own confusion, the country will not do well" (van Soldt 1995:103.)

Tablet 28(29), lines 70–84. "If in [Ajaru] the sun is surrounded by a halo in the morning: Adad will beat down the late crop, 'rust' will affect the barley, variant: the sesame will thrive" (van Soldt 1995:104.)

ENŪMA ANU ENLIL: *PLANETARY OMENS, CONSTELLATIONS ETC.: TABLETS 50 AND 51*

Tablets 50–51 of the *Enūma Anu Enlil* were translated and published in 1981 by Erika Reiner in collaboration with David Pingree. They offer additional points of comparison for later omen documents, including zodiacal data and predictions of weather. Reiner's Text XIII also furnishes many features similar to those of the *Brontosocopic Calendar.*

A sampling from Enūma Anu Enlil, tablet 50 (Reiner and Pingree 1981:35–47 et passim)

Text I, omen 5: "The Wild Sheep is for pestilence."

Text I, omen 17: "The Star of the sunset is for the raining of rain."

Text II, omen 12e: "The Goat reaches the Harros: in that year there will be an epidemic among cattle."

Text III, omens 13, 13a, 13b: "Ninsi is for the end of the dynasty. Ninsi is Mars. 'If the Stranger approaches Jupiter: the king of Akkad will die, and the harvest of the land will prosper.'"

Text IV, omens 11, 11a: "Scintillating stars are for the rising of wind. 'If the stars flare up again and again: a strong wind, variant: king, will (not) rise.'"

Enūma Anu Enlil, tablet 51; Reiner's Text XIII selected lines (after Reiner and Pingree 1981:67)

4. If in month VII EN.TE.NA.BAR.HUM rises heliacally: the king's days will be long.
5. If in month I the Demon with the Gaping Mouth rises heliacally: for five years in Akkad at the command of Irra there will be plague, but it will not affect cattle.
6. If in month I the Fish rises heliacally: there will be drought in the land.
7. If in month V the Bow rises heliacally: flax will prosper.
8. If the Bow comes close to UD.AL.TAR (Jupiter): Elam will eat fine food.

9. If in month VI the Kidney rises heliacally: the cultivated barley field will prosper.
10. [If...] is seen, fish and birds will be scarce [lit. cease], fish in the river [will not] spawn, birds in the sky [will lay no eggs].

TABLE B.3. *The meteorological series of the* Enūma Anu Enlil

Tablet	Topic	Incipit	Reference
40	Begins series with a *mušelû*-storm		Weidner 1968–69:71
41	"A darkness spreads and Adad throws his voice [i.e., thunder]"	[– – att]alû iššakan-ma ᵈadad rigim-šú iddi-ma	Weidner 1968–69:71
42	"The day darkens and a storm rises ..." also mentions the *mušelû*-storm		Weidner 1968–69:71
43	"a lightning bolt strikes in the South," discusses lightning and earthquake	ina šâri i birqu ib-riq	Weidner 1968–69:71–72
44	tally of 115 lightning and thunder *omina*; compares thunder with animal voices, count of thunderclaps; types of clouds from which thunder emerges "In Nisan Adad throws his voice"	ina itunisanni ᵈadad rigim-šu iddi	Weidner 1968–69:72
45	thunderclaps in different seasons, counting of thunderclaps, color of the rainbow, thunder at unexpected times "At sunrise Adad throws his voice"	ᵈadad ina nipih šamši rigim-šu iddi	Weidner 1968–69:72–73
46	"Adad throws his voice into the midst of the Seven-deities (the Pleiades)"	ᵈadad ina qabal ᵈimina-bi rigim-šu iddi-ma	Weidner 1968–69:73
47/48	(assorted fragments preserved) lightning, rain, winds, clouds, sunrise-sunset skies, earthquake, etc.		Weidner 1968–69:73
49/50	only fragments preserved		Weidner 1968–69:73
50	ends the Adad/thunder series, and celestial/astronomical *omina* resume *ina*^{itu}*nisanni šâr šûti illik* "In Nisan a South wind blows"		Weidner 1968–69:74

THE MUL.APIN ASTRONOMICAL TEXT

TABLE B.4. *Topics in the MUL.APIN (after Hunger and Pingree 1989:13)*

MUL.APIN Tablet, section	Topic	Reference: Hunger and Pingree 1989:	Cross-reference to other texts
I, i1–ii 35	Catalog of stars	137ff	
I, ii 36–iii 12	Dates of heliacal risings	139ff	
I, iii 13–33	Simultaneous risings and settings	140–41	
I, iii 34–48	Time intervals between dates of heliacal risings	141	
I, iv 1–30	*Ziqpu*-stars	141ff	
I, iv 31– II, i 8	Path of the Moon	144	
II, i 9–24	First intercalation scheme	150–51	
II, i 25–37	Heliacal risings, wind directions	152	
II, i 38–67	Planetary theory	146ff	*Enūma Anu Enlil*, tablet 56
II, i 68–71	Heliacal risings, wind directions	152	
II, A, 1–ii 20	Second intercalation scheme	152–53	
II, ii 21–42	Shadow table	153–54	
II, ii 43–iii 15	Water clock	154	*Enūma Anu Enlil*, tablet 14
II, iii 16–iv 12	Omens	145–46, 150	*Enūma Anu Enlil*, tablet 55, etc.

A SAMPLING OF OMENS FROM THE MUL.APIN, TABLETS II, III (AFTER HUNGER AND PINGREE 1989:108–23)

II, iii, 22–25: If the *U.RI.RI*-star becomes visible: rain and flood.

If the *U.RI.RI*-star approaches the Chariot: horses will die [lit., "mountain donkeys"]. If the *U.RI.RI*-star approaches the Furrow: the same.

II, iii, 43: If a star flares up from the West and enters the Yoke: there will be revolution.

II, iii, 53: If [...] ...: the king of the West will prosper.

II, iii, Gaps B1–B7 (Hunger and Pingree 1989:117–18):

[If] the star of Marduk reaches the Stars: in this year the Storm god will devastate.

[If] the star of Marduk reaches the Raven: early sesame will prosper. [If the star] of Marduk sees the body of a man: epilepsy[?] will seize him.

If a man takes a bath in front of the star of Marduk: there will be guilt.

If the star of Marduk is dark when it becomes visible: in this year there will be *asakku*-disease.

If the Yoke is dim when it comes out: the late flood will come.

THE *ŠUMMA ALU INA MĒLÊ ŠAKIN*

SAMPLE ENTRIES FROM THE ŠUMMA ĀLU, *TABLET 2, SELECTED OMENS (AFTER FREEDMAN 1998:71)*

72. If potsherds are habitually standing up at an angle in the street, that city will go to ruin.
77. If it rains sherds – uprising of the city and its environs.
79. If it rains frogs in a city, there will be plague in that city.
80. If dust storms whirl [...] – there will be violent deaths in the land.

ŠUMMA ĀLU, *TABLET 12, SELECTED OMENS (AFTER FREEDMAN 1998:195, 199)*

29. If fungus is seen on the south wall [of a house], the mistress of the house will die.
30. If fungus appears in a man's house in the middle of the north wall, the owner of the house will die and that house will be dispersed.
34. If fungus is seen on an exterior wall, someone mortally ill among the servants will die
43. If there is black fungus in a man's house, there will be profit in the man's house; the man's house will come rich.
44. If there is black fungus on the south, profit will be regular.
45. If there is black fungus on the north – the same.
46. If there is black fungus on the east – the same.
47. If there is black fungus on the west – the same.

THE *ŠUMMA IZBU* OMEN SERIES

TABLE B.5. *Topics covered in the* Šumma izbu *series*

Tablets	Category	Title
1–4	Human births	*Šumma sinništu arātma* ("If a woman is pregnant")
5	Sheep	Oldest of tablets in this series (does not use formula with *izbu*)
6–17	Any malformed newborn	*Šumma izbu*
18	Goats	*Šumma izbu*
19	Cattle	*Šumma izbu*
20–21	Horses	*Šumma izbu*
22	Pigs	*Šumma izbu*
23	Dogs	*Šumma izbu*
24	Gazelles	*Šumma izbu*

SAMPLE OF ENTRIES IN THE ŠUMMA IZBU *TABLET SERIES, SELECTED TO ILLUSTRATE COMMON TYPES AND ALSO TO SHOW ENTRIES FOR COMPARISON WITH ETRUSCAN DOCUMENTS*

Tablet 1, omen 7. "If a woman gives birth to a dog – the owner of the house will die, and his house will be scattered; the land will go mad; pestilence." (Leichty 1970:32)

Tablet 1, omen 8. "If a woman gives birth to a pig – a woman will seize the throne." (Leichty 1970:32)

Tablet 1, omen 9. "If a woman gives birth to a bull – there will be a despotic king in the land." (Leichty 1970:33)

Tablet 5, omen 1. "If a ewe gives birth to a lion – the weapons (that were) abandoned will be raised; the king will have no opponent." (Leichty 1970:73)

Tablet V, omens 29–30. "If a ewe gives birth to a lion, and it has four horns on the right and left – the prince will rule the [four] quarters. If a ewe gives birth to a lion, and it has no right ear – bad weather will destroy the cattle." (Leichty 1970:76)

Tablet 5, omens 52–54. "If a ewe gives birth to a lion, and it has the face of a ram – the progeny of the herd will not prosper. If a ewe gives birth to a lion, and it has the face of an ass – there will be severe famine in the land. If a ewe gives birth to a lion, and it has the face of a dog – pestilence." (Leichty 1970:78)[2]

Tablet 5, omen 69. "If a ewe gives birth to a lion, and its tongue covers its nostrils – [...] the land will rebel against the king." (Leichty 1970:79)

Tablet 5, omen 90. "If a ewe gives birth to a wolf – there will be plague in the land; madness; destruction of the land; disease of the herds." (Leichty 1970:81)

Tablet 10, omen 46′. "If an anomaly's eyes are on its forehead and bumps cover its eyes – disease of the cattle." (Leichty 1970:125)

Tablet 11, omen 72′. "If an anomaly's ears are turned around and face its back – the prince will slip through (bad) advice and the land will go mad." (Leichty 1970:136)

Tablet 20, omen 41′. "If a mare gives birth to one [foal] and its tail is white and its forehead is black [...] – [...] will revolt against the prince in his palace; lightning [...]" (Leichty 1970:183)

[2] The last omen might refer to the goddess of health, Gula, who was symbolized with a dog.

Tablet 20, omen 42′. "If a mare gives birth to one [foal] and its tail is black and its forehead is white [...] – [...] the prince ... a *muškênu*-class person will seize the throne." (Leichty 1970:183)

Tablet 21, omen 18. "If the anomaly of a mare has its feet on its chest – the notables will become lowly." (Leichty 1970:186)

THE *IQQUR ĪPUŠ* OMEN COMPILATION

TABLE B.6. *Breakdown of subjects in the* Iqqur īpuš

Sections	Topics	Sources
§ 1–28 § 50–53	Work on house, absence of owner, Hearth and home	*Šumma ālu* - -
§ 29–33	Building, restoration of Sacred buildings, objects	*Šumma ālu*
§ 34–41	Ritual activities	*Šumma ālu*
§ 42–49 § 60	Agricultural activities (dig well, plant garden, clear land)	*Šumma ālu* - -
§ 55–57	Purging, fever, hallucinations	*Šumma ālu*
§ 58–59	Meeting an ill omen (snake) Warning not to go out	*Šumma ālu*
§ 61–64	Marriage and birth	- -
§ 65–66	Fire	- -
§ 67–68	Conjunction of sun and moon; new moon	*Enūma Anu Enlil*
§ 69–81	Lunar and solar eclipses; haloes	*Enūma Anu Enlil*
§ 82–86	Rising, visibility of Venus	*Enūma Anu Enlil*
§ 87	Zodiacal light	*Enūma Anu Enlil*
§ 88–94 94	Thunderstorms, sometimes with other phenomena	*Enūma Anu Enlil* *Enūma Anu Enlil* *Adad*, text IV, lines 34–36 (Gehlken 2005:241)
§ 95–99	Rain, fog, south wind	*Enūma Anu Enlil*
§ 100–101	Earthquake	*Enūma Anu Enlil*
§ 102	Mudslides	*Enūma Anu Enlil*
§ 103–104	River floods	*Šumma ālu*
§ 105	List of gods and their months	(cf. *Enbu bêl arhim*)

SIMPLIFIED VERSIONS OF IQQUR ĪPUŠ, *SECTIONS 88–89: SELECTED SECTIONS FEATURING METEOROLOGICAL AND SEISMOLOGICAL EVENTS (BASED ON LABAT'S FRENCH TRANSLATION, 1965:172–75)*

§ 88. If in the month of Nisan, Adad hurls his rumbling [i.e., thunders]: the reign will change; Adad with his storms will ravage the wheat.

If in the month of Ajar: there will be deaths [variant: hostilities] in the country.

If in the month of Siwan: there will be rebellion in the country.

If in the month of Du'uzu: the harvest of the country will be prosperous.

If in the month of Arahsamna: the country will know blockades.

§ 89. If in the month of Ajar [thunder rumbles constantly]: wheat and chick-peas will be abundant.

If in the month of Arahsamna: the rains which were lacking will arrive, but the sesame and dates will not do well.

If in the month of Kislev: there will be rains and the locusts will appear suddenly and devour the (harvest of) the land.

If in the month of Shebat: the South wind will rise (it will sweep away the trees and) the houses will be destroyed.

SCHOLARS' RESEARCH REPORTS TO THE ASSYRIAN KING: SAMPLE EXCERPTS

REPORT FOR MARCH 15 (ADDARU 16) 669 BC (BAMAYA)

(SAA 8 169, quoted from Koch-Westenholz 1995:181; note that Hunger 1992:100, gives "date plantations" at line 3, and that he translates "Amurru" as "the Westland"):

> *If the moon is surrounded by a halo and Mars stands in it:*
> loss of cattle in all lands,
> the cultivated land and the dates will not thrive,
> ditto: Amurru will grow weak.
> If the sun stands in the halo of the moon:
> In all lands truth will be spoken,
> The son will speak truthfully to his father.
> If two stars stand in the halo of the moon:
> A reign of long duration.
> From Bamaja.

REPORT FROM ISSAR-ŠUMU-EREŠ: THUNDER IN AB

(Hunger 1992:17–18, no. 31, lines 4–6 = SAA 8 31)

"If in Ab (V) Adad thunders, the day becomes cloudy, it rains, a rainbow stretches, lightning flashes: the floodings will become scarce at the sources.

If it rains on a day without clouds: the land will be reduced by famine.

If Adad shouts on a day without clouds: there [will] be darkness and famine in the land.

If a storm [blows] from the west: fall of Elam."

REPORT FROM ISSAR-ŠUMU-EREŠ: THUNDER IN TISHRI

(Hunger 1992:18, no. 33 = SAA 8 33)

"If in Tishri (VII) Adad thunders, the day becomes cloudy, it rains, a rainbow stretches, lightning flashes: the gods will have mercy on the land.

From the Chief Scribe."

REPORT FROM TABIYA: THUNDER IN TISHRI

(Hunger 1992:250, no. 444 = SAA 8 444)

"If Adad thunders in Tishri (VII): there will be hostility in the land.

If it rains in Tishri (VII): fall of sick people and cattle, variant: fall of the enemy.

From Tabiya."

TEXTS ON HARUSPICY AND RELATED PROCEDURES

A sampling of haruspical tablets is drawn from a set of texts on the Secrets of Extispicy, published by Koch (2005), who offers a thorough presentation of parallel texts in Akkadian and so on. (Some tablets are attributed to the famous 8th-c. scholar Nabû-zuqup-kēnu, but its core texts had already been serialized and consulted in the second half of the second millennium.) (Note that the terminology of sheep livers appears here: thus, the Presence [*manzāzu*] = groove on the left lobe; the Increment [*sibtu*] = *processus papillaris*, etc.)[3]

MUTĀBILTU, *TABLET 1: SAMPLE OMEN TEXTS (KOCH 2005:93)*

Omen 14: Thickness means peaceful dwelling as in: "If the Increment is thick: The crops will thrive."

[3] For maps and detailed lists of terms and conditions, see Koch-Westenholz 2000:38–70; Koch 2005:74–83. For more information on liver anatomy, see Meyer 1987 and Leiderer 1990.

Omen 15: [...] means reconciliation as in: "If the front of the Gall Bladder is placed to the right: Reconciliation of god with man, the king will love his land."

Omen 17: To bend downwards means success as in: "If the Presence (is bent) downwards like a sickle: You will surround the enemy land in a siege."

MUTĀBILTU, *TABLETS 12–13, TEXT 1: SAMPLE OMEN TEXTS (KOCH 2005:172–76, PASSIM)*

Omen 4′: If the x]s of the Finger are atrophied three times: There will be murders [in the land].

Omen 14′: If the Gate of the Entrail's Cover is atrophied downwards: A terrible famine will afflict the land.

Omen 20′: If it is atrophied upwards: There will be want of barley, hay, and straw in the land.

Omen 65: If the top of the Rise of Wind next to the base of the Presence is atrophied downwards: The enemy will destroy my fortress, Adad will ravage the crop of the land.

MANZĀZU, *TABLET 3: DIFFERENT REGIONS OF THE LIVER ARE MARKED (KOCH-WESTENHOLZ 2000:94–95 AND PASSIM)*

Fall of the mass of the army.
Adad will flood the threshing floors.
Rise of storm, lightning and cold.
A snake will bite a man.
The gods of your army will forsake it at its destination.
Death of cattle, sheep and donkeys; alternatively a mortally ill person in the man's circle of acquaintances will die.
The king will bring about the fall of his nobles.

PĀN TĀKALTI, *TABLET 5: ONE OF THE RARE TEXTS CITING SLAVES (KOCH-WESTENHOLZ 2000:339–41 AND PASSIM)*

Escape of a famous slave.
Escape of a slave.
The slave who ran away will return.
A year of prosperity.
Lions will attack the campaign.
Snakes will attack the campaign.
Time to heap up earth and build a city wall.
They will rebel against the prince in his palace.
A favorable word will turn into a word of severe hostility.

APPENDIX C

OTHER *BRONTOSCOPIA* IN THE CLASSICAL TRADITION: SAMPLE TEXTS WITH ENGLISH TRANSLATION

For comparison with the Etruscan *Brontoscopic Calendar*, a short sampling of texts, preserved in Greek manuscripts, is offered: some sections have been excerpted only. All except Text C.2 are my own translations.

TEXT 1. LYDUS *DE OSTENTIS*, SECTIONS 39–41: *BRONTOSCOPION* ATTRIBUTED TO FONTEIUS CAPITO[1]

ΒΡΟΝΤΟΣΚΟΠΙΑ ΕΚ ΤΩΝ ΦΩΝΤΗΙΟΥ ΤΟΥ ΡΩΜΑΙΟΥ *ΚΑΘ' ΕΡΜΗΝΕΙΑΝ ΠΡΟΣ ΛΕΞΙΝ*

39. Σελήνης ἐχούσης τὸν αἰγόκερον εἰ ἐν ἡμέρᾳ βροντὴ γένηται, τύραννον ἐπαναστήσεσθαι ἀπειλεῖ τοῖς ἀπὸ τῆς στενῆς θαλάσσης ἄχρι Νείλου χωρίοις, ἐκπεσεῖν δὲ ὅμως αὐτὸν τῆς ἐγχειρήσεως. οὐχ ἥκιστα δὲ καὶ τῶν ἐπιτηδείων ἔσται ἐλάττωσις. ὑπονοστήσει δὲ καὶ ὁ Νειλος, καὶ διχόνοια ἔσται τέκνοις πρὸς γονεῖς, καὶ ταραχαὶ τῶν ἀρχόντων ἐνίοις. Πέρσαι δὲ καὶ οἱ τῆς Εὐρώπης τὸ δυτικὸν ἔχοντες ἠρεμήσουσιν. εἰ δὲ νυκτός, ἔθνη βάρβαρα κατ' ἀλλήλων ἐξενεχθήσεται, καὶ δογμάτων ἕνεκα τὰ τῆς Ῥωμαϊκῆς εἰρήνης σαλευθήσεται, καὶ τόπους δέ τινας τοῦ πολιτεύματος πολέμιοι πρὸς βραχὺ καθέξουσιν. ἄρχοντες ἐκ τῶν δυτικῶν ὁρμώμενοι ἡγήσονται παρανόμως τῶν κοινῶν. δυσδαίμονες δὲ ἔσονται οἱ τῶν ἀνθρώπων πλείους, καὶ χειμῶνες πικροί, καὶ ναυάγια, καὶ σάλοι σφαλεροί. Σελήνης ὑδρχόῳ οὔσης εἰ ἐν ἡμέρᾳ βροντήσῃ, κλεινῶν ἀνθρώπων καταλύσεις ἐμπρησμούς τε βαρεῖς. καὶ ὁ μὲν ἄμητος ἱλαρός, ἀκρίσι δὲ καὶ θηρίοις ἐκδοθήσεται, ἡ δὲ σταφυλὴ σφιγξῖ. καὶ τὸ πάντων

[1] Cf. Greek edition, Bekker 1837:332–36. Compare Italian translation of Domenici and Maderna 2007:99–101.

σχετλιώτερον, φάσματα τοῖς σώμασιν ἐνοχλήσει μετὰ νόσων, καὶ τῶν ὑδάτων διαμαρτήσει τὸ νόστιμον. καὶ οἱ μὲν ἐλάττονες τὰς τύχας ἐπὶ τὸ βέλτιον ἁρπασθήσονται, οἱ δὲ μείζους ἐπὶ τὸ χεῖρον ἐκπεσοῦνται.
ἀμετρία τε τῶν νόμων τὴν εἰρὴνην ταρήξει, καὶ ἐθνικῶν ἔσται ἐπὶ τῆς πολιτείας πάροδος. εἰ δὲ νυκτός βροντήσῃ, ἀπολεγχτω μὲν τῇ εἰρήνῃ τὸ δυτικόν, ὁ δὲ ἔμητος τῆς Ἀσίας εὐκταιότατος, καὶ οἱ ἄρχοντες παρὰ τὸ κρατοῦν ἐν αὐτοῖς οὐ σφόδρα τοὺς πηκόους λυπήσουσι.
Σελήνης ἰχθύσιν οὔσης ἐὰν βροντὴ ἐν ἡμέρᾳ γένηται, ὑδάτων ἐπίδοσιν δηλοῖ οὐ ποταμίων μόνον ἀλλὰ καὶ θαλαττίων, ἀχλύν τε τῷ ἀέρι, ἀλλὰ καὶ ἀφανισμὸν τοῖς ἰχθύσιν ἀπειλεῖ. τὰ δὲ θηρία πρὸς τὰς παγίδας αὐτομολήσει. φυτεύειν δ' ὅμως φυτὰ ἐπευκταῖον. καὶ νοσῶδες μὲν τὸ κατὰστημα, οἱ δὲ ἄρχοντες ἀνωμάλως διακείσονται.
40. Σελήνης οὔσης κριῷ ἐὰν βροντὴ γένηται ἐν ἡμέρᾳ, φυγὴν μὲν Ἀραβίοις, λοιμὸν δὲ Πέρσαις ἀπειλεῖ. εἰ δὲ νυκτί, Ἀσίαν καὶ Εὐρώπην ταράξουσι πολέμιοι, οἵ τε ναυτιλλόμενοι οὐκ ἔξω κινδύνων ἔσονται. καὶ ἀφθονία μὲν τῶν καρπῶν ἔσται, δυνατοῦ δὲ προσώπου Πτῶσις. καὶ τὰ κοινὰ ἐγγὺς εὐθηνίας ἔσται.
Σελήνης οὔσης ταύρῳ εἰ ἐν ἡμέρᾳ βροντήσῃ, ἐπομβρίαν δηλοῖ καὶ χαλαζώδη τὸν ἀέρα, καὶ ποταμῶν παροχεύματα. καὶ ὁ μὲν ἰχθὺς ἄφθονος, οἱ δέ περὶ τοὺς θαλαττὶους πόρους ἀγρυπνοῦντες λυπηρότατοι. καὶ ὁ μὲν ἄμητος ἀγαθός, τοῖς δὲ κτήνεσι λοιμός. εἰ δὲ νύκτωρ, καρπῶν μὲν ἔνδειαν, ὑετῶδες δὲ καὶ χαλαζῶδες τὸ θερινὸν κατάστημα. ἄμβλωσίς τε γυναικῶν συμβήσεται, καὶ τὸ νὸστιμον ὕδωρ ἐπιλείψει καὶ κινήσεις μὲν στρατευμάτων λυπηραί, τῇ δὲ ἀνατολῇ, τύραννος οὐκ εὐτυχὴς ἐνοχλήσει.
Σελήνης οὔσης διδύμοις εἰ ἐν ἡμέρᾳ βροντήσῃ, θεοσεβέστατα ἔσται τὰ κοινά εἰ δὲ νυκτί, τὰ ἐναντία.
Σελήνης οὔσης καρκίνῳ εἰ ἐν ἡμέρᾳ βροντὴ γένηται καὶ νεκρὰ συμβαίη, ταραχαὶ τοῖς κοινοῖς πράγμασιν ἔσονται καινοτομίαι τε [τοῖς πράγμασιν] ἐπὶ τὸ χεῖρον. οἱ δὲ καρποὶ ψεύσονται. ἀλλὰ μὴν καὶ ἡ θάλασσα τοῖς ναυτιλλομένοις ὀλεθρία ἔσται, καὶ φθορὰ μὲν ἀνθρώπων, ἀφθονία δὲ ἰχθύων. εἰ δὲ νυκτί, Αἰθίοψι καὶ Πέρσαις ὄλεθρος ἐνσκήψσει. τὰ δὲ δυτικὰ οὕτως λιμώξει ὡς καὶ εἰς Αἴγυπτον δι' ἔνδειαν καταφυγεῖν.
Σελήνης οὔσης λέοντι ἐὰν βροντὴν ἐν ἡμέρᾳ γενέσθαι συμβαίη, δόλον κατὰ τῆς βασιλείας μελετᾶσθαι δηλοῖ. καινισμοί τε νόσων ἔσονται, καὶ θάνατοι τοῦ πλήθους ἐκ τιμωριῶν. καὶ ὁ μὲν ἄμητος ἄφθονος ἔσται, ὑπὸ δὲ πολεμίων διαρπαγήσονται, καὶ διαφερόντως κατὰ τὸ δυτικόν. κατισχύσουσί τε καὶ ἀδικήσουσιν οἱ ὑπερέχοντες τὸ ὑπήκοον. εἰ δὲ νυκτός, καὶ οὕτως ἡ βασιλεία ἐπιβουλευθήσεται, καὶ μέρος τῆς ἀνατολῆς ὑπὸ βαρβάροις τελέσει. κατόμβρισις δὲ τοῖς ὑγροῖς καρποῖς λυμανεῖται. ἐπισυρήσονται δὲ τοῖς κτήνεσιν ἑρπετὰ θανατηφόρα. δυνατοὶ δὲ τῆς πολιτείας πεσοῦνται.

41. Σελήνης οὔσης παρθένῳ εἰ ἐν ἡμέρᾳ βροντήσῃ, ὑγρότερον ἔσται τὸ τοῦ ἀέρος κατάστημα, ἀφθονία δὲ τῶν ξηρῶν καρπῶν, ὡς ἠρέμα πως ἐπιρρεπεστέρους γενέσθαι τοὺς ἀνθρώπους ἐπὶ τὰ πταίσματα. ἔσται δὲ καὶ φθορὰ τοῦ γυναικείου ἔθνους. εἰ δὲ νυκτός, λοιμὸς ἐνσκήψει τοῖς δυτικοῖς. τὰ δὲ θηρία οὕτως ἐπελεύσεται τοῖς ἀνθρώποις ὡς φωλεοῖς αὐτοῖς ἐγκατακρύπτεσθαι καὶ δυνατῶν δὲ ἀνδρῶν ἔσται πτῶσις, καὶ ἔνδεια ξηρῶν καρπῶν.

Σελήνης οὔσης ζυγῷ εἰ ἐν ἡμέρᾳ βροντήσῃ, ἀγαθὰ ἐκδεχέσθωσαν ἄνθηρωποι, ἐξῃρημένων Αἰγυπτίων. εἰ δὲ νυκτός, ἡ μὲν νεολαία ἐν ὅπλοις τελέσει καὶ ἐν πολέμοις φθαρήσεται, ἀφθονία γε μὴν τῶν καρπῶν ἔσται, ὑπὸ δὲ βαρβάρων δαπανηθήσεται.

Σελήνης οὔσης σκορπίῳ ἐὰν ἐν ἡμέρᾳ βροντὴν ἢ ἀστραπὴν γενέσθαι συμβαίη, Ἀράβιοι ταραχθήσονται, ἔνδεια δὲ τῶν καρπῶν ἔσται, καὶ οἱ ἄνθρωποι κατ᾽ ἀλλήλων χωρήσουσιν. Ἀσσίρια γε μὴν πιεσθήσεται λιμῷ. εἰ δὲ νυκτός, πυρκαϊαὶ μὲν ἔσονται, πτώσεις δὲ πόλεων παραλίων καὶ διαφθορὰ καρπῶν καὶ τετραπόδων, θηρίων δὲ παντοίων ἐπιδρομή ὅθεν εὔχεσθαι δεῖ, ὅπως μὴ καὶ κεραυνῶν ἐνσκήψεσιν ἀναλωθῶσι.

Σελήνης οὔσης τοξότῃ ἐὰν ἐν ἡμέρᾳ βροντήσῃ, Πέρσαις ἀφανισμὸν ἀπειλεῖ, καὶ ὁ βασιλεὺς δὲ αυτῶν δολοφονηθήσεται, εὐκαρπία δὲ ἔσται καὶ χειμῶνες εὔδιοι ἔσονται, φθορὰ μὲν ἀνθρώπων, ἀνδραποδισμοὶ δὲ πόλεων. εἰ δὲ νυκτί, ἐπομβρίαν δηλοῖ, καὶ κάματον δὲ τοῖς ἐγκύμοσι καὶ νόσους καὶ θανάτους αἰφνιδίους, ὡς περιττεύειν μὲν τοῖς καρπούς, ἀπολείπεσθαι δὲ τοὺς δαπανῶντας. ἔσονται δὲ σάλοι καὶ ἄνεμοι βίαιοι, καὶ ἀφανισθήσεται ἀνὴρ δυνατὸς καὶ βαρὺς τοῖς πράγμασιν.

TRANSLATION OF THE BRONTOSCOPIA *FROM THE WRITINGS OF FONTEIUS THE ROMAN ACCORDING TO A WORD-FOR-WORD TRANSLATION*

With the moon being in Capricorn, if thunder happens during the day, it threatens an attempt on the throne by a usurper in the coastal regions that extend toward the Nile, but his attempt will fail. Nevertheless there will be a dearth of necessities. The Nile will drop, there will be dissensions of children against parents and confusion among those in power. But the Persians and the peoples of western Europe will remain at peace. If, however [the thunder is] at night, barbarian peoples will conspire among themselves, and because of certain judgments, the areas in which peace had been established by the Romans will lapse into instability, and in a short while the enemies will occupy certain parts of the state. Chiefs from the western regions will control the state illegally. The greater part of the population will be struck with misfortune, there will be biting winters, and shipwrecks and perilous earthquakes.

With the moon being in Aquarius, if it thunders in daytime, the downfall of famous men and serious fires. The harvest will be good, but will be devoured by locusts and pests, likewise the grapes [will be destroyed] by wasps. And further, the most dreadful of all, apparitions accompanied by diseases will damage bodies, and the drinking water will not be wholesome. The lower classes will improve their situation but for the more affluent their affairs will deteriorate. The inequity of the laws will upset the peace, and there will be admission of foreigners into the citizenship. If instead it thunders at night, the west will renounce the peace, the harvest in Asia will be even greater than one could hope, and the rulers of the state, during their tenure, will not excessively harass the subject allies.

With the moon being in Pisces, if thunder should happen during the day, it warns of a rise in the water level not only of rivers but also of the seas, and threatens a cloudy sky but also slaughter for the fish. Wild animals will fall into traps all by themselves. Yet it is the right time to plant trees. And the summer will be unhealthy, yet the rulers will survive the ups and downs of fortune.

With the moon being in Aries, if thunder should happen during the day, it threatens flight for the Arabs, plague for the Persians. If at night, the enemies will bring confusion to Asia and Europe, and those sailing will encounter dangers. And there will be an abundance of crops and the ruin of a powerful personage, and the community will be close to prosperity.

With the moon being in Taurus, if it thunders during the day, it signals heavy rain, and the air hail-bringing, and rivers will flood. And although there will be an abundance of fish, yet those who make their way upon the seas will be greatly tormented. And although the market will be good, yet there will be a plague upon the flocks and herds. But if [the thunder] is nocturnal, then a dearth of crops and the summer season will bring rain and hail. Women will miscarry and there will be no pure water. And the movements of armies will be distressing, yet in the east the tyrant will not succeed in annoying.

With the moon being in Gemini, if it thunders during the day, the community will be especially god-fearing – but if at night, then the opposite.

With the moon being in Cancer, if thunder happens during the day and it comes to pass that it dies down, it signals that there will be disturbances in public affairs and attempted revolts that will go bad. The crops will not ripen. But the sea will be deadly for those sailing and there will be ruin for men yet an abundance of fish. But if by night, misfortune will strike the Ethiopians and Persians. But in western regions they shall suffer hunger so that they will seek to escape from their want even as far as Egypt.

With the moon being in Leo, if thunder should happen to occur during the day, it announces that a plot against the royal government will be squelched. There will be new diseases and deaths by vengeance of the common people. But although the harvest will be plentiful, yet it will be pillaged by the enemies, especially those living toward the west. The victors will seize power and treat the subject allies unjustly. If at night, then also the kingdom will be the object of plots, and a portion of the eastern region will be seized by barbarians. Heavy rains will damage the soft fruits. Lethal snakes will slither among the flocks. Powerful men in the government will fall.

With the moon being in Virgo, if it thunders during the day, there will be a more humid condition to the air but an abundance of grains, so that men will indulge their own propensity to make false steps. But there will be ruin among the class of women. But if at night, a plague will break out among the westerners. Wild animals will attack men so often that they will resort to hiding in caves. And there will be the downfall of powerful men, and a shortage of cereals.

With the moon being in Libra, if it thunders during the day, men will await good things, excepting the Egyptians. But if at night, young men will go under arms and will perish in wars; although there will be an abundance of crops, yet they will be consumed by the barbarians.

With the moon being in Scorpio, if thunder or lightning should happen to occur during the day, the Arabs will be disturbed, there will be a dearth of crops, and men will attack one another in turn. Assyria will certainly be oppressed by famine. But if at night, there will be fires, the downfall of coastal cities, and the destruction of crops and the four-footed, and incursions of all sorts of wild animals, so that it will be necessary to beseech [the gods] lest men be destroyed even by the lightning bolts.

With the moon being in Sagittarius, if it should thunder during the day, it indicates destruction of the Persians, and their king will be slain by treachery, there will be a fruitful harvest and mild winters, yet ruin of men, and the enslavement of cities. But if at night, it announces rainy weather and heavy labor for pregnant women and diseases and sudden deaths, and that the crops will be plentiful, but they will not find markets. Then there will be earthquakes and strong winds, and a powerful man important to the state will perish.

TEXT 2. EXCERPTS FROM HEPHAISTION'S *APOTELESMATIKA* 1.21

Portions of Fragment 6 Riess = Pingree 1973:52–61, §§ 1.21.1–2, and 7–10 (Greek text), quoting the modern translation of Naphtali Lewis.[2]

<Κα′> Ἀποτελέσματα τῶν γινομένων ἐκλείψεων καὶ τῶν κομητῶν

Ἀνέγραψαν δὲ οἱ παλαιοὶ Αἰγύπτιοι τὰ ἀποτελέσματα τὰ ἐκ τῶν ἐκλείψεων γινόμενα ὧν παρατιθέμεθα καὶ τούτων τὰ πλεῖστα συντόμως οὕτως ἐπὶ

[2] Lewis 1996:139–44 et passim.

τῶν τελείων ἐκλείψεων τὸ μὲν χρῶμα τὸ μέλαν θάνατον τοῦ ἄρχοντος καὶ ταπείνωσιν καὶ λιμὸν καὶ μεταβολὴν σημαίνει τὸ δὲ ἐρυθρὸν τῆς χώρας κάκωσιν, τὸ δὲ ὑπόλευκον λιμὸν καὶ θάνατον τοῖς κτήνεσι καὶ ἐμπόροις τό δέ ἰοειδὲς πόλεμον καὶ λιμόν, τὸ δὲ χρυσοειδὲς λοιμὸν καὶ θάνατον.

Σελήνης δὲ οὔσης ἐν τῇ ἐκλείψει ὅταν διάττων ἀστὴρ ἐπιδραμὼν ἔλθῃ εἰς αὐτήν, μηνύσει τύραννον μέγαν ταπεινωθήσεσθαι καὶ ταραχὴν ἕξειν τὸν οἶκον αὐτοῦ εἰ δὲ ἅλως περὶ τὴν Σελήνην γένοιτο ἐκλελοιπυῖαν, ἔσται τειχήρης ὁ τύραννος πολιορκούμενος ὑπὸ πολεμίων ἐὰν δὲ ἐξέλθῃ ἀστήρ, ὑπ᾽ αὐτῶν αἰχμάλωτος ἔσται ἐκλελοιπυίας <δὲ> αὐτῆς βροντὴ γενομένη καθαίρεσιν τυράννου καὶ ταπείνωσιν τῆς χώρας σημαίνει, ἐκλείποντος δὲ Ἡλίου ἢ Σελήνης ὑετὸς γινόμενος θάνατον σημαίνει. ἐὰν δὲ ἅμα τῇ ἀνατολῇ γένηται ἔκλειψις καὶ διακρατήσῃ εἰς τρεῖς ὥρας ἄχρι τῆς ἀνακαθάρσεως, τὰ σημαινόμενα ἔσται μετὰ τὴν ἔκλειψιν τριμήνου παρελθούσης· ἐὰν δὲ ἀπὸ τετάρτης ὥρας γένηται ἡ ἔκλειψις καὶ μείνῃ ὥρας δύο ἕως ἂν καθαρθῇ, ἑξαμήνου παρελθούσης ἔσται τὰ σημαινόμενα καὶ ἕως δωδεκάτης ὥρας ὁμοίως συγκρίνεται ὁ χρόνος.

Ὁλοσχερῶς δέ φασι τὴν Σελήνην ἐκλείπουσαν ἐν μὲν τῷ Κριῷ ἀφορίαν πάντων ἔσεσθαι σημαίνειν, ἐν δὲ Ταύρῳ σίτου φθοράν, ἐν δὲ Διδύμοις καὶ Καρκίνῳ σίτου δαψίλειαν καὶ οἴνου, ἐν δὲ Λέοντι σίτου πλῆθος κατὰ τὰ πεδία, ἐν δὲ Παρθένῳ προβάτων φθορὰν καὶ δένδρων, ἐν δὲ Ζυγῷ σίτου φθορὰν ὑπὸ ἀκρίδος καὶ νόσου, ἐν δὲ Σκορπίῳ λοιμόν, ἐν δὲ Τοξότῃ ὕδωρ καὶ χάλαζαν, ἐν <δὲ> Αἰγοκέρωτι σίτον ἐν τοῖς πεδίοις ἔσεσθαι καὶ ἀνέμους μείζονας, ἐν δὲ Ὑδροχόῳ εὐθηνίαν καὶ σίτου φθορὰν ἐν τοῖς πεδίοις, ἐν δὲ τοῖς Ἰχθύσι πρωΐμους τοὺς σιτικοὺς καρπούς.

THE RESULTS OF ECLIPSES AND COMETS ON HUMAN DESTINY

The ancient Egyptians recorded the effects of eclipses on human destiny. Following is a concise exposition of the main points. In full eclipses the black color portends death of a ruler, humiliation, famine and reversal, the red prophesies suffering in the land, the whitish, famine and death to cattle and merchants, the green, war and famine, the gold, plague and death.

...

When the moon is in eclipse, a shooting star rushing toward it indicates that a great tyrant will be humbled and his house will be in turmoil.... Thunder occurring during the moon's eclipse presages overthrow of a tyrant and abasement of the land. Rain occurring during an eclipse of the sun or moon portends death. If an eclipse coincides with rising and lasts for three hours until clearing, the indicated consequences will occur three months after the eclipse. If the eclipse occurs four hours after rising and lasts two hours to clearing, the indicated consequences will occur after six months have passed. The time is interpreted similarly up to the twelfth hour.

The portent of a full eclipse of the moon is generally held to be as follows: in Aries, a dearth of everything; in Taurus, damage to the grain crop; in Gemini and Cancer, abundance of grain and wine; in Leo, plenty of grain in the fields; in Virgo, damage to flocks and trees; in Libra, damage to the grain crop from locusts and disease; in Scorpio, plague; in Sagittarius, rain and hail; in Capricorn, grain in the fields and strong winds; in Aquarius, prosperity and damage to the grain crop in the fields; in Pisces, early grain harvests.

TRANSLATIONS OF SAMPLE GREEK ASTROLOGICAL TEXTS

These representative samples convey the tone and formatting of most of the comparable, Late Antique or Byzantine manuscripts that dealt with thunder-omens or related topics, such as earthquakes. The two selected here have a high proportion of apodoses similar to those of the Etruscan Brontoscopic Calendar, yet it is readily apparent that the texts as a whole are not derived from the same sources, and that the formation of these Greek texts occurred later than the Etruscan *Brontoscopic Calendar* recorded by Figulus.

TEXT 3.MANUSCRIPT, GREEK TEXT NO. 20: EL ESCORIAL CODEX 1 (F. 165) – βροντολόγιον[3]

Βροντολόγιον

Κριός. –Ἐὰν βροντήσῃ εἰς τὰ ἀνατολικὰ μέρη θανατικὸν καὶ ἀκαταστασία εἰς πολλοὺς κακοὺς καὶ ἀνδρῶν σοφῶν ἀπώλεια. εἰ δὲ καὶ σεισμός, λιμὸς ἔσται.
Ταῦρος. –Ἐὰν βροντήσῃ, ἔνθα παραδράμῃ ἡ βροντή, χαρὰ γίνεται εἰς τοὺς βασιλεῖς, εἰς δὲ τὴν ἀνατολὴν ἀκαταστασία καὶ ἀλληλοσφαγία καὶ ἄκουσμα βασιλέως καὶ ἀρροστία πολλὴ καὶ ἀπώλεια. εἰ δὲ καὶ σεισμός, πολλὰ ἐναντία σημαίνει.
Δίδυμος. –Ἐὰν βροντήσῃ, ἀγῶνες καὶ σάλος εἰς τοὺς ἀνθρώπους, σίτου λεῖψις καὶ κλύδωνες ἐν τῇ θαλάσσῃ καὶ εἰς τὰ ὑψηλὰ ἐλαφρὸς χειμὼν καὶ καρπῶν λεῖψις καὶ ἄκουσμα βασιλέως εἰς τὴν οἰκουμένην, καὶ ἑρπετῶν ἀπώλεια. ἔνθα δὲ παραδρᾶμῃ ἡ βροντή, ταράσσεται ἡ χώρα ἐκείνη. εἰ δὲ σεισμός, ταραχαὶ πολλαί.
Καρκῖνος. –Ἐὰν βροντήσῃ, ἄνεμοι πολλοί, σίτου καὶ κριθῆς καρπὸς πολὺς καὶ πῦρ ἐν τῇ πόλει καὶ ψόφος βοῶν, χειμὼν καλός,

[3] Greek text following *CCAG*:vol. 11, pt. 1; *Codices Hispanienses*, ed. C. O. Zuretti (1932), 155–57.

ὄσπρια πολλὰ καὶ εἰς τὸν βορρᾶν κατάλυσις καὶ ἄκουσμα βασιλέως. εἰ δὲ καὶ σεισμός, νοσήματα πολλά.

Λέων. – Ἐὰν βροντήσῃ, σίτου καὶ οἴνου καὶ παντὸς ἀγαθοῦ λεῖψις καὶ ἐν τῇ δύσει ἀλλαξοβασιλεία, χειμὼν ποικίλος, πῦρ ἐν τῇ πόλει καὶ πτῶσις βοῶν, κίνδυνος πρὸς τὸν βορρᾶν καὶ δαιμόνων ἐμφάνεια. εἰ δὲ καὶ σεισμός, σοφοὶ ἄνδρες ἀπολοῦνται.

Παρθένος. – Ἐὰν βροντήσῃ, βασιλέως καὶ λαοῦ πολλοῦ ἀπώλεια καὶ εἰς τὰ μέρη τῆς Αἰγύπτου ἀκαταστασία, καὶ τὰ πλησίον κινδυνεύσουσι, λεῖψις δὲ παντὸς εἴδους εἰ δὲ σεισμὸς γένηται, ἄνεμοι πολλοὶ καὶ βυθοστρέμματα ἐν τῇ θαλάσσῃ, χειμγν δὲ βαρὺς καὶ θανατικὸν καὶ νοσήματα εἰς τόπους.

Ζυγός. – Ἐὰν βροντήσῃ, πόλεμος καὶ ἀλληλοσφαγίαι, καὶ τὰ ἄγρια ζῷα ἀπολοῦνται, καὶ καύματα φανήσονται, καὶ χειμὼν καλός, καὶ πόλεμοι γενήσονται καὶ ἀπὸ νόσου χαμὸς ἀνθρώπων καὶ ἐν τῇ δύσει λιμὸς καὶ περίστασις καὶ μετάθεσις ἀνθρώπων, τετραπόδων καὶ ἑρπετῶν ἀπώλεια. εἰ δὲ σεισμός, θανατικὸν κατὰ τόπους.

Σκόρπιος. – Ἐὰν βροντήσῃ, χειμὼν καὶ ἀλληλοσφαγίαι καὶ ὀρνέων ἀπώλεια καὶ ἀρχόντων θάνατος, ὀσπρίου λεῖψις καὶ εἰς τὸν βορρᾶν κατάλυσις καὶ κοῦρσα καὶ ἀλλαξοβασιλεῖαι καὶ ἀρχόντων περιστάσεις. εἰ δὲ καὶ σεισμός γένηται, σίτου χαρὰ καὶ πολλὴ ἀγαθοσύνη, νοσήματα δέ.

Τοξότης. – Ἐὰν βροντήσῃ, ἀγῶνες καὶ ἀκαταστασία. ἔνθα Παραδράμῃ ἡ βροντή, βλάβην σημαίνει τῇ χώρᾳ ἐκείνῃ. εὐφορία παντὸς εἴδους, καὶ ἐν τῇ ἀνατολῇ κατάλυσις χειμὼν ὄψιμος καὶ ἀκρίδων ἐμφάνεια πρὸς κατάλυσιν τῶν καρπῶν. εἰ δὲ σεισμός, προβάτων πτῶσις.

Αἰγόκερως. – Ἐὰν βροντήσῃ, ὄμβρον σημαίνει ἡμέρας τεσσαράκοντα. καὶ βασιλέως ἀτιμία καὶ ταραχὴ καὶ εἰς τὰς βασιλικὰς αὐλὰς ἄλλου βασιλέως ἄκουσμα, ἐν ἀνατολῇ καὶ εἰς ἄλλους τόπους λιμὸς καὶ θανατικὸν καὶ χάλαζα εἰς τόπους καὶ πλοίων ἀπώλεια καὶ ἀνδρῶν μεγάλων ἀφανισμός. εἰ δὲ καὶ σεισμός γένηται, μαλακισμοὺς καὶ νόσους ἐπικινδύνους δηλοῖ.

Ὑδροχόος. – Ἐὰν βροντήσῃ, πόλεμον δηλοῖ καὶ κατὰ τόπους νόσον, κατὰ τόπους χαρά εὐφορία σίτου καὶ κριθῆς καὶ ὀσπρίων παντοίων. ἔνδοξος χειμὼν καὶ ποικίλος, καὶ λαῶν χαμός. εἰ δὲ σεισμός γένηται, πόλεμον δηλοῖ.

Ἰχθύς. – Ἐὰν βροντὴσῃ, σίτου λεῖψις εἰς τόπους καὶ κίνδυνος ὀλίγος καὶ ταραχὴ εἰς τὴν χώραν ἐκείνην καὶ εἰς τοὺς κατοικοῦντας βασιλεῖς, καὶ λιμὸς ἔσται καὶ βασιλέως ἀπώλεια, ἔνδοξος χειμών, καὶ τὰ πρώιμα καὶ ὥριμα καλὰ ἔσονται. εἰ δὲ σεισμός γένηται, νόσος εἰς τοὺς ἀνθρώπους.

TREATISE ON THUNDER-DIVINATION

Aries. – If in any way it should thunder in the southern region, a deadly (time) and instability for many bad (men) and destruction of wise men, but if there is also an earthquake, there will be famine.

Taurus. – If in any way it should thunder, wherever the thunder rolls overhead, joy for the kings, but in the south, anarchy and people slaughtering each other, and sedition about the king and great weakness, and destruction. But if there is an earthquake, it indicates many reversals.

Gemini. – If in any way it should thunder, struggles and restlessness for men, lack of wheat and turbulence of the sea and in the heavens, mild winter and a lack of harvests and sedition about the king, to the inhabited world, and destruction of reptiles. Wherever the thunder rolls overhead, that land will be in a state of disorder. But if there is an earthquake, many disturbances.

Cancer. – If in any way it should thunder, many winds, abundant harvest of wheat and barley and fire in the city and bellowing of cattle, a good winter, abundant pulses and to the northern region dissolution and sedition about the king. But if an earthquake happens, many diseases.

Leo. – If in any way it should thunder, a lack of wheat and wine and of much good, and in the place where the sun sets (the west) a change of regime, winter unstable, fire in the city, and plague of cattle, danger to the north, and manifestation of spirits. But if there is an earthquake, wise men will perish.

Virgo. – If in any way it should thunder, destruction of the king and of many of the people, and in the region of Egypt, instability, and the neighboring region will be exposed to danger, but a dearth of all commodities. But if there is an earthquake, many winds, and deep waves in the sea, but winter will be harsh and deadly and diseases in places.

Libra. – If in any way it should thunder, war and people slaughtering each other, and there will be a dearth of wild animals and burning heat will appear, and a good winter, and wars will happen, and from disease men will be muzzled (constrained), and in the neighboring region to the west, famine, and misfortune and death of men, destruction of the four-footed and of reptiles. But if an earthquake occurs, deadly for some regions.

Scorpio. – If in any way it should thunder, winter, and people slaughtering each other, and destruction of birds and the death of rulers, a dearth of pulses and in the northern region weakness, and marauding parties, and regime changes, and misfortune for the rulers. And if an earthquake happens, joy (abundance) of wheat and much kindness, but diseases.

Sagittarius. – If in any way it should thunder, strife and instability. Wherever the thunder rolls overhead, it means damage in that land. Abundance of all commodities, and in the east dissolution. Winter late, and appearance of locusts, to the loss of crops. But if there is earthquake, plague of sheep.

Capricorn. – If in any way it should thunder, it signals cloudiness for forty days. And disgrace of the king and political tumult and for the royal palace, rumor of another king, famine in the east and also in other regions, and something deadly, and hail in some regions and destruction of ships and extermination of great men. But if an earthquake happens, it indicates effeminacy [weakness?] and dangerous diseases.

Aquarius. – If in any way it should thunder, it means war and disease in some regions, in other regions joy. Abundance of wheat and barley and all kinds of pulses. A glorious winter although changeable, and a muzzle for the people [oppression?]. But if an earthquake happens, it indicates war.

Pisces. – If in any way it should thunder, a dearth of wheat in some regions and little danger, and disorder in that land, and in the neighboring kingdoms, and there will be famine and the destruction of the king, a glorious winter, and the early fruits and the late [ripe] fruits will be good. But if an earthquake happens, disease for men.

TEXT 4. MANUSCRIPT GREEK TEXT NO. 21, FROM: *GEOPONICA* 1.10: *BRONTOLOGION* ATTRIBUTED TO ZOROASTER[4]

Σημείωσις τῶν ἀποτελουμένων ἐκ τῆς πρώτης βροντῆς καθ' ἕκαστον ἔτος, μετὰ τὴν τοῦ κυνὸς ἐπιτολήν. Ζωροάστρου.

1. Ἐκείνην χρὴ πρώτην βροντὴν ἡγεῖσθαι καθ' ἕκαστον ἐνιαυτὸν, τὴν μετὰ τὴν ἀνατολὴν τοῦ κυνὸς γινομένην. δεῖ οὖν παρατηρεῖσθαι ἐν ποίῳ οἴκῳ τοῦ ζωδιακοῦ κύκλου τῆς σελήνης οὔσης ἡ βροντὴ ἡ πρώτη γίνεται. 2. ἐπὰν ἐν κριῷ τῆς σελήνης οὔσης ἡ βροντὴ ἡ πρώτη γένηται, σημαίνει πτοηθῆναί τινας κατὰ τὴν χώραν, καὶ γενέσθαι ἀγωνίαν καὶ πυγὴν ἀνθρώπων, ὕστερον δέ κατάστασιν 3. ἐν ταύρῳ τῆς σελήνης οὔσης ἐὰν βροντήσῃ, φθορὰν σίτου κριθῆς σημαίνει, καὶ ἀκρίδθν ἔφοδον. ἐν δὲ βασιλικῇ αὐλῇ χαράν. τοῖς δὲ πρὸς ανατολὰς θλίψιν καὶ λιμὸν 4. ἐν διδύμοις ἐὰν βροντήσῃ, ταραχὴν καὶ νόσον δηλοῖ, σίτου φθοράν, καὶ Ἀράβων ἀπώλειαν. 5. ἐν καρκίνῳ, κριθῆς φθοράν, καὶ ἀνομβρίαν, καὶ βοῶν θάνατον. περὶ δὲ Μάρτιον καὶ Ἀπρίλλιον πολυομβρίαν. 6. ἐν λέοντι, σίτου καὶ κριθῆς φθορὰν ἐν τῇ ὀρεινῇ, καὶ κνησμοὺς καὶ λειχῆνας. 7. ἐν παρθένῳ, βασιλέως ἀπώλειαν καὶ ἄλλον βασιλεύειν τῆς χώρας. τοῖς ναυτιλλομένοις κίνδυνον, περὶ δὲ τὰ σπόριμα ἐρυσίβην. 8. ἐν ζυγῷ, πολέμους καὶ πληγῶν πλησμονὰς καὶ καρπῶν φθοράν. 9. ἐν σκορπίῳ, λιμὸς ἔσται, τὰ δὲ πετεινὰ πληθυνθήσονται. 10. ἐν τοξότῃ, σάλον ἐν τῇ χώρᾳ. ἐν τῇ ὀρεινῇ σίτου εὐφορίαν. ἐν τῇ πεδιάδι φθοράν. 11. ἐν αἰγοκὲρωτι, ὄμβρον ἔσεσθαι σημαίνει ἐπὶ ἡμέρας πεντήκοντα, βασιλέων δόλον, καὶ ψόγον, καὶ ἀπρεπη λαλιάν, καὶ ἐπιφάνειαν ἄλλου βασιλέως ἀπ' ἀνατολῆς, ὃς ἐπικρατήσει πάσης τῆς

[4] *Geoponica sive Cassiani Bassi scholastici de re rustica eclogae*, ed. H. Beckh (Stuttgart and Leipzig: Teubner, original edition 1895, reprinted 1994), 19–20.

οἰκουμένης. εὐκαρπία δὲ ἔσται, καὶ ἐνδόξων ἀνδρῶν θάνατος, καὶ προβάτων γονή. 12. ἐν ὑδροχόῳ, πολέμους ἰσχυροὺς ἐν τῇ παράλῳ. καρπῶν δὲ τῶν ἄλλων φθοράν, ὀσπρίων δὲ ἀπώλειαν. 13. ἐν ἰχθύσι, σίτου φθορὰν ὀλίγην, καὶ θάνατον ἀνδρὸς δυνατοῦ.

Interpretation of the outcomes [of the omens] of the first thunder of each year, after the first appearance of the dog (Sirius). Of Zoroaster.

1. It is necessary therefore to observe closely in which house of the zodiac's circuit the moon is when thunder first happens in the year. 2. If, when the moon is in Aries, thunder first occurs, it indicates that some will be distraught in the land and suffering and flight of men will take place, but afterwards a respite. 3. If, with the moon in Taurus it should thunder, dearth of wheat and barley and arrival of locusts; in the royal palace joy, for those in the east, affliction and famine. 4. In Gemini, if it thunders, disturbance and disease it indicates, and lack of wheat and destruction of Arabs. 5. In Cancer, lack of barley and lack of rain and death of cattle. Around March and April, excessive rain.

6. In Leo, lack of wheat and barley in the hill country, itching, eruption of the skin on animals. 7. In Virgo, destruction of kings and another will reign in the land. Danger for sailors, the sown fields will suffer from rust. 8. In Libra, wars and they will be surfeited with plagues and lack of fruits. 9. In Scorpio, famine and winged ones (birds) shall multiply. 10. In Sagittarius, restlessness in the land, abundance of wheat in the hill country, famine in the plains. 11. In Capricorn, rain for fifty days and cunning plans of kings and censure and unseemly chatter and the appearance of another king from the east, who will overcome the whole world. Abundance of crops, death of a glorious man, and offspring of sheep [plentiful]. 12. In Aquarius, difficult wars in the coastland. Dearth of other crops and destruction of pulses. 13. In Pisces, a slight lack of wheat, and death of a powerful man.

TIMELINE OF PERSONAGES, CULTURAL PHASES, AND HISTORICAL PERIODS

Neolithic period (Italy) (ca. 6000–2800 BC): environmental conditions, foods, lifestyles established from which later cultures developed

Third-millennium BC Mesopotamia (Early Dynastic period at Ur, etc.): earliest traces of divination literature and rituals

Bronze Age (Italy) (2300–ca. 1000 BC): technologies, lifestyles established that were ancestral to those of Etruscan, Italic cultures

Enmeduranki: antediluvian hero, scribe, sage, king of Sippar; shared secrets of divination with the men of Mesopotamia

Avellino eruption of Vesuvius (ca. 1780 BC): catastrophic ("plinian") destruction around Bay of Naples

Zimrilim, king of Mari in Syria (destroyed ca. 1775 BC): his court acquired texts of Mesopotamian divination

Samsuiluna king of Babylon (1749–1712 BC): son of Hammurabi I, mandated use of the Mesopotamian Standard Calendar

Final Bronze Age (Protovillanovan period, Italy) (ca. 1200–1000 BC): cultural changes in Italy set the stage for historical Etruscan cities; distinctive material culture, burial customs

Ugarit (modern Ras Shamra): Canaanite kingdom on Syrian coast, source of tablet texts, interchange with Mesopotamia. Destroyed by "Sea Peoples" invaders, 1190–1185 BC

Tiglath-Pileser I (1114–1076 BC): Assyrian king, waged famous campaigns toward the West

Esagil-kin-apli (1068–1047 BC): the famous Assyrian divination scholar/scribe under King Adad-aplu-iddina

Iron Age (Villanovan period, Italy) (ca. 1000–750/700 BC): distinctive culture in areas that would develop Etruscan cities (Tyrrhenian, Upper Adriatic, Bay of Naples), characterized by metalworking

Hallstatt Minimum (ca. 850–800 BC, effects continued until ca. 500 BC): cold, stormy, wet climatic period affected cultural development in Europe and the Mediterranean, including Italy; today prevents reliable radiocarbon dating of this era

Ashurnasirpal II (883–859 BC): Assyrian king, military empire builder, founded the city of Kalḫu (Nimrud), which preserves a wealth of art, artifacts, and texts

Shalmaneser III (858–824 BC): Assyrian king known for many military campaigns, construction of capital, palaces at Kalḫu (Nimrud)

Late Geometric period (era of Greek colonization in the Italian archipelago) (8th c. BC)

Pithekoussai (early 8th c. BC): founding of large, early colony-city of mixed Greek, Levantine, and Italian settlers; fosters technological, economic, artistic exchanges

Tiglath-Pileser III (744–727 BC): Assyrian king, restructured kingdom and army, used deportation as means of control of captive regions

Sargon II (ruled 722–705 BC): Assyrian king; his dynasty supported major scholarly initiatives; built capital Dur-Sharrukin (Khorsabad)

Orientalizing phase (of Etruscan culture) (ca. 740/720–680 BC, continued into 6th c. BC): urbanization, profound Eastern and Greek influences on Etruscan culture, especially art

Era of Princely Tombs (7th c. BC): ostentatious tombs, offerings associated with presumed rulers of Etruria, Latium, and Campania, some of them female (e.g., Larthia Velthurus of Caere, Vetusia of Praeneste)

Sennacherib (704–681 BC): king of Assyria, many military campaigns, attacked Judah, built capital of Nineveh and Palace "Without Rival"

Nabû-zuqup-kēnu of Kalḫu (Nimrud): great scholar during reigns of Sargon II and Sennacherib

Esarhaddon (680–669 BC): Assyrian king during difficult times: he promoted learning, and relied heavily upon diviners

Ashurbanipal (668–627 BC): Assyrian king, highly educated (the first king to read and write), and the last warrior-king of Assyria

Library of Ashurbanipal (destroyed 612 BC): in North West palace, Nineveh; royal archive that contained large collection of texts, including all major divination categories

Demaratus (ca. 650 BC): exiled from Corinth after expulsion of Bacchiad king; he emigrated to Tarquinia and is recorded as having stimulated development of art and technology in Etruria; father of Tarquinius Priscus, Etruscan king of Rome

Archaic period (Etruria)(ca. 600–475/450 BC)

510/509 BC: Rome expels last (Etruscan) king, forms Republic

ca. 500 BC: end of Hallstatt climatic minimum, climate in Italy similar to modern-day

Classical period (Etruria) (475/450–300 BC)

Hellenistic period (ca. 331–30 BC): from conquests of Alexander to death of Cleopatra, last Hellenistic Greek ruler.
Late Republican period (Italy) – ends as Octavian is named Augustus and "*princeps*" of Rome
73–71 BC: slave revolt of Spartacus in Italy defeated by Pompey and Crassus
63 BC: Cicero consul in Rome, executes Catilinarian conspirators, supported by Nigidius Figulus and others
60 BC: Pompey, Crassus, and Julius Caesar form the "First Triumvirate"
58 BC: Publius Clodius tribune, institutes grain dole in Rome
56 BC: noises in heavens above Latium prompt Rome to invite Etruscan haruspices to interpret the omen
50s BC: Publius Nigidius Figulus publishes works on Etruscan religion
45 BC: Figulus dies in exile, having supported Pompey
47–44 BC: Caesar dictator; assassinated on the Ides of March of the new "Julian" calendar
Late Republican period (1st c. BC): scholars, political figures, authors include Publius Nigidius Figulus, Aulus Caecina, Marcus Tullius Cicero, Varro, Diodorus Siculus, Dionysios of Halikarnassos
Early Empire (1st c. AD): scholars whose writings give evidence of Etruscan culture, and religion; Seneca, Pliny the Elder (d. 79 AD), Columella
Byzantine period (6th c. AD): period of Justinian and the career of Johannes Lydus ("John the Lydian"), civil servant in the palace at Constantinople, preserves ancient documents relating to pagan (including Etruscan) religion

GLOSSARY

Apkallū – the Seven Sages of mythical Mesopotamia (before the Flood)

apodosis – the prediction clause(s) of an omen text, follows the protasis

Apsu – the living, freshwater under the earth, domain to which the Apkallu sages return

Bārûtu – Mesopotamian set of texts covering divination by extispicy (examining victim's entrails); *bārû* = haruspex, diviner using entrails, especially liver

Brontoscopion – Greek title for a text concerning divination by thunder (*brontē*)

Capua tile (ca. 480 BC) – terracotta tablet, part of an Etruscan text of a calendar listing dates for performance of religious rituals, found in the necropolis of Capua

Chaldei ("Chaldeans") – Roman name for itinerant soothsayers/diviners for hire in Late Republican Rome, implying Babylonian origins

Enūma Anu Enlil – great cuneiform text series of celestial divination, sections named by modern scholars after pertinent gods: Shamash (sun), Adad (weather), Ishtar (planets), Sin (moon).

etrusca disciplina – the "teachings of the Etruscans"; the texts (revealed by Tages) comprising rituals necessary for Etruscan religion, how to found cities, expiate sins, and cultivate the gods and the dead

extispicy – the practice of divination by excising the entrails (especially liver) of a sacrificed animal

fulgur (L. "lightning bolt") – Roman religion specialized in divination by lightning (as opposed to thunder), and places, persons, objects struck by lightning had to be dealt with according to special rituals. A *fulguriator* was a priest skilled in this form of divination

haruspicy – the practice of divination by scrutinizing the victim's liver; the officiant is a *haruspex* (Latin) or *netśvis* (Etruscan)

hemerology – a work of divination arranged according to the day on which an omen occurs

Hermes Trismegistus – "Hermes Thrice-great"; originated in the syncretistic religious philosophy of Ptolemaic Egypt, but later was understood by scholars as a famous personage, to whom they attributed a variety of writings, including divination texts

incipit – the opening words of a text, often used today as the "title" for an ancient text

Iqqur īpuš – Mesopotamian divination text formatted by months or days; omens compiled from many other categories of divination

menology – a work of divination arranged according to the month in which an omen occurs

MUL.APIN – early and long-popular Mesopotamian text for divination by celestial, especially astronomical, phenomena; influenced many other works

Namburbû ("the omen's undoing") – Mesopotamian rituals by which a harmful prediction might be averted

nethśrac, netśvis – Etruscan terms for haruspicy, haruspex

palafitte ("platform villages") – known in Europe and Italy from the Neolithic period through the Iron Age. A settlement type known especially in Italian Bronze and Iron ages for sophisticated woodwork

Petosiris – an Egyptian scholar, often named by Classical scholars; with a (7th c. BC) pharaoh "Nechepso" as the author of texts on divination and astrology (said texts were probably Ptolemaic or later, not to be confused with a Hellenistic-era priest Petosiris, whose tomb has been excavated at Tuna-el-Gebel)

Piacenza liver – Etruscan bronze model of a sheep's liver, marked with names of gods who preside over different areas of a victim's liver, hence the messages from those gods (Hellenistic period, perhaps ca. 100 BC)

plinian (volcanic eruption) – term for an extremely violent volcanic eruption, such as that of 79 AD in which Roman scholar and admiral Pliny the Elder was killed

protasis – the opening conditional clause of an omen statement: "If it thunders ..." preceding the apodosis, the actual prediction

puls – Latin term for a porridge made from grains: the staple of primitive diet, rather than bread. Etruscans may have been called *pultiphagi*, "porridge eaters" by other cultures

Seismologion – Greek title for a text concerning divination by earthquake

Selindromia – Greek title for a text concerning divination by phases of the moon

Šumma ālu (*šumma* = "if") – "If a city is built on a hill ..."; incipit of a famous Mesopotamian text series for divination by phenomena in city life

Šumma izbu – "If the deformed newborn ..."; incipit of a text for teratological divination (by birth defects or unnatural behaviors)

Tages – legendary prophetic being, resembling a child and an old man, who emerged from a plowed furrow and recited/sang the text of the Etruscan scriptures (*etrusca disciplina*) recorded in the proto-urban period

Tarchon – legendary founder of the city of Tarquinii, to whom (in some versions of the legend) Tages revealed the religious teachings, the *etrusca disciplina*

Tonitruale – Latin title for a text concerning divination by thunder

tupšarru – Akkadian (Assyrian) term for "scholar"

Zagreb *Liber linteus* – Etruscan "Linen book" found re-used as mummy wrappings brought from Egypt to Zagreb in the nineteenth century; written in ink (2nd c. BC), it is a calendar of religious rituals, one of the longest extant Etruscan documents

Zilath mechl rasnal (*zilaθ meχl rasnal*) (Latin *praetor Etruriae*) – high-ranking magistrate of the Etruscan people

Zoroaster – seminal prophet of the Persian religion; his life is little known and variously dated (either ca. 1000 BC or 6th c. BC); much later works of divination were attributed to him

BIBLIOGRAPHY

Acha, P. N., and Szyfres, B. 2003. *Zoonoses and Communicable Diseases Common to Man and Animals*. Vol. 3, *Parasitoses*, 3rd ed. Washington DC: Pan American Health Organization.

Agostiniani, L., and F. Nicosia. 2000. *Tabula Cortonensis*. Rome: L'Erma di Bretschneider.

Aigner Foresti, L. 2000. "Orientalische Elemente im etruskischen Königtum?" In Prayon and Röllig, 275–86.

Albore Livadie, C. 2002. "A First Pompeii: The Early Bronze Age Village of Nola-Croce del Papa (Palma Campania phase)." *Antiquity* 76: 941–42.

Albore Livadie, C., and C. Cicirelli. 2003. "L'insediamento protostorico in località Longola di Poggiomarino." *La Parola del Passato* 58: 88–128.

Albore Livadie, C., and G. Vecchio. 2005. "Un Villaggio del Bronzo Antico a Nola-Croce del Papa (Campania)." In P. Attema, A. Nijboer, and A. Zifferero, eds., *Papers in Italian Archaeology VI: Communities and Settlements from the Neolithic to the Early Medieval Period*, 2:581–87. (Proceedings of the Sixth Conference of Italian Archaeology held at the University of Groningen, Groningen Institute of Archaeology, The Netherlands, April 15–17, 2003 = BAR Int. Ser. 1452[II].) Oxford: Archaeopress.

Alimentazione = Various authors. 1987. *L'alimentazione nel mondo antico: Gli Etruschi*. Ministero per I Beni Culturali e Ambientali, Giroanta Mondiale dell'Alimentazione, 1987. Rome: Istituto Poligrafico e Zecca dello Stato.

Amadasi-Guzzo, M. G. 1990. *Iscrizioni fenicie e puniche in Italia*. Rome: Istituto di Studi del Vicino Oriente.

Amann, P. 2000. *Die Etruskerin: Geschlechterverhältnis und Stellung der Frau im frühen Etrurien (9.-5. Jh. v. Chr.)*. Vienna: Österreichischen Akademie der Wissenschaften.

Ambos, C., and I. Krauskopf. 2010. "The Curved Staff in the Ancient Near East as a Predecessor of the Etruscan *lituus*." In L. B. van der Meer, ed., *Material Aspects of Etruscan Religion*, 126–53. (Proceedings of the International Colloquium Leiden, May 29 and 30, 2008.) Leuven: Peeters.

Ampolo, C. 1976–77. "Demarato: Osservazioni sulla mobilità sociale arcaica." *Dialoghi di Archeologia* 9–10: 333–45.

1990–91. "Lotte sociali in Italia centrale. Un documento controverso: Il calendario brontoscopico attribuito a Nigido Figulo." *Opus* 9–10: 185–97.

Ancillotti, A., and R. Cerri. 1996A. *Le tavole di Gubbio e La Civiltà degli Umbri*. Perugia: Edizioni Jama.

1996B. *The Tables of Iguvium* (English translation, abridged version of *Le tavole di Gubbio e La Civiltà degli Umbri*). Perugia: Edizioni Jama.

Anderson, R., and R. May. 1992. *Infectious Diseases of Humans*. Oxford: Oxford University Press.

Andrews, S. J. 1999. "The Life Cycle of *Fasciola hepatica*." In J. P. Dalton, ed., *Fasciolosis*, 1–29. Wallingford-Cambridge: CABI.

Angel, J. Lawrence. 1971. *Lerna*. Vol. 2, *The People*. Princeton, N.J.: American School of Classical Studies.

Aranguren, B., C. Bellini, M. Mariotti Lippi, M. Mori Secci, and P. Perazzi. 2007. "L'avvio della coltura della vite in Toscana: L'esempio di San Lorenzo a Greve (Firenze)." In A. Ciacci, P. Rendini, and A. Zifferero, eds., *Archeologia della vite e del vino in Etruria*, 88–97. (Atti del Convegno Internazionali di Studi, Scansano Teatro Castagnoli, 9–10 settembre 2005.) Siena: Ci.Vin.

Armstrong, G. L., J. Hollingsworth, and J. G. Morris, Jr. 1998. "Bacterial Foodborne Disease, Medical Costs and Productivity Losses." In Evans and Brachman, 109–38.

Arietti, F. 2005. "Intervento." In Bartoloni and Delpino, 638–41.

Aspetti della cultura di Volterra etrusca fra l'età del ferro e l'età ellenistica. 1997. (Atti del XIX Convegno di Studi Etruschi ed Italici, Volterra 1995.) Florence: Istituto di Studi Etruschi.

Aspöck, H., H. Auer, and O. Picher. 1996. "*Trichuris trichiura* Eggs in the Neolithic Glacier Mummy from the Alps." *Parasitology Today* 12: 255–56.

1999. "Parasites and Parasitic Diseases in Prehistoric Human Populations in Central Europe." *Helminthologia* 36: 139–45.

Aubet, M. E. 2001. *The Phoenicians and the West*, 2nd ed. Trans. M. Turton. Cambridge: Cambridge University Press.

Aubet Semmler, M. E. 2002. "The Phoenician Impact on Tartessos: Spheres of Interaction." In M. R. Bierling, ed. and trans., *The Phoenicians in Spain: An Archaeological Review of the Eighth–Sixth Centuries B.C.E.*, 199–224. Winona Lake, Ind.: Eisenbrauns.

Aufderheide, A. C., and C. Rodríguez-Martín. 1998. *The Cambridge Encyclopedia of Human Paleopathology*. Cambridge: Cambridge University Press.

Aygon, J.-P. 2003. "Météorologie et épidémie: le topos de la peste chez les poètes latins." In C. Cusset, ed., *La Météorologie dans l'Antiquité: entre science et croyance*, 275–89. (Actes du Colloque internationale interdisciplinaire de Toulouse, 2–4 mai 2002.) Saint-Étienne: Université de Saint-Étienne.

Azzaroli, A. 1972. "Il cavallo domestico in Italia dall'età del Bronzo agli Etruschi." *Studi Etruschi* 40: 273–306.

1998. "Canal Bianco Tomb of Adria, and Large Sized Horses in the Ancient World." In *Proceedings of the XIII Congress, International Union of Prehistoric and Protohistoric Science*, vol. 6, tome I, 83–85, Forlí, Italy, 8–14 September 1996.

Babbi, A. 2002. "Appliques e pendenti nuragici dalla Raccolta Comunale di Tarquinia." In O. Paoletti and L. Tamagno Perna, eds., *Etruria e Sardegna centro-settentrionale tra l'età del Bronzo Finale e l'arcaismo*, 433–52. (Atti del XXI Convegno di Studi Etruschi ed Italici, Sassari-Alghero-Oristano-Torralba, 13–17 ottobre 1998.) Pisa and Rome: Istituti Editoriali Poligrafici Internazionali.

Babbi, A., and A. Piergrossi. 2005. "*Per una definizione della cronologia relativa e assoluta del Villanoviano veiente e tarquiniese.*" In Bartoloni and Delpino, 293–318.

Bacchielli, L. 1986. "I piatelli Genucilia." In J. Swaddling, ed., *Italian Iron Age Artefacts in the British Museum*, 375–80. Papers of the Sixth British Museum Classical Colloquium 1982. London: British Museum.

Baggieri, G. 2005. "Paleopatologia scheletrica degli individui della necropoli di Monterozzi e di San Giovenale." In O. Paoletti and G. Camporeale, eds., *Dinamiche di Sviluppo delle Città nell'Etruria Meridionale: Veio, Caere, Tarquinia, Vulci*, 2: 703–13. (Atti del XXIII Convegno di Studi Etruschi ed Italici, Roma, Veio, Cerveteri/Pyrgi, Tarquinia, Tuscania, Vulci, Viterbo, 1–6 ottobre 2001.) Pisa and Rome: Istituti Editoriali e Poligrafici Internazionali.

Bagnasco Gianni, G. 1996. *Oggetti iscritti di epoca orientalizzante in Etruria (Biblioteca di Studi Etrusci* 30). Florence.

Baldwin, B. 1991. "John Lydos." In A. P. Kazhdan, ed., *The Oxford Dictionary of Byzantium*, 2: 1061–62. Oxford: Oxford University Press.

Balista, C., and G. Leonardi. 1996. "Gli abitati di ambiente umido nel Bronzo Antico dell'Italia settentrionale." In M. Cocchi Genick, ed., *L'Antica Età del Bronzo*, 199–228. (Atti del Congresso di Viareggio, 9–12 gennaio 1995.) Florence: F. Cantini.

Barber, E. W. 1994. *Women's Work: The First 20,000 Years*. New York: Norton.

Barber, E. W., and P. T. Barber. 2004. *When They Separated Earth from Sky: How the Human Mind Shapes Myth*. Princeton, N.J.: Princeton University Press.

Barbieri, G. 1987. "L'alimentazione carnea degli Etruschi." In *L'alimentazione nel mondo antico: Gli Etruschi*, 49–53. Ministero per I Beni Culturali e Ambientali, Giroanta Mondiale dell'Alimentazione, 1987. Rome: Istituto Poligrafico e Zecca dello Stato.

Barbieri, G., and M. J. Becker. 1996–97. "Tomba della donna con i sandali." *Notizie degli Scavi* 1996–97: 343–55.

Bardinet, T. 1988. "Remarques sur les maladies de la peau, la leper, et le châtiment divin dans l'Egypte ancienne." *Revue d'Egyptologie* 39: 3–36.

Barker, G., and T. Rasmussen. 1998. *The Etruscans*. Oxford: Blackwell.

Bartoli, F., F. Mallegni, and G. Fornaciari. 1997. "Le risorse alimentari nel mondo etrusco: aspetti della paleodieta in due gruppi umani a cultura etrusca." In *Aspetti della cultura di Volterra etrusca*, 477–88.

Bartoli, F., A. D'Alessandro, and F. Mallegni. 1999. "Aspetti paleonutrizionali di campioni osteologici umani paleoveneti provenienti da Piovego ed Este." In *Protostoria e storia del "Venetorum Angulus,"* 217–25. (Atti del XX Convegno di Studi Etruschi ed Italici, Portogruaro, Quarto d'Altino, Este, Adria, 16–19 ottobre 1996.) Pisa: Istituti Editoriali e Poligrafici Internazionali.

Bartoloni, G. 1987. "Le comunità dell'Italia centrale tirrenica e la colonizazzione greca in Campania." In M. Cristofani, ed., *Etruria e Lazio arcaico* (*Quaderni AEI* 15), 37–53. Rome: Consiglio Nazionale delle Ricerche.

1991. "Populonia: Characteristic Features of a Port Community in Italy during the First Iron Age." In E. Herring, R. Whitehouse, and J. Wilkins, eds., *Papers of the Fourth Conference of Italian Archaeology [1990]*, Part 2, *Archaeology of Power*, 101–15. London: Accordia Research Center.

2000. "La donna del principe." In G. Bartoloni, F. Delpino, C. Morigi Govi, and G. Sassatelli, eds., *Principi etruschi tra Mediterraneo ed Europa*, 271–306. (Mostra, Bologna, Museo Civico Archeologico.) Bologna: Marsilio.

2002. *La cultura villanoviana: all'inizio della storia etrusca*, 2nd ed. Rome: Carocci.

2005. "Veio, Piazza d'Armi." In O. Paoletti and G. Camporeale, eds., *Dinamiche di Sviluppo delle Città nell'Etruria Meridionale: Veio, Caere, Tarquinia, Vulci*, 1:73–84. (Atti del XXIII Convegno di Studi Etruschi ed Italici, Roma, Veio, Cerveteri/Pyrgi, Tarquinia, Tuscania, Vulci, Viterbo, 1–6 ottobre 2001.) Pisa and Rome: Istituti Editoriali e Poligrafici Internazionali.

2006. "L'inizio del processo di formazione urbana in Etruria: Analogie e differenze venute in luce nei recenti scavi." In M. Bonghi Jovino, ed., *Tarquinia e le civiltà del Mediterraneo*, 49–82. (Convegno internazionale, Milano, 22–24 giugno 2004.) Milan: Cisalpino, Istituto Editoriale Universitario.

2007. "Véies: Recherches récentes sur la 'citadella' de Piazza d'Armi." *Les Dossiers d'archéologie* 322: 22–29.

2009. "The Earliest Etruscan Toast: Considerations on the Earliest Phases of Populonia." In M. Gleba and H. Becker, eds., *Votives, Places and Rituals in Etruscan Religion: Studies in Honor of Jean MacIntosh Turfa*, 159–70. Leiden: Brill.

Bartoloni, G., et al. 1997. "Il complesso rurale di Campassini (Monteriggioni): considerazioni sull'alimentazione nell'Etruria settentrionale nell'VIII e VII sec. A.C." In *Aspetti della cultura di Volterra etrusca*, 93–128.

Bartoloni, G., A. Berardinetti, and L. Drago. 2000. "Le communità della bassa valle tiberina e il Mediterraneo orientale prima della colonizazzione greca." In F. Krinzinger, ed., *Die Ägäis und das westliche Mittelmeer: Beziehungen und Wechselwirkungen 8. bis 5. Jh. v. Chr.*, 525–33. (Akten des Symposions Wien, 24. bis 27. März 1999.) Vienna: Österreichische Akademie der Wissenschaften.

Bartoloni, G., and F. Delpino, eds. 2005. *Oriente e Occidente: metodi e discipline a confronto: Riflessioni sulla cronologia dell'età del ferro in Italia*. (Atti dell'Incontro di studi, Roma, 30–31 ottobre 2003.) (Series *Mediterranea*, I, 2004 [2005] *Quaderni annuali dell'Istituto di Studi sulle Civiltà Italiche e del Mediterraneo antico del Consiglio Nazionale delle Ricerche*, formerly *Quaderni di Archeologia Etrusco-Italica*). Pisa and Rome: Istituti Editoriali e Poligrafici Internazionali.

Barton, T. 1994. *Ancient Astrology*. London: Routledge.

Basch, L. 1987. *Le musée imaginaire de la marine antique*. Athens: Institute Hellénique pour la Préservation Nautique.

Bass, G. F. 1967. "VI. The Bronzes." In G. F. Bass, ed., *Cape Gelidonya: A Bronze Age Shipwreck (Transactions of the American Philosophical Society*, n.s. vol. 57, pt. 8: 84–121). Philadelphia: American Philosophical Society.

Baumgarten, A. I. 1981. *The Phoenician History of Philo of Byblos*. Leiden: Brill.

Baxter, P. J. 1990. "Medical Effects of Volcanic Eruption." *Bulletin of Volcanology* 52: 532–44.

Beard, M. 1991. "Ancient Literacy and the Function of the Written Word in Roman Religion." In M. Beard, ed., *Literacy in the Roman World, Journal of Roman Archaeology*, Suppl. 3, Ann Arbor Michigan, 35–58.

2004. "Writing and Religion." In S. I. Johnston, ed., *Religions of the Ancient World: A Guide*, 127–38. Cambridge, Mass.: Harvard University Press.

Beard, M., J. North, and S. Price. 1998. *Religions of Rome*. Vol. 1, *A History*; Vol. 2, *Texts*. Cambridge: Cambridge University Press.

Becker, H. W. 2007. "Production, Consumption, and Society in Northern Etruria during the Archaic and Classical Periods: The World of Lars Porsenna." Ph.D diss., University of North Carolina, Chapel Hill.

Becker, M. J. 1995. "Human Skeletal Remains from the Pre-Colonial Greek Emporium of Pithekoussai on Ischia (NA): Culture Contact in Italy from the Early VIII to the II Century BC." In Christie, 273–81.

1998. "The Cremations in the Calabrese Urn from Cerveteri and From Other Cinerary Containers in the Museo Gregoriano Etrusco, Vatican Museum." *Bollettino – Monumenti, Musei e Gallerie Pontifiche* XVIII: 57–73.

1999. "Bioarchaeology." *American Journal of Archaeology* 103: 528–31.

Becker, M. J., and A. Donadio. 1992. "A Summary of the Analysis of Cremated Human Skeletal Remains from the Greek Colony of Pithekoussai at Lacco Ameno, Ischia, Italy." *Old World Archaeology Newsletter* 16(1): 15–23.

Becker, M. J., and Salvadei, L. 1992. "Analysis of the Human Skeletal Remains from the Cemetery of Osteria dell'Osa." In A. M. Bietti Sestieri, ed., *La Necropoli Laziale di Osteria dell'Osa*, 53–191. Rome: Quasar.

Becker, M. J., J. M. Turfa, and B. Algee-Hewitt. 2009. *Human Remains from Etruscan and Italic Tomb Groups in the University of Pennsylvania Museum* (*Biblioteca di Studi Etruschi* 48). Pisa and Rome: Istituto di Studi Etruschi ed Italici, Fabrizio Serra Editore.

Beckman, G. 1983. "Mesopotamians and Mesopotamian Learning at Hattuša." *Journal of Cuneiform Studies* 35: 97–114.

Bedini, E. 1997. "I resti faunistici." In M. Bonghi Jovino and C. Chiaramonte Treré, eds., *Tarquinia: Testimonianze archeologiche e ricostruzione storica*, 103–44. Rome: L'Erma di Bretschneider.

Behm, C. A., and M. C. Sangster. 1999. "Pathology, Pathophysiology and Clinical Aspects." In J. P. Dalton, ed., *Fasciolosis*, 185–224. Wallingford-Cambridge: CABI.

Bekker, I. 1837. *Ioannes Lydus. (Corpus Scriptorum Historiae Byzantinae, editio emendatior et copiosior, consilio B. G. Nieburhrii C. F., instituta, auctoritate Academiae Litterarum Regiae Borussicae continuata.)* Bonn: Ed. Weber.

Benelli, E. 1994. *Le iscrizioni bilingii etrusco-latine. (Biblioteca di Studi Etruschi 27)*. Florence: Olschki.

1996. "Sui cosidetti penesti etruschi." *La Parola del Passato* 51: 335–44.

2000. "Iscrizione bilingue da Pesaro." In M. Torelli, ed., *Gli Etruschi*, 638, no. 329. Milan: Bompiani.

2003. "Una misconosciuta nota di Gustav Herbig e l'etrusco etera*." *Miscellanea Etrusco-Italica* 3: 209–21.

Berkin, J. 2003. *Orientalizing Bucchero Pottery from the Lower Building at Poggio Civitate (Murlo)*. Philadelphia: AIA Monograph, University Museum Press.

Bernabò Brea, M., and A. Mutti. 1994. "... *Le Terremare si scavono per concimare i prati* ..." Parma: Silva Editore.

Bernabò Brea, M., A. Cardarelli, and M. Cremaschi, eds. 1997. *Le Terremare più antica civiltà padana* (Catalogo della mostra). Modena: Electa.

Bernicchia, A., M. A. Figazzola, V. Gemelli, B. Mantovani, A. Lucchetti, M. Cesari, and E. Speroni, 2006. "DNA Recovered and Sequenced from an Almost 7000 Year-old Neolithic Polypore, *Daedaleopsis tricolor*." *Mycological Research* 110(1): 14–17.

Bertani, M. G. 1995. "Il 'banchetto dei morti' in Etruria padana (IX–IV sec. A.C.): Risorse del territorio e alimentazione nelle testimonianze funerari." In L. Quilici and S. Quilici Gigli, eds., *Agricoltura e commerci nell'Italia antica*, 41–64. Rome: L'Erma di Bretschneider.

Bettelli, M. 2004. "'Divinità migranti'? Possibili influssi dal Mediterraneo orientale nell'iconografia dell'età del Bronzo italiana." In N. Negroni Catacchio, ed., *Preistoria e Protostoria in Etruria*, 1: 247–60. (Atti del Sesto Incontro di Studi. Miti Simboli Decorazioni Ricerche e scavi, Pitigliano – Valentano 13–15 settembre 2002.) Milan: Centro Studi di Preistoria e Archeologia.

Bezold, C. 1911. *Astronomie, Himmelschau und Astrallehre bei den Babyloniern = Sitzungsberichte der Heidelberger Akademie der Wissenschaften*, Philosophisch-historische Klasse 1911, Abhandlung 2. Heidelberg: Carl Winter's Universitätsbuchhandlung.

Bezold, C., and F. Boll. 1911. *Reflexe astrologischer Keilschriften bei griechischen Schriftstellern = Sitzungsberichte der Heidelberger Akademie der Wissenschaften*, Philosophisch-historische Klasse 1911, Abhandlung 7. Heidelberg: Carl Winter's Universitätsbuchhandlung.

Bierling, M. R., ed. and trans. 2002. *The Phoenicians in Spain. An Archaeological Review of the Eighth–Sixth Centuries* BCE. *A Collection of Articles translated from Spanish.* Winona Lake, Wis.: Eisenbrauns.

Bietti Sestieri, A. M. 1973. "The Metal Industry of Continental Italy, 13th–11th cent., and Its Aegean Connections." *Proceedings of the Prehistoric Society* 39: 383–424.

1986. "Italian Swords and Fibulae in the Late Bronze and Early Iron Ages." In J. Swaddling, ed., *Italian Iron Age Artefacts in the British Museum*, 3–23. Papers of the Sixth British Museum Classical Colloquium 1982. London: British Museum.

1992A. *The Iron Age Community of Osteria dell'Osa: A Study of Socio-political Development in Central Tyrrhenian Italy*. Cambridge: Cambridge University Press.

Bietti Sestieri, A. M., ed. 1992B. *La Necropoli Laziale di Osteria dell'Osa*. Rome: Quasar.

Bietti Sestieri, A. M. 1995. "The Role of Interregional Contact in the Development of Latial Society in the Early Iron Age." In Christie, 353–64.

1997. "Italy in Europe in the Early Iron Age." *Proceedings of the Prehistoric Society* 63: 371–402.

1998. "L'Italia in Europa nella prima età del Ferro: una proposta di ricostruzione storica." *Archeologia Classica* 50: 1–67.

2005. "A Reconstruction of Historical Processes in Bronze and Early Iron Age Italy Based on Recent Archaeological Research." In P. Attema, A. Nijboer, and A. Zifferero, eds., *Papers in Italian Archaeology VI: Communities and Settlements from the Neolithic to the Early Medieval Period*, 2:9–24. (Proceedings of the Sixth Conference of Italian Archaeology held at the University of Groningen, Groningen Institute of Archaeology, The Netherlands, April 15–17, 2003 = BAR Int. Ser. 1452[II].) Oxford: Archaeopress.

2010. *L'Italia nell'età del bronzo e del ferro: Dalle palafitte a Romolo (2200–700 a.C.)*. Rome: Carocci.

Bietti Sestieri, A. M., A. De Santis, and A. La Regina. 1989–1990. "Elementi di tipo cultuale e doni personali nella necropoli laziale di Osteria dell'Osa." *Scienze dell'Antichità* 3–4: 65–88.

Biggs, R. D. 1967. "More Babylonian 'Prophecies.'" *Iraq* 29: 117–32.

Bjorkman, J. K., 1973. "Meteors and Meteorites in the Ancient Near East." *Meteoritics* 8: 91–132. (Also published as a booklet by Center for Meteorite Studies, publication no. 12, Arizona State University, June 30, 1973.)

Black, F. 1975. "Infectious Diseases in Primitive Societies." *Science* 187: 515–18.

Black, J., and A. Green. 1992. *Gods, Demons and Symbols of Ancient Mesopotamia: An Illustrated Dictionary*, with illustrations by T. Rickards. Austin: University of Texas and British Museum.

Blanck, H., and G. Proietti. 1986. *La Tomba dei Rilievi di Cerveteri*. Rome: De Luca.

Blomé, B., C. Nylander, and I. Pohl. 1996. "An Etruscan Earthquake c. 550 B.C.?" *Opuscula Romana* 20: 261–63.

Blong, R. J. 1984. *Volcanic Hazards: A Sourcebook on the Effects of Eruptions*. Sydney: Academic Press.

Bobrova, L., and A. Militarev. 1993. "From Mesopotamia to Greece: To the Origin of Semitic and Greek Star Names." In H. D. Galter, ed., *Die Rolle der Astronomie in den Kulten Mesopotamiens*, 307–29. Beiträge zum 3. Grazer Morgenländischen Symposion (23.–27. September 1991). Graz: GrazKult.

Boccolini, P. 2004. "Elementi di corredo: un'ipotesi di ricostruzione dello *status* della donna a Tarquinia nel IX secolo a.C." In N. Negroni Catacchio, ed., *Preistoria e Protostoria in Etruria*, 2: 581–90. (Atti del Sesto Incontro di Studi. Miti Simboli Decorazioni Ricerche e scavi, Pitigliano - Valentano 13–15 settembre 2002.) Milan: Centro Studi di Preistoria e Archeologia.

Bodson, L., and D. Orr. 2002. "Amphibians and Reptiles: Evidence from Wall Paintings, Sculpture, Skeletal Remains and Ancient Authors." In Jashemski and Meyer, 327–50.

Boese, J., R. Echt, and A. Miron, eds. 1983. *Frühe Phöniker im Libanon. 20 Jahre deutsche Ausgrabungen in Kāmid el-Lōz*. Mainz/Rhein: Philipp von Zabern.

Boiardi, A., and P. von Eles. 1997. "La necropoli Lippi di Verucchio: Ipotesi preliminare per un analisi delle strutture sociali." *Archeologia dell'Emilia Romagna* 1(1): 23–40.

Boitani, F. 2001. "[Veio. Le necropoli.] I.G.8. La tomba principesca n. 5 di Monte Michele." In A. M. Moretti Sgubini, ed., *Veio, Cerveteri, Vulci: Città d'Etruria a Confronto*, 113–18. (Exhibition, Rome, 1 ottobre–30 dicembre 2001.) Rome: L'Erma di Bretschneider.

2004. "La tomba del guerriero AA1 dalla necropoli dei Quattro Fontanili di Veio." In A. M. Sgubini Moretti, ed., *Scavo nello scavo: Gli Etruschi non visti*, 128–36 (137–49 catalog). (Ricerche e 'riscoperte' nei depositi dei musei archeologici dell'Etruria meridionale: 5 marzo 2004–30 giugno 2004, Viterbo, Fortezza Giulioli.) Rome: Soprintendenza per I Beni Archeologici per l'Etruria Meridionale/Union Printing Edizioni.

Boiy, T. 2004. *Late Achaemenid and Hellenistic Babylon*. Leuven: Peeters.

Bonfante, L. 1981. *Out of Etruria: Etruscan Influence North and South*. BAR Int.Ser. 1732. Oxford: B.A.R.

1990. *Reading the Past: Etruscan*. London: British Museum.

2006A. "33. Etruscan Inscribed Bucchero Aryballos." In L. Bonfante and B. Fowlkes, eds., *Classical Antiquities at New York University*, 111–13. Rome: L'Erma di Bretschneider.

2006B. "Etruscan Inscriptions and Etruscan Religion." In N. T. de Grummond and E. Simon, eds. 2006. *The Religion of the Etruscans*, 9–26. Austin: University of Texas Press.

Bonfante, L., and G. Bonfante. 2002. *The Etruscan Language: An Introduction*, 2nd ed. Manchester: The University Press.

Bonfante, L., and B. Fowlkes, eds. 2006. *Classical Antiquities at New York University*. Rome: L'Erma di Bretschneider.

Bonghi Jovino, M. 1991. "Osservazioni sui sistemi di costruzione a Tarquinia: techniche locali ed impiego del 'muro a pilastri' fenicio." *Archeologia Classica* 43: 171–91.

2000. "Funzione, simboli e potere: I 'Bronzi' del 'Complesso' tarquiniese." In Prayon and Röllig, 287–98.

2010. "The Tarquinia Project: A Summary of 25 Years of Excavation." *American Journal of Archaeology* 114: 161–80.

Bonghi Jovino, M., and C. Chiaramonte Treré. 1987. *Tarquinia: Ricerche, Scavi e Prospettive.* (Mostra, Milan 1986.) Milan: Edizioni ET.

Bonghi Jovino, M., and C. Chiaramonte Treré, eds., 1997. *Tarquinia: Testimonianze archeologiche e ricostruzione storica.* Rome: L'Erma di Bretschneider.

Bonghi Jovino, M., F. Malegni, and L. Usai. 1997. "Una morte violenta: Sul rinvenimento di uno scheletro nell'area del 'complesso sacro-istituzionale' della Civita di Tarquinia." In *Aspetti della cultura di Volterra etrusca*, 489–98.

Bonghi Jovino, M., et al. 2001. "'Area sacra/complesso monumentale' della Civita." In A. M. Moretti Sgubini, ed., *Tarquinia etrusca: Una nuova storia*, 21–44. (Catalogo della mostra, Tarquini, 2001.) Rome: L'Erma de Bretschneider.

Borgognini Tarli, S. M. 1992. "Aspetti antropologici e paleodemografici dal Paleoliticio superiore alla prima età del ferro." In A. Guidi and M. Piperno, eds., *Italia preistorica*, 238–73. Rome: Laterza.

Borgongino, M. 2006. *Archeobotanica: Reperti vegetali da Pompei e del territorio vesuviano.* Rome: L'Erma di Bretschneider.

Boscato, P., C. Carioni, A. Brandolini, L. Sadori, and M. Rottoli. 2008. "Molecular Markers for the Discrimination of *Triticum turgidum* L. subsp. *dicoccum* (Schrank ex Schubl.) Thell. and *Triticum timopheevii* (Zhuk.) Zhuk. Subsp. *Timopheevii.*" *Journal of Archaeological Science* 35(2): 239–46.

Botta, P. E., and E. Flandin. 1849–1850. *Monument de Ninive.* Paris: Imprimerie Nationale.

Botto, M. 2005. "Per una riconsiderazione della cronologia degli inizi della colonizzazione fenicia nel Mediterraneo centro-occidentale." In Bartoloni and Delpino, 579–628.

Bouchet, E. 1997. "Helminth Eggs: Trace Elements of Neolithic and Paleolithic Parasitosis on French Sites." *Comptes Rendus de la Societé de Biologie et de ses Filiales* 191: 529–36.

Bouma, J. W., P. A. J. Attema, A. J. Beijer, A. J. Nijboer, and R. A. Olde Dubbelink. 1995. "The Economy of an Early Latin Settlement, Borgo Le Ferriere-*Satricum*, 800–200 BC." In Christie, 183–95.

Boura, F., et al., 1992. "Le gisement de Vandières Les Grandes Corvées (Meurthe et Moselle)." In M.-J. Roulière-Lambert, ed., *Un Monde Villageois: habitat et milieu naturel en Europe de 2000 à 500 av. J.-C.*, 73–82. Lons-le-Saunier: Cercle Giradot (1992).

Bowmer, M., and T. Mollesen. 1986. "Appendix: Identification of Human Remains from the Hut Urns." In G. Bartoloni, ed., "24. Le urne a capanna: ancora sulle prime scoperte nei Colli Albani," in J. Swaddling, ed., *Italian Iron Age Artefacts* [pp. 235–248]. London: British Museum, 238–239.

Brachman, P. S., and A. R. Kaufmann. 1998. "Anthrax." In Evans and Brachman, 95–108.

Brack-Bernsen, L. 2007. "The 360-Day Year in Mesopotamia." In J. M. Steele, ed., *Calendars and Years: Astronomy and Time in the Ancient Near East*, 83–100. Oxford: Oxbow.

Briquel, D. 1984. *Pélasges en Italie: recherches sur l'histoire de la légende.* Rome: Bibliothèque de l'École Française de Rome.

1991. *L'origine lydienne des Etrusques, histoire du theme dans la literature antique.* Rome: École Française de Rome.

1996. "I riti di fondazione." In M. Bonghi Jovino and C. Chiaramonte Treré, eds., *Tarquinia: Ricerche, Scavi e Prospettive*, 171–90. (Mostra, Milan 1986.) Milan: Edizioni ET.

1997. *Chrétiens et haruspices: La religion étrusque, dernier rampart du paganisme romain.* Paris: École Normale Supérieure.

1999A. "Cornelius Labeo et la reaction paienne." In D. Briquel and C. Guittard, eds., *La divination dans le monde étrusco-italique.* Vol. 8, *Les écrivains du troisiéme siécle et l'etrusca disciplina. Des Sévères à Constantin.* (Actes de la table ronde tenue à l'ENS les 24 et 25 octobre 1997), *Caesarodunum* Suppl. 96: 51–62.

1999B. "Les presages de royauté de la divination étrusque." In E. Smadja and E. Geny, eds., *Pouvoir, divination, & prédestination dans le monde antique*, 185–204. Paris: Presses Universitaires Franc-Comtoises.

2009. "Les inscriptions votives du sanctuaire de Portonaccio à Véies." In M. Gleba and H. Becker, eds., *Votives, Places and Rituals in Etruscan Religion: Studies in Honor of Jean MacIntosh Turfa*, 43–67. Leiden: Brill.

Britton, J. P. 2002. "Treatments of Annual Phenomena in Cuneiform Sources." In J. M. Steele and A. Imhausen, eds., *Under One Sky: Astronomy and Mathematics in the Ancient Near East* 21–78. Münster: Ugarit-Verlag.

Brocato, P., 2000. *La necropoli etrusca della Riserva del Ferrone: Analisi di una communità arcaica dei Monti della Tolfa*. Rome: Edizioni Quasar.

Brown, D. 2000. *Mesopotamian Planetary Astronomy – Astrology*. Groningen: Styx.

Brown, P. J. 1993. "Favism." In Kiple, 722–24. Cambridge: Cambridge University Press.

Brown, Shelby. 1991. *Late Carthaginian Child Sacrifice and Sacrificial Monuments in their Mediterranean Context*. Sheffield: Sheffield Academic Press.

Bruce-Chwatt, L., and J. de Zulueta. 1980. *The Rise and Fall of Malaria in Europe: A Historico-Epidemiological Study*. Oxford: Clarendon.

Bruni, S. 1998. *Pisa etrusca: Anatomia di una città scomparsa*. Milan: Longanesi.

2001. "Analisi botaniche." In A. M. Moretti Sgubini, ed., *Tarquinia etrusca: Una nuova storia*, 59–61. (Catalogo della mostra, Tarquinia, 2001.) Rome: L'Erma de Bretschneider.

Brusadin Laplace, D., and S. Patrizi Montoro. 1977–78. "L'imbarcazione monossile della necropoli di Caolino al Sasso di Furbara." *Origini* 11: 355–79.

Bryce, T., 2003. *Letters of the Great Kings of the Ancient Near East: The Royal Correspondence of the Late Bronze Age*. London: Routledge.

Bucher, G. S. 1987–88. The *Annales Maximi* in the Light of Roman Methods of Keeping Records." *American Journal of Ancient History* 12–13: 1–61.

Burkert, W. 1985. "Itinerant Diviners and Magicians: A Neglected Element in Cultural Contacts." In R. Hägg, ed., *The Greek Renaissance of the Eighth Century* B.C.: *Tradition and Innovation*, 115–19. (Proceedings of the Second International Symposium at the Swedish Institute in Athens, 1–5 June, 1981.) Stockholm: Svenska Institutet i Athen.

1992. *The Orientalizing Revolution: Near Eastern Influence on Greek Culture in the Early Archaic Age*. Trans. M. E. Pinder and W. Burkert. Cambridge, Mass.: Harvard University Press.

1995. "Lydia between East and West, or How to Date the Trojan War: A Study in Herodotus." In J. B. Carter and S. P. Morris, eds., *The Ages of Homer: A Tribute to Emily Townsend Vermeule*, 139–48. Austin: University of Texas Press.

2004. *Babylon Memphis Persepolis: Eastern Contexts of Greek Culture*. Cambridge, Mass.: Harvard University Press.

2005. "Signs, Commands, and Knowledge: Ancient Divination Between Enigma and Epiphany." In S. I. Johnston and P. T. Struck, eds., *Mantikē: Studies in Ancient Divination*, 29–49. Leiden: Brill.

Buranelli, F., and M. Sannibale. 2005. "Non più solo 'Larthia': Un documento epigrafico inedito dalla Tomba Regolini-Galassi di Caere." In B. Adembri, ed., *ΑΕΙΜΝΗΣΤΟΣ: Miscellanea di Studi per Mauro Cristofani*, 1:220–31. Florence: Centro Di.

Camporeale, G. 1967. *La Tomba del Duce*. Florence: Olschki.

1969. *I commerci di Vetulonia in età orientalizzante*. Florence: Sansoni.

1984. *La caccia in Etruria*. Rome: G. Bretschneider.

Camporeale, G., ed. 1997. *L'abitato etrusco dell'Accesa: Il quartiere B*. Rome: G. Bretschneider.

Camporeale, G. 2000. "Sopravvivenze vilıanoviane nell'orientalizzante vetuloniese." In Prayon and Röllig, 153–69.

Camporeale, G., ed. 2001. *Gli Etruschi fuori Etruria*. Verona: Arsenale.

Camporeale, G., and S. Giuntoli. 2000. *The Accesa Archaeological Park at Massa Marittima*. Trans. J. Denton. Follonica: Editrice Leopoldo II.

Canciani, F. 1981. "Griechische und orientalische Handwerker in Mittelitalien." In *Die Aufnahme fremder Kultureinflüsse in Etrurien und das Problem des Retardierens in der etruskischen Kunst*, 53–59. (Symposium Mannheim, 8.-10.2. 1980 = Schriften des Deutschen Archäologen-Verbandes V.) Mannheim: Archäologischer Seminar der Universität Mannheim.

Capasso, L. 1986. "Resti dell' 'ultima pasta' in una inumazione dello millennio a.C. in Abruzzo (Celano, Bacino del Fucino)." In *Quaderni di Protostoria 1. Atti dell'Incontro di Acquasparta 1985. "Gli insediamenti perilacustri dell'età del bronzo e della prima età del ferro: il caso dell'antico Lacus Velinus, Perugia: Cattedra di Protostoria europea,"* 357–59. Perugia: Istituto di Archeologia.

1999. "Brucellosis at Herculaneum." *International Journal of Osteoarchaeology* 9: 277–88.

2001. *I Fuggiaschi di Ercolano: Paleobiologia delle vittime dell'eruzione vesuviana del 79 d.C.* Rome: L'Erma di Bretschneider.

Capasso, L., and G. Di Tota. 1997. "Le malattie infettive degli Etruschi." In G. Maetzke and L. Tamagno Perna, eds., *Aspetti della cultura di Volterra Etrusca fra l'età del ferro e l'età ellenistica e contributi della Ricerca Antropologica alla conoscenza del popolo etrusco,* 551–57. (Atti del XIX Convegno di Studi Etruschi ed Italici, Volterra 15–19 ottobre 1995.) Florence: Olschki.

1998. "Lice Buried under the Ashes of Herculaneum." *Lancet* 351(9107): 992.

1999. "Tuberculosis in Herculaneum." In G. Palfi, O. Dutour, and I. Deaks, eds., *Tuberculosis Past and Present*, 461–67. Budapest: Golden Book Tuberculosis Foundation.

Capdeville, G. 1989. "Les dieux fulgurants dans la doctrine étrusque." In *Atti di Secondo Congresso Internazionale Etrusco* (Firenze 26 maggio–2 giugno 1985*)*, 3:1171–90. (Supplemento di *Studi Etruschi*) Rome: Giorgio Bretschneider.

1992. "Le tre manubie di Tinia." *Studi Etrusci* 58: 155–70.

1996. "Les dieux de Martianus Capella." *Revue de l'histoire des religions* 213: 251–300.

1997. "Les livres sacrés des Étrusques." In J.-G. Heintz, ed., *Oracles et prophétes dans l'antiquité*, 457–508. (Actes du Colloque, Strasbourg 1995.) Paris: De Boccard.

Capozza, M. 1966. *Movimenti servili nel mondo romano in età repubblicana.* Vol. 1, *Dal 50 al 184 a.Cr.* Rome: Università degli Studi di Padova.

Carman, J., ed. 1997. *Material Harm: Archaeological Studies of War and Violence.* Glasgow: Cruithne.

Carmichael, A. G. 1993A. "Diphtheria." In Kiple, 680–83. Cambridge: Cambridge University Press.

1993B. "Bubonic Plague." In Kiple, 628–31.

1993C. "The Plague of Athens." In Kiple, 934–37.

Carslaw, K. S., R. G. Harrison, and J. Kirkby. 2002. "Cosmic Rays, Clouds, and Climate." *Science* 298: 1732–37.

Cataldi, M. 2001. "Tomba a fossa 'Dei Due Giovinetti' (6337)." In A. M. Moretti Sgubini, ed., *Tarquinia etrusca: Una nuova storia*, 95–99. (Catalogo della mostra, Tarquinia, 2001.) Rome: L'Erma de Bretschneider.

Cateni, G. 1991. *Volterra: Guarnacci Museum*. Pisa: Pacini Editore.

Caubet, A., and J.-C. Courtois. 1986 "Une modele de foie d'Enkomi." *Report of the Department of Antiquities, Cyprus* 1986: 72–77.

CCIR 1963 = (International Telecommunication Union), International Radio Consultative Committee, *Report 322*, 10th Plenary Assembly, Geneva 1963. (Cited by Vincent 1985: 34–4, fig. 2.)

Celenza, A., et al. 1996. "Thunderstorm Associated Asthma: A Detailed Analysis of Environmental Factors." *British Medical Journal* 312(7031): 604–8.

Cerchiai, L. 1985. "Una tomba principesca del periodo Orientalizzante Antico a Pontecagnano." *Studi Etrusci* 53: 27–42.

Chadwick, R. 1989. "Comets and Meteors in the Last Assyrian Empire." In A. F. Aveni, ed., *World Archaeoastronomy*, 186–94. (Selected papers from the 2nd Oxford International Conference on Archaeoastronomy Held at Merida, Yucatan, Mexico, 13–17 January 1986.) Cambridge: Cambridge University Pres.

Chauveau, M. 2000. *Egypt in the Age of Cleopatra*. Trans. D. Lorton. Ithaca, N.Y.: Cornell University Press.

Cherici, A. 1989. "Keraunia." *Archeologia Classica* 41: 329–76.

Chiaramonte Treré, C. 1988. "I depositi all'ingresso dell'edificio tarquiniese: Nuovi dati sui costumi rituali etruschi." *Mélanges de l'École Française à Rome: Antiquité* 100(2): 565–600.

1995. "Seppellimenti in abitato: Il caso di Tarquinia." In N. Negroni Catacchio, ed., *Preistoria e protostoria in Etruria*, 241–48. (Atti del secondo Incontro di Studi. Tipologia delle necropoli e rituali di deposizione. Ricerche e scavi, Farnese, 21–23 maggio 1993.) Milan: Istituto italiano di preistoria e protostoria.

Chiesa, F. 2001. "Il settore M: la deposizione femminile e il deposito delle olle." In A. M. Moretti Sgubini, ed., *Tarquinia etrusca: Una nuova storia*, 38–40. (Catalogo della mostra, Tarquinia, 2001.) Rome: L'Erma de Bretschneider.

Christie, N., ed. 1995. *Settlement and Economy in Italy 1500 BC to AD 1500*. Oxford: Oxbow.

Ciafaloni, D. 2006. "Nota sulle tipologie architettoniche e murarie tarquiniesi: Ulteriori corrispondenze con il Vicino Oriente antico." In M. Bonghi Jovino, ed., *Tarquinia e le civiltà del Mediterraneo*, 145–61. (Convegno internazionale, Milano, 22–24 giugno 2004.) Milan: Cisalpino, Istituto Editoriale Universitario.

Cianferoni, G. C. 2005. "Vino e olio 'principi' della tavola." In *Cibi e sapori nel mondo antico*, 79–91. (Exhibition, Florence, 18 March 2005–15 January 2006). Livorno: sillabe (Soprintendenza per i Beni Archeologici della Toscana).

2007. "La vite e il vino in Etruria: I problemi aperti." In A. Ciacci, P. Rendini, and A. Zifferero, eds., *Archeologia della vite e del vino in Etruria*, 19–20. (Atti del Convegno Internazionali di Studi, Scansano Teatro Castagnoli 9–10 settembre 2005.) Siena: Ci. Vin.

Ciarallo, A. 2002. "About an Ancient Mixture Found in Pompeii." In J. Renn and G. Castagnetti, eds., *Homo faber: Studies on Nature, Technology, and Science at the Time of Pompeii*, 153–67. (Presented at a conference at the Deutsches Museum, Munich, 21–22 March 2000.) Rome: L'Erma di Bretschneider.

Cioni, R. 1995. "Note sulla evoluzione geologica del Distretto Vulcanico Vulsino." In P. Tamburini, ed., *Un abitato villanoviano perilacustre: Il "Gran Carro" sul lago di Bolsena (1959–1985)*, 6–10. Rome: G. Bretschneider.

Cliff, A., P. Haggett, and M. Smallman-Raynor. 1993. *Measles: An Historical Geography of a Major Human Viral Disease*. Oxford: Oxford University Press.

Cocchi Genick, D. 1998. *L'antica età del bronzo nell'Italia centrale: profilo di un'epoca e di un'appropriata strategia metodologica*. Florence: Octavo.

Cocchi Genick, D., ed. 1996. *L'antica età del Bronzo*. (Atti del congresso di Viareggio, 9–12 gennaio 1995.) Florence: Octavo.

Coles-Brooks, J. 1967. "A Villanovan Belt from Euboea." *Bulletin of the Institute for Classical Studies* 14: 22–24.

Cohen, M. E. 1993. *The Cultic Calendars of the Ancient Near East*. Bethesda, Md.: CDL Press.

Cohen, M. N. 1989. *Health and the Rise of Civilization*. New Haven, Conn.: Yale University Press.

Coldstream, J. N. 1993. "Mixed Marriages at the Frontiers of the Early Greek World." *Oxford Journal of Archaeology* 12: 89–107.

Colonna, G. 1966. "*REE*." *Studi Etruschi* 34: 321–22.

1970. "Una nuova iscrizione etrusca del VII secolo e appunti sull'epigrafia ceretana dell'epoca." *Mélanges de l'École Française à Rome: Antiquité* 82: 637–72.

1975. "Firme arcaiche di artefici nell'Italia centrale." *Römische Mitteilungen* 82: 181–92.

1976. "Scriba cum rege sedens." In *Mélanges offerts à Jacques Heurgon: L'Italie préromaine et la Rome républicaine*, 187–95. Rome: École Française de Rome.

1977. "Nome gentilizio e società." *Studi Etruschi* 45: 175–92.

1988. "Una nuova dedica alla etrusca Uni." *Bolletino d'Arte* ser. 6 vol. 73 fasc. 48: 23–26.

1991. "Gli scudi bilobati dell'Italia centrale e l' *Ancile* dei Salii." *Archeologia Classica* 43: 55–122.

1993. "Ceramisti e donne padrone di bottega nell'Etruria arcaica." In G. Meiser, ed., *Indogermanica et Italica: Festschrift fur Helmut Rix zum 65. Geburtstag*, 61–68. Innsbruck: Institut fur Sprachwissenschaft.

1993 (1994). "A proposito degli dei del Fegato di Piacenza." *Studi Etruschi* 59: 123–36, 137–39.

1995. "Etruschi a Pitecusa nell'Orientalizzante antico." In A. Storchi Marino, ed., *L'incidenza dell'Antico: Studi in memoria di E. Lepore*, 1:325–42. Naples: Luciano Editore.

1997. "Il possesso del carro e il significato della sua deposizione nella tomba: culture a confronto. L'Italia antica: l'Italia centrale." In A. Emiliozzi, ed., *Carri da Guerra e principi etruschi*, 15–23. (Mostra, Viterbo 1997–98.) Rome: L'Erma di Bretschneider.

2002. "REE 71." *Studi Etruschi* 65–68: 351–57.

2004. "Comments Recorded in Symposium." In Bartoloni and Delpino, 478–83.

2006. "A proposito della presenza etrusca nella Gallia meridionale." In S. Gori and M. C. Bettini, eds., *Gli Etruschi da Genova ad Ampurias*, 2:657–78. (Atti del XXIV Convegno di Studi Etruschi ed Italici, Marseille-Lattes, 26 settembre–1 ottobre 2002.) Pisa: Istituti Editoriali e Poligrafici Internazionali.

2007. "L'Apollo di Pyrgi, Śur/Śuri (il 'Nero') e l'Apollo Sourios." *Studi Etruschi* 73: 100–34.

Colonna, G., and F.-W. von Hase. 1986. "Alle origini della statuaria etrusca: La Tomba delle Statue presso Ceri." *Studi Etruschi* 52: 13–59.

Como 1986 = *Como fra Etruschi e Celti: la città preromana e il suo ruolo commerciale* (Catalogo della mostra, Como 1986). [Como?]: Società archeologica comense.

Comstock, G. W., and R. J. O'Brien. 1998. "Tuberculosis." In Evans and Brachman, 777–804.

Copenhaver, B. F. 1995. Hermetica*: The Greek* Corpus Hermeticum *and the Latin* Asclepius *in a new English translation, with notes and introduction*. Cambridge: Cambridge University Press.

Corcoran, T. H. 1971/1999. *Seneca. Naturales Quaestiones Books I–III, with an English translation by Thomas H. Corcoran*. Cambridge, Mass.: Harvard University Press.

Cornell -2000 = Web site: "Aegean Dendrochronology Project December 2000 Project Report," Cornell University Dendrochronology Laboratory (Peter Kuniholm): www.arts.cornell.edu/dendro/2000news/2000adp.html (accessed July 1, 2008).

Cornell, T. 1991. "The Tyranny of the Evidence: A Discussion of the Possible Uses of Literacy in Etruria and Latium in the Archaic Age." In M. Beard, ed., *Literacy in the Roman World*, 7–33. (*Journal of Roman Archaeology* Suppl. 3.) Ann Arbor, Mich.

Cornell, T. J. 1995. *The Beginnings of Rome*. London: Routledge.

Corridi, C. 1994. "Naturalistica. Fauna." In L. Donati, ed. 1994. *La Casa dell'Impluvium: Architettura etrusca a Roselle*, 143–57. Rome: G. Bretschneider.

Costantini, L., L. Costantini Biasini, and S. Scali. 1987. "Bolsena-Gran Carro." In *L'alimentazione nel mondo antico: Gli Etruschi*, 61–70. (Ministero per I Beni Culturali e Ambientali, Giroanta Mondiale dell'Alimentazione, 1987.) Rome: Istituto Poligrafico e Zecca dello Stato.

Costantini, L., L. Costantini Biasini, and M. Hopf. 1995. "I resti vegetali del villaggio del 'Gran Carro,' Bolsena (VT): scavo 1974." In P. Tamburini, ed., *Un abitato villanoviano perilacustre: Il "Gran Carro" sul lago di Bolsena (1959–1985)*, 325–38. Rome: G. Bretschneider.

Cristofani, M. 1985. *I Bronzi degli Etruschi*. Novara: Istituto Geografico De Agostini.

1993. "Un etrusco a Egina." *Studi Etruschi* 59: 159–62.

1995. *Tabula Capuana: un calendario festivo di età arcaica*. Florence: Olschki.

1996. *Etruschi e altre genti nell'Italia preromana Mobilità in età arcaica*. Rome: G. Bretschneider.

Cristofani, M., and P. Fronzaroli. 1971. "Un'iscrizione cuneiforme su un vaso bronzeo da una tomba di Faleri." *Studi Etruschi* 39: 313–31.

Crosby, A. W. 1993A. "Smallpox." In Kiple, 1008–13.

1993B. "Influenza." In Kiple, 807–11.

Culican, W. 1973. "Phoenician Jewellery in New York and Copenhagen." *Berytus* 22: 31–47.

Cumont, F., F. Boll et al. 1898–1953. *Catalogus codicum astrologorum graecorum [CCAG]* (Brussels 1889–1911). Brussels: Henri Lamertin.

Cupitò, M. 2005. "Addenda interpretative sul sistema figurativa del carello di Bisenzio." In P. Attema, A. Nijboer, and A. Zifferero, eds., *Papers in Italian Archaeology VI: Communities and Settlements from the Neolithic to the Early Medieval Period*, 2:739–41. (Proceedings of the Sixth Conference of Italian Archaeology held at the University of Groningen, Groningen Institute of Archaeology, The Netherlands, April 15–17, 2003 = BAR Int. Ser. 1452[II].) Oxford: Archaeopress.

Curtis, J. E., and J. E. Reade, eds. 1995. *Art and Empire: Treasures from Assyria in the British Museum*. New York: Metropolitan Museum of Art.

Cygielman, M., and L. Pagnini. 2002. "Presenze sarde a Vetulonia: alcune considerazioni." In O. Paoletti and L. Tamagno Perna, eds., *Etruria e Sardegna centro-settentrionale tra l'età del Bronzo Finale e l'arcaismo*, 387–410. (Atti del XXI Convegno di Studi Etruschi ed Italici. Sassari-Alghero-Oristano-Torralba, 13–17 ottobre 1998.) Pisa and Rome: Istituti Editoriali Poligrafici Internazionali.

2006. *Tomba del Tridente a Vetulonia*. Pisa: Istituti Editoriali e Poligrafici Internazionali.

D'Agostino, B. 1990. "Relations between Campania, Southern Etruria, and the Aegean in the Eighth Century BC." In J.-P. Descoeudres, ed., *Greek Colonists and Native Populations*, 73–85. Oxford: Oxford University Press.

Dalley, S. 1998. "Occasions and Opportunities. I. To the Persian Conquest." In S. Dalley, ed., *The Legacy of Mesopotamia*, 9–33. Oxford: Oxford University Press.

Dalton, J. P., ed. 1999. *Fasciolosis*. Wallingford-Cambridge: CABI.

Damgard Andersen, H. 1993. "The Etruscan Ancestor Cult – Its Origin and Development and the Importance of Anthropomorphization." *Analecta Romana* 21: 7–66.

Damiani, I., S. Festuccia, and A. Guidi. 1998. "Le Caprine." In N. Negroni Catacchio, ed., *Preistoria e Protostoria in Etruria*, 203–14. (Atti del Terzo Incontro di Studi. Protovillanoviani e/o Protoetruschi. Ricerche e Scavi, Manciano – Farnese 12–14 maggio 1995.) Octavo/Franco Cantini.

Davidson, A., ed. 1999. *The Oxford Companion to Food*. Oxford: Oxford University Press. (All citations are to entries written by Davidson.)

Davidson, A. C., et al. 1996. "A Major Outbreak of Asthma Associated with a Thunderstorm: Experience of Accident and Emergency Departments and Patients' Characteristics." *British Medical Journal* 312(7031): 601–4.

Debiasi, A. 2000. "Le tre eruzioni dell'Etna in Tucidide." *Hesperia* 12: 227–33.

De Grossi Mazzorin, J. 1989. "Testimonianze di allevamento e caccia nel Lazio antico tra l'VIII e il VII secolo a.C." *Dialoghi di Archeologia*, 3rd series, 7(1): 125–42.

1995A. "Economie di allevamento in Italia centrale dalla media età del Bronzo alla fine dell'età del Ferro." In Christie, 167–77.

1995B. "I reperti paleontologici." In P. Tamburini, ed., *Un abitato villanoviano perilacustre: Il 'Gran Carro' sul lago di Bolsena (1959–1985)*, 339–51. Rome: G. Bretschneider.

1998. "Analisi dei resti faunistici da alcune strutture di Sorgenti della Nova." In N. Negroni Catacchio, ed., *Preistoria e Protostoria in Etruria*, 169–80. (Atti del Terzo Incontro di Studi. Protovillanoviani e/o Protoetruschi. Ricerche e Scavi, Manciano – Farnese 12/14 maggio 1995.) Octavo/ Franco Cantini.

2004. "Some Considerations about the Evolution of Animal Exploitation in Central Italy from the Bronze Age to the Classical Period." In B. Santillo Frizell, ed., *Pecus: Man and Animal in Antiquity*, 38–49. (Proceedings of the Conference at the Swedish Institute in Rome, September 9–12, 2002.) (The Swedish School in Rome. Projects and Seminars, 1.) Rome: Swedish Institute.

De Grossi Mazzorin, J., A. Riedel, and A. Tagliacozzo. 1998. "Horse Remains in Italy from the Eneolithic to the Roman Period." In *Proceedings of the XIII Congress, International Union of Prehistoric and Protohistoric Science*, 6:87–92 (tome I). Forlí, Italy, 8–14 september 1996 (published 1998, Forlí).

2004. "L'evoluzione delle popolazioni animali e dell'economia nell'età del bronzo recente." In D. Cocchi Genick, ed., *L'etá del bronzo recente in Italia*, 227–32. (Atti del Congresso Nazionale di Lido di Camaiore, 26–29 ottobre 2000.) Lucca: Mauro Baroni editore.

de Grummond, N. T. 2000. "Mirrors and Manteia: Themes of Prophecy on Etruscan Mirrors." In M. D. Gentili, ed., *Aspetti e problemi*, 27–67. Rome: Aracne.

2002. "Mirrors, Marriage and Mysteries." *Journal of Roman Archaeology* 47: 63–85.

2006A. *Etruscan Myth, Sacred History and Legend.* Philadelphia: University Museum Press.

2006B. "Prophets and Priests." In de Grummond and Simon, 27–44.

de Grummond, N. T., and E. Simon, eds. 2006. *The Religion of the Etruscans.* Austin: University of Texas Press.

De Juliis, E. M. 2004. "La prima età del ferro in Puglia." In Bartoloni and Delpino, 453–66.

Delattre, D. 2006. *La Villa des Papyrus et les rouleuax d'Herculaneum: la bibliothèque de Philodème.* Liège: CEDOPAL.

Della Casa, A. 1962. *Nigidio Figulo.* Rome: Edizioni dell'Ateneo.

del Olmo Lete, G. 1984. *Interpretación de la Mitología Cananea: Estudios de semántica ugarítica.* Valencia: Institución San Jerónimo.

2004. *Canaanite Religion According to the Liturgical Texts of Ugarit.* Trans. W. G. E. Watson. Winona Lake, Ind.: Eisenbrauns.

Delpino, F. 2000. "Il principe e la cerimonia del banchetto." In G. Bartoloni, F. Delpino, C. Morigi Govi, and G. Sassatelli, eds., *Principi etruschi tra Mediterraneo ed Europa*, 193–95. (Mostra, Bologna, Museo Civico Archeologico.) Bologna: Marsilio.

2002. "Brocchette a collo oblique dall'area etrusca." In O. Paoletti and L. Tamagno Perna, eds., *Etruria e Sardegna centro-settentrionale tra l'età del Bronzo Finale e l'arcaismo*, 363–85. (Atti del XXI Convegno di Studi Etruschi ed Italici, Sassari-Alghero-Oristano-Torralba, 13–17 ottobre 1998.) Pisa and Rome: Istituti Editoriali Poligrafici Internazionali.

2005. "Dinamiche sociali e innovazioni rituali a Tarquinia villanoviana: le tombe I e II del sepolcreto di Poggio dell'Impiccato." In O. Paoletti and G. Camporeale, eds., *Dinamiche di Sviluppo delle Città nell'Etruria Meridionale: Veio, Caere, Tarquinia, Vulci*, 1:343–58. (Atti del XXIII Convegno di Studi Etruschi ed Italici, Roma, Veio, Cerveteri/Pyrgi, Tarquinia, Tuscania, Vulci, Viterbo, 1–6 ottobre 2001.) Pisa and Rome: Istituti Editoriali e Poligrafici Internazionali.

2007. "Viticoltura, produzione e consumo del vino nell'Etruria protostorica." In A. Ciacci, P. Rendini, and A. Zifferero, eds., *Archeologia della vite e del vino in Etruria*, 133–46. (Atti del Convegno Internazionali di Studi, Scansano Teatro Castagnoli 9–10 settembre 2005.) Siena: Ci.Vin.

Delpino, F., and G. Bartoloni. 2000. "Il principe: stile di vita e manifestazione del potere." In G. Bartoloni, F. Delpino, C. Morigi Govi, and G. Sassatelli, eds., *Principi etruschi tra Mediterraneo ed Europa*, 221–70. (Mostra, Bologna, Museo Civico Archeologico.) Bologna: Marsilio.

De Lucia Brolli, M. A. 2004. "Una tomba a cremazione entro custodia litica dall'Agro falisco." In A. M. Moretti Sgubini, ed., *Scavo nello scavo: gli Etruschi non viste*, 109–11 (112–17 catalog). (Ricerche e 'riscoperte' nei depositi dei musei archeologici dell'Etruria meridionale: 5 marzo 2004–30 giugno 2004, Viterbo, Fortezza Giulioli.) Rome: Soprintendenza per I Beni Archeologici per l'Etruria Meridionale.

DeMarco, M. 1985. "9.4. Deposito votivo di Villa Marchi a Fiesole." In G. Colonna, ed., *Santuari d'Etruria*, 168–69. Milan: Electa.

De Marinis, R. C. 1988. "Liguri e Celto-Liguri." In G. Pugliese Carratelli, ed., *Italia omnia terrarium alumna*, 157–259. Milan: Scheiwiller.

De Marinis, R. C., and F. M. Gambari. 2005. "La cultura di Golasecca dal X agli inizi del VII secolo a.C.: cronologia relative e correlazioni con alter aree culturali." In Bartoloni and Delpino, 197–225.

de Martino, F. 1993. "Sull'alimentazione degli schiavi." *La Parola del Passato* 48: 401–27.

De Puma, R. D. 1973. "The Dioskouroi on Four Etruscan Mirrors in Midwestern Collections." *Studi Etruschi* 41: 159–70.

Dergachev, V. A., O. M. Raspopov, B. van Geel, and G. I. Zaitseva. 2004. "The 'Sterno-Etrussia' Geomagnetic Excursion around 2700 BP and Changes of Solar Activity, Cosmic Ray Intensity, and Climate." *Radiocarbon* 46(2): 661–81.

Dershowitz, N., and E. M. Reingold. 2008. *Calendrical Calculations*, 3rd ed. Cambridge: Cambridge University Press.

De Santis, A. 1995. "Contatti fra Etruria e Lazio Antico alla Fine dell'VIII sec. a.C.: La Tomba del Guerriero di Osteria dell'Osa." In Christie, 365–75.

2005. "Da capi guerrieri a principi: La strutturazione del potere politica nell'Etruria protourbana." In O. Paoletti and G. Camporeale, eds., *Dinamiche di Sviluppo delle Città nell'Etruria Meridionale: Veio, Caere, Tarquinia, Vulci*, 2: 615–31. (Atti del XXIII Convegno di Studi Etruschi ed Italici, Roma, Veio, Cerveteri/Pyrgi, Tarquinia, Tuscania, Vulci, Viterbo, 1–6 ottobre 2001.) Pisa and Rome: Istituti Editoriali e Poligrafici Internazionali.

De Santis, A., R. Merlo, and J. De Grossi Mazzorin. 1998. *Fidene: Una casa dell'età del ferro* (Ministero dei Beni Culturali e Ambientali, and Soprintendenza Archeologica di Roma.) Milan: Electa.

Deshayes, J., Sznycer, M., and P. Garelli. 1981. "Remarques sur les monuments de Karatepe." *Revue Archéologique* 75: 31–60.

de Simone, C. 1981. "Fremde Gentilnamen in Etrurien in archaischer Zeit." In *Die Aufnahme fremder Kultureinflüsse in Etrurien und das Problem des Retardierens in der etruskischen Kunst*, 89–93. (Symposium Mannheim, 8.-10.2. 1980 = Schriften des Deutschen Archäologen-Verbandes V). Mannheim: Archäologischer Seminar der Universität Mannheim.

De Simone, C. 1993. "Le iscrizioni chiusine arcaiche." In *La civiltà di Chiusi e del suo territorio*, 25–38. (Atti del XVII Convegno di Studi Etruschi ed Italici, Chianciano Terme, 28 maggio–1 guigno 1989.) Florence: Olschki.

Devoto, G. 1948. *Le Tavole di Gubbio*. Florence: G. C. Sansoni (1967 repr).

DeVries, K. 2002. "The Throne of Midas?" Paper delivered at 103rd Annual Meeting, Archaeological Institute of America, Philadelphia, January 3–6, 2002. Abstract: *American Journal of Archaeology* 106: 275.

2005. "Greek Pottery in Gordion." In L. Kealhofer, ed., *The Archaeology of Midas and the Phrygians*, 36–55. Philadelphia: University Museum Press.

DeVries, K., P. I. Kuniholm, G. K. Sams, and M. M. Voigt in *Antiquity* 77 (2003). http://www.antiquity.ac.uk/projgall/devries296.

Dickson, J. H., K. Oeggl, T. G. Holden, L. L. Handley, T. C. O'Connell, and T. Preston. 2000. "The Omnivorous Tyrolean Iceman: Colon Contents (Meat, Cereals, Pollen, Moss and Whipworm) and Stable Isotope Analyses." *Philosophical Transactions of the Royal Society London*, (Biol. Sci.) Series B 355: 1843–49.

Diodorus/Oldfather 1939 = *Diodorus of Sicily*, with an English translation by C. H. Oldfather, vol. 3 (Loeb Classical Library). Cambridge, Mass.: Harvard University Press.

Dirks, R. 1993. "Famine and Disease." In Kiple, 157–63.

Dittmar, K., and W. R. Teegen. 2003. "The Presence of *Fasciola hepatica* (liver-fluke) in Humans and Cattle from a 4,500-Year-old Archaeological Site in the Saale-Unstrut Valley, Germany." *Memorias do Instituto Oswaldo Cruz* 98 (Suppl. 1): 141–43.

Docter, R. F. 2006. "Etruscan Pottery: Some Case Studies in Chronology and Context." In S. Gori and M. C. Bettini, eds., *Gli Etruschi da Genova ad Ampurias*, 1: 233–40. (Atti del XXIV Convegno di Studi Etruschi ed Italici, Marseille-Lattes, 26 settembre – 1 ottobre 2002.) Pisa: Istituti Editoriali e Poligrafici Internazionali.

Docter, R. F., H. G. Niemeyer, A. J. Nijboer, and H. van der Plicht. 2004. "Radiocarbon Dates of Animal Bones in the Earliest Levels of Carthage." In Bartoloni and Delpino, 557–77.

Dolfini, A. 2004. "Le simbologie ornitomorfe in Italia durante il Bronzo Finale: prospettive di analisi." In N. Negroni Catacchio, ed., *Preistoria e Protostoria in Etruria*, 1:279–305. (Atti del Sesto Incontro di Studi. Miti Simboli Decorazioni Ricerche e scavi, Pitigliano – Valentano 13–15 settembre 2002.) Milan: Centro Studi di Preistoria e Archeologia.

2005. "La fase di transizione Bronzo-Ferro nel territorio di Vulci: elementi di continuità e discontinuità nella cultura materiale." In *Dinamiche di Sviluppo delle Città nell'Etruria Meridionale: Veio, Caere, Tarquinia, Vulci*, 1: 509–21. (Atti del XXIII Convegno di Studi Etruschi ed Italici, Roma, Veio, Cerveteri/Pyrgi, Tarquinia, Tuscania, Vulci, Viterbo, 1–6 ottobre 2001.) Pisa and Rome: Istituti Editoriali e Poligrafici Internazionali.

Domenici, I. 2001. "L'eroe con l'aratro: A proposito di un'urnetta inedita di Heidelberg." *Archäologischer Anzeiger* 2001: 79–90.

Domenici, I., and E. Maderna. 2007. *Giovanni Lido: Sui segni celesti*. Ed. and introduced Ilaria Domenici; trans. Erika Maderna. Milan: Medusa.

Donati, L., ed. 1994. *La Casa dell'Impluvium: Architettura etrusca a Roselle*. Rome: G. Bretschneider.

Donati, A., and P. Pasini, eds. 1997. *Pesca e pescatori nell'antichità*. Milan: Leonardo Arte.

Donati, L. 2000. "Architettura civile, sacra e domestica." In M. Torelli, ed., *Gli Etruschi*, 313–33. (Mostra, Palazzo Grassi, Venice, 2000.) Milan: Bompiani.

2005. "La coppia di figure sedute incise sui cinerari biconici: Gli esempi di Traquinia." In Paoletti and Camporeale, 1: 371–82.

2005A. "Due cippi etruschi nell'Abbazia di Sant'Antimo." In B. Adembri, ed., *ΑΕΙΜΝΗΣΤΟΣ: Miscellanea di Studi per Mauro Cristofani*, 2: 678–83. Florence: Centro Di.

Dougherty, R. 1928. "Writing upon Parchment and Papyrus among the Babylonians and the Assyrians." *Journal of the American Oriental Society* 48: 109–35.

Drescher-Schneider, R., J.-L. de Beaulieu, M. Magny, A.-V. Walter-Simonnet, G. Bossnet, L. Millet, E. Brugiapaglia, and A. Drescher. 2007. "Vegetation History, Climate and Human Impact over the Last 15,000 Years at Lago dell'Accesa (Tuscany, Central Italy)." *Vegetation History and Archaeobotany* 16: 279–99.

Drews, R. 1991. "The Coming of the City to Central Italy." *American Journal of Ancient History* 6: 133–65.

Dunn, F. L. 1993. "Malaria." In Kiple, 855–62.

DuPont, H. L. 1993. "Diarrheal Diseases (Acute)." In Kiple, 676–80.

Ebbinghaus, S. 2008. "Patterns of Elite Interaction: Animal-Headed Vessels in Anatolia in the Eighth and Seventh Centuries BC." In B. J. Collins, M. R. Bachvarova, and I. C. Rutherford, eds., *Anatolian Interfaces: Hittites, Greeks and Their Neighbors*, 181–90. (Proceedings of an International Conference on Cross-Cultural Interaction, September 17–19, 2004, Emory University, Atlanta, Ga.) Oxford: Oxbow.

Eck, W. 1997. "Cornelius [II 19] Labeo." *Der Neue Pauly* 3, col. 193.

Eder, W. 2005. "The Political Significance of the Codification of Law in Archaic Societies: An Unconventional Hypothesis." In K. A. Raaflaub, ed., *Social Struggles in Archaic Rome: New Perspectives on the Conflict of the Orders*, 239–67. Oxford: Blackwell.

Edlund-Berry, I. E. M. 1991. "Power and Religion: How Social Change Affected the Emergence and Collapse of Power Structures in Central Italy." In E. Herring, R. Whitehouse, and J. Wilkins, eds., *Papers of the Fourth Conference of Italian Archaeology [1990]. Part 2, Archaeology of Power*, 161–71. London: Accordia Research Center.

1992. "Etruscans at Work and Play: Evidence for an Etruscan Calendar." In H. Frohning et al., eds., *Kotinos: Festschrift für Erika Simon*, 330–38. Mainz/Rhein: von Zabern.

2006. "Ritual Space and Boundaries in Etruscan Religion." In de Grummond and Simon, 116–31.

Elayi, J. 1980. "Remarques sur un type de mur phénicien." *Rivista di Studi Fenici* 8: 165–80.

Emiliozzi, A., ed. 1997. *Carri da guerra e principi etruschi*. (Catalogo della mostra, Viterbo Palazzo dei Papi, 24 maggio 1997–31 gennaio 1998.) Rome: L'Erma di Bretschneider.

Emiliozzi, A. 2001. "Technical Problems concerning Orientalizing Vehicles in Cyprus and Etruria." In L. Bonfante and V. Karageorghis, eds. *Italy and Cyprus in Antiquity 1500–450 B.C.*, 315–34. Nicosia: Costakis and Leto Severis Foundation.

Engelbrecht, A. 1887. *Hephaestio von Theben und sein astrologisches Compendium*. Vienna: Carl Konegen.

Esposito, A. M., ed. 1999. *Principi guerrieri: La necropoli etrusca di Casale Marittimo*. Milan: Electa.

Etxeberria, F. 1994. "Vertebral Epiphysitis: Early Symptoms of Brucellar Disease." *Journal of Paleopathology* 6(1): 41–49.

Fedele, F., and P. P. Petrone, eds. 2000. *Un'eruzione vesuviana 4000 anni fa: Reperti provenienti del sito archeologico di San Paolo Belsito*. (Mostra, Università degli Studi di Napoli,

Federico II. Centro Musei delle Scienze Naturali. Museo di Antropologia.) Naples: Fridericiana editrice universitaria.

Evans, A. S. 1998. "Epidemiological Concepts." In Evans and Brachman, 13–64.

Evans, A. S., and P. S. Brachman, eds. 1998. *Bacterial Infections of Humans: Epidemiology and Control*, 3rd ed. New York: Plenum.

Ewald, P. 1994. *Evolution of Infectious Disease*. Oxford: Oxford University Press.

Facchetti, G. M. 2000. *Frammenti di diritto privato etrusco*. Florence: Oschki.

2002. "L'appellativo etrusco *etera*." *Studi Etruschi* 65–68: 225–35.

Facchini, F., and P. Brasili. 1997. "I reperti villanoviani ed etruschi del territorio Bolognese: Sintesi antropologica." In *Aspetti della cultura di Volterra etrusca*, 423–33.

Falchi, I. 1908. "Vetulonia – Nuove scoperte nella necropoli." *Notizie Scavi* 1908: 419–37.

Falconer, W. A. 1923/1996. *Cicero. De Senectute. De Amicitia. De Divinatione*, with an English translation by William Armistead Falconer. Cambridge, Mass: Harvard University Press.

Farello, P. 1994. "L'insediamento di Castenaso: allevamento, caccia e pesca in un sito villanoviano." In Forte and von Eles, 218–23.

2002. "Analisi dei resti alimentari." In P. von Eles, ed., *Guerriero e sacerdote. Autorità e comunità nell'età del ferro a Verucchio. La Tomba del Trono*, 314–15. (Quaderni di Archeologia dell'Emilia Romagna 6.) Florence: all'Insegna del Giglio.

Fedele, F., and P. P. Petrone, eds. 2000. *Un'eruzione vesuviana 4000 anni fa: Reperti provenienti del sito archeologico di San Paolo Belsito*. (Mostra, Università degli Studi di Napoli, Federico II. Centro Musei delle Scienze Naturali. Museo di Antropologia.) Naples: Fridericiana editrice universitaria.

Ferranti, F. 2004. "L'orizzonte tardo-geometrico enotrio alla vigilia delle fondazioni coloniali greche." In Bartoloni and Delpino, 441–52.

Feruglio, A. E. 1991. "Un carro da parata da Castel San Mariano di Corciano presso Perugia." In L. Bonfante and F. Roncalli, eds., *Gens antiquissima Italiae: Antichità dell'Umbria a New York*, 135–40. (Exhibition, New York 1991.) Milan: Electa.

Fenelli, M. 1984. "Lavinium." *Archeologia Laziale* 6: 325–44.

Finkel, I. L. 1988. "Adad-apla-iddina, Esagil-kīn-apli, and the Series SA.GIG." In E. Leichty, M. de J. Ellis, and P. Gerardi, eds., *A Scientific Humanist: Studies in Memory of Abraham Sachs*, 143–59. Philadelphia: University Museum Press.

Fletcher, R. N. 2007. *Patterns of Imports in Iron Age Italy*. BAR Int.Ser. 1732. Oxford: Archaeopress.

Forbes, H. A. 1976. *We Have a Little of Everything: The Ecological Basis of Some Agricultural Practices in Methana, Trizinia*. New York: New York Academy of Sciences.

Formicola, V. Q., C. Milanesi, and C. Scarsini. 1987. "Evidence of Spinal Tuberculosis at the Beginning of the Fourth Millennium BC from Arena Candida Cave (Liguria, Italy)." *American Journal of Physical Anthropology* 72: 1–7.

Fornaciari, G. 1997. "Paleopatologia di gruppi umani a cultura etrusca: il caso di Pontecagnano, Salerno (VII–IV sec. a.C.)." In *Aspetti della cultura di Volterra etrusca*, 467–75.

Fornaciari, G., and F. Mallegni. 1997. "I resti paleoantropologici." In M. Bonghi Jovino and C. Chiaramonte Treré, eds., *Tarquinia: Testimonianze archeologiche e ricostruzione storica*, 100–102. Rome: L'Erma di Bretschneider.

Forni, G. 2007. "Quando e come sorse la viticoltura in Italia." In A. Ciacci, P. Rendini, and A. Zifferero, eds., *Archeologia della vite e del vino in Etruria*, 69–81. (Atti del Convegno Internazionali di Studi, Scansano Teatro Castagnoli 9–10 settembre 2005.) Siena: Ci. Vin.

Forte, M. 1994. "Lo scavo dell'insediamento di Castenaso – Via Gramsci: Le strutture (1981)." In Forte and von Eles, 193–99.

Forte, M., and P. von Eles, eds. 1994. *La Pianura Bolognese nel Villanoviano: Insediamenti della prima età del Ferro*. (Ministero Beni Culturaii e Ambientali, and Soprintendenza Archeologica dell'Emilia Romagna. Studi e documenti di archeologia, *Quaderni* 5.)

Fowden, G. 1993. *The Egyptian Hermes: A Historical Approach to the Late Pagan Mind*. Princeton, N.J.: Princeton University Press.

Fowler, W. Warde. 1899/1969. *The Roman Festivals of the Period of the Republic.* Port Washington, N.Y.: Kennikat.

Frankfort, T. 1959. "Les classes serviles en Étrurie." *Latomus* 18: 3–22.

Freedman, S. M. 1998. *If a City Is Set on a Height: The Akkadian Omen Series* Šumma Alu ina Mēlê Šakin. Vol. 1, *Tablets 1–21*. Philadelphia: Samuel Noah Kramer Fund, University of Pennsylvania Museum.

Fricker, C., M. Spigelman, and E. Fricker. 1997. "The Detection of Escherichia coli DNA in the Ancient Remains of Lindow Man Using the Polymerase Chain Reaction." *Letters in Applied Microbiology* 24: 351–54.

Furlani, G. 1931. "Fulmini mesopotamici, hittiti, greci ed etruschi." *Studi Etruschi* 5: 203–31.

Galeotti, L. 1986–88. "Considerazioni sul carro a due ruote nell'Etruria e nel *Latium vetus.*" *Archeologia Classica* 38–40: 94–104.

Galli, E. 1912. "Il sepolcreto visentino delle 'Buccacce.'" *MonAnt* 21: 409–98.

Gambari, F. M., and G. Colonna. 1988. "Il bicchiere con iscrizione arcaica da Castelletto Ticino e l'adozione della scrittura nell'Italia nord-occidentale." *Studi Etruschi* 54: 119–64.

Garbini, G. 1980. "Gli 'Annali di Tiro' e la storiografia fenicia." In R. Y. Ebeid and M. J. L. Young, eds., *Oriental Studies Presented to B. S. J. Isserlin*, 112–27. Leiden: Brill.

Garnsey, P. 1999. *Food and Society in Classical Antiquity.* Cambridge: Cambridge University Press.

Gastaldi, P. 1998. *Pontecagnano II.4. La necropoli di Pagliarone, AION ArchStAnt, Quad.* 10, Naples.

Gaultier, F., J. Gran-Aymerich, and D. Briquel. 1989–90. *REE* 56: 73 = *Studi Etruschi* 56: 350–56, no. 73.

Gehlken, E. 2005. "Die Adad-Tafeln der Omenserie *Enūma Anu Enlil.* Teil 1: Einführung." *Baghdader Mitteilungen* 36: 235–73.

Geller, M. J. 1998. "New Documents from the Dead Sea: Babylonian Science in Aramaic." In M. Lubetski, C. Gottlieb, and S. Keller, *Boundaries of the Ancient Near Eastern World. A Tribute to Cyrus Gordon*, 224–29. (Journal for the Study of the Old Testament, Suppl. Series 273.) Sheffield: Sheffield Academic Press.

Gentili, G. V. 2003. *Verucchio villanoviana: Il sepolcreto in località Le Pegge e la necropoli al piede della Rocca Malatestiana.* Rome: G. Bretschneider.

Geppert, K. 2004. "Importfunde des spätheteitischen Raumes in Mittelitalien und ihre Wirkung auf das einheimische Kunsthandwerk." In M. Novák, F. Prayon, and A.-M. Wittke, eds., *Die Aussenwirkung des späthethitischen Kulturraumes. Güteraustausch – Kulturkontakt – Kulturtransfer*, 63–83. (Akten des zweiten Forschungstagung des Graduiertenkollegs "Anatolien und seine Nachbarn" der Eberhard-Karls-Universität Tübingen, 20. bis 22. November 2003). Münster: Ugarit-Verlag.

Geppert, K., and F. Gaultier. 1999. "Zwei Pasticci und ihre Folgen: Die Bildmotive der Caeretaner Pyxiden D 150 und D 151 im Louvre." In Prayon and Röllig, 211–18.

Giardino, C., C. Belardelli, and A. Malizia. 1991. "Power and the Individual in Funerary Ideology: The Emergence of the Aristocracy in the Villanovan Period in the Bologna Region." In E. Herring, R. Whitehouse, and J. Wilkins, eds., *Papers of the Fourth Conference of Italian Archaeology [1990].* Part 2, *Archaeology of Power*, 9–19. London: Accordia Research Center.

Giglioli, G. Q. 1929. "L'oinochoe di Tragliatella." *Studi Etruschi* 3: 111–59.

Giulierini, P. 2005. "Cibo e alimentazione in Etruria: produzione e consumi." In *Cibi e sapori nel mondo antico*, 66–78. (Exhibition, Florence, 18 March 2005–15 January 2006.) Livorno: sillabe (Soprintendenza per i Beni Archeologici della Toscana).

Giusberti, G. 1994. "Nota antropologica relativa ad uno scheletro di epoca villanoviana rinvenuto nel territorio di Quarto Inferiore (metà VI sec. A.C.)." In Forte and von Eles, 306–9.

Gleba, M. 2000. "Textile Production at Poggio Civitate (Murlo) in the 7th c. BC." In D. Cardon and M. Feugère, eds., *Archéologie des textiles des origines au Ve siecle*, 75–80. (Actes du colloque de Lattes, octobre 1999. Monographies *Instrumentum* 14.) Montpellier: Éditions Monique Mergoil.

2008. *Textile Production in Pre-Roman Italy* (Ancient Textiles 4). Oxford: Oxbow.

Gleba, M., and Turfa J. M. 2005. "Digging for Archaeological Textiles in Museums: 'New' Finds in the University of Pennsylvania Museum of Archaeology and Anthropology." In A. Rast-Eicher and R. Windler, eds., *Proceedings of the 9th North European Symposium for Archaeological Textiles, NESAT IX – 18–21 Mai, 2005, Braunwald*, 35–40. Ennenda: Archetex.

Goedicke, H. 1975. *The Report of Wenamon.* Baltimore: Johns Hopkins University Press.

Gooch, P. S. 1983. "Helminth Parasites." In G. D. Hart, ed., *Disease in Ancient Man: An International Symposium [March 5–7, 1979]*, 205–11. Toronto: Clarke Irwin.

Graf, F. 1998. "Fonteius [I6 and I9]." In *Der Neue Pauly*, vol. 4. cols. 587–88.

Gran-Aymerich, J. 2006. "La diffusion des vases étrusques en Méditerranée nord-occidentale: l'exception gauloise." In S. Gori and M. C. Bettini, eds., *Gli Etruschi da Genova ad Ampurias*, 1: 205–19. (Atti del XXIV Convegno di Studi Etruschi ed Italici, Marseille-Lattes, 26 settembre–1 ottobre 2002.) Pisa: Istituti Editoriali e Poligrafici Internazionali.

2009. "'Gli Etruschi fuori d'Etruria': Dons et Offrandes Étrusques en Méditerranée Occidentale et dans l'Ouest de l'Europe." In M. Gleba and H. Becker, eds., *Votives, Places and Rituals in Etruscan Religion: Studies in Honor of Jean MacIntosh Turfa*, 15–41. Leiden: Brill.

Gras, M. 1985. *Traffics tyrrhéniens archäiques.* Rome: École Française de Rome.

Green, A. R. W. 2003. *The Storm-God in the Ancient Near East.* (Biblical and Judaic Studies 8.) Winona Lake, Ind.: Eisenbrauns.

Greenblatt, C. L., and M. Spigelman. 2003. *Emerging Pathogens: The Archaeology, Ecology, and Evolution of Infectious Disease.* Oxford: Oxford University Press.

Greenfield, J. C., and M. Sokoloff. 1989. "Astrological and Related Omen Texts in Jewish Palestinian Aramaic." *Journal of Near Eastern Studies* 48: 201–14.

2005. "4Q318 (4Qzodiology and Brontology ar) [translation of Qumran Cave 4 text 318]." In D. W. Parry and E. Tov, eds., *The Dead Sea Scrolls Reader.* Part 6, *Additional Genres and Unclassified Texts*, 224–26. Leiden: Brill.

Grmek, M. D. 1989/91. *Diseases in the Ancient Greek World.* Trans. M. Muellner and L. Muellner (French edition 1983). Baltimore: Johns Hopkins University Press.

Gromatici veteres/Corpus agrimensorum from K. Lachmann and A. Rudorff, *Die Schriften der römischen Feldmesser* I, Berlin 1848. Reprinted 1967, Hildesheim: G. Olms.

Grove, A. T., and O. Rackham. 2001. *The Nature of Mediterranean Europe: An Ecological History.* New Haven, Conn.: Yale University Press.

Guidoboni, E., A. Comastri, and G. Traina. 1994. *Catalogue of Ancient Earthquakes in the Mediterranean Area up to the 10th Century.* Trans. B. Phillips. Rome: Istituto Nazionale di Geofisica.

Guillaumont, F. 1986. "Ciceron et les techniques de l'haruspicine." In *La divination dans le monde étrusco-italique*, vol. 2. (Actes de la Table ronde tenue le 22 mars 1986, à l'École normale supérieure) = *Caesarodunum* Suppl. 54: 121–135.

Guinan, A. K. 1989. "The Perils of High Living: Divinatory Rhetoric in the Šumma Ālu." In H. Behrens, D. Loding, and M. T. Roth, eds., DUMU-E$_2$-DUB-BA-A: *Studies in honor of Åke W. Sjöberg*, 227–35. (Occasional Publications of the Samuel Noah Kramer Fund, 11.) Philadelphia: University Museum Press.

1990. "The Human Behavioral Omens: On the Threshold of Psychological Inquiry." *Bulletin of the Canadian Society for Mesopotamian Studies* 19: 9–14.

2002. "A Severed Head Laughed: Stories of Divinatory Interpretation." In L. Ciraolo and J. Seidel, eds., *Magic and Divination in the Ancient World*, 7–40. Leiden: Brill/Styx.

Guittard, C. 1990. "Contribution des sources littéraires a notre conaissance de l'*etrusca disciplina*: Tarquitius Priscus et les *Arbores infelices*." In H. Heres and M. Kunze, eds., *Die Welt der Etrusker*, 91–99. (Internationales Kolloquium 24.-26. Oktober 1988 in Berlin.) Berlin: Akademe-Verlag.

1997. "Questions sur la divination étrusque: les formulas dans la tradition latine." In F. Gaultier and D. Briquel, eds., *Les Étrusques, les plus religieux des hommes: État de la recherche sur la religion étrusque*, 399–412. (Actes du colloque international, Galeries nationales du Grand Palais, 17–18–19 novembre 1992.) Paris: École du Louvre.

2003. "Les calendriers brontoscopiques dans le monde étrusco-romain." In C. Cusset, ed., *La Météorologie dans l'Antiquité: entre science et croyance*, 455–66. (Actes du Colloque internationale interdisciplinaire de Toulouse, 2–3–4 mai 2002.) Saint-Étienne: Université de Saint-Étienne.

2005. "L'*etrusca disciplina* dans le *Liber prodigorum* de Julius Obsequens." In *La divination dans le monde étrusco-italique*. Vol. 9, *Les écrivains du IVe siécle: L'*etrusca disciplina *dans un monde en mutation*. (Actes de la table ronde de Clermont-Ferrand 17 et 18 septembre 1999), *Caesarodunum* Suppl. 67: 41–48.

Gundel, W., and H. G. Gundel. 1966. *Astrologoumena: Die astrologische Literatur in der Antike und ihre Geschichte*. Wiesbaden: Franz Steiner Verlag.

Haack, M.-L. 2006. *Prosopographie des haruspices romains*. (Biblioteca di *Studi Etruschi* 42.) Pisa and Rome: Istituto Nazionale di Studi Etruschi ed Italici, Istituti Editoriali e Poligrafici Internazionali.

Hadjichristodoulou, C., et al. 2000. "Surveillance of Communicable Diseases in the Balkans." *Lancet* 355(9213): 1465.

Hakkert, A. M., ed. 1964. *Publius Nigidius Figulus, Operum Reliquiae*. Amsterdam. = Another edition of Swoboda.

Hall, J. F. 1984. "The Municipal Aristocracy of Etruria and Their Participation in Politics at Rome B.C. 91–A.D. 14." Ph.D. dissertation, University of Pennsylvania.

1996. "From Tarquins to Caesars: Etruscan Governance at Rome." In J. F. Hall, ed., *Etruscan Italy: Etruscan Influences on the Civilizations of Italy from Antiquity to the Modern Era*, 149–89. Provo, Utah: Museum of Art, Brigham Young University.

Hannah, R. 2005. *Greek and Roman Calendars: Constructions of Time in the Classical World*. London: Duckworth.

2009. *Time in Antiquity*. London: Routledge.

Hansen, W. 1996. *Phlegon of Tralles' Book of Marvels*, translated with an introduction and commentary. Exeter: University of Exeter Press.

Harden, V. A. 1993. "Typhus, Epidemic." In Kiple, 1080–84.

Harding, A. 1999. "Warfare: A Defining Characteristic of Bronze Age Europe?" In J. Carman and A. Harding, eds., *Ancient Warfare: Archaeological Perspectives*, 157–73. Phoenix Mill, UK: Sutton.

Harris, W. V. 1971. *Rome in Etruria and Umbria*. Oxford: Clarendon.

Harrison, A., and E. M. Bartels. 2006. "A Modern Appraisal of Ancient Etruscan Herbal Practices." *American Journal of Pharmacology and Toxicology* 1(1): 21–24.

Harrison, A., I. Cattani, and J. M. Turfa. 2009. "Metallurgy, Environmental Pollution and the Decline of Etruscan Civilization." *Environmental Safety and Pollution Research International (ESPR)* 31 March 2009. Online: DOI 10.1007/s11356-009-0141-5.

Harrison, R. J. 1988. *Spain at the Dawn of History: Iberians, Phoenicians and Greeks*. London: Thames and Hudson.

Hawass, Z., Y. Z. Gad, S. Ismail, R. Khairat, D. Fathalla, N. Hasan, A. Ahmed, H. Elleithy, M. Ball, F. Gaballah, S. Wasef, M. Fateen, H. Amer, P. Gostner, A. Selim, A. Zink, and C. M. Pusch, 2010. "Ancestry and Pathology in King Tutankhamun's Family." *Journal of the American Medical Association* 303(7): 638–47.

Hawkins, J. D. 1992. "What Does the Hittite Storm-God Hold?" In D. J. W. Meijer, ed., *Natural Phenomena: Their Meaning, Depiction and Description in the Ancient Near East*, 53–82. (Proceedings of the Colloquium, Amsterdam U89.) Amsterdam: Royal Netherlands Academy of Arts and Sciences.

Haynes, S. 1985. *Etruscan Bronzes*. London: Sothebys.

2000. *Etruscan Civilization: A Cultural History*. Los Angeles: J. Paul Getty Museum.

Helbaek, H. 1981. "Appendix." In E. Berggren and K. Berggren, eds., *San Giovenale vol. II, fasc. 2, Skrifter XXVI:II, 2, Excavations in Area B, 1957–1960*. Lund: C. W. K. Gleerup, 57.

Hencken, H. 1968. *Tarquinia and Etruscan Origins*. London: Thames and Hudson.

Henneberg, M., and R. J. Henneberg. 1998. "Biological Characteristics of the Population Based on Analysis of Skeletal Remains." In J. C. Carter, ed., *The Chora of Metaponto: The Necropoleis* II, 503–56. Austin: University of Texas Press.

Herring, E., R. Whitehouse, and J. Wilkins, eds. 1991. *Papers of the Fourth Conference of Italian Archaeology (1990) 2. The Archaeology of Power,* Pts. 1 and 2. London: Accordia Research Center.

Herrmann, B., and S. Hummel. 2003. "Ancient DNA Can Identify Disease Elements." In C. L. Greenblatt and M. Spigelman, eds., *Emerging Pathogens: The Archaeology, Ecology, and Evolution of Infectious Disease,* 143–49. Oxford: Oxford University Press.

Herrmann, H.-V. 1983. "Altitalisches und Etruskisches in Olympia (NeueFunde und Forschungen)." In *Grecia, Italia e Sicilia nell'VIII e VII secolo a.C., Atti del convegno internazionale Atene 1979 = Annuario della Scuola italiana ad Atene* 61, 1983 (1984): 271–94.

Heurgon, J. 1953. "Tarquitius Priscus et l'organisation de l'ordre des haruspices sous l'empereur Claude." *Latomus* 12: 402–17.

1957. "L'état étrusque." *Historia* 6: 63–97.

1959. "The Date of Vegoia's Prophecy." *Journal of Roman Studies* 49: 41–45.

1964. *Daily Life of the Etruscans.* Trans. J. Kirkup. New York: Macmillan.

1970. "Classes et ordres chez les Étrusques." In *Recherches sur les structures sociales dans l'antiquité classique,* 29–41. (Caen 25–26 avril 1969.) Paris: CNRS.

Hirn, M. 1988. "The Sound of Thunder." *Scientific American* 59: 201.

Hoeppli, R. J. 1971. "Early Navigation and the Spread of Parasitic Diseases." *Episteme* 5: 201–17.

Holden, P., and N. Purcell. 2000. *The Corrupting Sea.* Oxford: Blackwell.

Holmes, S. J. 1997. "Mumps." In A. S. Evans and R. A. Kaslow, eds., *Viral Infections of Humans: Epidemiology and Control,* 4th ed., chap. 18. New York: Plenum.

Hopf, M. 1995. "Determinazione dei resti vegetali del villaggio del 'Gran Carro' (Lago di Bolsena, Viterbo): scavo 1980." In P. Tamburini, ed., *Un abitato villanoviano perilacustre: Il 'Gran Carro' sul lago di Bolsena (1959–1985),* 336–38. Rome: G. Bretschneider.

Hostetler, K. L. 2004–2007. "Serpent Iconography." *Etruscan Studies* 10: 203–9.

Huber, P. T. 1987. "Dating by Lunar Eclipse Omens with Speculations on the Birth of Omen Astrology." In J. L. Bergren and B. R. Goldstein, eds., *From Ancient Omens to Statistical Mechanics: Essays on the Exact Sciences Presented to Asger Aaboe,* 3–13. Copenhagen: University Library.

Hunger, H. 1992. *Astrological Reports to Assyrian Kings* (State Archives of Assyria VIII). Helsinki: Helsinki University Press.

Hunger, H., and D. Pingree. 1989. MUL.APIN. *An Astronomical Compendium in Cuneiform. Archiv für Orientforschung Beiheft* 24. Horn: Ferdinand Berger und Söhne.

1999. *Astral Sciences in Mesopotamia.* Leiden: Brill.

Hunter, P. R. 1997. *Waterborne Diseases: Epidemiology and Ecology.* New York: Wiley.

Inghirami, F. 1825. *Monumenti Etruschi o di etrusco nome, disegnati, incise, illustrate e pubblicati dal cavaliere Francesco Inghirami,* vol. 3. Fiesole: Poligrafia fiesolana.

Jacobs, J. 2010. "Traces of the Omen Series *Šumma izbu* in Cicero, *De divinatione.*" In A. Annus, ed., *Divination and Interpretation of Signs in the Ancient World,* 317–39. Chicago: Oriental Institute.

Jacobsen, T. 1988. "The *Asakku* in Lugal-e." In E. Leichty, M. deJ. Ellis, and P. Gerardi, eds., *A Scientific Humanist: Studies in Memory of Abraham Sachs,* 225–32. Philadelphia: University Museum.

Jannot, J. R. 1984. *Les reliefs archäiques de Chiusi.* Rome: École Française de Rome.

1993. "*Insignia potestatis*: Les signes de pouvoir dans l'iconographie de Chiusi." In *La civiltá di Chiusi e del suo terriorio,* 217–37. (Atti del XVII Convegno di Studi Etruschi ed Italici, Chianciano Terme, 28 maggio–1 giugno1989.) Florence: Olschki.

Jashemski, W. F., and F. G. Meyer, eds. 2002. *The Natural History of Pompeii.* Cambridge: Cambridge University Press.

Jeske-Janicka, M., and P. K. Janicki. 1998. "Appendix 11A.5. Detection of Treponema Pallidum Antigens." In J. C. Carter, ed., *The Chora of Metaponto: The Necropoleis* II, 557–59. Austin: University of Texas.

Jeyes, U. 1997. "Assurbanipal's *bārûtu*." *Compte rendu, rencontre assyriologique internationale* 39: 61–65.

Jeziorska, M. 2008. "Palaeopathology at the Beginning of the New Millennium: A Review of the Literature." In R. David, ed., *Egyptian Mummies and Modern Science*, 83–98. Cambridge: Cambridge University Press.

Johnston, W. D. 1993. "Tuberculosis." In Kiple, 1059–68.

Johnston, A., and M. Pandolfini. 2000. *Gravisca. Scavi nel santuario Greco: Le iscrizioni*. Bari: Edipuglia.

Jones, A. H. M., J. R. Martindale, and J. Morris, eds. 1980. *The Prosopography of the Later Roman Empire*, vol. 2. Cambridge: Cambridge University Press.

Jones, M. 2001. *The Molecule Hunt: Archaeology and the Search for Ancient DNA*. London: Allen Lane.

Jones, W. H. S. 1923/1995. *Hippocrates*, vol. 1, with an English translation by W. H. S. Jones. Cambridge, Mass.: Harvard University Press.

Jurgeit, F. 2000. "Die Fussbänke vom Typ Ceri." In Prayon and Röllig, 219–24.

Kassianidou, V. 2001. "Cypriot Copper to Sardinia: Yet Another Case of Bringing Coals to Newcastle?" In L. Bonfante and V. Karageorghis, eds., *Italy and Cyprus in Antiquity: 1500–450 BC*, 97–119. Nicosia: Severis Foundation.

Kim-Farley, R. J. 1993. "Mumps." In Kiple, 887–89.

King, A. 2002. "Mammals: Evidence from Wall Paintings, Sculpture, Mosaics, Faunal Remains, and Ancient Literary Sources." In Jashemski and Meyer, 401–50.

Kiple, K. F., ed. 1993. *Cambridge World History of Human Disease*. Cambridge: Cambridge University Press.

Klingshirn, W. E. 2006. "Inventing the *sortilegus*: Lot Divination and Cultural Identity in Italy, Rome, and the Provinces." In P. Harvey and C. Schultz, eds., *Religion in Republican Italy*, 137–61. *Yale Classical Studies* XXXIII. Cambridge: Cambridge University Press.

Klotz, A. 1927. "7) Joannes Laurentius Lydos." In *Paulys Real-Encyklopädie* 26: 2210–17.

Koch, U. S. 2005. *Secrets of Extispicy: The Chapter Mutābiltu of the Babylonian Extispicy Series and Nisirti ba-rûti Texts mainly from Assurbanipal's Library*. Münster: Ugarit-Verlag.

Koch-Westenholz, U. 1995. *Mesopotamian Astrology: An Introduction to Babylonian and Assyrian Celestial Divination*. Copenhagen: Museum Tusculanum Press.

2000. *Babylonian Liver Omens: The Chapters Manzāzu, Padānu and Pān tākalti of the Babylonian Extispicy Series Mainly from Aššurbanipal's Library*. Copenhagen: Museum Tusculanum Press.

Kochavi, M. 1998. "The Eleventh Century BCE Tripartite Pillar Building at Tel Hadar." In S. Gitin, A. Mazar, and E. Stern, eds., *Mediterranean Peoples in Transition: In Honor of Professor Trude Dothan*, 468–78. Jerusalem: Israel Exploration Society.

Kornmüller, M. 2006. *Etrusca Disciplina: Manuale teorico-pratico di divinazione etrusca*. Rome: Irradiazioni.

Kossack, G. 1999. *Religiöses Denken in dinglicher und bildlicher Überlieferung Alteuropas aus der Spätbronze- und frühen Eisenzeit (9.-6. Jahrhundert v. Chr. Geb.)*. Munich: Bayerischen Akademie der Wissenschaften.

Kourou, N. 2005. "Greek Imports in Early Iron Age Italy." In Bartoloni and Delpino, 497–515.

Krauss, H., A. Weber, M. Appel, B. Enders, H. D. Isenberg, H. G. Schiefer, W. Slenczka, A. von Graevenitz, and H. Zahner. 2003. *Zoonoses: Infectious Diseases Transmissible from Animals to Humans*, 3rd ed. Washington, DC: ASM Press.

Kristiansen, K. 1999. "The Emergence of Warrior Aristocracies in Later European History and Their Long-term History." In J. Carman and A. Harding, eds., *Ancient Warfare: Archaeological Perspectives*, 175–89. Phoenix Mill, UK: Sutton.

Kroll, W. 1932. "Tarquitius Priscus (7)." Paully-Wissowa *Realencyklopädie* IVA, cols. 2392–94. Stuttgart.

Kudlien, F. 1991. *Sklaven-Mentalität im Spiegel antiker Wahrsagerei.* (Forschungen zur antiken Sklaverei, XXIII.) Stuttgart: Franz Steiner.

Kuhnke, L. 1993. "Disease Ecologies of the Middle East and North Africa." In Kiple, 453–63.

Kuniholm, Peter. 2007. Personal Communication.

Kunze, E. 1951–53. "Etruskische Bronzen in Griechenland." In *Studies Presented to D. M. Robinson*, 1:736–46. St. Louis: Washington University Press.

Kyrieleis, H., and W. Röllig. 1988. "Ein altorientalischer Pferdeschmuck aus dem Heraion von Samos." *Athenische Mitteilungen* 103: 37–75.

Labat, R. 1939. *Hémérologies et Ménologies d'Aššur.* Paris: Librairie d'Amérique et d'Orient Adrien-Maisonneuve.

1965. *Un calendrier babylonien des travaux, des signes et des mois (séries iqqur īpuš).* Paris: Librairie Honoré Champion.

1972–75. "Hemerologien." In *Reallexikon der Assyriologie* 4: 317–23.

Lacheman, E. R. 1937. "An Omen Text from Nuzi." *Revue d'Assyriologie* 34: 1–8.

Lachmann, K. and A. Rudorff. 1848/1967. *Die Schriften der römischen Feldmesser I*, Berlin 1848 (reprinted 1967). Hildesheim: G. Olms.

Lambert, W. G. 1962. "A Catalogue of Texts and Authors." *Journal of Cuneiform Studies* 16: 59–77.

Lambrinoudakis, V., and S. Petrounakos. 2005. "3b. Consecration. IV. Dedication of Humans, Death by Lightning." In *Thesaurus Cultus et Rituorum Antiquorum III, Divination – Prayer – Veneration – Hikesia – Asylia – Oath – Malediction – Profanation – Magic Rituals*, 330–31. Los Angeles: J. Paul Getty Museum.

Landsberger, B. 1915. *Der kultische Kalender der Babylonier und Assyrier.* Leipzig: J. C. Hinrich.

Larew, H. G. 2002. "Insects: Evidence from Wall Paintings, Sculpture, Mosaics, Carbonized Remains, and Ancient Authors." In Jashemski and Meyer, 315–26.

LeBaron, C. W., and D. N. Taylor. 1993. "Typhoid Fever." In Kiple, 1071–77.

Lee, H. D. P. 1952/1987. *Aristotle. VII. Meteorologica*, with an English translation by H. D. P. Lee. Cambridge Mass.: Harvard University Press.

Legrand, L. 1930. *Publius Nigidius Figulus, philosophe néopythagoricien orphique.* Paris: Éditions de l'oevre d'Auteuil.

Leichty, E. 1960. "Teratological Omens in Ancient Mesopotamia." Ph.D. diss., University of Chicago.

1970. *The Omen Series Šumma izbu.* (Texts from Cuneiform Sources series vol. 4.) Locust Valley, N.Y.: J. J. Augustin.

1993. "The Origins of Scholarship." In H. D. Galter, ed., *Die Rolle der Astronomie in den Kulten Mesopotamiens*, 11–29. (Beiträge zum 3. Grazer Morgenländischen Symposion, 23.-27. September 1991.) Graz: GrazKult.

Leiderer, R. 1990. *Anatomie der Schafsleber im babylonischen Leberorakel: Eine makroskopisch-analytische Studie.* Munich: W. Zuckschwerdt.

Lemaire, A. 1992. "Divination." In E. Lipinski et al., eds., *Dictionnaire de la Civilisation phénicienne et punique*, 131–32. Turnhout: Brepols.

Lerner, R. E. 2008. "Flies: Some Scratchy Issues Concerning the Black Death." *Journal of the Historical Society* 8(2): 205–28.

Levine, M. M. 1998. "Typhoid Fever." In Evans and Brachman, 839–58.

Lewis, N. 1996. *The Interpretation of Dreams and Portents in Antiquity.* Wauconda Ill.: Bolchazy-Carducci.

Lieberman, S. 1990. "Canonical and Official Cuneiform Texts: Towards an Understanding of Ashurbanipal's Personal Tablet Collection." In T. Abusch, J. Huehnergard, and P. Steinkeller, eds., *Lingering Over Words: Studies in Ancient Near Eastern Literature in Honor of William L. Moran*, 305–36. (Harvard Semitic Studies 37.) Atlanta: Scholars Press.

Linderski, J. 2005. "Religious Aspects of the Conflict of the Orders: The Case of *confarreatio*." In K. A. Raaflaub, ed., *Social Struggles in Archaic Rome: New Perspectives on the Conflict of the Orders*, 223–38. Oxford: Blackwell.

Lipiński, E. 2004. *Itineraria Phoenicia.* (*Studia Phoenicia* XVIII.) Leuven: Peeters.

Liuzzi, D. 1983. *Nigidio Figulo, astrologo e mago: Testimonianze e frammenti.* Lecce: Milella.

Livingstone, A. 1993. "The Case of the Hemerologies: Official Cult, Learned Formulation, and Popular Practice." In E. Matsushima, ed., *Official Cult and Popular Religion in the Ancient Near East*, 96–113. (Papers of the First Colloquium on the Ancient Near East – The City and its Life held at the Middle Eastern Culture Center in Japan (Mitaka, Tokyo), March 20–22, 1992.) Heidelberg: Universitätsverlag C. Winter.

1993–97. "Menologie." In *Reallexikon der Assyriologie* 8:59–60.

Livy/Schlesinger 1959 = *Livy*, vol. 14, Trans. A. C. Schlesinger. Cambridge Mass.: Harvard University Press = Loeb Classical Library Livy, vol. 14.

Longrigg, J. 2000. "Death and Epidemic Disease in Classical Athens." In V. M. Hope and E. Marshall, eds., *Death and Disease in the Ancient City*, 55–64. London: Routledge.

Lorimer, H. L. 1950. *Homer and the Monuments.* London: Macmillan.

Lo Schiavo, F. 2001. "Late Cypriot Bronzework and Bronzeworkers in Sardinia, Italy and Elsewhere in the West." In L. Bonfante and V. Karageorghis, eds., *Italy and Cyprus in Antiquity: 1500–450 BC*, 131–52. Nicosia: Severis Foundation.

2002. "Osservazioni sul problema dei rapporti fra Sardegna ed Etruria in età nuragica – II." In O. Paoletti and L. Tamagno Perna, eds., *Etruria e Sardegna centro-settentrionale tra l'età del Bronzo Finale e l'arcaismo*, 51–70. (Atti del XXI Convegno di Studi Etruschi ed Italici. Sassari-Alghero-Oristano-Torralba, 13–17 ottobre 1998.) Pisa and Rome: Istituti Editoriali Poligrafici Internazionali.

Losi, A. 1994. "La sepoltura di Quarto Inferiore." In Forte and von Eles, 301–5.

Maas, M. 1992. *John Lydus and the Roman Past.* New York: Routledge.

MacBain, B. 1982. *Prodigy and Expiation: A Study on Religion and Politics in Republican Rome.* (Collection Latomus 177.) Brussels: Latomus.

Macfarlane, J. T. 1996. "*Tyrrhena Regum Progenies:* Etruscan Literary Figures from Horace to Ovid." In J. F. Hall, ed., *Etruscan Italy. Etruscan Influences on the Civilizations of Italy from Antiquity to the Modern Era*, 241–65. Provo, Utah: Museum of Art, Brigham Young University.

MacIntosh, J. 1974. "Representations of Furniture on the Frieze Plaques from Poggio Civitate (Murlo)." *Römische Mitteilungen* 81:15–40.

Macnamara, E. 2001. "Evidence and Influence of Cypriot Bronzework in Italy from the 8th to 6th centuries B.C." In L. Bonfante and V. Karageorghis, eds., *Italy and Cyprus in Antiquity: 1500–450 bc*, 291–313. Nicosia: Severis Foundation.

Maggiani, A. 1973. "Coppa fenicia da una tomba villanoviana di Vetulonia." *Studi Etruschi* 41: 73–95.

1989. "Immagini di aruspici." In *Atti del Secondo Congresso Internazionale Etrusco, Firenze 26 maggio – 2 giugno 1985* (Suppl. *Studi Etruschi*), 3:1557–63.

1992. "L'uomo e il sacro nei rituali e nella religione etrusca." In *Le Civiltà del Mediterraneo e il sacro*, 191–209. Milan: Jaca Book-Massimo.

1997. "Réflexions sur la religion étrusque 'primitive': de l'époque villanovienne à l'époque archaïque." In F. Gaultier and D. Briquel, eds., *Les Étrusques, les plus religieux des hommes: État de la recherche sur la religion étrusque*, 431–47. (Actes du colloque international, Galeries nationals du Grand Palais, 17–18–19 novembre 1992.) Paris: École du Louvre.

1998. "Appunti sulle magistrature etrusche." *Studi Etruschi* 62: 95–138.

2002. Una brocchetta bronzea da Vetulonia." In O. Paoletti and L. Tamagno Perna, eds., *Etruria e Sardegna centro-settentrionale tra l'età del Bronzo Finale e l'arcaismo*, 411–18. (Atti del XXI Convegno di Studi Etruschi ed Italici. Sassari-Alghero-Oristano-Torralba, 13–17 ottobre 1998.) Pisa and Rome: Istituti Editoriali Poligrafici Internazionali.

2005. "La divinazione in Etruria." In *Thesaurus Cultus et Rituum Antiquorum*, ed. V. Lambrinoudakis and J. C. Balty. Los Angeles and Basel, (2004–), 3: 52–78.

Magness, J. 2001. "A Near Eastern Ethnic Element Among the Etruscan Elite?" *Etruscan Studies* 8: 79–117.

Magny, M. 1995. *Une histoire du climat: Des derniers mammouths au siècle de l'automobile.* Paris: Editions Errance.

Mainardi, S., and E. Pacciani. 1994. "Nota antropologica sullo scheletro umano rinvenuto a Castenaso- Via Gramsci nella struttura η." In Forte and von Eles, 213–17.

Makkay, J. 1983. "Metal Forks as Symbols of Power and Religion." In *Archäologischer Anzeiger* 35: 313–44.

Mallegni, F., and A. Vitiello. 1997. "Le ricerche antropologiche sui gruppi umani a cultura etrusca." In *Aspetti della cultura di Volterra etrusca*, 17–32.

Malnati, L., and V. Manfredi. 1991. *Gli Etruschi in Val Padana.* Milan: Mondadori.

Mancinelli, D., et al. 1997. "Inquadramento paleodemografico di alcuni gruppi a cultura etrusca nell'ambito delle popolazioni dell'Italia centro-meridionale nel I millennio a.C." In *Aspetti della cultura di Volterra etrusca*, 499–513.

Manthe, U. 1979. "Ein etruskischer Schiedsspruch: Zur Interpretation des Cippus Perusinus." *Revue Internationale des Droits de l'Antiquité* 26: 261–305.

Marcattili, F. 2005. "[1a. Cult Places.] Bidental." *Thesaurus Cultus et Rituum Antiquorum*, ed. V. Lambrinoudakis and J. C. Balty. Los Angeles and Basel. (2004–) 3: 201–2.

Marchesi, M. 2000. "Rhyton configurato a protome di leone." In G. Bartoloni, F. Delpino, C. Morigi Govi, and G. Sassatelli, eds., *Principi etruschi tra Mediterraneo ed Europa*, 128–29, no. 78. (Mostra, Bologna, Museo Civico Archeologico.) Bologna: Marsilio.

Marchesini, M., and S. Marvelli. 2002. "7.4 Analisi botaniche del contenuto del vaso biconico (Cat. N. 8)." In P. von Eles, ed., *Guerriero e sacerdote. Autorità e comunità nell'età del ferro a Verucchio. La Tomba del Trono*, 299–307. (Quaderni di Archeologia dell'Emilia Romagna 6.) Florence: all'Insegna del Giglio.

Marchesini, S. 1997. *Studi onomastici e sociolinguistici sull'Etruria arcaica: Il caso di Caere.* (*Biblioteca di Studi Etruschi* 32.) Florence: Olschki.

Markoe, G. 1985. *Phoenician Bronze and Silver Bowls from Cyprus and the Mediterranean.* Berkeley: University of California Press.

Martelli, M. 1994. "Sulla produzione di vetri orientalizzanti." In M. Martelli, ed., *Tyrrhenoi Philotechnoi*, 75–97. (Atti della giornata di studio, Viterbo 1990.) Rome: Gruppo editoriale internazionale.

Martelli, M. C. 1973. "Documenti di arte orientalizzante da Chiusi." *Studi Etruschi* 41: 97–120.

Martelli, M., B. Adembri, F. Canciani, M. Cristofani, and M. A. Rizzo. 1987. *La ceramica degli Etruschi:. La pittura vascolare.* Novara: Istituto Geografico de Agostini.

Marzatico, F. 1992. "Les habitats préhistoriques successifs de la Tourbiére de Fiavé (Trentin, Italie)." In M.-J. Roulière-Lambert, ed., *Un Monde Villageois: habitat et milieu naturel en Europe de 2000 à 500 av. J.-C.*, 185–90. (Papers presented 1990, published 1992). Lons-le-Saunier: Cercle Giradot.

2004. "150 Years of Lake-Dwelling Research in Northern Italy." In F. Menotti, ed., *Living on the Lake in Prehistoric Europe: 150 Years of Lake-Dwelling Research*, 83–97. London: Routledge.

Marzatico, F., and P. Gleirscher, eds. 2004. *Guerrieri Principi ed Eroi fra il Danubio e il Po dalla Preistoria all'Alto Medioevo.* (Catalogo della Mostra tenuta a Trento dal 19 giugno al 7 novembre 2004.) Trento: Castello del Buonconsiglio, monumenti e collezioni provinciali.

Massa-Pairault, F.-H. 1986. "Les jeux équestres de Poggio Civitate: Représentation et sociètè." *Ktema: Civilisations de l'Orient, de la Grece et de Rome antiques* (University of Strasbourg) 11: 179–87.

1999. "Mito e miti nel territorio volsiniese." In *Annali della Fondazione per il Museo "Claudio Faina" (Orvieto)* 6: 77–108.

Mastandrea, P. 1979. *Un neoplatonico latino Cornelio Labeone: (testimonianze e frammenti).* Leiden: Brill.

Mastrolorenzo, G., P. Petrone, L. Pappalardo, and M. F. Sheridan. 2006. "The Avellino 3780-yr-B.P. Catastrophe as a Worst-case Scenario for a Future Eruption of Vesuvius." *Proceedings of the National Academy of Sciences* 103(12): 4366–70.

Matsushima, E., ed. 1993. *Official Cult and Popular Religion in the Ancient Near East.* Heidelberg: C. Winter.

Maul, S. M. 1994. "Zukunftsbewältigung: Eine Untersuchung altorientalischen Denkens anhand der babylonisch-assyrischen Löserituale (Namburbi)." *Baghdader Forschungen* 18. Mainz am Rhein: Philipp von Zabern.

2009. "Divination Culture and the Handling of the Future." In G. Leick, ed., *The Babylonian World*, 361–72. New York: Routledge.

Mavrakis, A., and S. Lykoudis. 2006. "Heavy Precipitation Episodes and Cosmic Rays Variation." *Advances in Geoscience* 7: 157–61.

Mayor, A. 2000. *The First Fossil Hunters*. Princeton, N.J.: Princeton University Press.

2003. *Greek Fire, Poison Arrows and Scorpion Bombs*. Woodstock, N.Y.: Overlook.

Mays, S. 1998. *The Archaeology of Human Bones*. London: Routledge.

Mazzarino, S. 1957. "Sociologia del mondo etrusco e problemi della tarda etruschicità." *Historia* 6: 98–122.

Mazzuoli, G. 2007. "Gli Etruschi e gli Ucelli: Uno studio ornitologico nell'iconografia tombale etrusca." http://www.canino.infoi/inserti/monografie/etruschi/vari/etruschi_uccelli/01/index.htm (accessed June 2008).

McGovern, P. E. 2007. "The Archaeological and Chemical Hunt for the Origins of Viniculture in the Near East and Etruria." In A. Ciacci, P. Rendini, and A. Zifferero, eds., *Archeologia della vite e del vino in Etruria*, 108–22. (Atti del Convegno Internazionali di Studi, Scansano Teatro Castagnoli 9–10 settembre 2005.) Siena: Ci.Vin.

Mckinney, J. D., W. R. Jacobs, Jr., and B. R. Bloom. 1998. "Persisting Problems in Tuberculosis." In R. Krause, J. I. Gallin, and A. S. Fauci, eds., *Emerging Infections*, 51–146. San Diego: Academic Press.

McNeill, W. H. 1976. *Plagues and Peoples*. New York: Anchor/Doubleday (reprint 1998.)

Melnick, J. L. 1997. "Poliovirus and Other Enteroviruses." In A. S. Evans and R. A. Kaslow, eds., *Viral Infections of Humans: Epidemiology and Control*, chap. 21. New York: Plenum Medical.

Melville, R. 1999. *Lucretius. On the Nature of the Universe*. Oxford: Oxford University Press.

Menichetti, M. 1994. *Archeologia del potere: re, imagini e miti a Roma e in Etruria in età arcaica*. Milan: Longanesi.

Menotti, E. M., ed. 2005. *Cibo: vita e cultura nelle collezioni del Museo Archeologico Nazionale di Mantova*. (Exhibition: *Cibi e Saproi nell'Italia Antica*, with English translation, *Food: Life and Culture*.) Mantova: Tre Lune Edizioni.

Meyer, J.-W. 1987. *Untersuchungen zu den Tonlebermodellen aus dem Alten Orient*. Kevelaer and Neukirchen-Vluyn: Butzon & Bercker.

Miari, M. 1995. "Offerte votive legate al mondo vegetale e animal nelle cavità naturali dell'Italia protostorica." In L. Quilici and S. Quilici Gigli, eds., *Agricoltura e commerci nell'Italia antica*, 11–29. Rome: L'Erma di Bretschneider.

Michels, A. K. 1967. *The Calendar of the Roman Republic*. Princeton, N.J.: Princeton University Press.

Milcent, P.-Y. 2006. "Les importations italiques au nord-ouest du Midi gaulois (milieu du Xe – debut du IVe s. av. J.-C.): inventaire et perspectives d'interprétation." In S. Gori and M. C. Bettini, eds., *Gli Etruschi da Genova ad Ampurias*, 1:319–55. (Atti del XXIV Convegno di Studi Etruschi ed Italici, Marseille-Lattes, 26 settembre – 1 ottobre 2002.) Pisa: Istituti Editoriali e Poligrafici Internazionali.

Miller, J. C. 1987. "Gathering Animals: Potential Health Hazard." *Museum Applied Science Center for Archaeology Journal* 4(3): 144–49.

Minetti, A. 2004. "La tomba della Quadriga Infernale di Sarteano." *Studi Etruschi* 70: 135–59.

Mingazzini, P. 1965. "*Fulgur conditum* e *bidental*, nonché l'etimologia del nome bidental." In Centro Studi Ciociaria (a cura di), *Gli archeologi italiani in onore di Amedeo Maiuri*, 315–36. Cava dei Tirreni: Di Mauro.

Minto, A. 1921. *Marsiliana d'Albegna: Le scoperte archeologiche del Principe Don Tommmaso Corsini*. Florence: Istituto di Edizioni Artistiche/Fratelli Alinari.

Momigliano, A. 2005. "The Rise of the Plebs in the Archaic Age of Rome." In K. A. Raaflaub, ed., *Social Struggles in Archaic Rome: New Perspectives on the Conflict of the Orders*, 168–84. Oxford: Blackwell.

Morandi, A. 1988. "Nuove osservazioni sul Fegato di Piacenza." *Mélanges de l'École Française à Rome: Antiquité* 100: 283–97.

1989. "Note di epigrafia etrusca veiente." *Mélanges de l'École Française à Rome: Antiquité* 101(2): 581–96.

1991. *Nuovi lineamenti di lingua etrusca*. Rome: Erre emme edizioni.

Moretti Sgubini, A. M., ed. 2004. *Eroi Etruschi e Miti Greci: Gli affreschi della Tomba François tornano a Vulci*. (Exhibition, Vulci, Castello della Badia, 26 giugno–26 settembre 2004.) Florence: Edizioni Cooperativa Archeologia.

Morigi Govi, C. 1971. "Il tintinnabulo della 'Tomba degli ori' dell'Arsenale militare di Bologna." *Archeologia Classica* 23: 212–35.

Morris, S. 1992. *Daidalos and the Origins of Greek Art*. Princeton, N.J.: Princeton University Press.

Moscati, S. 1999. "Sarcophagi." In S. Moscati, ed., *The Phoenicians*, 355–59. (U.S. edition of exhibition volume, published 1988, Milan: Bompiani.) New York: Rizzoli.

Moser, M. E. 1996. "The Origins of the Etruscans: New Evidence for an Old Question." In J. F. Hall, ed., *Etruscan Italy: Etruscan Influences on the Civilizations of Italy from Antiquity to the Modern Era*, 29–43. Provo, Utah: Museum of Art, Brigham Young University.

Muss, U. 2000. "Das Artemision von Ephesos – Wege von und nach Westen." In F. Krinzinger, ed., *Die Ägäis und das westliche Mittelmeer. Beziehungen und Wechselwirkungen 8. bis 5. Jh.v.Chr.*, 149–55. (Akten des Symposions Wien, 24. bis 27. März 1999.) Vienna: Österreichische Akademie der Wissenschaften.

Naso, A. 2000A. "Le aristocrazie etrusche in period orientalizzante: cultura, economia, relazioni." In M. Torelli, ed., *Gli Etruschi*, 111–29. (Mostra, Palazzo Grassi, Venice, 2000.) Milan: Bompiani.

2000B. "Etruscan and Italic Artifacts from the Aegean." In D. Ridgway, F. R. Serra Ridgway, M. Pearce, E. Herring, R. D. Whitehouse, and J. B. Wilkins, eds., *Ancient Italy in its Mediterranean Setting: Studies in Honour of Ellen Macnamara*, 193–207. London: Accordia Research Institute, University of London.

2000C. "Commercio e scambi tra Oriente e Occidente." In *Magna Grecia e Oriente Mediterraneo prima dell'età ellenistica*, 125–85. (Atti del trentanovesimo convegno di Studi sulla Magna Grecia, Taranto, 1–5 ottobre 1999.) Naples: Istituto per la Storia e l'Archeologia della Magna Grecia.

2000D. "Etruskische und italische Weihungen in griechischen Heiligtümern: altbekannte und neue Funde." In F. Krinzinger, ed., *Die Ägäis und das westliche Mittelmeer. Beziehungen und Wechselwirkungen 8. bis 5. Jh.v.Chr.*, 157–63. (Akten des Symposions, Wien, 24. bis 27. März 1999.) Vienna: Österreichische Akademie der Wissenschaften.

Nataro, J. P., O. C. Stine, J. B. Kaper, and M. M. Levine. 2003. "The Archaeology of Enteric Infection: *Salmonella, Shigella, Vibrio*, and Diarrheagenic *Escherichia coli*." In C. L. Greenblatt and M. Spigelman, eds., *Emerging Pathogens: The Archaeology, Ecology, and Evolution of Infectious Disease*, 167–74. Oxford: Oxford University Press.

Negroni Catacchio, N., ed. 1982. *Sorgenti della Nova: Una communità protostorica ed il suo territorio nell'Etruria Meridionale*. (Mostra, Arezzo, Grosseto, 1982–83.) Florence: Vision:Viella.

Negroni Catacchio, N., ed. 1995. *Sorgenti della Nova: L'abitato del Bronzo Finale*. Florence: Istituto italiano di preistoria e protostoria.

Negroni Catacchio, N., and L. Domanico. 2001. "L'abitato protourbano di Sorgenti della Nova: dagli spazi dell'abitare all'organizazzione sociale." In J. R. Brandt and L. Karlsson, eds., *From Huts to Houses: Transformations of Ancient Societies*, 40, LVI, 337–59. (Proceedings of an International Seminar organized by the Norwegian and Swedish Institutes in Rome, 21–24 September 1997.) Stockholm: Paul Åström/Skrifter utgivna av Svenska Institutet i Rom.

Negroni Catacchio, N., and M. Miari. 1995. "La Ricostruzione Grafica di Alcune Strutture Residenziali e di Servizio in Etruria: Problemi e Metodi." In Christie, 521–30.

Neumann, J. 1985. "Climatic Change as a Topic in the Classical Greek and Roman Literature." *Climatic Change* 7: 441–54.

Niebuhr, B. G., ed. 1837. *Corpus Scriptorum Historiae Byzantinae. Johannes Lydus*. Bonn: E. Weber.

Nielsen, E. 1987. "Some Preliminary Thoughts on Old and New Terracottas." *Opuscula Romana* 16: 91–119.

Niemeyer, H. G. 1986. "El yacimiento fenicio de Toscanos: urbanistica y función." In G. del Olmo Lete and M. E. Aubet Semmler, eds., *Los Fenicios en la peninsula iberica* I, 109–26. Barcelona: Editorial AUSA.

1990. "The Phoenicians in the Mediterranean: A Non-Greek Model or Expansion and Settlement in Antiquity." In J.-P. Descoeudres, ed., *Greek Colonists and Native Populations*, 469–89. Oxford: Oxford University Press.

2002. "The Phoenician Settlement at Toscanos: Urbanization and Function." In M. R. Bierling, ed. and trans., *The Phoenicians in Spain: An Archaeological Review of the Eighth–Sixth Centuries* B.C.E., 31–48. Winona Lake, Ind.: Eisenbrauns.

Nijboer, A. J. 1998. *From Household Production to Workshops*. Groningen: University of Groningen, Department of Mediterranean Archaeology.

2006. "The Iron Age in the Mediterranean: A Chronological Mess or 'Trade Before the Flag,' Part II." In *The Iron Age in the Mediterranean*, 255–77. (*Ancient West and East* vol. 4, no. 2.) Leiden: Brill.

Nijboer, A. J., J. van der Plicht, A. M. Bietti Sestieri, and A. De Santis. 1999/2000. "A High Chronology for the Early Iron Age in Central Italy." *Palaeohistoria* 41/42:163–76.

Nissinen, M. 2000. "The Socioreligious Role of the Neo-Assyrian Prophets." In M. Nissinen, ed., *Prophecy in its Ancient Near Eastern Context: Mesopotamian, Biblical, and Arabian Perspectives*, 89–114. Atlanta: Society of Biblical Literature.

Nizzo, V. 2005. "Intervento." In Bartoloni and Delpino, 645–46.

Oaks, S. C., et al. 1991. *Malaria. Obstacles and Opportunities*. Washington DC: National Academy Press.

Ocean Studies Board, National Research Council. 1999. *From Monsoons to Microbes: Understanding the Ocean's Role in Human Health*. Washington DC: National Academy Press.

Ogilvie, R. M. 1958. "Livy, Licinius Macer and the libri lintei." *Journal of Roman Studies* 48: 40–46.

Onisto, N. 2002. "7.1.Analisi morfologica e metrica dei resti ossei umani." In P. von Eles, ed., *Guerriero e sacerdote. Autorità e comunità nell'età del ferro a Verucchio. La Tomba del Trono*, 277–89. (Quaderni di Archeologia dell'Emilia Romagna 6.) Florence: all'Insegna del Giglio.

Oppenheim, A. L. 1969. "Divination and Celestial Observation in the Last Assyrian Empire." *Centaurus* 14: 97–135.

1974. "A Babylonian Diviners Manual." *Journal of Near Eastern Studies* 33: 197–220.

Orr, D. 2002. "Snakes on Pompeian Household Shrines." In Jashemski and Meyer, 350–56.

Ortner, D. J., and G. Theobald. 1993. "Diseases in the Pre-Roman World." In Kiple, 247–61.

Pacciarelli, M. 1998. "Rito funerario e società' nel Bronzo Finale dell'Etruria meridionale." In N. Negroni Catacchio, ed., *Preistoria e Protostoria in Etruria*, 35–46. (Atti del Terzo Incontro di Studi. Protovillanoviani e/o Protoetruschi. Ricerche e Scavi, Manciano - Farnese 12–14 maggio 1995.) Florence: Franco Cantini Editore.

Pagnini, L. 2000. "281. Tridente." In G. Bartoloni, F. Delpino, C. Morigi Govi, and G. Sassatelli, eds., *Principi etruschi tra Mediterraneo ed Europa*, 242. (Exhibition, Bologna, 2000–2001.) Bologna: Marsilio.

Pairault-Massa, F.-H. 1985. "La divination en Étrurie: Le IVe siècle, période critique:" In Ch. Guittard, ed., *La divination dans le monde étrusco-italique*, vol. 1. (Actes de la Table ronde, 23 février 1985), *Caesarodunum* Suppl. 52: 56–115.

Pallottino, M. 1948–49. "Sulla lettura e sul contenuto della grande iscrizione di Capua." *Studi Etruschi* 20: 159–96.

1952. "Un ideogramma araldico etrusco?" *Archeologia Classica* 4: 245–47.

1979. *Saggi di antichità*. (Anthology of publications, 3 vols.) Rome: G. Bretschneider.

1987. "Il fregio dei Vibenna e le sue implicazioni storiche." In F. Buranelli, ed., *La Tomba François di Vulci*, 225–43. (Mostra, Città del Vaticano, Braccio di Carlo Magno, 20 marzo–17 maggio 1987.) Rome: Quasar.

Pallottino, M., et al. 1986. *Liber Linteus Zagabrensis = Vjesnik. Arheoloskog Museja u Zagrebu* (3rd series) 19: 1–106.

et al. 1992. *Gli Etruschi e l'Europa*. Milan: Fabbri Editori.

Palm, J. 1952. "Vejian Tomb Groups in the Museo Preistorico, Rome." *Opuscula Archaeologica* 7: 50–86.

Palmer, R. E. A. 1996. "Locket Gold, Lizard Green." In J. F. Hall, ed., *Etruscan Italy: Etruscan Influences on the Civilizations of Italy from Antiquity to the Modern Era*, 17–27. Provo, Utah: Museum of Art, Brigham Young University.

Park, K. 1993. "Black Death." In Kiple, 612–16.

Parke, H. W. 1988. *Sibyls and Sibylline Prophecy in Classical Antiquity*. London: Routledge.

Parpola, S. 1970–1983. *Letters from Assyrian Scholars to the Kings Esarhaddon and Assurbanipal, Parts I-III* (AOAT 5/1 and 5/2). Neukirchen-Vluyn: Verlag Butzon & Bercker Kevelaer.

1993A. *Letters from Assyrian and Babylonian Scholars* (*SAA* 10). Helsinki: Helsinki University Press.

1993B. "Mesopotamian Astrology and Astronomy as Domains of the Mesopotamian 'Wisdom.'" In H. D. Galter, ed., *Die Rolle der Astronomie in den Kulten Mesopotamiens*, 47–59. (Beiträge zum 3. Grazer Morgenländischen Symposion, 23.–27. September 1991.) Graz: GrazKult.

1997. *Assyrian Prophesies* (*SAA* 9). Helsinki: Helsinki University Press.

Pascucci, P. 1998. "L'insediamento costiero della prima età del Ferro de 'La Mattonara' (Civitavecchia)." *Archeologia Classica* 50: 69–115.

Patitucci Uggeri, S., and G. Uggeri. 1993. "La topografia della città." In F. Berti and P. G. Guzzo, eds., *Spina: Storia di una città tra Greci ed Etruschi*, 21–31. (Exhibition, Ferrara, 1993–1994.) Ferrara: Comitato Ferrara Arte.

Patrick, A. 1967. "Disease in Antiquity: Ancient Greece and Rome." In D. Brothwell and A. T. Sandison, eds., *Diseases in Antiquity*, 238–48. Springfield, Ill.: C. C. Thomas.

Patterson, K. D. 1993A. "Tapeworm." In Kiple, 1035–36.

1993B. "Ascariasis, and Echinococcosis (Hydatidosis)." In Kiple, 603–4, 703.

Patterson, J. R. 2000. "On the Margins of the City of Rome." In V. M. Hope and E. Marshall, eds., *Death and Disease in the Ancient City*, 85–103. London: Routledge.

Pearce, L. E. 1995A. "The Scribes and Scholars of Ancient Mesopotamia." In J. M. Sasson, J. Barnes, G. Beckman, and K. S. Rubinson, eds., *Civilizations of the Ancient Near East*, 4: 2265–78. New York: Charles Scribner's Sons.

1995B. "SEPĪRU and [LÚ]A.BA: Scribes of the Late First Millennium." In K. van Lerberghe and G. Voet, eds., *Languages and Cultures in Contact: At the Crossroads of Civilizations in the Syro-Mesopotamian Realm*, 355–68. (Proceedings of the 42th [*sic*] *RAI*.) Leuven: Peeters.

2004. "Sacred Texts and Canonicity: Mesopotamia." In S. I. Johnston, ed., *Religions of the Ancient World: A Guide*, 626–28. Cambridge, Mass.: Harvard University Press.

Pedroni, L. 1998. "Ipotesi sull'evoluzione del calendario arcaico di Roma." *Papers of the British School at Rome* 66: 39–55.

Pellegrini, E. 1998. "Aspetti della metallurgia protovillanoviana in Etruria." In N. Negroni Catacchio, ed., *Preistoria e Protostoria in Etruria*, 23–34. (Atti del Terzo Incontro di Studi. Protovillanoviani e/o Protoetruschi. Ricerche e Scavi, Manciano – Farnese 12–14 maggio 1995.) Florence: Franco Cantini Editore.

Pellicer Catalán, M. 2002. "Phoenician and Punic Sexi." In M. R. Bierling, ed. and trans., *The Phoenicians in Spain: An Archaeological Review of the Eighth–Sixth Centuries* B.C.E., 49–77. Winona Lake, Ind.: Eisenbrauns.

Perkins, P. 1999. *Etruscan Settlement, Society and Material Culture in Central Coastal Etruria*. Oxford: John and Erica Hedges.

Peroni, R. 1989. *Protostoria dell'Italia continentale.* (Popoli e civiltà dell'Italia antica.) Rome: Ente per la diffusione e l'educazione storica.

Peroni, R., and A. Vanzetti. 2004. "Intervento." In G. Bartoloni and F. Delpino, 658–60.

Pfiffig, A. J. 1964. *Religio iguvina: Studien zi den Bronzetafeln von Gubbio.* (Öst., Akad. der Wiss. Phil.-hist. Klasse Denkschriften 84.) Vienna: Kommissionsverlag der Österreichischen Akademie der Wissenschaften.

1969. *Die etruskische Sprache.* Graz: Akademische Druck- und Verlagsanstalt.

1975. *Religio etrusca.* Graz: Akademische Druck- und Verlagsanstalt.

Piccaluga, G. 1994. "La specifità dei libri lintei." *Scrittura e civiltà* 18: 5–22.

Pielou, E. C. 2001. *The Energy of Nature.* Chicago: University of Chicago Press.

Pietrangeli, C. 1949–51. "Bidentalia." *Rendiconti della Pontifica Accademia* 25–6: 37–52.

Piganiol, A. 1951. "Sur le calendrier brontoscopique de Nigidius Figulus." In P. R. Coleman-Norton et al., eds., *Studies in Roman Economic and Social History in Honor of Allan Chester Johnson*, 79–87. Princeton N.J.: Princeton University Press.

Pingree, D., ed. 1973. *Hephaestionis Thebani Apotelesmaticorum Libri tres*, vol. 1. Leipzig: Teubner.

Pingree, D. 1974. "Petosiris." In C. Coulston Gillispie, ed., *Dictionary of Scientific Biography*, 10:547–49. New York: Scribner.

1982. "Mesopotamian Astronomy and Astral Omens in Other Civilizations." In H.-J. Nissen and J. Renger, eds., *Mesopotamien und seine Nachbarn: Politische und kulturelle Wechselbeziehungen im Alten Vorderasien vom 4. bis 1. Jahrtausend v. Chr.*, 613–31. (XXV. Rencontre Assyriologique, 1978.) Berlin: Dietrich Reimer.

1998. "Legacies in Astronomical and Celestial Omens." In S. Dalley, ed., *The Legacy of Mesopotamia*, 123–37. Oxford: Oxford University Press.

Pinza, G. 1905. *Monumenti primitivi di Roma e del Lazio antico = Monumenti Antichi 15.* Milan: Hoepli.

Pleiner, R. 2000. *Iron in Archaeology: The European Bloomery Smelters.* Prague: Archeologický ústav AV ČR.

Poccetti, P. 1987. "Etrusco *feluske* = *faliscus*? Note sull'iscrizione della stele arcaica di Vetulonia." *Studi Etruschi* 63: 281–90.

Pohl, I. 1987. "San Giovenale." In *L'alimentazione nel mondo antico: Gli Etruschi*, 71–73. (Ministero per I Beni Culturali e Ambientali, Giroanta Mondiale dell'Alimentazione.) Rome: Istituto Poligrafico e Zecca dello Stato.

Poland, J. D., and D. T. Dennis. 1998. "Plague." In Evans and Brachman, 545–58.

Pongratz-Leisten, B. 1999. *Herrschaftswissen in Mesopotamien. Formen der Kommunikation zwischen Gott und König im 2. und 1. Jahrtausend v.Chr.* (SAAS vol. 10). Helsinki: University of Helsinki.

Potter, D. 1990. "Sibyls in the Greek and Roman World." *Journal of Roman Archaeology* 3: 471–83. (Review article on Parke 1988.)

Poultney, J. W. 1959. *The Bronze Tables of Iguvium.* Baltimore: American Philological Association.

Prayon, F. 1977. "L'Oriente e la statua etrusca arcaica." *Colloqui del Sodalizio* 5: 165–72.

2001. "Near Eastern Influences of Early Etruscan Architecture?" In L. Bonfante and V. Karageorghis, eds., *Italy and Cyprus in Antiquity: 1500–450 BC*, 335–50. Nicosia: Severis Foundation.

2004. "Überlegungen zur Monumentalität frühetruskischer Plastik und Architektur." In M. Novák, F. Prayon, and A.-M. Wittke, eds., *Die Aussenwirkung des späthethitischen Kulturraumes. Güteraustausch - Kulturkontakt - Kulturtransfer*, 85–105. (Akten des zweiten Forschungstagung des Graduiertenkollegs "Anatolien und seine Nachbarn" der Eberhard-Karls-Universität Tübingen, 20. bis 22. November 2003.) Münster: Ugarit-Verlag.

Prayon, F., and W. Röllig, eds. 2000. *Der Orient und Etrurien.* (Akten des Kolloquiums, Tübingen 1997, Zum Phänomen des 'Orientalisierens' im westlichen Mittelmeeraum, 10.-6. Jh.v.Chr.) Pisa and Rome: Istituti Editoriali e Poligrafici Internazionali.

Prosdocimi, A. L. 1984. *Le tavole iguvine*. I. Florence: Olschki.

Pulak, C. 1998. "The Uluburun Shipwreck: An Overview." *International Journal of Nautical Archaeology* 27: 188–224.

2001. "The Cargo of the Uluburun Ship and Evidence for Trade with the Aegean and Beyond." In L. Bonfante and V. Karageorghis, eds., *Italy and Cyprus in Antiquity: 1500–450 BC*, 13–60. Nicosia: Severis Foundation.

Raaflaub, K. 1991. "City-State, Territory and Empire in Classical Antiquity." In A. Molho, K. Raaflaub, and J. Emlen, eds., *City States in Classical Antiquity and Medieval Italy*, 565–88. Ann Arbor: University of Michigan Press.

Rackham, H. 1938/1991. *Pliny. Natural History, Preface and Books 1–2*, with an English translation by H. Rackham. Cambridge, Mass.: Harvard University Press.

Rallo, A., ed. 1989. *Le donne in Etruria*. Rome: L'Erma di Bretschneider.

Ramelli, I. 2003. *Cultura e religione etrusca nel mondo romano: La cultura etrusca dalla fine dell'indipendenza*. San Donato Milanese: Edizioni dell'Orso.

Rasenna = M. Pallottino, G. Pugliese Caratelli, et al. 1986. *Rasenna: Storia e civiltà degli Etruschi*. Milan: Libri Schweiwiller.

Rasmussen, S. W. 2000. "Cicero's Stand on Prodigies: A Non-Existent Dilemma?" In R. Lorsch Wildfang and J. Isager, eds., *Divination and Portents in the Roman World*, 9–24. Odense: Odense University Press.

Rasmussen, T. B. 2005. "Urbanization in Etruria." In R. Osborne and B. Cunliffe, eds., *Mediterranean Urbanization, 800–600 BC*, 71–90. Oxford: Oxford University Press.

Rathje, A. 1989. "Alcune considerazioni sulle lastre da Poggio Civitate con figure femminili." In A. Rallo, ed., *Le donne in Etruria*, 75–84. Rome: L'Erma di Bretschneider.

2000. "'Princesses' in Etruria and Latium Vetus?" In D. Ridgway, F. R. Serra Ridgway, M. Pearce, E. Herring, R. D. Whitehouse, and J. B. Wilkins, eds., *Ancient Italy in its Mediterranean Setting: Studies in Honour of Ellen Macnamara*, 295–300. London: Accordia Research Institute, University of London.

2006. "Il sacro e il politico: Il deposito votivo di Tarquinia." In M. Bonghi Jovino, ed., *Tarquinia e le civiltà del Mediterraneo*, 103–18. (Convegno internazionale, Milano, 22–24 giugno 2004.) Milan: Cisalpino, Istituto Editoriale Universitario.

Rawlings, L. 1999. "Condottieri and Clansmen: Early Italian Raiding, Warfare and the State." In K. Hopwood, ed., *Organised Crime in Antiquity*, 97–127. Swansea: Duckworth.

Rawson, E. 1985. *Intellectual Life in the Late Roman Republic*. Baltimore: Johns Hopkins University Press.

1978/1991. "Caesar, Etruria, and the *Disciplina Etrusca*." *Journal of Roman Studies* 68: 132–52. (Repr. in *Roman Culture and Society: Collected Papers*, 289–323. Oxford: Clarendon, 1991).

1971/1991. "Prodigy Lists and the Use of the *Annales Maximi*." *Classical Quarterly* 21(1): 158–69. (Repr. in *Roman Culture and Society: Collected Papers*, 1–15. Oxford: Clarendon, 1991).

Reese, D. S. 2002A. "Fish: Evidence from Specimens, Mosaics, Wall Paintings, and Roman Authors." In Jashemski and Meyer, 274–91.

2002B. "Marine Invertebrates, Freshwater Shells, and Land Snails: Evidence from Specimens, Mosaics, Wall Paintings, Sculpture, Jewelry, and Roman Authors." In Jashemski and Meyer, 292–314.

Reiner, E. 1995. *Astral Magic in Babylonia*. (*Transactions of the American Philosophical Society* 85, pt. 4.) Philadelphia: American Philosophical Society.

1999. "Babylonian Celestial Divination." In N. M. Swerdlow, ed., *Ancient Astronomy and Celestial Divination*, 21–37. Cambridge, Mass.: MIT Press.

Reiner, E., with D. Pingree. 1975. *Babylonian Planetary Omens*. Part 1, *The Venus Tablet of Ammisaduqa*, vol. 2, fasc. 1. Malibu, Calif.: Undena.

with D. Pingree. 1981. *Babylonian Planetary Omens*. Part 2, *Enūma Anu Enlil, Tablets 50–51*, vol. 2, fasc. 2. Malibu, Calif.: Undena Publications.

with D. Pingree. 2005. *Babylonian Planetary Omens*, Part 4. Leiden: Brill/Styx.

Rendeli, M. 1993. *Città aperte: Ambiente e paesaggio rurale organizzato nell'Etruri meridionale costiera durante l'età orientalizzante e arcaica.* Rome: Gruppo editoriale internazionale.

Ricciardi, M., and G. G. Aprile. 1988. "Identification of Some Carbonized Plant Remains from the Archaeological Area of Oplontis." In *Studia Pompeiana et Classica in Honor of W. F. Jashemski*, 317–25. New Rochelle, N.Y.: A. D. Caratzas.

Richards, E. G. 1998. *Mapping Time: The Calendar and Its History.* Oxford: Oxford University Press.

Richardson, E. H. 1984. "The Lady at the Fountain." In *Studi di antichità in onore di Guglielmo Maetzke*, 2:447–54. Rome: G. Bretschneider.

Richardson, W. F. 2004. *Numbering and Measuring in the Classical World: An Introductory Handbook*, rev. 2nd ed. Bristol: Bristol Phoenix Press.

Ridgway, D. 1992. *The First Western Greeks.* Cambridge: Cambridge University Press.

1994. "Phoenicians and Greeks in the West: A View from Pithekoussai." In G. R. Tsetskhladze and F. de Angelis, eds., *The Archaeology of Greek Colonisation: Essays Dedicated to Sir John Boardman*, 35–46. Oxford: Oxford University School of Archaeology.

1997 (2000). "The Orientalizing Phenomenon in Campania: Sources and Manifestations." In Prayon and Röllig, 233–44.

2000. "The First Western Greeks Revisited." In D. Ridgway, F. R. Serra Ridgway, M. Pearce, E. Herring, R. D. Whitehouse, and J. B. Wilkins, eds., *Ancient Italy in its Mediterranean Setting: Studies in Honour of Ellen Macnamara*, 179–91. London: Accordia Research Institute, University of London.

Ridgway, D., and F. R. S. Ridgway. 1994. "Demaratus and the Archaeologists." In R. D. De Puma and P. Small, eds., *Murlo and the Etruscans: Art and Society in Ancient Etruria*, 6–15. Madison: University of Wisconsin Press.

Ridgway, F. R. S. 1990. "Etruscans, Greeks, Carthaginians: the Sanctuary at Pyrgi." In J.-P. Descoeudres, ed., *Greek Colonists and Native Populations*, 511–30. Oxford: Oxford University Press.

2003. "The University of Milan's Final Reports on Tarquinia, and Two Exhibitions on Tarquinia and Her Neighbors." *Journal of Roman Archaeology* 16: 463–68.

2005. "Rediscovered Etruscans: Two Exhibitions at Viterbo and Vulci." *Journal of Roman Archaeology* 18: 466–71.

Riess, E. 1892. "*Nechepsonis et Petosiridis Fragmenta Magica.*" *Philologus* Suppl.6: 325–94.

Rix, H. 1963. *Das etruskische Cognomen.* Wiesbaden: Harrassowitz.

1984. "Etr. *meχ rasnal* = Lat. *res publica.*" In *Studi di antichità in onore di Guglielmo Maetzke*, 2: 455–68. Rome: G. Bretschneider.

1994. *Die Termini der Unfreiheit in den Sprachen Alt-Italiens.* Stuttgart: Franz Steiner.

1997. "Les priers du *liber linteus* de Zagreb." In F. Gaultier and D. Briquel, eds., *Les Étrusques, les plus religieux des hommes: État de la recherche sur la religion étrusque*, 391–98. (Actes du colloque international, Galeries nationals du Grand Palais, 17–18–19 novembre 1992.) Paris: École du Louvre.

Robb, J. 1997. "Violence and Gender in Early Italy." In D. L. Martin and D. W. Frayer, eds., *Troubled Times: Violence and Warfare in the Past*, 111–44 (*War and Society* 3.) Amsterdam: Overseas Publishers Association.

Roberts, C., and K. Manchester. 1995. *The Archaeology of Disease*, 2nd ed. Ithaca, N.Y.: Cornell University Press.

Robson, E. 2008. *Mathematics in Ancient Iraq: A Social History.* Princeton, N.J.: Princeton University Press.

Rochberg, F. 1999. "Babylonian Horoscopy: The Texts and Their Relations." In N. M. Swerdlow, ed., *Ancient Astronomy and Celestial Divination*, 39–59. Cambridge, Mass.: MIT Press.

2004. *The Heavenly Writing: Divination, Horoscopy, and Astronomy in Mesopotamian Culture.* Cambridge: Cambridge University Press.

Rochberg-Halton, F. 1988. *Aspects of Babylonian Celestial Divination: The Lunar Eclipse Tablets of* Enūma Anu Enlil = *Archiv für Orientforschung Beiheft* 22. Horn: Ferdinand Berger und Söhne.

Rolandi, G., P. Petrosino, and J. McGeehin. 1998. "The Interplinian Activity at Somma-Vesuvius in the Last 3500 Years." *Journal of Volcanology and Geothermal Research* 82: 19–52.

Röllig, W. 2004. "Sprachen und Schriften der Levante in Anatolien." In M. Novák, F. Prayon, and A.-M. Wittke, eds., *Die Aussenwirkung des späthethitischen Kulturraumes. Güteraustausch – Kulturkontakt – Kulturtransfer*, 205–17. (Akten des zweiten Forschungstagung des Graduiertenkollegs "Anatolien und seine Nachbarn" der Eberhard-Karls-Universität Tübingen, 20. bis 22. November 2003.) Münster: Ugarit-Verlag.

Rollinger, R. 2001. "The Ancient Greeks and the Impact of the Ancient Near East: Textual Evidence and Historical Perspective (ca. 750–650 BC)." In R. M. Whiting, ed., *Mythology and Mythologies: Approaches to Intercultural Influences*, 233–64. (Helsinki, The Neo-Assyrian Text Corpus Project, Melammu Symposia II, 2001.) Helsinki: Neo-Assyrian Text Corpus Project.

Romagnoli, M. 1997. "I cremati della collezione del Museo Guarnacci di Volterra (PI)." In *Aspetti della cultura di Volterra etrusca*, 447–57.

Roncalli, F. 1980. "*Carbasinis uoluminibus implicati libri*: Osservazioni sul *liber linteus* di Zagabria." *Jahrbuch des deutschen Instituts* 95: 227–64.

1981. "Die Tracht des Haruspex als frühgeschichtliches Relikt in historischer Zeit." In *Die Aufnahme fremder Kultureinflüsse in Etrurien und das Problem des Retardierens in der etruskischen Kunst*, 124–32. (Schriften des deutschen Archäologen-Verbands 5.) Mannheim.

Roncalli, F., ed. 1985. *Scrivere etrusco: dalla leggenda alla conoscenza. Scrittura e letteratura nei massimi documenti della lingua etrusca, Catalogo della mostra, Perugia, 1985*, ed. C. Piravano. Milan: Electa.

Roncalli, F. 1990. "La pietra come '*instrumentum scriptorium*' e il cippo di Perugia." In *Annali della Fondazione per il Museo "Claudio Faina" (Orvieto)* 4: 11–20.

2010. "Between Divination and Magic: Role, Gesture and Instruments of the Etruscan *Haruspex*." In L. B. van der Meer, ed., *Material Aspects of Etruscan Religion*, 117–26. (Proceedings of the International Colloquium, Leiden, May 29 and 30, 2008.) Leuven: Peeters.

Rosen, A. M. 2007. *Civilizing Climate: Social Responses to Climate Change in the Ancient Near East*. Lanham, Md.: AltaMira.

Rossoni, G. 1997. "I tridenti metallici nel Vicino Oriente antico tra uso pratico e simbologia: Proposte di interpretazione di una particolare classe di materiali." In *Contributi e materiali di archeologia orientale*, 561–90. (Vol. 7 = Studi in memoria di Henri Frankfort, 1897–1954.) Rome: Università degli Studi "La Sapienza."

Rottoli, M. 1997. "I resti vegetali." In M. Bonghi Jovino and C. Chiaramonte Treré, eds., *Tarquinia: Testimonianze archeologiche e ricostruzione storica*, 92–99. Rome: L'Erma di Bretschneider.

Rubini, M., F. Bartoli, E. Bonafede, S. Mogliazza, and L. Moreschini. 2000. "Appendice: Studio antropologico sugli inumati della necropoli etrusca del Ferrone (VII–VI sec. a.C.)." In P. Brocato, ed., *La necropoli etrusca della Riserva del Ferrone: Analisi di una communita' arcaica dei Monti della Tolfa*, 499–523. Rome: Edizioni Quasar.

Ruiz Mata, D. 2002. "The Ancient Phoenicians of the 8th and 7th Centuries B.C. in the Bay of Cádiz: State of the Research." In M. R. Bierling, ed. and trans., *The Phoenicians in Spain: An Archaeological Review of the Eighth–Sixth Centuries B.C.E.*, 155–98. Winona Lake, Ind.: Eisenbrauns.

Russo, F. 2005. "I *Carmina Marciana* e le tradizioni sui Marcii." *La Parola del Passato* 60: 5–32.

Sachs, A. J., and H. Hunger. 1988–2001. *Astronomical Diaries and Related Texts from Babylonia*, 5 vols. Vienna: Verlag der Österreichischen Akademie der Wissenschaften.

Säflund, G. 1939. *Le Terremare delle provincie di Modena, Reggio Emilia, Parma, Piacenza*. (Svenska institutet i Rom. Skrifter 4,7.) Lund: Gleerup.

1986. "'Hieros Gamos'-Motive in der etruskischen Sepulkralkunst." In J. Swaddling, ed., *Italian Iron Age Artefacts in the British Museum*, 471–78. London: British Museum.

Sallares, R. 1991. *The Ecology of the Ancient Greek World*. Ithaca, N.Y.: Cornell University Press.

2002. *Malaria and Rome*. Oxford: Oxford University Press.

Salvesen, A. 1998. "The Legacy of Babylon and Nineveh in Aramaic Sources." In S. Dalley, ed., *The Legacy of Mesopotamia*, 139–61. Oxford: Oxford University Press.

Sannibale, M. 2006. "Tra cielo e terra: Considerazioni su alcuni aspetti della religione etrusca a Vulci." *Studi Etruschi* 72 (2006 = 2009): 117–47.

2008. "Iconografie e simboli orientali nelle corti dei principi etruschi." *Byrsa* 7(1/2): 85–123.

Santillo Frizell, B. 2004. "Curing the Flock: The Use of Healing Waters in Roman Pastoral Economy." In B. Santillo Frizell, ed., *Pecus: Man and Animal in Antiquity*, 80–90. (Proceedings of the Conference at the Swedish Institute in Rome, September 9–12, 2002.) (The Swedish School in Rome. Projects and Seminars, 1.) Rome: Swedish Institute.

Santuari d'Etruria 1985= G. Colonna, ed., *Santuari d'Etruria*. (Exhibition, Arezzo, Museo archeologico C. Cilnio Mencenate, 19 maggio–20 ottobre 1985). Milan: Electa. (Some catalog entries attributed to "*redattore*.")

Sassatelli, G. 1999. "The Diet of the Etruscans." In J.-L. Flandrin and M. Montanari, eds., *Food: A Culinary History from Antiquity to the Present*, 106–12. (English edition of *Histoire de l'alimentation* [1996], ed. A. Sonnenfeld, trans. C. Botsford.) New York: Columbia University Press.

2000. "Il principe e la pratica della scrittura." In G. Bartoloni, F. Delpino, C. Morigi Govi, and G. Sassatelli, eds., 2000. *Principi etruschi tra Mediterraneo ed Europa*, 309–17. (Mostra, Bologna, Museo Civico Archeologico.) Bologna: Marsilio.

Scarpa, G. 1986. "La fauna." In R. De Marinis, ed., *Gli Etruschi a nord del Po*, 1: 184–92. Mantova: Regione Lombarda, Provincia commune di Mantova.

Scarth, A. 1994. *Volcanoes*. College Station: Texas A&M Press.

Schäfer, T. 1989. *Imperii Insignia. Sella Curulis* und *Fasces: Zur Repräsentation römischer Magistrate*. (RM Ergänzungsheft 29.) Mainz/Rhein: Phillip von Zabern.

Schaefer, V. J., and J. A. Day. 1981. *A Field Guide to the Atmosphere*. Boston: Houghton Mifflin.

Scheffer, C. 1987. "Acquarossa." In *L'alimentazione nel mondo antico: Gli Etruschi*, 75–76. (Ministero per i Beni Culturali e Ambientali, Giroanta Mondiale dell'Alimentazione, 1987.) Rome: Istituto Poligrafico e Zecca dello Stato.

Scheidel, W. 1996. *Measuring Sex, Age and Death in the Roman Empire* (= *Journal of Roman Archaeology* Suppl. Ser. 21). Portsmouth, RI: *Journal of Roman Archaeology*.

Schlesinger, A. C. 1959. *Livy with an English translation. XIV. Summaries, Fragments, and Obsequens*, with a general index to Livy by Russel M. Geer. Cambridge, Mass.: Harvard University Press.

Schmidt, P. L. 2000. "Nigidius Figulus, P." *Der Neue Pauly* 8: 890–91. See also his entry in *Brill's New Pauly: Antiquity*, vol. 9, eds. H. Cancik and H. Schneider, Leiden, 2006: 751–52.

Schmitz, P., 1995. "The Phoenician Text from the Sanctuary at Pyrgi." *Journal of the American Oriental Society* 115: 559–75.

2009. "Phoenician KRNTRYŠ, Archaic Greek *ΚΟΡΥΝΗΤΗΡΙΟΣ, and the Storm God of Aleppo." In *KUSATU* 11: 119–60.

Schomer, P. D. 1985. "Electroacoustics." In E. C. Jordan, ed., *Reference Data for Engineers: Radio, Electronics, Computer, and Communications*, 7th ed., 40–1 to 40–25. Indianapolis, Ind.: Howard W. Sams.

Schultz, C. 2006. "Juno Sospita and Roman Insecurity in the Social War." In P. Harvey and C. Schultz, eds., *Religion in Republican Italy*, 207–27. (*Yale Classical Studies* XXXIII.) Cambridge: Cambridge University Press.

Sciacca, F. 2003. "Nota sul *rhyton* a protome di leone da Veio: confronti e produzione." *Archeologia Classica* 54: 301–19.

2004. "Per una nuova interpretazione del tridente in bronzo dal Circolo del Tridente di Vetulonia." *Archeologia Classica* 55: 269–82.

Scrivere etrusco 1985 = *Scrivere etrusco: dalla leggenda alla conoscenza. Scrittura e letteratura nei massimi documenti della lingua etrusca, Catalogo della mostra, Perugia, 1985*, F. Roncalli et al., ed. C. Piravano. Milan: Electa.

Scullard, H. H. 1981. *Festivals and Ceremonies of the Roman Republic*. London: Thames & Hudson.

Shackleton Bailey, D. R. 2000. *Valerius Maximus: Memorable Doings and Sayings*. Cambridge, Mass.: Harvard University Press.

Shaw, B. D. 1996. "'Seasons of Death': Aspects of Mortality in Imperial Rome." *Journal of Roman Studies* 86: 100–38.

Sider, D. 2005. *The Library of the Villa dei Papiri at Herculaneum*. Los Angeles: J. Paul Getty Museum.

Simkin, T., and L. Siebert. 1994. *Volcanoes of the World*. Tucson, Ariz.: Geoscience Press.

Simon, E. 2006. "Gods in Harmony: The Etruscan Pantheon." In de Grummond and Simon, 45–65.

Sjöberg, Å. 1973. "Der Vater und sein misratener Sohn." *Journal of Cuneiform Studies* 25: 105–69.

Skeates, R. 1994. "A Radiocarbon Date-List for Prehistoric Italy (c. 46,400 BP – 2450 BP/400 cal. BC)." In R. Skeates and R. Whitehouse, eds., *Radiocarbon Dating and Italian Prehistory*, 147–288. London: Accordia Research Centre.

Small, J. P. 1986. "The Tragliatella Oinochoe." *Römische Mitteilungen* 93: 63–96.

Smit, H. J. M., et al. 1999. "Atmospheric Pressure Changes and Outdoor Temperature Changes in Relation to Spontaneous Pneumothorax." *Chest* 116: 676–81.

Söderlind, M. 2004. "Man and Animal in Antiquity: Votive Figures in Central Italy from the 4th to 1st Centuries B.C." In B. Santillo Frizell, ed., *Pecus: Man and Animal in Antiquity*, 277–94. (Proceedings of the Conference at the Swedish Institute in Rome, September 9–12, 2002.) (The Swedish School in Rome. Projects and Seminars, 1.) Rome: Swedish Institute.

Sophocles, E. A. 1887. *Greek Lexicon of the Roman and Byzantine periods (from B.C. 146 to A.D. 1100)*. New York: C. Scribner's Sons.

Sorrentino, C. 1981A. "Appendix: La fauna." In B. Olinder and I. Pohl, eds., *San Giovenale* vol. II, fasc. 4. *The Semi-Subterranean Building in Area B*, Skrifter 40, XXVI:II, 4. Lund: C. W. K. Gleerup, 85–89.

1981B. "Appendix: La fauna." In E. Bergren and K. Berggren, eds., *San Giovenale* vol. II, fasc. 2, Skrifter XXVI:II, 2, *Excavations in Area B*, 1957–1960. Lund: C. W. K. Gleerup, 58–64.

2004. "Il materiale osteologico animale del tumulo di via San Jacopo." In S. Bruni, T. Caruso, and M. Massa eds., *Archaeologica Pisana: Scritti per Orlando Pancrazi*, 366–69. Pisa: Giardini.

Speranza, A., B. van Geel, and J. van der Plicht. 2003. "Evidence for Solar Forcing of Climate Change at ca. 850 cal BC from a Czech Peat Sequence." *Global and Planetary Change* 35 (1–2): 51–65.

Speranza, F. 1974. *Scriptorum Romanorum De Re Rustica Reliquiae*, vol. 1. (Biblioteca di Helikon, Testi e Studi 8.) Messina: Università degli Studi.

Stannard, J. 1993. "Diseases in Western Antiquity." In Kiple, 262–70.

Starr, I. 1990. *Queries to the Sungod: Divination and Politics in Sargonid Assyria* (with J. Aro, S. Parpola, and J. Reade). Helsinki: Helsinki University Press.

Stary, P. 1979. "Foreign Elements in Etruscan Arms and Armor: 8th to 3rd Centuries B.C." *Proceedings of the Prehistoric Society* 45: 79–206.

1980A. "Zum Bedeutung und Funktion zweirädrigen Wagen während der Eisenzeit in Mittelitalien." *Hamburger Beiträge zur Archäologie* 7: 7–21.

1980B. "Orientalische und griechische Einflüsse in der etruskischen Bewaffnung und Kampfesweise." In *Die Aufnahme fremder Kultureinflüsse in Etrurien und das Problem des Retardierens in der etruskischen Kunst*, 25–40. (Symposium Mannheim, 8.–10.2. 1980 = Schriften des Deutschen Archäologen-Verbandes V) Mannheim: Archäologischer Seminar der Universität Mannheim, 1981.

1981. *Zur eisenzeitlichen Bewaffnung und Kampfesweise in Mittelitalien.* (*Marburger Studien* 3.) Mainz/Rhein: Philipp von Zabern.

2000. "Early Iron Age Armament and Warfare: Near Eastern Influences from the Aegean via Etruria to Andalucia." In D. Ridgway, F. R. Serra Ridgway, M. Pearce, E. Herring, R. D. Whitehouse, and J. B. Wilkins, eds., *Ancient Italy in its Mediterranean Setting: Studies in Honour of Ellen Macnamara*, 209–20. London: Accordia Research Institute, University of London.

Steinbauer, D. H. 1999. *Neues Handbuch des Etruskischen.* St. Katharinen: Scripta Mercaturae.

Steiner, R. C. 2009. Paper, "H>H: On the Diffusion of an Assyro-Aramaic Sound Change to Babylonia," delivered 3/30/2009, at *Tablet and Torah: Mesopotamia and the Biblical World. A conference in honor of Dr. Barry L. Eichler*, University of Pennsylvania Museum, Philadelphia (to be published by CLC Press.)

Steingräber, S. 1986. *Etruscan Painting. Catalogue Raisonné of Etruscan Wall Paintings* (eds. D. Ridgway and F. R. Serra Ridgway). New York: Harcourt, Brace Jovanovich.

2000. "L'urbanistica etrusca." In M. Torelli, ed., *Gli Etruschi*, 291–311. (Mostra, Palazzo Grassi, Venice, 2000.) Milan: Bompiani.

2001. "The Process of Urbanization of Etruscan Settlements from the Late Villanovan to the Late Archaic Period." *Etruscan Studies* 8: 7–33.

2002. "Ahnenkult und bildliche Darstellungen von Ahnen in etruskischen und unteritalischen Grabgemälden aus vorrömischer Zeit." In J. M. Højte, ed., *Images of Ancestors*, 127–58. Aarhus: Aarhus University Press.

Steuernagel, D. 2005. "Vom Wissen zur Wissenschaft: Etruskische Eingeweideschauer und ihre bildliche Repräsentation." In T. L. Kienlin, ed., *Die Dinge als Zeichen: Kulturelles Wissen und materielle Kultur*, 383–94. (Internationale Fachtagung and der Johann Wolfgang Goethe-Universität Frankfurt am Main 3.–5. April 2003.) Bonn.

Stieglitz, R. 1998. "The Phoenician-Punic Menology." In M. Lubetski, C. Gottlieb, and S. Keller, eds., *Boundaries of the Ancient Near Eastern World: A Tribute to Cyrus Gordon*, 211–21. (*Journal for the Study of the Old Testament*, Suppl. Series 273.) Sheffield: Sheffield Academic Press.

Stoddart, S. 1989. "Divergent Trajectories in Central Italy, 1200–500 B.C." In T. C. Champion, ed., *Centre and Periphery: Comparative Studies in Archaeology*, 88–101. London: Unwin Hyman.

Stol, M. 1993. *Epilepsy in Babylonia.* Groningen: Styx.

Stopponi, S. 2000. "I rapporti con i popoli vicini." In M. Torelli, ed., *Gli Etruschi*, 181–89. (Mostra, Palazzo Grassi, Venice, 2000.) Milan: Bompiani.

Stothers, R. B., and M. R. Rampino. 1983. "Volcanic Eruptions in the Mediterranean Before A.D. 630 from Written and Archaeological Sources." *Journal of Geophysical Research* 88 (B8): 6357–71.

Streck, K. M. 1916. *Assurbanipal und die letzte assyrischen Könige bis zum Untergange Ninevehs*, vol. 2. Leipzig: J. C. Hinrichs.

Stuart-Macadam, P. L. 1989. "Nutritional Deficiency Diseases: A survey of Scurvy, Rickets and Iron-deficiency Anemia." In M. Y. Iscan and A. R. Kennedy, *Reconstruction of Life from the Skeleton*, 201–22. New York: Alan R. Liss.

1998. "Iron Deficiency Anemia: Exploring the Difference." In A. L. Grauer and P. L. Stuart-Macadam, eds., *Sex and Gender in Paleopathological Perspective*, 45–63. Cambridge: Cambridge University Press.

Suetonius/Rolfe 1998 = *Suetonius*, vol. 1, with an English translation by J. C. Rolfe. Cambridge, Mass.: Harvard University Press.

Svensmark, H. 2000. "Cosmic Rays and Earth's Climate." *Space Science Reviews* 93: 155–66.

Swerdlow, N. M. 1998. *The Babylonian Theory of the Planets.* Princeton, N.J.: Princeton University Press.

Swoboda, A., ed. 1889. *Publii Nigidii Figuli, Operum Reliquiae.* Prague and Vienna: Tempsky.

Tabanelli, M. 1962. *Gli Ex-Voto Polyviscerali Etruschi e Romane.* Florence: Olschki.

1963. *La medicina nel mondo degli Etruschi.* Florence: Olschki.

Tafuri, M. A. 2005. *Tracing Mobility and Identity: Bioarchaeology and Bone Chemistry of the Bronze Age Sant'Abbondio Cemetery (Pompeii, Italy).* (BAR Int.Ser. 1359.) Oxford: Archaeopress.

Tamburini, P. 1991. "Pietra ovoide con fulmine in rilievo." In L. Bonfante and F. Roncalli, eds., *Gens antiquissima Italiae: Antichità dell'Umbria a New York*, 273–76. (Exhibition, New York 1991.) Milan: Electa.

1995A. "I 'prodigi' volsiniesi nelle fonti classiche." In P. Tamburini, ed., *Un abitato villanoviano perilacustre: Il "Gran Carro" sul lago di Bolsena (1959–1985)*, 11–12. Rome: G. Bretschneider.

1995B. "Conclusioni." In Tamburini, *Un abitato villanoviano perilacustre*, 353–65.

Tassi Scandone, E. 2001. *Verghe, scuri e fasci littori in Etruria.* (*Biblioteca di Studi Etruschi* 36.) Pisa and Rome: Istituti Editoriali e Poligrafici Internazionali.

Taub, L. C. 2003. *Ancient Meteorology.* New York: Routledge.

Teegen, W.-R. 1995. "Grave Dimensions as a Diagnostic Tool for Palaeodemography and Social Ranking: The Example of Veio-Quattro Fontanili." In Christie, 261–71.

Terrosi Zanco, O. 1961. "Ex-voto allungati dell'Italia centrale." *Studi Etruschi* 29: 423–59.

Thesleff, H. 1965. Review of Della Casa (1962). *Gnomon* 37: 44–48.

Thompson, R. C. 1900. *The Reports of the Magicians and Astrologers of Nineveh and Babylon in the British Museum.* Vol. 2, *English Translations, Vocabulary, etc.* London: Luzac & Co. (Vol. 1, *Cuneiform Facsimiles*, is not cited here.)

Thulin, C. O. 1906. *Die etruskische Disciplin.* Vol. 3, *Die Haruspicin.* Göteborg: Wald. Zachrissons Boktryckeri.

Tinnefeld, F. 1999. "Lydos [3] Iohannes Lydos." In *Der Neue Pauly* 7: 550–51.

Tinner, W., et al. 2003. "Climatic Change and Contemporaneous Land-use Phases North and South of the Alps 2300 BC to 800 AD." *Quaternary Science Reviews* 22(14): 1447–60.

Tirelli, M. 1981. "La rappresentazione del sole nell'arte etrusca." *Studi Etruschi* 49: 41–50.

Torelli, M. 1969. "Senatori etruschi della tarda repubblica e dell'impero." *Dialoghi di Archeologia* 3: 285–363.

1971. "Il santuario di Hera a Gravisca." *La Parola del Passato* 26: 44–67.

1975. *Elogia Tarquiniensia.* Florence: Sansoni.

1977. "Il santuario Greco di Gravisca." *La Parola del Passato* 32: 398–458.

1982. "Per la definizione del commercio Greco-orientale: Il caso di Gravisca." *La Parola del Passato* 37: 302–26.

1986. "Un *templum augurale* di età repubblicana a Bantia." *Rendiconti dell'Accademia dei Lincei* 8th ser., 21: 293–94.

1987. *La società etrusca. L'età arcaica, l'età classica.* Rome: La Nuova Italia scientifica.

1988. "*Etruria principes disciplinam doceto*: Il mito normativo dello specchio di Tuscania." In *Studia Tarquiniensia*, 109–18. (*Archeologica Perusina* 9.) Rome: G. Bretschneider.

1997. *Il rango, il rito e l'immagine: alle origini della rappresentazione storica romana.* Milan: Electa. (Originally in *Ostraka* 5(2) 1996: 333–68.)

2000. "Le regie etrusche e laziali tra orientalizzante e arcaismo." In G. Bartoloni, F. Delpino, C. Morigi Govi and G. Sassatelli, eds., *Principi etruschi tra Mediterraneo ed Europa*, 67–78. (Mostra, Bologna, Museo Civico Archeologico.) Bologna: Marsilio.

2009. *Storia degli Etruschi*, 8th ed. (original 1981). Rome and Bari: Laterza.

Tromp, S. W. 1980. *Biometeorology: The Impact of Weather and Climate on Humans and Their Environment (Animals and Plants).* London: Heyden.

Trucco, F., D. De Angelis, C. Iaia, and R. Vargiu 2005. "Nuovi dati sui rituali funerari della prima età del Ferro a Tarquinia." In O. Paoletti and G. Camporeale, eds., *Dinamiche di Sviluppo delle Città nell'Etruria Meridionale: Veio, Caere, Tarquinia*, Vulci, 1:359–69. (Atti del XXIII Convegno di Studi Etruschi ed Italici, Roma, Veio, Cerveteri/Pyrgi, Tarquinia, Tuscania, Vulci, Viterbo, 1–6 ottobre 2001.) Pisa and Rome: Istituti Editoriali e Poligrafici Internazionali.

Trump, D. H. 1966. *Central and Southern Italy before Rome.* London: Thames & Hudson.

Turcan, R. 1976. "Encore la prophétie de Végoia." In *L'Italie préromaine et la Rome républicaine: Mélanges offerts à Jacques Heurgon*, 2:1009–19. (Collection de l'École Française de Rome.) Rome: École Française de Rome.

Turfa, J. M. 1986. "III. International Contacts: Commerce, Trade, and Foreign Affairs." In L. Bonfante, ed., *Etruscan Life and Afterlife: A Handbook of Etruscan Studies*, 66–91. Detroit: Wayne State University Press.

1994. "Anatomical Votives and Italian Medical Traditions." In R. D. De Puma and J. P. Small, eds., *Murlo and the Etruscans*, 224–40. Madison: University of Wisconsin Press.

1999. "Parasols in Etruscan Art." *Source* 18(2): 15–24.

2001. "The Etruscans and the Phoenicians of Cyprus: The 8th to 6th Centuries B.C." In L. Bonfante and V. Karageorghis, eds., *Italy and Cyprus in Antiquity 1500–450 B.C.*, 271–90. Nicosia: Severis Foundation.

2004. "[Weigeschenke: Altitalien und Imperium Romanum. 1. Italien.] B. Anatomical votives." In *Thesaurus Cultus et Rituum Antiquorum*. Vol. 1, *Processions – Sacrifices – Libations – Fumigations – Dedications*, 359–68. Los Angeles and Basel.

2005. *Catalogue of the Etruscan Gallery of The University of Pennsylvania Museum*. Philadelphia: University Museum Press.

2006A. "Votive Offerings in Etruscan Religion." In de Grummond and Simon, 90–115.

2006B. "Etruscan Religion at the Watershed: Before and After the Fourth Century BC." In P. Harvey and C. Schultz, eds., *Religion in Republican Italy*, 62–89. (*Yale Classical Studies* XXXIII.) Cambridge: Cambridge University Press.

2006C. Review of Haack 2006. *Bryn Mawr Classical Review* 2006.12.34.

2006D. "Appendix A: The Etruscan *Brontoscopic Calendar*." In de Grummond and Simon, 173–90.

2006E. "Was There Room for Healing in the Healing Sanctuaries?" *Archiv fur Religionsgeschichte* 8: 63–80.

2006F. Review of Bartoloni and Delpino 2004. *Bryn Mawr Classical Review* 2006.08.10.

2007. "The Etruscan *Brontoscopic Calendar* and Modern Archaeological Discoveries." In J. Swaddling and P. Perkins, eds., *Acts of Conference "Etruscans Now!" Proceedings of the Symposium held at the British Museum, 2002. Part II = Etruscan Studies* 10 (2004–2007), 163–73.

2008A. Review of Domenici and Maderna 2007. *Bryn Mawr Classical Review* 2008.07.14.

2008B. Review of van der Meer 2007. *Bryn Mawr Classical Review* 2008.05.37.

2009. "From the Tombs of Bisenzio." In S. Bruni, ed., *Etruria e Italia Preromana: Studi in Onore di Giovannangelo Camporeale*, 2:525–33. Pisa and Rome: Fabrizio Serra Editore.

2011. "Telmu and Petrui: A Rediscovered Romance?" *Expedition* 53(1): 7–8.

(in press). "'If it thunders, it signals plague, but not especially life-threatening': Health and Disease in the *disciplina etrusca*." In C. Beer, ed., *Proceedings of Conference on Archaeology and History of Medicine* (Stockholm, 8–9 November 2001). Stockholm: Medelhavsmuseet and Karolinska Institutet.

Turfa, J. M., and S. Gettys. 2009. "The Skill of the Etruscan Haruspex: A Biological Basis for Successful Divination?" *BABESCH* 84: 41–52.

Turfa, J. M., and A. G. Steinmayer, Jr. 1993. "In Defence of Patroklos: A Plea to Common Sense." *Antichthon* 27: 1–12.

1996. "The Comparative Structure of Greek and Etruscan Monumental Buildings." *Papers of the British School at Rome* 64: 1–39.

1999. "The Earliest Foresail, on Another Etruscan Vase." *International Journal of Nautical Archaeology* 28(3): 292–96.

2001. "Sewn Hulls and Self-defense." *International Journal of Nautical Archaeology* 30(1): 122–27.

2002. "Interpreting Early Etruscan Structures: The Question of Murlo." *Papers of the British School at Rome* 70:1–28.

T-W-Fiennes, R. N. 1978. *Zoonoses and the Origins and Ecology of Human Disease*. London and New York: Academic.

Uman, M. A. 1984. *Lightning*. New York: Dover.

Urbainczyk, T. 2008. *Slave Revolts in Antiquity*. Berkeley: University of California Press.

Vagnetti, L. 2000. "Western Mediterranean Overview: Peninsular Italy, Sicily and Sardinia at the Time of the Sea Peoples." In E. D. Oren, ed., *The Sea Peoples and Their World: A Reassessment*, 305–26. (University Museum Symposium Series 11.) Philadelphia: University Museum Press.

Valvo, A. 1988. *La "Profezia di Vegoia": Proprietà fondiaria e aruspicina in Etruria nel I secolo A.C.* Rome: Istituto italiano per la storia antica.

van der Meer, L. B. 1979. "*Iecur Placentinum* and the Orientation of the Etruscan Haruspex." *BABESCH* 57: 87–99.

1987. *The Bronze Liver of Piacenza: Analysis of a Polytheistic Structure*. Amsterdam: J. C. Gieben.

2007. *Liber Linteus Zagrabensis. The Linen Book of Zagreb. A Comment on the Longest Etruscan Text*. Leuven: Peeters.

van Geel, B., and B. E. Berglund. 2000. "A Causal Link between a Climatic Deterioration around 850 cal BC and a Subsequent Rise in Human Population Density in NW-Europe?" *Terra Nostra* 7: 126–30.

van Geel, B., J. Buurman, and H. T. Waterbolk. 1996. "Archaeological and Palaeoecological Indications of an Abrupt Climate Change in The Netherlands, and Evidence for Climatological Teleconnections around 2650 BP." *Journal of Quaternary Science* 11(6): 451–60.

van Geel, B., O. M. Raspopov, J. van der Plicht, and H. Renssen. 1998. "Solar Forcing of Abrupt Climate Change around 850 Calendar Years BC." In B. J. Peiser, T. Palmer and M. E. Bailey, eds., *Natural Catastrophes During Bronze Age Civilisations: Archaeological, Geological, Astronomical and Cultural Perspectives*, 162–68. (BAR IS 728.) Oxford: Archaeopress.

van Soldt, W. H. 1995. *Solar Omens of Enūma Anu Enlil: Tablets 23 (24) – 29 (30)*. Den Haag: Nederlandsch Historisch-Archaeologisch Instituut te Istanbul.

van Strydonck, M. 2004. "Radiocarbon and Archaeological Evidence for a Possible Climate Induced Cultural Change on the Balaeric Islands around 2700 BP." In T. Higham, C. Bronk Ramsay and C. Owen eds., *Radiocarbon and Archaeology*, 247–62. (Fourth International Symposium, St. Catherine's College, Oxford, 9–14 April 2002.) Oxford: Oxford University School of Archaeology.

Vargiu, R., and M. J. Becker. 2005. "Appendice: Studio antropologico dei resti scheletrici umani," 395–409 in Cataldi, M., *Sulle 'tombe a buca' di Tarquinia*, in *Dinamiche di sviluppo delle città nell'Etruria Meridionale. Veio, Caere, Tarquinia, Vulci. Atti del XXIII Convegno di Studi Etruschi ed Italici (1–6 ottobre 2001)*. Pisa-Rome, Istituti Editoriali e Poligrafici Internazionali, vol 1: 409–11.

Vidal-Naquet, P. 1970. "Esclavage et gynécocratie dans la tradition, le mythe, l'utopie." In *Recherches sur les structures sociales dans l'antiquité classique*, 63–80. (Caen 25–26 avril 1969.) Paris: CNRS.

Villard, F. 1994. "Les sièges de Syracuse et leurs pestilences." In P. Ginouvés et al., eds., *L'eau, la santé et la maladie dans le monde grec. Actes du Colloque…Paris…du 25 au 27 novembre 1992* = *BCH* Suppl. 28: 337–44.

Vincent, W. R. 1985. "34. Radio Noise and Interference." In E. C. Jordan, ed., *Reference Data for Engineers: Radio, Electronics, Computer, and Communications*, 7th ed., 34–1 to 34–13. Indianapolis, Ind.: Howard W. Sams.

Virolleaud, C. 1905–12. *L'astrologie chaldéene, le livre intitulé <enuma (Anu) iluBêl> publié, transcript et traduit, fascicules 1–14*. Paris: P. Geuthner.

Vitaliano, D. B. 1976. *Legends of the Earth: Their Geologic Origins*. Bloomington: Indiana University Press.

von der Osten-Sacken, E. 2002. "Zur Göttin auf dem Burneyrelief." In S. Parpola and R. M. Whiting, eds., *Sex and Gender in the Ancient Near East*, 479–87. (Proceedings of the XLVIIe Rencontre Assyriologique Internationale, Helsinki, 2.) Helsinki: The Neo-Assyrian Text Corpus Project.

von Eles, P., ed. 2002. *Guerriero e sacerdote. Autorità e comunità nell'età del ferro a Verucchio. La Tomba del Trono.* (Quaderni di Archeologia dell'Emilia Romagna 6.) Florence: all'Insegna del Giglio.

von Hase, F.-W. 1981. "Zum Beginn des Fernhandels von und nach Etrurien unter besonderer Berücksichtigung der frühesten mittelitalischen Funde in Griechenland." In *Die Aufnahme fremder Kultureinflüsse in Etrurien und das Problem des Retardierens in der etruskischen Kunst* (Symposium Mannheim, 8.-10.2. 1980 = Schriften des Deutschen Archäologen-Verbandes V). Mannheim: Archäologischer Seminar der Universität Mannheim, 9–24.

1997. "Presences etrusques et italiques dans les sanctuaries grecs (VIIIe–VIIe siecle av. J.-C.)." In F. Gaultier and D. Briquel, eds., *Les Étrusques, les plus religieux des hommes: État de la recherche sur la religion étrusque*, 293–323. (Actes du colloque international, Galeries nationals du Grand Palais, 17–18–19 novembre 1992.) Paris: École du Louvre.

Wachsmuth, C. 1897. *Johannnes Lydus De ostentis*. Leipzig: Bibliotheca scriptorum Graecorum et Romanorum Teubneriana (original edition 1863).

Wainwright, G. A. 1932. "Iron in Egypt." *Journal of Egyptian Archaeology* 18: 3–15.

Wallace, R. E. 2008. *Zikh Rasna: A Manual of the Etruscan Language and Inscriptions.* Ann Arbor, Mich.: Beech Stave.

Warden, P. G. 1983. "*Bullae*, Roman Custom and Italic Tradition." *Opuscula Romana* 14: 69–75.

Warmington, E. H. 1967. *Remains of Old Latin. III. Lucilius. The Twelve Tables.* Cambridge, Mass.: Harvard University Press.

Watson, G. E. 2002. "Birds: Evidence from Wall Paintings, Mosaics, Sculpture, Skeletal Remains, and Ancient Authors." In Jashemski and Meyer, 357–400.

Watts, N. H., 1923. *Cicero. The Speeches, with an English translation. Pro Archia ... Pro Plancio.* Cambridge, Mass.: Harvard University Press (reprinted 1961).

Weidemann, T. 1981. *Greek and Roman Slavery.* Baltimore: Johns Hopkins University Press (reprinted 1997).

Weidner, E. 1927. "Eine Beschreibung des Sternenhimmels aus Assur." *Archiv für Orientforschung* 4: 73–85.

1968–69. "Die astrologische Serie Enûma Anu Enlil." *Archiv für Orientforschung* 22: 65–75.

Weinberg, E. 1992. "Iron Withholding in Prevention of Disease." In P. L. Stuart-Macadam and S. Kent, eds., *Diet, Demography and Disease: Changing Perspectives on Anemia*, 105–50. New York: Aldine De Gruyter.

Weinstock, S. 1950. "C. Fonteius Capito and the *Libri Tagetici*." *Papers of the British School at Rome* 8: 44–49.

1951. "*Libri fulgurales.*" *Papers of the British School at Rome* 19: 122–53.

West, M. L. 1997. *The East Face of Helicon: West Asiatic Elements in Greek Poetry and Myth.* Oxford: Clarendon.

Wetter, E. 1969, "Ricerche topografiche nei territori circostanti Acqua Rossa." *Opuscula Romana* 7: 109–37.

White, K. D. 1970. *Roman Farming.* Ithaca, N.Y.: Cornell University Press.

Wiggerman, F. 1992. "Mythological Foundations of Nature." In D. J. W. Meijer, ed., *Natural Phenomena: Their Meaning, Depiction and Description in the Ancient Near East*, 279–306. Amsterdam: Koninklijke Nederlandse Akademie van Wetenschappen Verhandelingen, Afd. Letterkunde, Niuewe Reeks deel 152.

Wikander, Ö. 1988. "Ancient Roof-tiles – Use and Function." *Opuscula Atheniensia* 17: 203–16.

1992. "Archaic Roof-tiles: The First (?) Generation." *Opuscula Atheniensia* 19: 151–61.

Wildfang, R. L. 2000. "*Fulgura* et *Fulmina*: Or What It Portends When a Family Tomb Is Struck by a *fulmen quod decussit*." In R. Lorsch Wildfang and J. Isager, eds., *Divination and Portents in the Roman World*, 67–78. Odense: Odense University Press.

Wilkins, J. B. 1994. "The Iguvine Tables: Problems in the Interpretation of Ritual Text." In C. Malone and S. Stoddart, eds., *Territory, Time and State: The Archaeological Development of the Gubbio Basin*, 152–77. Cambridge: Cambridge University Press.

Wilkins, J., et al., eds. 1995. *Food in Antiquity*. Exeter: Exeter University Press.

Wilkinson, L. 1993A. "Anthrax." In Kiple, 582–83.

1993B. "Brucellosis." In Kiple, 625–28.

Williams, C. K., II. 1980. "Demaratus and Early Corinthian Roofs." In *Stēlē: tomos eis mnēmēn Nikolaou Kontoleontos*, 245–50. Athens: To Sōmateion tōn philōntōn Nikolaou Kontoleontos.

Williams, C. K., II, and J. Fisher. 1973. "Corinth, 1972: The Forum Area." *Hesperia* 42: 1–44.

Wise, M. O. 1994. *Thunder in Gemini and Other Essays on the History, Language and Literature of Second Temple Palestine*. Sheffield: JSOT.

Wiseman, D. J. 1955. "Assyrian Writing-Boards." *Iraq* 17: 3–13.

Wiseman, D. J., and J. A. Black. 1996. *Literary Texts from the Temple of Nabû*. London: British School of Archaeology in Iraq.

Wiseman, T. P. 1995. *Remus: A Roman Myth*. Cambridge: Cambridge University Press.

Wissowa. 1900. "Clodius Tuscus." *Pauly-Wissowa Realencyclopädie* vol. 4/col. 104, no. 61.

Wood, A. R. 2002. "Under the Weather? A Once-scorned Science May Have Answers." *Philadelphia Inquirer*, June 23, 2002, A1, A14.

Woytowitsch, E. 1978. *Die Wagen der Bronze- und frühen Eisenzeit in Italien*. (Prähistorische Bronzefunde XVII, 1.) Munich: C. H. Beck.

Wulf, S., M. Kraml, A. Brauer, J. Keller, and J. F. W. Negendank. 2004. "Tephrochronology of the 100 ka Lacustrine Sediment Record of Lago Grande di Monticchio (Southern Italy)." *Quaternary International* 122: 7–30.

Wünsch, R., ed. 1898. *De mensibus*. Leipzig: Leskii.

Wyatt, H. V. 1993. "Poliomyelitis." In Kiple, 942–50.

Xella, P. 1982. "L'influence babylonienne à Ougarit, d'après les texts alphabétiques rituals et divinatoires." In H.-J. Nissen and J. Renger eds., *Mesopotamien und seine Nachbarn: Politische und kulturelle Wechselbeziehungen im Alten Vorderasien vom 4. bis 1. Jahrtausend v. Chr.*, 321–38. (XXV. Rencontre Assyriologique, 1978.) Berlin: Dietrich Reimer.

1999. "La figure du 'Prince prédestiné' au Proche-Orient ancient: destin des puissants et volonté des dieux." In E. Smadja and E. Geny, eds., *Pouvoir, divination, & prédestination dans le monde antique*, 159–73. Paris: Presses Universitaires Franc-Comtoises.

York, M. 1986. *The Roman Festival Calendar of Numa Pompilius*. New York: P. Lang.

Young, E. J. 1983. "Human Brucellosis." *Reviews of Infectious Diseases* 5: 821–42.

Young, E. J., and W. H. Hall. 1998. "Brucellosis." In Evans and Brachman, 155–68.

Yung, A. P., and M. L. Grayson. 1988. "Psittacosis – A Review of 135 Cases." *Medical Journal of Australia* 148(5): 228–33.

Zamarchi Grassi, P. 1985. "10.2. La stipe della Fonte Veneziana." In G. Colonna, ed., *Santuari d'Etruria*, 174–79. Milan: Electa.

Zifferrero, A. 1991. "Miniere e metallurgia estrattiva in Etruria Meridionale: Per una lettura critica di alcuni dati archeologici e minerari." *Studi Etruschi* 57: 201–41.

1995. "Rituale funerario e formazione delle aristocrazie nell'Etruria protostorica: osservazioni sui corredi femminili e infantili di Tarquinia." In N. Negroni Catacchio, ed., *Tipologia delle necropoli e rituali di deposizione*, 257–65. (Atti del Secondo Incontro di studi, Preistoria e Protostoria in Etruria, Farnese 1993.) Milan: Centro Studi di Preistoria e Archeologia.

2004. "Simbolismo astrale e segnalazione del rango nell'aristocrazia tirrenica: Ipotesi sul signifcato e sull'impiego della bulla etrusca e latina." In N. Negroni Catacchio, ed., *Preistoria e Protostoria in Etruria*, 1: 327–37. (Atti del Sesto Incontro di Studi. Miti Simboli Decorazioni Ricerche e scavi, Pitigliano – Valentano, 13–15 settembre 2002). Milan: Centro Studi di Preistoria e Archeologia.

Zohary, D., and M. Hopf. 1993. *Domestication of Plants in the Old World: The Origin and Spread of Cultivated Plants in West Asia, Europe and the Nile Valley*, 2nd ed. Oxford: Oxford University Press.

INDEX

For EU product safety concerns, contact us at Calle de José Abascal, 56–1°, 28003 Madrid, Spain or eugpsr@cambridge.org.

www.ingramcontent.com/pod-product-compliance
Lightning Source LLC
LaVergne TN
LVHW062109200926
848484LV00007B/1625